Springer Series in Operations Research

Editors:
Peter W. Glynn   Stephen M. Robinson

Springer
New York
Berlin
Heidelberg
Barcelona
Hong Kong
London
Milan
Paris
Singapore
Tokyo

Springer Series in Operations Research

Peter J. Haas

Stochastic Petri Nets

Modelling, Stability, Simulation

With 71 Illustrations

Springer

Peter J. Haas
IBM Research Division
San Jose, CA 95120-6099
USA
peterh@almaden.ibm.com

Series Editors:
Peter W. Glynn
Department of Management Science
    and Engineering
Terman Engineering Center
Stanford University
Stanford, CA 94305-4026
USA

Stephen M. Robinson
Department of Industrial Engineering
University of Wisconsin–Madison
Madison, WI 53706-1572
USA

Library of Congress Cataloging-in-Publication Data
Haas, Peter J. (Peter Jay)
    Stochastic Petri nets : modelling, stability, simulation / Peter J. Haas.
      p. cm. — (Springer series in operations research)
    Includes bibliographical references and index.
    ISBN 978-1-4419-3001-9   e-ISBN 978-0-387-21552-5
    1. Petri nets.   2. Stochastic analysis.   I. Title.   II. Series.
  QA267 .H3 2002
  511.3—dc21   2002019559

Printed on acid-free paper.

Manufacturing supervised by Jerome Basma.
Camera-ready copy prepared from the author's LaTeX2e files using Springer's macros.

Printed in the United States of America.

9 8 7 6 5 4 3 2 1

Springer-Verlag   New York Berlin Heidelberg
A member of BertelsmannSpringer Science+Business Media GmbH

To the Memory of
Walter H. Haas

Preface

This book was motivated by a desire to bridge the gap between two important areas of research related to the design and operation of engineering and information systems. The first area concerns the development of mathematical tools for formal specification of complex probabilistic systems, with an eye toward subsequent simulation of the resulting stochastic model on a computer. The second area concerns the development of methods for analysis of simulation output.

Research on modelling techniques has been driven by the ever-increasing size and complexity of computer, manufacturing, transportation, workflow, and communication systems. Many engineers and systems designers now recognize that the use of formal models has a number of advantages over simply writing complicated simulation programs from scratch. Not only is it much easier to generate software that is free of logical errors, but various qualitative system properties—absence of deadlock, impossibility of reaching catastrophic states, and so forth—can be verified far more easily for a formal model than for an ad-hoc computer program. Indeed, certain system properties can sometimes be verified automatically.

Our focus is on systems that can be viewed as making state transitions when events associated with the occupied state occur. More specifically, we consider *discrete-event systems* in which the stochastic state transitions occur only at an increasing sequence of random times. The "Bedienungsprozess" (service process) framework, developed by König, Matthes, and Nawrotzki in the 1960s and early 1970s, provided the first set of building blocks for formal modelling of general discrete-event systems. The modern incarnation of the Bedienungsprozess is the "generalized semi-Markov process" (GSMP). Although useful for a unified theoretical treatment of discrete-event stochastic systems, the GSMP framework is not always well suited to practical modelling tasks. In particular, the modeller is forced to specify the "state of the system" directly as an abstract vector of random variables. Such a specification can be highly nontrivial: the system state definition must be as concise as possible for reasons of efficiency, but must also contain enough information so that (1) a sequence of state transitions and transition times can be generated during a simulation run and (2) the system characteristics of interest can be determined from such a sequence. Stochastic Petri nets (SPNs), introduced in the 1980s, are very appealing in that they not only have the same modelling power as GSMPs (see Chapter 4) but also admit a graphical representation that is well suited to top-down and bottom-up modelling of complex systems.

In parallel to these advances in modelling, a rigorous theory of simulation output analysis has been developed over the past 25 years. Much of this theory pertains to the problem of obtaining point estimates and confidence intervals for long-run performance measures of interest. Such point and interval estimates are typically used to compare alternative system designs or operating policies. These estimates also form the basis for simulation-based optimization procedures. Confidence intervals can be particularly difficult to obtain, but are necessary to distinguish true differences in system behavior from mere random fluctuations. The basic idea is to view each simulation run as the sample path of a precisely defined stochastic process. Point estimates and confidence intervals are then established by appealing to limit theorems for such processes.

Unfortunately, many of the results in the output-analysis literature have not been provided in a form that is directly useful to practicing simulation analysts. Typically, a specified estimation or optimization procedure is shown to produce valid results if the output process of the simulation has specified stochastic properties—for example, obeys specified limit theorems or has a sequence of regeneration points. Verification of the required properties for a specific (and usually complicated) simulation model often turns out to be a formidable task. Indeed, when studying the long-run performance of a specified system, it is often hard even to establish that the simulation problem at hand is well posed in that the system is stable and long-run performance measures actually exist.

This book is largely concerned with making a connection between modelling practice and output-analysis theory. We illustrate the use of the SPN building blocks for modelling and discuss the basic principles that underlie estimation procedures such as the regenerative method and the method of batch means. Tying these topics together are verifiable conditions on the building blocks of an SPN under which the net is stable over time and specified estimation procedures are valid. Our treatment highlights perhaps the most appealing aspect of SPNs: the formalism is powerful enough to permit accurate modelling of a wide range of real-world systems and yet simple enough to be amenable to stability and convergence analysis.

When studying the literature related to SPNs, one quickly encounters a multitude of SPN variants as well as a variety of other frameworks for modelling discrete-event systems. Partly for this reason, we provide—in addition to our other results—methods for comparing the modelling power of different discrete-event formalisms. Although we emphasize the comparison of SPNs with GSMPs, our general approach provides a means for making principled choices between alternative modelling frameworks. Our methodology can also be used to extend recurrence results and limit theorems from one framework to another. This latter application of our modelling-power theorems both simplifies the proofs of certain results for SPNs and makes the material in this book relevant not only to SPNs but also to the general study of discrete-event systems. Indeed, this book can be viewed as a survey of some fundamental stability, convergence, and estimation issues for discrete-event systems, using SPNs as a convenient and appealing framework for the discussion.

Our view of SPNs differs from many in the literature in that we focus on the close relationship between SPNs and GSMPs. To some extent this viewpoint is necessary: because we allow completely arbitrary clock-setting distributions, the underlying marking process of an SPN is not, in general, a Markov or semi-Markov process. Our viewpoint also is advantageous, in that it lets us exploit the many powerful results that have been established for both GSMPs and their underlying general state-space Markov chains. We emphasize, however, that SPNs have unique features that require extension—rather than straightforward adaptation—of results for GSMPs. The prime example is given by "immediate transitions," which have no counterpart in the GSMP model and lead to a variety of mathematical complications.

The presentation is self-contained. Knowledge of basic probability theory, statistics, and stochastic processes at a first-year graduate level is needed to understand the theory and examples. We occasionally use results from the theory of Markov chains on a general state space—most of the technical complexities for such chains can safely be glossed over in our setting, and the results we use are directly analogous to classical results for chains with finite or countably infinite state spaces. The Appendix summarizes the key mathematical results used in the text. To increase accessibility, we suppress measure-theoretic notation whenever possible—the Appendix contains a discussion of basic measure-theoretic concepts and their relation to the terminology used in the text. The more applied reader will wish to focus primarily on the discussion of modelling techniques and on specific estimation methods. These topics are covered primarily in Chapter 1, Chapter 2, Section 3.1.3, Section 6.3, Sections 7.2.2–7.2.4 and 7.3.3–7.3.5, Sections 8.1, 8.2.2–8.2.4, 8.3.2, and 8.3.3, and Sections 9.1 and 9.3.

I am grateful to the IBM Corporation for support of this work and for the resources of the Almaden Research Center. I also wish to thank Thomas Kurtz and the Center for the Mathematical Sciences at the University of Wisconsin–Madison for hospitality during the 1992–1993 academic year. I have benefitted from conversations with many colleagues over the years, including Sigrun Andradottir, James Calvin, Donald Iglehart, Sean Meyn, Joseph Mitchell, William Peterson, Karl Sigman, and Mary Vernon. Thanks also are due to the students of the graduate course on simulation that I taught at Stanford University during the 1998–1999 and 2000–2001 academic years. Shane Henderson provided valuable feedback on an initial version of the manuscript. As is apparent from the notes and references in the text, I am deeply indebted to Gerald Shedler, who introduced me to both SPNs and stochastic simulation and who has co-authored most of the papers I have written on these topics. Perhaps less apparent, but equally important, are the technical insights and general encouragement that I have received from Peter Glynn. The staff of Springer-Verlag has been exceedingly helpful throughout the production of this book—special thanks go to Achi Dosanjh for her help in jump-starting the project and to Kristen Cassereau for her meticulous copyediting. Finally, I wish to thank my wife, Laura, and my children, Joshua and Daniel, for their love, patience, and support.

San Jose, California   Peter J. Haas
                                                                            March 2002

Contents

List of Figures

Selected Notation

$s \to s'$ Marking s' can be reached from marking s in one step (see Definition 4.9 in Chapter 4)

$s \rightsquigarrow s'$ Marking s' can be reached from marking s in a finite number of steps (see Definition 4.9 in Chapter 4)

1_A Indicator of the set A

$|A|$ Number of elements in the set A

$x \wedge y$ Minimum of x and y

$x \vee y$ Maximum of x and y

$C_n = (C_{n,1}, \ldots, C_{n,M})$ Vector of clock readings just after the nth marking change

$C(s)$ Set of possible clock-reading vectors when the marking is s

$C[0,1]$ Space of continuous real-valued functions on $[0,1]$

$C^l[0,1]$ Space of continuous $\Re^l$-valued functions on $[0,1]$

$D = \{d_1, \ldots, d_L\}$ Set of places

E	Set of transitions
E'	Set of immediate transitions
$E(s)$	Set of transitions enabled in marking s
$E^*(s,c)$	Set of transitions—starting with marking s and clock-reading vector c—that trigger the next marking change
$E_n^* = E^*(S_n, C_n)$	Set of transitions that trigger the $(n+1)$st marking change
$\bar{\phi}$	Recurrence measure for the underlying chain of an SPN that satisfies Assumption PD (see Section 5.1.2)
$F(\,\cdot\,; s', e', s, E^*)$	Clock-setting distribution for new transition e' after a marking change from s to s' triggered by the firing of the transitions in E^*
$F_0(\,\cdot\,; e, s)$	Initial clock-setting distribution for transition e when the initial marking is s
$\gamma(n)$	Index of nth marking change at which the new marking is timed
G	Marking set
$G(e)$	Set of markings in which transition e is enabled
h_q	Function used in drift criterion for stability: $h_q(s,c) = \exp\left(q \max_{1 \leq i \leq M} c_i\right)$
H_b	Set of states of the underlying chain such that each clock reading is bounded above by b
i.i.d.	Independent and identically distributed
$I(e)$	Set of normal input places for transition e
$J(e)$	Set of output places for transition e
L	Number of places
$L(e)$	Set of inhibitor input places for transition e
μ	Initial distribution of the underlying chain

μ^+	Initial distribution of the embedded chain
M	Number of transitions
ν_0	Initial-marking distribution
$N(s'; s, E^*)$	Set of new transitions at marking change from s to s' triggered by the firing of the transitions in E^*
o.i.d.	One-dependent and identically distributed
o.d.s.	One-dependent and stationary
$O(s'; s, E^*)$	Set of old transitions at marking change from s to s' triggered by the firing of the transitions in E^*
$\psi(s)$	Number of ongoing delays when the marking is s
P_μ	Probability law of the underlying chain when the initial distribution is μ
$P_{(s,c)}$	Probability law of the underlying chain when the initial state is (s, c)
$P((s,c), A)$	Transition kernel of the underlying chain: $P((s,c), A) = P_{(s,c)} \{ (S_1, C_1) \in A \}$
$P^r((s,c), A)$	r-step transition kernel of the underlying chain: $P^r((s,c), A) = P_{(s,c)} \{ (S_r, C_r) \in A \}$
$\mathcal{P}(e)$	Priority of transition e
$r(s, e)$	Speed of clock for transition e when marking is s
$\Re^l$	l-dimensional Euclidean space ($\Re = \Re^1$ denotes the set of real numbers)
$\Re_+$	The set of nonnegative real numbers
Σ	State space of the underlying chain
Σ^+	State space of the embedded chain
S	Timed marking set
S'	Immediate marking set

$s = (s_1, s_2, \ldots, s_L)$ — Fixed marking of an SPN

$|s|$ — Total number of tokens when the marking is s

$S_n = (S_{n,1}, \ldots, S_{n,L})$ — Marking of the SPN just after the nth marking change

$\{(S_n, C_n): n \geq 0\}$ — Underlying chain of the marking process

$\{(S_n^+, C_n^+): n \geq 0\}$ — Embedded chain of the marking process: $(S_n^+, C_n^+) = (S_{\gamma(n)}, C_{\gamma(n)})$

τ_Δ — Lifetime of the marking process

$t^*(s, c)$ — Time—starting with marking s and clock-reading vector c—until the next marking change (holding-time function)

$\{X(t): t \geq 0\}$ — Marking process of an SPN

ζ_n — Time of the nth marking change

1
Introduction

Predicting the performance of a computer, manufacturing, telecommuni-
cation, workflow, or transportation system is almost always a challenging
task. Such a system usually comprises multiple activities or processes that
proceed concurrently. In a typical computer workstation, for example, the
storage subsystem writes data to a disk while, at the same time, one or
more CPUs perform computations and a keyboard transmits characters to
a buffer. Activities often have precedence relationships: assembly of a part
in a manufacturing cell does not begin until assembly of each of its subparts
has completed. Moreover, specified activities may be synchronized in that
they must always start or terminate at the same time. Activities frequently
compete for limited resources, and one activity may have either preemptive
or nonpreemptive priority over another activity for use of a resource. To
further complicate matters, many of the component processes of a system—
such as the arrival process of calls to a telephone network—are random in
nature. Because of this complexity and randomness, developing mathematical models of the system under study is usually nontrivial. The standard
"network of queues" modelling framework, for example, can fail to capture
complex synchronization behavior or precedence constraints. Assessment of
system performance is equally difficult. Models that are accurate enough to
adequately represent system behavior often cannot be analyzed using, for
example, methods based on the theory of continuous-time Markov chains
on a finite or countably infinite state space.

This book is about stochastic Petri nets (SPNs), which have proven to be a
popular and useful tool for modelling and performance analysis of complex
stochastic systems. We focus on some fundamental issues that arise when

modelling a system as an SPN and studying the long-run behavior of the resulting SPN model using computer simulation. Specifically, we consider the following questions:

- How can SPNs be used in practice to model computer, manufacturing, and other systems of interest to engineers and managers?

- How large a class of systems can be modelled within the SPN framework? To what degree do various SPN building blocks enhance modelling power?

- Under what conditions on the building blocks is an SPN model stable over time, so that long-run simulation problems are well posed?

- What simulation-based methods are available for estimating long-run performance characteristics? How can the validity of a given estimation method be established for a particular SPN model?

We address the first question by providing numerous examples of both SPN models and modelling techniques. To address the remaining questions, we study in detail the various stochastic processes associated with an SPN.

1.1 Modelling

It is frequently useful to view a complex stochastic system as evolving over continuous time and making state transitions when events associated with the occupied state occur. Often the system is a *discrete-event* system in that the stochastic state transitions occur only at an increasing sequence of random times. In a discrete-event system, each of the several events associated with a state competes to trigger the next state transition and each of these events has its own stochastic mechanism for determining the next state. At each state transition, new events may be scheduled and previously scheduled events may be cancelled.

The SPN framework provides a powerful set of building blocks for specifying the state-transition mechanism and event-scheduling mechanism of a discrete-event stochastic system. An SPN is specified by a finite set of *places* and a finite number of *transitions* along with a *normal input function,* an *inhibitor input function,* and an *output function* (each of which associates a set of places with a transition). A *marking* of an SPN is an assignment of *token* counts (nonnegative integers) to the places of the net. A transition is *enabled* whenever there is at least one token in each of its normal input places and no tokens in any of its inhibitor input places; otherwise, it is *disabled.* An enabled transition *fires* by removing one token per place from a random subset of its normal input places and depositing one token per place in a random subset of its output places. An *immediate* transition fires the instant it becomes enabled, whereas a *timed* transition fires

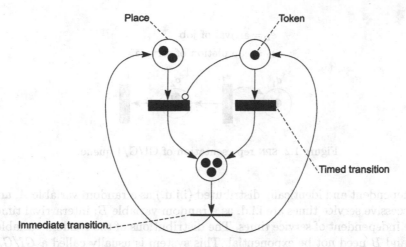

Figure 1.1. SPN building blocks.

after a positive (and usually random) amount of time. In the context of
discrete-event systems, the marking of the SPN corresponds to the state of
the system, and the firing of a transition corresponds to the occurrence of
an event. In general, for a given marking, some transitions are enabled and
others are not, reflecting the fact that some events can occur and others
cannot possibly occur when a discrete-event system is in a given state—for
example, a "departure of customer" event cannot occur if the state is such
that no customers are in the system.

SPNs have a natural graphical representation (see Figure 1.1) that facilitates modelling of discrete-event systems. This bipartite graph of the
places and transitions of an SPN determines the event-scheduling mechanism. In the graphical representation of an SPN, places are drawn as circles,
immediate transitions as thin bars, and timed transitions as thick bars. Directed arcs connect transitions to output places and normal input places to
transitions; arcs terminating in open dots connect inhibitor input places to
transitions. Tokens are drawn as black dots. In Figure 1.1, for example, the
place containing a single token is an inhibitor input place for the leftmost
of the two timed transitions and a normal input place for the rightmost of
the two timed transitions; the place containing three tokens is an output
place for each of the timed transitions. Observe that the leftmost timed
transition is not enabled (because there is a token in the inhibitor input
place) and the other two transitions are both enabled.

EXAMPLE 1.1 (GI/G/1 queue).   Consider a service center at which jobs
arrive one at a time for processing by a single server. The jobs queue for
service and are served one at a time in arrival order, that is, according to a
first-come, first-served service discipline. The server is never idle when jobs
are in the system. The times between successive arrivals to the system are

e_1 = arrival of job

e_2 = completion of service

Figure 1.2. SPN representation of GI/G/1 queue.

independent and identically distributed (i.i.d.) as a random variable A, and successive service times are i.i.d. as a random variable B; interarrival times are independent of service times. The distributions of the random variables A and B need not be exponential. This system is usually called a *GI/G/1 queue.* Here the "GI" stands for "general and independent" interarrival times, the "G" denotes a "general" service-time distribution, and the "1" denotes the number of servers.

An SPN representation of this system is displayed in Figure 1.2. In this SPN the tokens in place d_2 correspond to the jobs in the system, the firing of timed transition e_1 corresponds to the event "arrival of job," and the firing of timed transition e_2 corresponds to the event "completion of service." There is always exactly one token in place d_1, so that transition e_1 is always enabled, reflecting the fact that the arrival process to the queue is always active.[1] Thus, the marking of the net in Figure 1.2—which we write as $(1, 3)$—corresponds to the scenario in which three jobs are in the system; one job is undergoing service and two jobs are waiting in queue. Transition e_2 is enabled if and only if place d_2 contains one or more tokens, reflecting the fact that the server is never idle when jobs are in the system and at least one job must be in the system for the server to be busy. Whenever transition e_1 = "arrival of job" fires, it deposits a token in place d_2; this token corresponds to the newly arrived job. Moreover, it removes a token from place d_1 and deposits a token in place d_1 (so that the token count remains unchanged). Whenever transition e_2 = "completion of service" fires, it removes a token from place d_2; this token corresponds to the job that has just completed service and left the system. Observe that, for this particular SPN model, tokens are removed and deposited in a deterministic manner: a transition removes exactly one token from each normal input place and deposits one token in each output place whenever it fires.

This SPN model is appropriate for studying performance characteristics such as the long-run average queue length or the long-run fraction of time that the server is busy; see Example 2.2 in the next subsection. Observe that

---

[1]Place d_1 is unnecessary if we adopt the convention that a transition with no input places is always enabled.

the model can also be used for studying these performance characteristics under other service disciplines such as random service order or nonpreemptive last-come, first-served. This flexibility results because the SPN model does not explicitly keep track of the arrival order of the jobs in the system. This lack of information leads to complications, however, when studying delay characteristics such as the long-run fraction of waiting times in the queue that exceed a specified value. In Chapter 8 we discuss techniques for estimating delays in SPNs such as the one in Figure 1.2.

Heuristically, an SPN changes marking in accordance with the firing of a transition enabled in the current marking (or with the simultaneous firing of two or more transitions enabled in the current marking). Here the new marking may coincide with the current marking. The times at which transitions fire are determined by a stochastic mechanism. Specifically, a *clock* is associated with each transition. The *clock reading* for an enabled transition indicates the remaining time until the transition is scheduled to fire. Clocks run down at marking-dependent *speeds*, and a marking change occurs when one or more clocks run down to 0. The transitions enabled in a marking therefore compete to change the marking: the transitions whose clocks run down to 0 first are the "winners."

At time 0 the initial marking and clock readings are selected according to an initial probability distribution. At each subsequent marking change there are three types of transitions:

1. A *new* transition is enabled in the new marking and either is not enabled in the old marking—so that no clock reading is associated with the transition just before the marking change—or is in the set of transitions that triggers the marking change—so that the associated clock reading is 0 just before the marking change. For such a transition, a new clock reading is generated according to a probability distribution that depends only on the old marking, the new marking, and the set of transitions that triggers the marking change.

2. An *old* transition is enabled in both the old and new markings and is not in the set of transitions that triggers the marking change. The clock for such a transition continues to run down (perhaps at a new speed).

3. A *newly disabled* transition is enabled in the old marking and disabled in the new marking. If the transition is not in the set of transitions that triggers the marking change, then it is "cancelled" and its clock reading is discarded. Otherwise, the clock associated with the transition has just run down to 0 and no new clock reading is generated.

As mentioned before, we distinguish between immediate transitions which always fire the instant they become enabled and timed transitions which fire only after a positive amount of time elapses. The clock reading generated for a new immediate transition is always equal to 0 with probability 1, whereas the clock reading generated for a new timed transition is always positive with probability 1. If at least one immediate transition is enabled in a marking—as in Figure 1.1—then the marking is immediate; otherwise, the marking is timed. An immediate marking vanishes the instant it is attained.

EXAMPLE 1.2 (GI/G/1 queue). For the SPN model in Figure 1.2, the set of enabled transitions is $\{e_1\}$ whenever the marking is $(1,0)$, that is, whenever there are no tokens in place d_2 and hence no jobs in the system. Thus, as expected, the only event that can occur when the system is empty is an arrival of a job. Whenever the marking is of the form $(1,n)$ with $n > 0$, the set of enabled transitions is $\{e_1, e_2\}$, reflecting the fact that either an arrival of a job or a completion of service can occur when one or more jobs are in the system; for such a marking the clock readings associated with transitions e_1 and e_2 determine which event occurs first. Whenever transition e_1 fires, corresponding to an "arrival of job" event, e_1 immediately becomes enabled again, and a new clock reading is generated as an independent sample from the distribution of the interarrival-time random variable A. The time at which the clock next runs down to 0—so that transition e_1 fires—corresponds to the next arrival of a job. Similarly, successive clock readings for transition e_2 are generated as mutually independent samples from the distribution of the service-time random variable B. Transition e_2 can become enabled in two different ways: (1) when the marking is $(1,n)$ with $n \geq 2$ and transition e_2 fires, and (2) when the marking is $(1,0)$ and transition e_1 fires. In the former scenario, a job completes service and another job immediately begins service, so that transition e_2—which is enabled in marking $(1,n)$—fires and immediately becomes enabled again in the new marking $(1, n-1)$. In the latter scenario, a job arrives to an empty system and immediately starts to undergo service, so that transition e_2—which is not enabled in marking $(1,0)$—becomes enabled in the new marking $(1,1)$ just after transition e_1 fires. Observe that whenever the marking is $(1,1)$ and transition e_2 fires, so that a job completes service and leaves behind an empty system, transition e_2 is not enabled in the new marking $(1,0)$, and so a new clock reading is not generated for e_2 at this marking change.

EXAMPLE 1.3 (Alternative model of GI/G/1 queue). An alternative SPN model of the GI/G/1 queue is given in Figure 1.3. Here we distinguish between a job undergoing service—represented by a token in place d_3—and jobs waiting in queue—represented by tokens in place d_2. Transitions e_1 and e_2 have the same interpretation and behavior as in the SPN in Figure 1.2. Transition e_3 = "start of service" is immediate, reflecting the fact that a job starts to undergo service at the same instant it is selected for service. Whenever transition e_3 fires, it removes a token from place d_2 and deposits

e_1 = arrival of job

e_2 = completion of service

e_3 = start of service

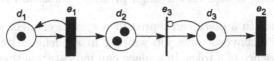

Figure 1.3. Alternative SPN representation of GI/G/1 queue.

a token in place d_3. Suppose, for example, that $n \geq 2$ jobs are in the system and transition e_2 fires, so that the marking changes from $(1, n - 1, 1)$ to $(1, n - 1, 0)$. Then transition e_3 becomes enabled and fires immediately, changing the marking to $(1, n - 2, 1)$. Similarly, whenever the system is empty and transition e_1 fires, the marking changes from $(1, 0, 0)$ to $(1, 1, 0)$; transition e_3 then becomes enabled and fires immediately, changing the marking to $(1, 0, 1)$. A marking of the form $(1, n, 0)$ with $n > 0$ is immediate, because transition e_3 is always enabled in such a marking. Observe that, due to the inhibitor arc, transition e_3 never fires when place d_3 contains a token, reflecting the fact that at most one job can undergo service at any time. Although the SPN in Figure 1.3 represents the service mechanism in greater detail, the SPN in Figure 1.2 is more convenient to work with in practice: the latter SPN has fewer places and transitions but can be used to study any performance characteristic that can be studied using the former SPN.

The timed transitions enabled in the current marking usually correspond to activities currently underway in the system, and the firing of a timed transition corresponds to the completion of an activity. SPNs are thus well suited to representation of

- *Concurrent activities*, because more than one transition can be enabled in a marking.

- *Synchronized activities*, because the firing of a transition can cause one or more transitions to become enabled (or disabled) simultaneously.

- *Activities with precedence relationships*, because a transition cannot become enabled until at least one token has been deposited in each of its normal input places and all tokens have been removed from each of its inhibitor input places. This deposit and removal of tokens typically occurs when one or more "preceding" transitions fire.

- *Priorities among activities*, because (1) a normal input place for a "high-priority" transition can also be an inhibitor input place for a "low-priority" transition, (2) at a marking change, a token representing a limited system resource can be "routed" to the normal input place for a "high-priority" transition, and (3) the clock for a "low-priority" transition can be made to run down at zero speed whenever the marking is such that a "high-priority" transition is enabled.

A token residing in a place can represent a system element such as a machine part on a conveyor or a job waiting in a queue. Alternatively, the presence or absence of a token in a place can indicate whether or not a logical condition holds. The token count in a place may be 0 or 1, for example, based on whether or not the number of vehicles on a specified stretch of road exceeds a given threshold. SPNs are conducive to both bottom-up and top-down modelling. In bottom-up modelling, a detailed subnet is developed for each component of a system, and then the subnets are combined to form the overall SPN model. In top-down modelling, a preliminary SPN model is developed that captures the main interactions between the components of the system without modelling each component in detail. Then the subnets corresponding to the system components are each progressively refined until the model is sufficiently detailed.

The *marking process* of an SPN records the marking as it evolves over continuous time. Formal definition of the process is in terms of a general state-space Markov chain that describes the SPN at successive marking changes. This underlying chain records the marking of the net together with the clock reading for each transition.

Many SPN formalisms have been proposed in the literature. Our particular choice of SPN model is motivated by several considerations:

1. *Modelling power*: As Chapter 4 shows, the class of SPNs we consider has the same modelling power as "generalized semi-Markov processes" (GSMPs). This means that a wide variety of discrete-event systems can be specified within our SPN framework.

2. *Simplicity*: The SPN formalism considered here, while powerful, consists of relatively few building blocks.

3. *Generality*: Our SPN model subsumes a number of models in the literature. The results in this book apply immediately to these latter models and often apply to other SPN models with minor modifications.

A problem sometimes encountered when modelling with SPNs is that the size of the SPN graph can become very large. One approach to this problem is to allow distinguishable tokens, so that the tokens in a place can convey more information about the state of the system than the token count alone imparts. The "colored SPNs" (CSPNs) considered in Chapter 9 are one such extension of the basic SPN model.

1.2   Stability and Simulation

Engineers and systems designers are often interested in performance characteristics such as the long-run average operating cost for a flexible manufacturing system, the long-run fraction of time a database is accessible, or the long-run utilization of a communications link. When the system of interest is modelled as an SPN, each of these characteristics typically can be specified as a time-average limit of the form

$$r(f) = \lim_{t\to\infty} \frac{1}{t} \int_0^t f\big(X(u)\big)\, du, \tag{2.1}$$

where f is a real-valued function and $X(t)$ denotes the marking of the net at time $t \geq 0$. Other performance measures of interest can be expressed as functions of such time-average limits or as (functions of) time-average limits of the underlying chain used to define the marking process.

EXAMPLE 2.2 (GI/G/1 queue).   Consider the SPN in Figure 1.2. For a marking s, write $s = (s_1, s_2)$, where s_i $(i = 1, 2)$ is the token count in place d_i. Then the long-run average number of jobs in the system is given by (2.1) with $f(s) = f(s_1, s_2) = s_2$. The long-run fraction of time that at least three jobs are in the system is given by (2.1) with

$$f(s) = \begin{cases} 1 & \text{if } s_2 \geq 3; \\ 0 & \text{otherwise,} \end{cases} \tag{2.3}$$

and the long-run fraction α of "busy time" (time when the system is nonempty) that at least three jobs are in the system is given by $r(f)/r(g)$, where $r(\cdot)$ is given by (2.1), f is defined as in (2.3), and

$$g(s) = \begin{cases} 1 & \text{if } s_2 \geq 1; \\ 0 & \text{otherwise.} \end{cases}$$

To see this, observe that

$$\alpha = \lim_{t\to\infty} \frac{\int_0^t f\big(X(u)\big)\, du}{\int_0^t g\big(X(u)\big)\, du} = \frac{\lim_{t\to\infty}(1/t)\int_0^t f\big(X(u)\big)\, du}{\lim_{t\to\infty}(1/t)\int_0^t g\big(X(u)\big)\, du} = \frac{r(f)}{r(g)}.$$

For $n \geq 0$, let $S_n = (S_{n,1}, S_{n,2})$ be the marking and $C_n = (C_{n,1}, C_{n,2})$ the vector of clock readings for transitions e_1 and e_2 just after the nth marking change. Also, set

$$\tilde{r}(\tilde{h}) = \lim_{n\to\infty} \frac{1}{n} \sum_{j=0}^{n-1} \tilde{h}(S_j, C_j)$$

for each real-valued function $\tilde{h}$ defined on the state space of the process $\{(S_n, C_n) \colon n \geq 0\}$. Then the long-run fraction α of jobs arriving to an empty system is given by $\tilde{r}(\tilde{f})/\tilde{r}(\tilde{g})$, where

$$\tilde{f}(s,c) = \tilde{f}(s_1,s_2,c_1,c_2) = \begin{cases} 1 & \text{if } s_2 = 0; \\ 0 & \text{otherwise} \end{cases}$$

and

$$\tilde{g}(s,c) = \tilde{g}(s_1,s_2,c_1,c_2) = \begin{cases} 1 & \text{if } s_2 = 0 \text{ or if } s_2 > 0 \text{ and } c_1 < c_2; \\ 0 & \text{otherwise.} \end{cases}$$

To see this, observe that $\tilde{g}(S_n, C_n) = 1$ if and only if e_1 is the next transition to fire, because either e_1 is the only transition enabled or both e_1 and e_2 are enabled, but the clock for e_1 runs down to 0 first. That is, $\tilde{g}(S_n, C_n) = 1$ if and only if the next event to occur is an arrival of a job for processing. Similarly, $\tilde{f}(S_n, C_n) = 1$ if and only if the system is empty, so that the next event to occur is an arrival of a job (to the empty system) for processing. Thus the quantity $\sum_{j=0}^{n-1} \tilde{g}(S_j, C_j)$ counts the number of arrivals to the system among the first n marking changes, and $\sum_{j=0}^{n-1} \tilde{f}(S_j, C_j)$ is the number of arrivals to an empty system among the first n marking changes. It follows that

$$\alpha = \lim_{n\to\infty} \frac{\sum_{j=0}^{n-1} \tilde{f}(S_j, C_j)}{\sum_{j=0}^{n-1} \tilde{g}(S_j, C_j)} = \frac{\lim_{n\to\infty}(1/n)\sum_{j=0}^{n-1} \tilde{f}(S_j, C_j)}{\lim_{n\to\infty}(1/n)\sum_{j=0}^{n-1} \tilde{g}(S_j, C_j)} = \frac{\tilde{r}(\tilde{f})}{\tilde{r}(\tilde{g})}$$

as asserted.

In Section 3.2 we discuss the formal specification of performance measures in more detail.

Under certain restrictions on the building blocks of an SPN, the marking process $\{\,X(t)\colon t \geq 0\,\}$ is a continuous-time Markov chain (CTMC) with finite or countably infinite state space; see Section 3.4. A variety of techniques is then available for determining whether the time-average limits of interest exist and, if so, for computing these limits either analytically or numerically. In general, however, the stochastic process $\{\,X(t)\colon t \geq 0\,\}$ is not a continuous-time Markov chain or even a semi-Markov process. Determining the existence of time-average limits then becomes a highly nontrivial task and the limits, if they exist, must be estimated using computer simulation.[2] We focus primarily on problems for which simulation is required,

---

[2]Even when the marking process is a CTMC, the chain's state space may be so large that simulation is the only practical means of assessing long-run behavior. Similarly, even when the performance measure of interest can be represented as a time-average limit of the underlying chain of the marking process—or as a function of such limits— simulation usually is required because the state space of the underlying chain is too complex to admit analytical or numerical solution methods.

and our discussion centers around stochastic process properties pertinent to estimation methods for SPNs.

The usual reason for estimating time-average limits is to compare alternative system designs or operating policies, and real differences must be distinguished from apparent differences caused by random fluctuations. It is therefore essential to provide not only an estimate of each time-average limit of interest, but also an assessment of the precision of each estimate. This assessment frequently takes the form of a confidence interval. In general, obtaining point estimates and confidence intervals for time-average limits is not an easy task, because successive observations of the marking process are usually far from being either independent or identically distributed. Indeed, the evolution of the marking process can depend heavily on the initial conditions of the simulation, even when the simulated time is large, so that the resulting estimates suffer from "initialization bias." To obtain meaningful estimates, effective methods are needed for selection of the number of runs, the length of each run, the initial conditions for each run, the quantities to be measured, and the form of the final estimates.

The estimation problem is simplified considerably when $\{ X(t) \colon t \geq 0 \}$ is a *regenerative process*, that is, when there exists an infinite sequence of random time points (called *regeneration points*) at which the process probabilistically restarts. The regeneration points decompose sample paths of the process into i.i.d. "cycles." Under mild regularity conditions, the regenerative property guarantees the existence of time-average limits. Moreover, the "regenerative method" for analysis of simulation output can be used to obtain strongly consistent point estimates and asymptotic confidence intervals for time-average limits; the method requires observation of only a finite portion of a single sample path of the marking process. It is often apparent that the marking process of an SPN probabilistically restarts whenever the net is in a specified marking and a specified transition fires, but it can be difficult to verify that such restarts occur infinitely often with probability 1. It is even harder to determine whether, as the method requires, both the expected time between regeneration points and the "regenerative variance constant" are finite. Establishing these properties often amounts to showing that the underlying chain hits a specified set of states infinitely often with probability 1 and that the times between successive hits have finite second moment. Thus stability properties such as recurrence are of central importance to our discussion.

The regenerative method is not applicable when a sequence of regeneration points cannot be found or when regenerations occur too infrequently. Sometimes, however, strongly consistent point estimates and asymptotic confidence intervals for time-average limits can be obtained nonetheless, using methods based on *standardized time series* (STS). Perhaps the best-known STS method is the method of *batch means* (with the number of batches independent of the simulation run length). A sufficient condition for the validity of STS methods is that the output process $\{ f(X(t)) \colon t \geq 0 \}$

obey a *functional central limit theorem* (FCLT). Roughly speaking, a stochastic process with time-average limit r obeys an FCLT if the associated cumulative (i.e., time-integrated) process, centered about the deterministic function $g(t) = rt$ and suitably compressed in space and time, converges in distribution to a Brownian motion as the degree of compression increases. The challenge, then, is to determine from an SPN's building blocks whether or not such an FCLT holds. As in the regenerative setting, this problem can be reformulated as a stability question for the underlying chain.

It may also be possible to obtain point estimates and confidence intervals for time-average limits using *consistent estimation methods* such as *variable* batch-means (in which the number of batches is an increasing function of the simulation run length) or *spectral* methods. These methods assume that the output process obeys an ordinary central limit theorem (CLT) and are based on consistent estimation of the variance constant that appears in the CLT. When applicable, consistent estimation methods yield confidence intervals that are asymptotically shorter and less variable than those STS methods provide. As with regenerative and STS methods, determining if a consistent estimation method is applicable to a specified SPN model amounts to analyzing the stability of the underlying chain.

The discussion so far has pertained to estimation of performance characteristics that can be expressed in terms of time-average limits of the marking process or underlying chain, such as long-run utilization, availability, and reliability. Frequently, however, assessment of delay phenomena also is of interest. Examples of delays include the time to produce an item in a flexible manufacturing system, the time to compute the answer to a query in a database management system, and the time to transmit a message from one node to another in a communication network. Typically, such delays correspond to lengths of certain "delay intervals" (random time intervals) determined by marking changes of an SPN, and the performance measures of interest can be expressed in the form $\lim_{n \to \infty} (1/n) \sum_{j=0}^{n-1} f(D_j)$, where f is a real-valued function and $\{ D_j : j \geq 0 \}$ is a sequence of delays. The limiting average delay $\lim_{n \to \infty} (1/n) \sum_{j=0}^{n-1} D_j$ can sometimes be estimated indirectly, that is, without measuring lengths of individual delay intervals. For general time-average limits of a sequence of delays, however, individual lengths must be measured and then combined to form point and interval estimates. Because there can be more than one ongoing delay at a time point and delays need not terminate in the order in which they start, measuring individual lengths is a nontrivial step of the simulation. A mechanism is needed to link the starts (left endpoints) and terminations (right endpoints) of individual delay intervals.

When the marking process is regenerative and there are no ongoing delays at any regeneration point, the sequence of delays is a regenerative process in discrete time. Strongly consistent point estimates and asymptotic confidence intervals for time-average limits can therefore be obtained using the regenerative method. When there are ongoing delays at each regeneration point, however, extensions of the standard regenerative method are needed to obtain point estimates and confidence intervals. When there is no apparent sequence of regeneration points or regenerations occur too infrequently, STS methods can be used to obtain point estimates and asymptotic confidence intervals for time-average limits, provided that the sequence of delays obeys an FCLT. Verifying that such an FCLT holds for a specific SPN model again amounts to establishing stability properties for the underlying chain.

1.3 Overview of Topics

The remainder of the book is organized as follows. We give a formal description of the SPN building blocks in Chapter 2 and, through a set of examples, illustrate the use of SPNs as models of discrete-event systems. Methods are provided for concise specification of SPN models in which more than one transition can fire simultaneously.

Chapter 3 focuses on basic properties of the marking process of an SPN. We give a formal definition of the marking process and show that this definition leads to an algorithm for generating sample paths. Through examples, we show that a wide variety of long-run performance measures can be represented as time-average limits of the marking process. Other performance measures can be represented as functions of time-average limits, where the limits are expressed in terms of either the marking process or the underlying chain. In this connection, we discuss some general relationships between limits in discrete and continuous time. Next, we show that a marking process can exhibit pathological behavior in which, with positive probability, an infinite number of marking changes occur in a finite time interval. Conditions that rule out such "explosions" are then developed. Finally, we give conditions under which the marking process is a continuous-time Markov chain with finite or countably infinite state space.

Modelling-power issues are explored in Chapter 4. We first show that for every GSMP there exists an SPN with a marking process that "strongly mimics" the GSMP; in this sense, SPNs have at least the modelling power of GSMPs. This result provides a justification for the SPN formulation introduced in Chapter 2. Indeed, since the SPN building blocks often are more convenient for modelling than the GSMP building blocks, the foregoing result establishes SPNs as an attractive general framework for modelling and simulation analysis of discrete-event systems. The methodology used to obtain the modelling-power result can also be used to assess the relative modelling power of different SPN formulations and the contribution of individual SPN building blocks to overall modelling power. For example, in contrast to a well-known result from the theory of ordinary (untimed, deterministic) Petri nets, we show that inclusion of inhibitor input places does not increase the modelling power of SPNs. We conclude the chapter by establishing the converse of our main modelling-power result: for every SPN there exists a GSMP that strongly mimics the marking process of the SPN. This result permits direct application to SPNs of results from the theory of GSMPs. Moreover, when establishing stability properties for SPNs, the converse result provides a useful tool for dealing with the various complications caused by the presence of immediate transitions.

In Chapter 5 we provide techniques for showing that specified subsets of the state space of the underlying chain are hit infinitely often with probability 1. Such recurrence arguments are needed to establish, for specific SPN models, both the existence of time-average limits and the applicability of various estimation methods. One approach to demonstrating recurrence is to show that the underlying chain "drifts" toward a specified compact subset of the state space whenever the chain lies outside this subset. We give "positive density" conditions on the clock-setting distributions under which a drift condition holds. An alternative approach that imposes less stringent constraints on the clock-setting distributions is based on a "geometric trials" recurrence criterion. This latter approach utilizes the detailed structure of the SPN model as well as properties of "GNBU" distributions.

Chapter 6 deals with estimation methods for SPNs in which the marking process or underlying chain is regenerative. After summarizing the relevant properties of regenerative processes, we give conditions on the building blocks of an SPN under which there exists a sequence of regeneration points both for the marking process and for the underlying chain. We then show how this regenerative structure can be used to obtain point estimates and confidence intervals for time-average limits. In addition to presenting the basic method, we discuss techniques for reducing the bias of the standard estimator, obtaining point estimates and confidence intervals for functions of time-average limits, and estimating gradients of time-average limits with respect to system parameters. We also describe extensions of the basic method that permit dependence between adjacent cycles.

Chapter 7 focuses on estimation methods that can be used when the regenerative method is inapplicable. We first consider methods based on standardized time series. The discussion covers the general theory of standardized time series, as well as the STS-area, STS-maximum, and batch-means methods. Based on stability results for general state-space Markov chains, conditions on the building blocks of an SPN are given under which the output process obeys an FCLT, so that STS methods are applicable. We then give conditions under which various consistent estimation methods can be applied. The idea is to first adapt results from the literature to obtain such conditions under the (unrealistic) assumption that the output process of the simulation is stationary. We then use a "coupling" argument to extend these results to the nonstationary setting usually found in practice. This development leads to conditions on the building blocks of an SPN under which a class of "quadratic-form" estimators are consistent for the asymptotic variance. Included in this class are estimators for the method of variable batch means and for various spectral methods.

Chapter 8 concerns delays in SPNs. We first introduce a recursively defined sequence of random vectors, called "start vectors," whose use provides a means both for specification of a sequence of delays $\{ D_j : j \geq 0 \}$ and for subsequent measurement of the delays during the course of a simulation run. When there exists a sequence of regeneration points for the underlying chain, the sequence of delays can be decomposed into one-dependent stationary (o.d.s.) cycles. Various extensions of the standard regenerative method can then be used to estimate general time-average limits—we compare the statistical efficiency of two such extensions. These estimation methods reduce to the standard regenerative method when there are no ongoing delays at any regeneration point. If the performance measure of interest is the limiting average delay, then a specialized estimation method can be used that does not require measurement of individual delays. When there is no apparent sequence of regeneration points for the underlying chain but the clock-setting distributions satisfy positive density conditions as in Chapter 5, it is still possible to decompose the sequence of delays into o.d.s. cycles. Although the random indices that decompose sample paths into such cycles cannot be determined explicitly, the mere existence of these points implies that, under mild regularity conditions, time-average limits are well defined and the output process $\{ f(D_j) : j \geq 0 \}$ obeys an FCLT. It then follows that STS methods such as batch means can be used to obtain strongly consistent point estimates and asymptotic confidence intervals for time-average limits.

Chapter 9 introduces colored stochastic Petri nets (CSPNs). A CSPN is similar to an ordinary SPN, except that tokens come in different "colors" and a transition fires "in a color." An "input incidence function" and an "output incidence function" determine the transitions enabled in a marking as well as the number of tokens of each color that are removed and deposited when a transition fires in a color. The primary appeal of CSPNs for modelling of discrete-event systems is that such nets permit concise specification, especially when there are many subsystems of similar structure or behavior. Virtually all the simulation-based estimation methodology developed for ordinary SPNs carries over to the CSPN setting. When the net exhibits "symmetry with respect to color," modifications of the standard regenerative method lead to shorter cycle lengths and—when estimating delays—to increased statistical efficiency.

Notes

Petri nets are named after Carl Adam Petri, who introduced the nets in his 1962 Ph.D. dissertation. At present, the literature contains over 7000 books, papers, and reports dealing with Petri nets and their extensions. Petri's original nets are deterministic and involve no notion of time. Overviews of the theory of such deterministic Petri nets can be found in the books of Peterson (1981) and Reisig (1985) and the survey paper of Murata (1989).

Symons (1978, 1980) proposed the use of transitions with random firing times together with transitions that take "an insignificant amount of time to fire" (that is, immediate transitions). Symons' work, together with that of Natkin (1980) and Molloy (1981), resulted in the first SPN models.

Ajmone Marsan et al. (1984, 1987) develop the "generalized SPN" (GSPN) model, a type of SPN in which each transition is either immediate or has exponentially distributed firing times. An introduction to GSPNs can be found in Ajmone Marsan et al. (1995).

The SPN formulation used in this book follows Haas and Shedler (1985b, 1989b). As indicated in Section 1.1, many of the results in the following chapters can be adapted to other SPN settings, for example, GSPNs.

In the literature, timed and immediate markings are also referred to as "tangible" and "vanishing" markings, respectively. The mechanism for scheduling the firing of transitions is sometimes called the "race model with enabling memory."

The stochastic-process viewpoint that is central to our approach can be traced back to the early work of Crane and Iglehart (1975), Whitt (1980), and Iglehart and Shedler (1983), among others. A useful, complementary view of SPNs and GSMPs can be based on the notion of "stochastic timed automata"—see Cassandras and LaFortune (1999) and Glasserman and Yao (1994) for examples of this approach.

A number of important topics pertinent to general simulation methodology lie outside the scope of our discussion. Such topics include choosing the level of detail for a simulation model, selecting input probability distributions, generating random numbers, choosing data structures and algorithms for generating sample paths, debugging a simulation model, and validating model output against real-world data. Banks (1998), Bratley et al. (1987), and Law and Kelton (2000), for example, discuss these aspects of simulation. These references and others also discuss more elaborate versions of the estimation methods given in this book—we focus on relatively simple versions of the various methods because their validity can be rigorously established for specific SPN models.

2
Modelling with Stochastic Petri Nets

Stochastic Petri nets (SPNs) are well suited to representing concurrency, synchronization, precedence, and priority. After presenting the basic SPN building blocks in Section 2.1, we give a series of examples in Section 2.2 that illustrates the use of SPNs for modelling discrete-event systems. We pay particular attention to complications that arise in the specification of *new-marking probabilities*. These probabilities determine the mechanism by which a transition removes tokens from a random subset of its normal input places and deposits tokens in a random subset of its output places when it fires. Consideration of a queueing system with batch arrivals shows that new-marking probabilities must be allowed to depend explicitly on the current marking; that is, the SPN formalism must include *marking-dependent* transitions. By means of an example, we show how new-marking probabilities for an SPN with marking-dependent transitions can be specified in a form suitable for processing by a computer program. Another complication arises when more than one transition can fire at a time point. In principle, new-marking probabilities must be defined for all possible sets of simultaneously firing transitions, and there can be an extremely large number of such sets. As shown in Section 2.3, concise specification of new-marking probabilities can be facilitated by assigning numerical "priorities" to transitions.

2.1 Building Blocks

The basic elements of an SPN "graph" are

- A finite set $D = \{ d_1, d_2, \ldots, d_L \}$ of *places*
- A finite set $E = \{ e_1, e_2, \ldots, e_M \}$ of *transitions*

- A (possibly empty) set $E' \subset E$ of *immediate transitions*

- Sets $I(e), L(e), J(e) \subseteq D$ of *normal input places, inhibitor input places*, and *output places*, respectively, for each transition $e \in E$

The transitions in $E - E'$ are called *timed transitions*. Denote by G the finite or countably infinite set of *markings*. For $s \in G$ we write $s = (s_1, s_2, \ldots, s_L)$, where s_j is the number of *tokens* in place $d_j \in D$.

Definition 1.1. An SPN is said to be *k-bounded* $(k \geq 1)$ if and only if

$$\max(s_1, s_2, \ldots, s_L) \leq k$$

for each $s = (s_1, s_2, \ldots, s_L) \in G$.

Thus an SPN is k-bounded if and only if the token count in a place never exceeds k.

Let $E(s)$ be the set of transitions that are *enabled* when the marking is s, that is, the set of transitions having at least one token in each normal input place and no tokens in any inhibitor input place:

$$E(s) = \{ e \in E : s_j \geq 1 \text{ for } d_j \in I(e) \text{ and } s_j = 0 \text{ for } d_j \in L(e) \}.$$

A transition $e \in E - E(s)$ is *disabled* when the marking is s. In a dual manner, set

$$G(e) = \{ s \in G : e \in E(s) \}$$

for $e \in E$, so that $G(e)$ is the set of markings in which transition e is enabled. Define the set S' of *immediate markings* by

$$S' = \{ s \in G : E(s) \cap E' \neq \varnothing \}$$

and the set S of *timed markings* by

$$S = G - S' = \{ s \in G : E(s) \cap E' = \varnothing \}.$$

According to this definition, an element of the marking set is an immediate marking if at least one immediate transition is enabled. Heuristically, an immediate marking vanishes the instant it is attained.

EXAMPLE 1.2 (GI/G/1 queue). For the SPN in Figure 1.3—see Example 1.3 in Chapter 1—we have $D = \{ d_1, d_2, d_3 \}$, $E = \{ e_1, e_2, e_3 \}$, and $E' = \{ e_3 \}$. The SPN graph is formally described by setting

- $I(e_1) = \{ d_1 \}$, $I(e_2) = \{ d_3 \}$, $I(e_3) = \{ d_2 \}$.

- $J(e_1) = \{ d_1, d_2 \}$, $J(e_2) = \varnothing$, $J(e_3) = \{ d_3 \}$.

- $L(e_1) = L(e_2) = \varnothing$, $L(e_3) = \{ d_3 \}$.

The set of markings is $G = \{1\} \times \{0, 1, 2, \dots\} \times \{0, 1\}$ (where $\times$ denotes Cartesian product), and the set of immediate markings is

$$S' = \{(s_1, s_2, s_3) \in G: s_2 > 0 \text{ and } s_3 = 0\}.$$

The sets of enabled transitions are given by

- $E((1, 0, 0)) = \{e_1\}$.

- $E((1, n, 1)) = \{e_1, e_2\}$ for $n \geq 0$.

- $E((1, n, 0)) = \{e_1, e_3\}$ for $n \geq 1$.

Similarly, $G(e_1) = G$, $G(e_2) = \{1\} \times \{0, 1, 2, \dots\} \times \{1\}$, and $G(e_3) = \{1\} \times \{1, 2, \dots\} \times \{0\}$.

The marking of an SPN changes when one or more enabled transitions *fire*. For $E^* \subseteq E(s)$, denote by $p(s'; s, E^*)$ the probability that the new marking is s' given that the marking is s and the transitions in the set E^* fire simultaneously. For each $s \in G$ and $E^* \subseteq E(s)$, the function $p(\cdot; s, E^*)$ is a probability mass function on G in that $\sum_{s' \in G} p(s', s, E^*) = 1$. Recall that a transition removes at most one token from each of its normal input places and deposits at most one token in each of its output places when it fires. We therefore permit $p(s'; s, E^*)$ to be positive only if $s = (s_1, s_2, \dots, s_L)$, $s' = (s'_1, s'_2, \dots, s'_L)$, and E^* satisfy

$$s_j - \sum_{e^* \in E^*} 1_{I(e^*)}(d_j) \leq s'_j \leq s_j + \sum_{e^* \in E^*} 1_{J(e^*)}(d_j) \tag{1.3}$$

for $1 \leq j \leq L$. Here 1_A denotes the indicator function of the set A, so that the quantity $\sum_{e^* \in E^*} 1_{I(e^*)}(d_j)$ is the number of transitions $e^* \in E^*$ for which d_j is a normal input place and $\sum_{e^* \in E^*} 1_{J(e^*)}(d_j)$ is the number of transitions $e^* \in E^*$ for which d_j is an output place. Observe that the token count of a place may increase or decrease by more than 1 when transitions fire simultaneously.

EXAMPLE 1.4 (Cyclic queues with feedback). Consider a closed network of queues with two single-server service centers and N (≥ 2) jobs; see Figure 2.1. With fixed probability $p \in (0, 1)$, a job that completes service at center 1 moves to center 2 and with probability $1 - p$ joins the tail of the queue at center 1. A job that completes service at center 2 moves to center 1. The queueing discipline at each center is first-come, first-served. Successive service times at center i ($i = 1, 2$) are i.i.d. as a random variable L_i having a continuous distribution function. Observe that, with probability 1, a service completion at center 1 and a service completion at center 2 never occur simultaneously.

An SPN model of this system is displayed in Figure 2.2. The tokens in place d_i ($i = 1, 2$) correspond to the jobs at center i (either waiting or

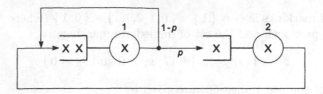

Figure 2.1. Cyclic queues with feedback (five jobs).

e_1 = service completion at center 1

e_2 = service completion at center 2

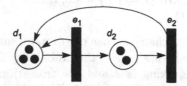

Figure 2.2. SPN representation of cyclic queues with feedback.

in service). Whenever transition e_2 fires, it removes a token from place d_2 and deposits a token in place d_1, reflecting the fact that a job that completes service at center 2 moves to center 1. Whenever transition e_1 fires, it removes a token from place d_1; moreover, it deposits a token in place d_2 with probability p and in place d_1 with probability $1 - p$. Equivalently, with probability p, transition e_1 removes a token from place d_1 and deposits a token in place d_2 and, with probability $1 - p$, removes and deposits no tokens when it fires. In this manner the SPN model captures the feedback mechanism in the network of queues. Formally, we have

$$p(s; s, \{e_1\}) = 1 - p,$$
$$p((s_1 - 1, s_2 + 1); s, \{e_1\}) = p,$$
$$p((s_1 + 1, s_2 - 1); s, \{e_2\}) = 1$$

for $s = (s_1, s_2) \in G$.

Now suppose that transitions e_1 and e_2 can fire simultaneously. This situation can arise, for example, if each service-time random variable L_i takes values in the set $\{1, 2, 3, \dots\}$. Whenever e_1 and e_2 fire simultaneously, the SPN changes marking as if one of the transitions fires immediately after the other (the order of the firings is immaterial). That is,

$$p((s_1 + 1, s_2 - 1); s, \{e_1, e_2\}) = 1 - p,$$
$$p(s; s, \{e_1, e_2\}) = p$$

for $s = (s_1, s_2) \in S$.

Often in applications the stochastic mechanism for removing and depositing tokens is degenerate and does not explicitly depend on the current marking.

Definition 1.5. A transition $e \in E$ is said to be *deterministic* if and only if, for all $s = (s_1, s_2, \ldots, s_L) \in G(e)$, we have $p(s'; s, \{e\}) = 1$, where s' is determined from s according to the relations

$$s'_j = \begin{cases} s_j - 1 & \text{if } d_j \in I(e) \cap (D - J(e)); \\ s_j + 1 & \text{if } d_j \in J(e) \cap (D - I(e)); \\ s_j & \text{otherwise} \end{cases}$$

for $1 \le j \le L$.

Thus a transition e is deterministic if, with probability 1, it removes exactly one token from each normal input place and deposits exactly one token in each output place whenever it fires (and no other transitions fire simultaneously).

EXAMPLE 1.6 (Deterministic transitions). For both the SPN in Figure 1.2 and the SPN in Figure 1.3, all transitions are deterministic. For the SPN in Figure 2.2, transition e_2 is deterministic but transition e_1 is not.

A *clock* is associated with each transition. The clock for an enabled transition records the remaining time until the transition is scheduled to fire. These clocks, along with the speeds at which the clocks run down, determine which of the enabled transitions trigger the next marking change. Denote by $r(s, e)$ (≥ 0) the *speed* (finite, deterministic rate) at which the clock associated with transition e runs down when the marking is $s \in G(e)$. The requirement that $r(s, e)$ be finite is needed to ensure that timed transitions never fire instantaneously. We require that $r(s, e) = 1$ for $e \in E'$ and $s \in G(e)$. In particular, this means that zero speeds are not allowed for immediate transitions; such transitions always fire the instant they become enabled. Typically in applications, all speeds for enabled transitions are equal to 1. There exist models, however, in which speeds other than 1 as well as state-dependent speeds are convenient. For example, zero speeds are needed for specification of queueing systems with service interruptions of the "preemptive-resume" type—see Example 2.3 in the following section. State-dependent speeds are needed for queueing systems in which the service effort is divided among the jobs receiving service (the "processor sharing" service discipline).

To avoid trivialities, we always assume without comment that

1. For each marking $s \in G$, there exists a transition $e \in E(s)$ with $r(s, e) > 0$.

2. For each transition $e \in E$, there exists a marking $s \in G(e)$ with $r(s, e) > 0$.

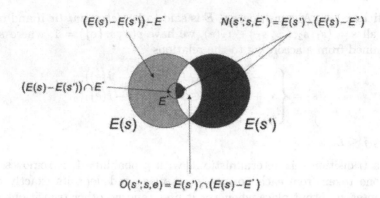

Figure 2.3. Sets of new, old, and newly disabled transitions.

The assumption in (2) implies that $L(e) \cap I(e) = \varnothing$ for $e \in E$, so that no place can be both a normal input place and an inhibitor input place for a transition.

The initial marking s_0 is selected according to an *initial-marking distribution* ν_0 defined on G. Then, for each enabled transition $e_i \in E(s_0)$, an initial clock reading is generated according to an *initial clock-setting distribution function* $F_0(\,\cdot\,; e_i, s_0)$. The distribution function ν_0 may be degenerate in the sense that $\nu_0(s) = 1$ for some $s \in G$.

At a subsequent marking change from s to s' triggered by the simultaneous firing of the transitions in the set E^*, a finite clock reading is generated for each *new transition* $e' \in N(s'; s, E^*) = E(s') - (E(s) - E^*)$. Denote the *clock-setting distribution function*—that is, the distribution function of such a new clock reading—by $F(\,\cdot\,; s', e', s, E^*)$. For $e' \in E'$, we require that $F(0; s', e', s, E^*) = 1$ for s, s', and E^*, so that immediate transitions always fire instantaneously. For $e' \in E - E'$, we require that $F(0; s', e', s, E^*) = 0$ for s, s', and E^*, so that timed transitions never fire instantaneously. For each *old transition* $e' \in O(s'; s, E^*) = E(s') \cap (E(s) - E^*)$, the old clock reading is kept after the marking change. A transition in the set $E(s) - E(s')$ is called a *newly disabled transition*, and we distinguish between two types of newly disabled transitions.

1. For $e' \in (E(s) - E(s')) - E^*$, transition e' (which was enabled before the transitions in E^* fired) is cancelled and the clock reading is discarded.

2. For $e' \in (E(s) - E(s')) \cap E^*$, the clock for transition e' has run down to 0 just before the marking change and no new clock reading is generated.

Figure 2.3 illustrates these definitions.

When the marking is s and the set E^* of transitions that trigger a marking change is a singleton set of the form $E^* = \{e^*\}$, we often write $p(s'; s, e^*)$ for $p(s'; s, \{e^*\})$, $O(s'; s, e^*)$ for $O(s'; e, \{e^*\})$, and so forth.

EXAMPLE 1.7 (GI/G/1 queue). For the SPN in Figure 1.2, all speeds are equal to 1. The clock-setting distribution functions are given by

$$F(x; s', e_1, s, E^*) \equiv F(x; e_1) = P\{A \leq X\}$$

and

$$F(x; s', e_2, s, E^*) \equiv F(x; e_2) = P\{B \leq X\},$$

where A and B are the interarrival-time and service-time random variables. Observe that whenever a job arrives, the next arrival event is scheduled immediately, so that e_1 is always a new transition at a marking change triggered by the firing of e_1. If place d_2 contains no tokens just before such a marking change—so that the job arrives to an empty system—then the arriving job immediately goes into service and a "completion of service" event is scheduled. That is, e_2 is also a new transition at such a marking change. Otherwise, if place d_2 contains one or more tokens, then a "completion of service" event has previously been scheduled, so that e_2 is an old transition rather than a new transition. Thus, for $s = (s_1, s_2) \in G$,

$$N(s'; s, e_1) = \begin{cases} \{e_1, e_2\} & \text{if } s_2 = 0; \\ \{e_1\} & \text{if } s_2 > 0. \end{cases}$$

Similarly,

$$N(s'; s, e_2) = \begin{cases} \varnothing & \text{if } s_2 = 1; \\ \{e_2\} & \text{if } s_2 > 1, \end{cases}$$

$$O(s'; s, e_1) = \begin{cases} \varnothing & \text{if } s_2 = 0; \\ \{e_2\} & \text{if } s_2 > 0, \end{cases}$$

and

$$O(s'; s, e_2) = \{e_1\}.$$

In each of these equations, s' denotes the unique new marking when the current marking is s and the specified transition fires. Suppose that at time 0 a job arrives to an empty system. Then the initial-marking distribution is

$$\nu_0(s) = \begin{cases} 1 & \text{if } s = (1, 1); \\ 0 & \text{otherwise} \end{cases}$$

or, equivalently, $\nu_0(s) = 1_{\{(1,1)\}}(s)$. For the initial marking $s_0 = (1,1)$, we have $E(s_0) = \{e_1, e_2\}$; the initial clock-setting distribution functions for transitions e_1 and e_2 are given by $F(x; e_1)$ and $F(x; e_2)$ as defined previously.

Observe that for each of transitions e_1 and e_2 in Example 1.7, the clock-setting distribution function does not explicitly depend on the new marking, old marking, or set of transitions that trigger the marking change. Such transitions frequently occur in practice and motivate the following definition.

Definition 1.8. A timed transition e' is said to be *simple* if there exists a distribution function $F(\cdot; e')$ such that

$$F(\cdot; s', e', s, E^*) \equiv F(\cdot; e')$$

and

$$F_0(\cdot; e', s) \equiv F(\cdot; e')$$

for all s', s, and E^*.

2.2 Illustrative Examples

The examples in this section illustrate the specification of discrete-event systems using the SPN building blocks. These examples demonstrate various modelling techniques and also highlight some important modelling issues.

2.2.1 Priorities: Producer–Consumer Systems

The activities in a system usually require various system resources. To process a part in an automated manufacturing system, for example, a suitable machine is needed. To transmit a voice conversation over a telephone system, a set of communication links must be available. When a resource is scarce, competition among activities for use of the resource usually is resolved by assigning relative priorities to the activities. The following examples show how immediate transitions, inhibitor input places, and zero speeds can be used to model a variety of preemptive and nonpreemptive priority schemes.

EXAMPLE 2.1 (Producer–consumer system with nonpreemptive priority). Consider a system consisting of two producers, two consumers, and two buffers, each numbered 1 and 2. The producers share a single channel for transmission (one at a time) of items to consumers. Producer i ($i = 1, 2$) creates items for consumer i one at a time; items created but not yet transmitted are placed in buffer i for transmission. Buffer i has finite capacity

e_1 = creation of item by producer 1

e_2 = start of transmission to consumer 1

e_3 = end of transmission to consumer 1

e_4 = creation of item by producer 2

e_5 = start of transmission to consumer 2

e_6 = end of transmission to consumer 2

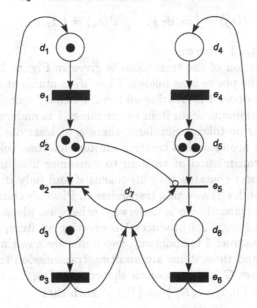

Figure 2.4. SPN representation of producer–consumer system with nonpreemptive priority and finite buffers.

$B_i > 0$; that is, an item created by producer i when the system already contains $B_i - 1$ items for consumer i causes the process of creation of items for consumer i to shut down. This process remains shut down until the first subsequent end of transmission to consumer i. Producer–consumer pair 1 has nonpreemptive priority over producer–consumer pair 2 for use of the channel. Items created by producer i are transmitted in the order in which they are created. The successive times required by producer i to create an item are i.i.d. as a positive random variable A_i with continuous distribution function, and the successive times to transmit an item to consumer i are i.i.d. as a positive random variable L_i with continuous distribution function. (All creation times and transmission times are mutually independent.)

This system can be specified as an SPN with deterministic timed and immediate transitions; see Figure 2.4 for $B_1 = 4$ and $B_2 = 3$. Let $D = \{ d_1, d_2, \ldots, d_7 \}$ be the set of places of the SPN, $E = \{ e_1, e_2, \ldots, e_6 \}$ be the set of transitions, and $E' = \{ e_2, e_5 \}$ be the set of immediate transitions.

Set

$$L(e_5) = \{ d_2 \}$$

and $L(e_j) = \varnothing$ otherwise. Also set

$$I(e_2) = \{ d_2, d_7 \}, \qquad I(e_5) = \{ d_5, d_7 \},$$

and $I(e_j) = \{ d_j \}$ otherwise. Finally, set

$$J(e_3) = \{ d_1, d_7 \}, \qquad J(e_6) = \{ d_4, d_7 \},$$

and $J(e_j) = \{ d_{j+1} \}$ otherwise.

The interpretation of the transitions is given in Figure 2.4, and the interpretation of the places is as follows. Place d_1 contains at least one token if and only if producer 1 is creating an item. Place d_3 contains one token if and only if transmission of an item to consumer 1 is underway; otherwise, place d_3 contains no tokens. Similarly, there is at least one token in place d_4 if and only if producer 2 is creating an item and one token in place d_6 if and only if transmission of an item to consumer 2 is underway. Place d_2 (resp., place d_5) contains k (≥ 0) tokens if and only if k items are in buffer 1 (resp., buffer 2) awaiting transmission. Place d_7 contains one token if and only if no transmission is underway; otherwise, place d_7 contains no tokens. Thus, in Figure 2.4 producer 1 is creating an item, a transmission of an item to consumer 1 is underway, two items are awaiting transmission to consumer 1, and three items are awaiting transmission to consumer 2.

The marking set G is the set of all elements $(s_1, s_2, \ldots, s_7) \in \{ 0, 1, \ldots, B_1 \}^2 \times \{ 0, 1 \} \times \{ 0, 1, \ldots, B_2 \}^2 \times \{ 0, 1 \}^2$ such that

1. $s_3 + s_6 + s_7 = 1$.

2. $s_1 + s_2 + s_3 = B_1$.

3. $s_4 + s_5 + s_6 = B_2$.

The first constraint reflects the fact that, at any time point, a transmission to consumer 1 is underway, a transmission to consumer 2 is underway, or the channel is idle. Thus the token that resides in place d_3, d_6, or d_7 represents the limited, shared channel resource. The second and third constraints reflect the fact that an item for consumer 1 (resp., consumer 2) is either "waiting to be produced," waiting to be transmitted, or undergoing transmission. The immediate marking set S' is given by

$$S' = \{ (s_1, s_2, s_3, s_4, s_5, s_6, s_7) \in G : s_7 = 1 \text{ and } s_2 + s_5 > 0 \}.$$

It can be shown that $|G| = 3B_1 B_2 + 2B_1 + 2B_2 + 1$, $|S| = 2B_1 B_2 + B_1 + B_2 + 1$, and $|S'| = B_1 B_2 + B_1 + B_2$. (Here, as elsewhere, $|A|$ denotes the number of elements in the set A.)

The new-marking probabilities are as follows. If $e^* = e_1 =$ "creation of item by producer 1," then the new-marking probability $p(s'; s, e^*) = 1$ when

$$s = (s_1, s_2, s_3, s_4, s_5, s_6, s_7) \text{ and } s' = (s_1 - 1, s_2 + 1, s_3, s_4, s_5, s_6, s_7).$$

If $e^* = e_2 =$ "start of transmission to consumer 1," then $p(s'; s, e^*) = 1$ when

$$s = (s_1, s_2, 0, s_4, s_5, 0, 1) \text{ and } s' = (s_1, s_2 - 1, 1, s_4, s_5, 0, 0).$$

If $e^* = e_3 =$ "end of transmission to consumer 1," then $p(s'; s, e^*) = 1$ when

$$s = (s_1, s_2, 1, s_4, s_5, 0, 0) \text{ and } s' = (s_1 + 1, s_2, 0, s_4, s_5, 0, 1).$$

If $e^* = e_4 =$ "creation of item by producer 2," then $p(s'; s, e^*) = 1$ when

$$s = (s_1, s_2, s_3, s_4, s_5, s_6, s_7) \text{ and } s' = (s_1, s_2, s_3, s_4 - 1, s_5 + 1, s_6, s_7).$$

If $e^* = e_5 =$ "start of transmission to consumer 2," then $p(s'; s, e^*) = 1$ when

$$s = (B_1, 0, 0, s_4, s_5, 0, 1) \text{ and } s' = (B_1, 0, 0, s_4, s_5 - 1, 1, 0).$$

If $e^* = e_6 =$ "end of transmission to consumer 2," then $p(s'; s, e^*) = 1$ when

$$s = (s_1, s_2, 0, s_4, s_5, 1, 0) \text{ and } s' = (s_1, s_2, 0, s_4 + 1, s_5, 0, 1).$$

All other new-marking probabilities of the form $p(s'; s, e^*)$ are equal to 0. It can be seen from the above specification that each transition $e \in E$ is deterministic. Observe that transitions never fire simultaneously because the random variables A_1, A_2, L_1, and L_2 have continuous distribution functions. Thus, new-marking probabilities of the form $p(s'; s, E^*)$ with $|E^*| > 1$ can be specified arbitrarily; in practice, this means that such probabilities need not be specified at all.

The clock-setting distribution functions for timed transitions e_1, e_3, e_4, and e_6 are $F(x; s', e_1, s, e) = P\{A_1 \leq x\}$, $F(x; s', e_4, s, e) = P\{A_2 \leq x\}$, $F(x; s', e_3, s, e) = P\{L_1 \leq x\}$, and $F(x; s', e_6, s, e) = P\{L_2 \leq x\}$, respectively—observe that each of these transitions is simple. All speeds for enabled transitions are equal to 1.

The sequence of marking changes illustrated in Figure 2.5 shows how the SPN model captures the nonpreemptive priority of producer–consumer pair 1 over producer–consumer pair 2 for use of the channel. When transition $e_3 =$ "end of transmission to consumer 1" fires, it deposits a token in place d_7, indicating that the channel is available for transmission of an

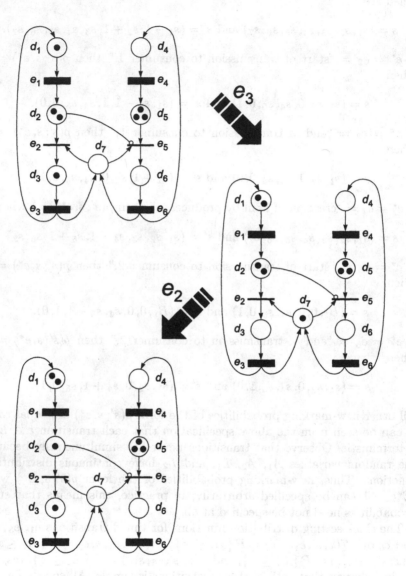

Figure 2.5. Marking changes for SPN representation of producer–consumer system with nonpreemptive priority and finite buffers.

e_1 = creation of item by producer 1
e_2 = end of transmission to consumer 1
e_3 = creation of item by producer 2
e_4 = end of transmission to consumer 2

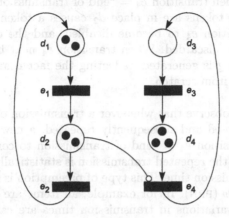

Figure 2.6. SPN representation of producer–consumer system with preemptive-repeat priority and finite buffers.

item. As shown in the figure, there are items awaiting transmission to consumer 1 and items awaiting transmission to consumer 2; these items are represented by the tokens in places d_2 and d_5, respectively. The presence of tokens in place d_2 causes immediate transition e_2 = "start of transmission to consumer 1" to fire while simultaneously inhibiting the firing of immediate transition e_5 = "start of transmission to consumer 2." When transition e_2 fires, it deposits a token in place d_3, causing transition e_3 to become enabled, and there is a start of transmission to consumer 1. Transition e_2 also removes a token from place d_7, indicating that the channel is now in use.

EXAMPLE 2.2 (Producer–consumer system with preemptive-repeat priority). Consider a producer–consumer system as in Example 2.1, but suppose that producer–consumer pair 1 has *preemptive-repeat* priority over producer–consumer pair 2. That is, whenever a transmission to consumer 2 is underway and producer 1 creates an item, the transmission to consumer 2 stops immediately and there is a start of transmission to consumer 1. The next time the channel becomes available to producer–consumer pair 2, the previously interrupted transmission to consumer 2 starts again from scratch. Figure 2.6 displays an SPN representation of this system. All transitions are deterministic and all speeds are equal to 1. The clock-setting distribution functions for timed transitions e_1, e_2, e_3, and e_4 are given by

$F(x; s', e_1, s, e) = P\{A_1 \leq x\}$, $F(x; s', e_3, s, e) = P\{A_2 \leq x\}$, $F(x; s', e_2, s, e) = P\{L_1 \leq x\}$, and $F(x; s', e_4, s, e) = P\{L_2 \leq x\}$, respectively. The preemptive-repeat priority of producer–consumer pair 1 over producer–consumer pair 2 is modelled by making d_2 an inhibitor input place for transition e_4. The idea is that the firing of transition $e_1 =$ "creation of item by producer 1" when transition $e_4 =$ "end of transmission to consumer 2" is enabled and no tokens are in place d_2 causes a token to be deposited in place d_2, transition e_4 to become disabled, and the clock reading for transition e_4 to be discarded. When transition e_4 next becomes enabled, a new clock reading is generated, reflecting the fact that transmission to consumer 2 starts from scratch.

In Example 2.2 observe that whenever a transmission of an item to consumer 2 is preempted and subsequently repeated, a new clock reading is generated for transition $e_4 =$ "end of transmission to consumer 2." That is, the duration of the repeated transmission is statistically independent of the original transmission time. This type of preemption is sometimes called *preempt-repeat new* (PRN). If, for example, all items are of the same size and the random variations in transmission times are caused by random delays in the transmission process, then the preemption mechanism can reasonably be modelled as PRN. Suppose, however, that the transmission process is deterministic and the random variations in transmission times are caused by random variations in the sizes of the items. Then, for a given item, the duration of the repeated transmission should be the same as the original transmission time. This latter type of preemption is called *preempt-repeat identical* (PRI). Although activities subject to PRI preemption cannot be modelled exactly within our SPN framework, they can be modelled approximately—see Example 2.8 in the next subsection.

EXAMPLE 2.3 (Producer–consumer system with preemptive-resume priority). Consider a producer–consumer system as in Example 2.1, but suppose that producer–consumer pair 1 has *preemptive-resume* priority over producer–consumer pair 2. That is, as in Example 2.2, creation of an item by producer 1 when a transmission to consumer 2 is underway always results in an interruption of the transmission. The next time the channel becomes available to producer–consumer pair 2, however, the transmission to consumer 2 resumes from the point at which it was interrupted. Figure 2.7 displays an SPN representation of this system. All transitions are deterministic and the clock-setting distributions are as in Example 2.2. Zero speeds are used to model preemptive-resume behavior as follows. For $s = (s_1, s_2, s_3, s_4) \in G(e_4)$, set $r(s, e_4) = 1$ if $s_2 = 0$ and $r(s, e_4) = 0$ otherwise. All other speeds are equal to 1. Thus the firing of transition $e_1 =$ "creation of item by producer 1" when transition $e_4 =$ "end of transmission to consumer 2" is enabled causes the clock for transition e_4 to stop running down. The clock resumes running down when the token count in

e_1 = creation of item by producer 1

e_2 = end of transmission to consumer 1

e_3 = creation of item by producer 2

e_4 = end of transmission to consumer 2

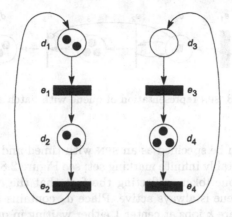

Figure 2.7. SPN representation of producer–consumer system with preemptive-resume priority and finite buffers.

place d_2 next becomes 0, that is, when the channel next becomes available to producer–consumer pair 2.

2.2.2 Marking-dependent Transitions

When a deterministic transition fires, the number of tokens it removes from each normal input place and deposits in each output place does not explicitly depend on the current marking. In general, however, transitions may exhibit "marking dependence." The following example shows that marking-dependent transitions are needed to model certain discrete-event systems.

EXAMPLE 2.4 (Queue with batch arrivals). Consider a queueing system consisting of one single-server center. Jobs arrive at the center in batches and are served one at a time. Whenever there is a completion of service and the queue is not empty, the server immediately starts a new service; the job to receive service is selected randomly and uniformly among the jobs waiting in queue. Successive batch sizes are i.i.d. as a discrete random variable B, successive service times are i.i.d. as a random variable L with continuous distribution function, and successive interarrival times between batches are i.i.d. as a random variable A with continuous distribution function. We assume that, for $i \geq 1$,

$$b_i \overset{\text{def}}{=} P\{B = i\} > 0.$$

e_1 = arrival of batch

e_3 = entry into queue of job in batch

e_4 = completion of service

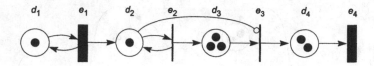

Figure 2.8. SPN representation of queue with batch arrivals.

This system can be specified as an SPN with timed and immediate transitions and a countably infinite marking set; see Figure 2.8. Place d_1 always contains exactly one token, reflecting the fact that the arrival process of batches to the queue is always active. Place d_4 contains k (≥ 0) tokens if and only if there are k jobs at center 1 either waiting in queue or receiving service. Transitions e_1, e_3, and e_4 are deterministic. Places d_2 and d_3 are used in conjunction with marking-dependent transition e_2 to "generate" the random size of each batch upon arrival.

The idea is that whenever transition e_1 = "arrival of batch" fires, it deposits a token in place d_2 and immediate transition e_2 becomes enabled. Transition e_2 then fires a random number of times in succession before becoming disabled, depositing a token in place d_3 each time it fires. The probability that e_2 fires exactly i times—so that exactly i tokens are deposited in place d_3—is equal to b_i for $i \geq 1$. When transition e_2 fires for the last time and becomes disabled, leaving a total of, say, k tokens in place d_3, it removes the token in place d_2 and (deterministic) transition e_3 becomes enabled. Transition e_3 then fires precisely k times in succession, removing all k tokens from place d_3 and depositing k tokens in place d_4. Thus, whenever transition e_1 fires, the net effect is to deposit a random number of tokens in place d_4; the distribution of the number of tokens deposited is the same as the distribution of the random variable B.

The foregoing marking-dependent firing mechanism for transition e_2 is specified as follows. Whenever place d_3 contains k (≥ 0) tokens and transition e_2 fires, a token is deposited in place d_3. With probability

$$p_k = \frac{b_{k+1}}{\sum_{i=k+1}^{\infty} b_i} = \frac{b_{k+1}}{1 - \sum_{i=1}^{k} b_i}, \tag{2.5}$$

a token also is removed from place d_2, and transition e_2 becomes disabled; with probability $1 - p_k$, a token is not removed from place d_2, and transition

e_2 remains enabled. Formally, $p(s'; s, e_2) = p_k$ when

$$s = (1, 1, k, m) \quad \text{with } k, m \geq 0$$
$$\text{and } s' = (1, 0, k + 1, m)$$

and $p(s'; s, e_2) = 1 - p_k$ when

$$s = (1, 1, k, m) \quad \text{with } k, m \geq 0$$
$$\text{and } s' = (1, 1, k + 1, m);$$

otherwise, $p(s'; s, e_2) = 0$. Observe that p_k is simply the conditional probability that $B = k + 1$, given that $B \geq k + 1$. A simple calculation shows that, for $k \geq 1$, the probability that transition e_2 fires exactly k times before becoming disabled is equal to b_k.

It does not appear possible to model the queue with batch arrivals without use of marking-dependent transitions.[1] The following two examples show that even when marking-dependent transitions are not needed, they can reduce the complexity of the SPN graph and the size of the marking set. The examples also highlight the fact that the SPN representation of a discrete-event system need not be unique.

EXAMPLE 2.6 (Token ring). Local area decentralized computer networks are usually configured in a ring or bus topology. Consider a unidirectional ring network having a fixed number of ports, labelled $1, 2, \ldots, N$ in the direction of signal propagation. At each port, message packets arrive according to a random process. A distinguished bit pattern, called a *ring token*, circulates around the ring from one port to the next. The time for the ring token to propagate from port j to the next port is a positive constant R_j. When a port observes the ring token and has a packet queued for transmission, the port converts the ring token to another distinguished bit pattern called a *connector* and transmits the packet followed by the ring token; the ring token continues to propagate if the port has no packet queued for transmission. Conceptually, the port "removes the token" from the ring at the start of a transmission, "holds the token" while the transmission is

---

[1] Some SPN variants associate a "multiplicity" with each arc between a place and a transition. The firing mechanism for a transition with N normal input places and M output places is as follows. Denote by n_i the multiplicity associated with the arc from the ith normal input place to the transition $(1 \leq i \leq N)$ and by m_j the multiplicity associated with the arc from the transition to the jth output place $(1 \leq j \leq M)$. Then the transition is enabled only if, for $1 \leq i \leq N$, the ith normal input place contains n_i tokens; whenever such a transition fires, it removes n_i tokens from the ith normal input place and deposits m_j tokens in the jth output place. It can be shown that the use of arc multiplicities does not increase the modelling power of the basic SPN formalism, so that this device is not sufficient to permit modelling of the queue with batch arrivals if the batch size is unbounded.

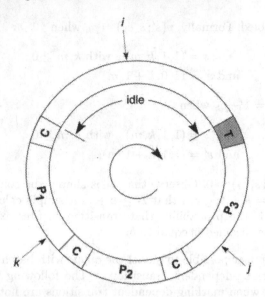

Figure 2.9. Token ring.

underway, and "releases the token" back onto the ring at the end of the transmission. By destroying the connector prefix the port "removes" the transmitted packet when it returns around the ring; see Figure 2.9. In the figure, i, j, and k denote three of the N ports; $\mathbf{T}$ denotes the ring token; $\mathbf{C}$ denotes a connector; and $\mathbf{P_1}$, $\mathbf{P_2}$, and $\mathbf{P_3}$ denote packets.

For simplicity, assume that at most one packet is awaiting transmission at any time at any particular port; the successive times from end of transmission by port j until the arrival of the next packet for transmission by port j are i.i.d. as a positive random variable A_j with continuous distribution function. Moreover, the successive times for port j to transmit a packet are i.i.d. as a positive random variable L_j with continuous distribution function.

This system can be specified as an SPN with marking-dependent transitions; see Figure 2.10 for $N = 2$. The set of places of the SPN is

$$D = \{\, d_{1,1}, d_{2,1}, d_{3,1}, d_{4,1}, \ldots, d_{1,N}, d_{2,N}, d_{3,N}, d_{4,N} \,\},$$

and the set of transitions is

$$E = \{\, e_{1,1}, e_{2,1}, e_{3,1}, \ldots, e_{1,N}, e_{2,N}, e_{3,N} \,\}.$$

All transitions are timed. (For clarity of exposition, we use double subscripts to index places, transitions, and token counts.)

Place $d_{1,j}$ contains one token if and only if port j either is transmitting a packet or has a packet queued for transmission. Place $d_{2,j}$ contains one token if and only if port j is not transmitting a packet and has no packet

$e_{1,j}$ = arrival of packet for transmission by port j

$e_{2,j}$ = end of transmission by port j

$e_{3,j}$ = observation of ring token by port j

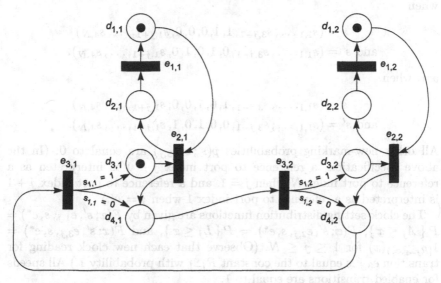

Figure 2.10. SPN representation of token ring (two ports).

queued for transmission. Place $d_{3,j}$ contains one token if and only if port j is transmitting a packet, and place $d_{4,j}$ contains one token if and only if the ring token is propagating from port j to the next port. Otherwise, a place contains no tokens.

The marking set G $(= S)$ is

$$G = \left\{ (s_{1,1}, s_{2,1}, \ldots, s_{4,N}) \in \{0,1\}^{4N} : s_{1,j} + s_{2,j} = 1 \text{ and} \right.$$
$$\left. s_{2,j} s_{3,j} = 0 \text{ for } 1 \leq j \leq N; s_{3,1} + s_{4,1} + \cdots + s_{3,N} + s_{4,N} = 1 \right\}.$$

It follows that $|G| = 3N2^{N-1}$. In any marking there are exactly $N+1$ tokens, and each place contains at most one token. Each of the disjoint sets of places $\{ d_{1,j}, d_{2,j} \}$ contains exactly one token indicating whether or not port j has a packet queued for transmission. The set of places $\{ d_{3,1}, d_{4,1}, d_{3,2}, d_{4,2}, \ldots, d_{3,N}, d_{4,N} \}$ contains exactly one token indicating the position and status of the ring token. There can never be tokens at places $d_{2,j}$ and $d_{3,j}$ simultaneously, reflecting the fact that there can be no arrival of a packet for transmission by port j during a transmission by port j.

Transitions $e_{1,j}$ and $e_{2,j}$ are deterministic for $1 \leq j \leq N$. Whenever transition $e_{3,j}$ = "observation of ring token by port j" fires, it removes a token from place $d_{4,j-1}$ and deposits a token either in place $d_{3,j}$ or in place

$d_{4,j}$, depending on whether $(s_{1,j}, s_{2,j})$ equals $(1, 0)$ or $(0, 1)$, respectively. Thus, when the ring token arrives at port j, either port j starts transmission or the ring token starts to propagate to the next port, depending on whether port j has a packet queued for transmission. Formally, $p(s'; s, e_{3,j}) = 1$ when

$$s = (s_{1,1} \ldots, s_{3,j-1}, 1, 1, 0, 0, 0, s_{1,j+1}, \ldots, s_{4,N})$$
$$\text{and } s' = (s_{1,1}, \ldots, s_{3,j-1}, 0, 1, 0, 1, 0, s_{1,j+1}, \ldots, s_{4,N}),$$

and when

$$s = (s_{1,1}, \ldots, s_{3,j-1}, 1, 0, 1, 0, 0, s_{1,j+1}, \ldots, s_{4,N})$$
$$\text{and } s' = (s_{1,1}, \ldots, s_{3,j-1}, 0, 0, 1, 0, 1, s_{1,j+1}, \ldots, s_{4,N}).$$

All other new-marking probabilities $p(s'; s, e_{3,j})$ are equal to 0. (In the above specification, a reference to port index $j - 1$ is interpreted as a reference to port index N when $j = 1$, and a reference to port index $j + 1$ is interpreted as a reference to port index 1 when $j = N$.)

The clock-setting distribution functions are given by $F(x; s', e_{1,j}, s, e^*) = P\{A_j \leq x\}$, $F(x; s', e_{2,j}, s, e^*) = P\{L_j \leq x\}$, and $F(x; s', e_{3,j}, s, e^*) = 1_{[R_{j-1}, \infty)}(x)$ for $1 \leq j \leq N$. (Observe that each new clock reading for transition $e_{3,j}$ is equal to the constant R_{j-1} with probability 1.) All speeds for enabled transitions are equal to 1.

As shown in the next example, the token ring of Example 2.6 can also be represented as an SPN with deterministic transitions; that is, no marking-dependent transitions are required. An advantage of this representation is that the SPN graph completely determines the state-transition mechanism of the net. A disadvantage is that the deterministic SPN has more places, transitions, and markings than the SPN of Example 2.6. This situation is typical; increasing the amount of information conveyed by the SPN graph usually increases the size and complexity of the graph.

EXAMPLE 2.7 (Alternative representation of token ring).   The system in Example 2.6 can be specified as an SPN with deterministic timed and immediate transitions and unit speeds; see Figure 2.11 for $N = 2$. Each place contains at most one token. The interpretations of places $d_{1,j}$, $d_{2,j}$, $d_{3,j}$, and $d_{4,j}$ are exactly as in Example 2.6. Place $d_{5,j}$ contains one token if and only if port j has just observed the ring token. The clock-setting distribution functions for timed transitions are as in Example 2.6.

The marking set G is

$$G = \big\{ (s_{1,1}, s_{2,1}, \ldots, s_{5,N}) \in \{0, 1\}^{5N} : s_{1,j} + s_{2,j} + s_{j,3} = 1$$
$$\text{for } 1 \leq j \leq N; s_{3,1} + s_{4,1} + s_{5,1} + \cdots + s_{3,N} + s_{4,N} + s_{5,N} = 1 \big\},$$

and

$$S' = \big\{ (s_{1,1}, s_{2,1}, \ldots, s_{5,N}) \in G : s_{5,j} = 1 \text{ for some } j \big\}$$

$e_{1,j}$ = arrival of packet for transmission by port j

$e_{2,j}$ = end of transmission by port j

$e_{3,j}$ = observation of ring token by port j

$e_{4,j}$ = start of transmission by port j

$e_{5,j}$ = start of propagation from port j

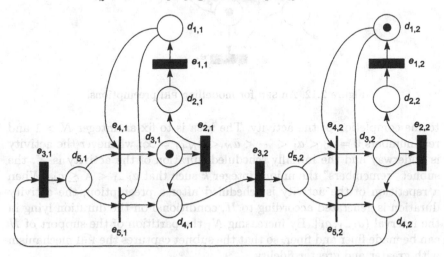

Figure 2.11. Deterministic SPN representation of token ring (two ports).

is the immediate marking set. For this SPN, $|G| = 5N2^{N-1}$, $|S| = 3N2^{N-1}$, and $|S'| = 2N2^{N-1}$. Thus the marking set G is larger than the marking set for the SPN of Example 2.6 by a factor of about 1.7.

Observe that when transition $e_{3,j}$ = "observation of ring token by port j" fires, it removes a token from place $d_{4,j-1}$ and deposits a token in place $d_{5,j}$; then immediate transition $e_{4,j}$ fires if $(s_{1,j}, s_{2,j})$ equals $(1,0)$ and immediate transition $e_{5,j}$ fires if $(s_{1,j}, s_{2,j})$ equals $(0,1)$. Thus, when the ring token arrives at port j, either port j starts transmission or the ring token starts to propagate to the next port, depending on whether or not port j has a packet queued for transmission.

Our next example shows how marking-dependent transitions can be used to approximately model the PRI preemption mechanism mentioned in the previous subsection.

EXAMPLE 2.8 (Modelling PRI preemption). Consider an activity that is subject to PRI preemption, and suppose that the duration of the activity has distribution function H; for concreteness, suppose that H has support on the nonnegative real line. Figure 2.12 shows a subnet that can be used to model the activity; for this subnet, the firing of transition e_1 corresponds

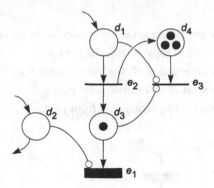

Figure 2.12. An SPN for modelling PRI preemptions.

to the completion of the activity. The idea is to fix an integer $N > 1$ and real numbers $0 = a_0 < a_1 < \cdots < a_N < a_{N+1} = \infty$; whenever the activity is underway and the initially scheduled duration of the activity is X, the subnet "remembers" the unique integer k such that $a_{k-1} < X \le a_k$. When a repetition of the activity is scheduled after a preemption, the activity duration is generated according to H, conditional on the duration lying in the interval $(a_{k-1}, a_k]$. By increasing N, the partition of the support of H can be made finer and finer, so that the subnet captures the PRI mechanism with greater and greater fidelity.

In more detail, the activity is initially scheduled when a token is deposited in place d_1—for simplicity, we assume that the set of places $\{ d_1, d_3 \}$ contains no more than one token at any time. Immediate transition e_2 then fires a random number of times in succession before becoming disabled, depositing a token in place d_4 each time it fires; the probability that e_2 fires exactly k times, so that exactly k tokens are deposited in place d_4, is $p_k = H(a_k) - H(a_{k-1})$ for $k \in \{ 1, 2, \dots, N+1 \}$. When e_2 fires for the last time and becomes disabled, it removes a token from place d_1 and deposits a token in place d_3, causing transition e_1 to become enabled. The precise specification of the new-marking probabilities that define this firing mechanism is similar to that given in Example 2.4 for the SPN model of a queue with batch arrivals.

Assuming that k tokens have been deposited in place d_4, a new clock reading for transition e_1 is generated according to the conditional distribution

$$
H_k(t) = \begin{cases} 0 & \text{if } t \le a_{k-1}; \\ \big(H(t) - H(a_{k-1})\big)/\big(H(a_k) - H(a_{k-1})\big) & \text{if } a_{k-1} < t \le a_k; \\ 1 & \text{if } t > a_k. \end{cases}
$$

A preemption of the activity occurs when a token is deposited in place d_2 and e_1 becomes disabled. A subsequent removal of the token in place d_2

causes e_1 to become reenabled, and the activity is repeated. At each such repetition, a new clock reading for e_1 is generated according to H_k. When the activity finally completes and transition e_1 fires—removing the token in place d_3—deterministic transition e_3 becomes enabled and fires k times in succession, removing all tokens from place d_4. The subnet is then ready for the next fresh start of the activity. This construction illustrates the utility of letting the clock-setting distribution depend explicitly on the new marking, since $F(x; s', e_1, s, E^*) \equiv F(x; s', e_1) = H_{s'_4}(x)$ for $s' = (s'_1, s'_2, s'_3, s'_4, \ldots) \in G$, $s \in G$, and $E^* \subseteq E(s)$.

We conclude our discussion of marking-dependent transitions by indicating how new-marking probabilities for an SPN with such transitions can be specified in a form suitable for processing by a computer program. In particular, we illustrate the specification of new-marking probabilities in SPSIM, a prototype software system developed at IBM for simulation of SPNs and other stochastic processes. The SPSIM system takes as input a model description, written in the SPSIM specification language, and automatically translates this description into an executable simulation program.

Consider first the SPN model, given in Example 2.4, of the queue with batch arrivals. The new-marking probabilities for this SPN can be specified by the following SPSIM statements:

```
\MARKING CHANGES
FOR (I* == 1) || (I* == 3) || (I* == 4)  DETERMINISTIC
FOR I* == 2
  IF TRUE THEN
    WITH PROB = P(S[3])      NEXT S'[2] = S[2] - 1;
                                  S'[3] = S[3] + 1
    WITH PROB = 1 - P(S[3]) NEXT S'[3] = S[3] + 1
```

The syntax of the SPSIM specification language is similar to that of the C programming language. In the above listing, we assume that a function P has been defined such that, for an integer-valued variable k, the expression P(k) evaluates to the probability p_k defined in (2.5). (The SPSIM system permits such user-defined functions.) The first line in the listing demarcates the section of the model specification in which new-marking probabilities are defined. The variable I* is a standard identifier that denotes the index of the transition that triggers the marking change. For example, if transition e_2 triggers the marking change, then I* is equal to 2. Similarly, S denotes the current marking and S' denotes the new marking. Brackets are used to specify components of a marking: S[3] denotes the third component of the current marking, that is, the token count in place d_3. The idea is that the logical expression in each FOR-clause is evaluated until a true expression is found. Each such logical expression has the same syntax as a logical expression in C and depends on the index I*. In the above listing, for example, the == and || operators are logical equality and logical OR operators as in C; the expression in the first FOR-clause is true if the trigger transition is equal to e_1, e_3, or e_4, and the expression in the second FOR-clause is true if the trigger transition is equal to e_2. For a well-specified SPN model, exactly one of the FOR-clauses contains a true expression. If this FOR-clause is followed by the DETERMINISTIC keyword, then the new marking S' is generated from the current marking S by decrementing (by 1) the token count in each normal input place of the trigger transition and incrementing the token count in each output place. Otherwise, the FOR-clause is followed by one or more IF-clauses, exactly one of which is assumed to contain a true logical expression. In the above listing, the logical expression in the displayed IF-clause consists of the keyword TRUE, so that the expression is always true. In general, the logical expression can depend on both I* and S. Associated with each IF-clause are one or more WITH-clauses. For the (unique) IF-clause that contains a true logical expression, one of the associated WITH-clauses is randomly chosen according to the specified probability, and the new marking S' is generated by the specified assignments to the components of S'. The components of S' for which no assignments are specified keep their values from the current marking S.

As a second example, consider the SPN model of Example 2.6 with $N = 5$ ports, and suppose that transitions never fire simultaneously. The new-marking probabilities for this SPN can be specified by the following SPSIM statements:

```
\REPLACEMENTS
N IS 5
J* IS I*[2]
J*MINUS1 IS ((J* - 2) MOD N) + 1

\MARKING CHANGES
FOR (I*[1] == 1) || (I*[1] == 2)  DETERMINISTIC
FOR I*[1] == 3
  IF S[1][J*] == 1 THEN
    WITH PROB = 1 NEXT S'[4][J*MINUS1] = S[4][J*MINUS1] - 1;
                      S'[3][J*] = S'[3][J*] + 1
  IF S[2][J*] == 1 THEN
    WITH PROB = 1 NEXT S'[4][J*MINUS1] = S[4][J*MINUS1] - 1;
                      S'[4][J*] = S'[4][J*] + 1
```

As illustrated by the above listing, both transitions and components of markings can have multiple indices. Brackets are used to specify a specific index. For example, if the transition that triggers a marking change is $e^* = e_{3,5}$, then the standard identifiers I*[1] and I*[2] are equal to 3 and 5, respectively. Similarly, if the current marking is $s = (s_{1,1}, s_{2,1}, s_{3,1}, s_{4,1}, \cdots, s_{1,5}, s_{2,5}, s_{3,5}, s_{4,5})$, then the standard identifier S[2][5] is equal to $s_{2,5}$. The first statement in the \REPLACEMENTS section of the model description specifies that every subsequent occurrence of the identifier N in the model description is to be replaced by the symbol 5 before translation of the model. The remaining statements in the \REPLACEMENTS section have a similar interpretation, and the statements in this section are executed in the opposite order in which they appear. Thus, for example, the identifier S[4][J*MINUS1] is equal to the token count $s_{4,4}$ whenever the current marking is $s = (s_{1,1}, \ldots, s_{4,5})$ and the trigger transition is $e^* = e_{3,5}$. To see this, observe that SPSIM generates the successive replacements

$$\text{S[4][J*MINUS1]} \Rightarrow \text{S[4][((J* - 2) MOD N) + 1]}$$
$$\Rightarrow \text{S[4][((I*[2] - 2) MOD N) + 1]}$$
$$\Rightarrow \text{S[4][((I*[2] - 2) MOD 5) + 1]}.$$

The rightmost expression is then evaluated with I*[2] equal to 5—here MOD is the standard modulo operator. Because $-1 \bmod n = n - 1$ for $n \geq 1$, it follows that, in general, J*MINUS1 is equal to N whenever J* is equal to 1 and to $j - 1$ whenever J* is equal to j with $1 < j \leq N$.

2.2.3 Synchronization: Flexible Manufacturing System

The following examples illustrate one way in which immediate transitions and marking-dependent transitions can be used to model synchronized activities, specifically, the synchronized unloading of manufactured parts. The first example also illustrates the utility of allowing the clock-setting distribution function for a transition to depend explicitly on the current and new markings.

EXAMPLE 2.9 (Flexible manufacturing system).   Consider a flexible manufacturing system that produces two types of parts and has three machines numbered 1, 2, and 3. Parts of type 1 require processing first by machine 1 and then by machine 2. Two processes can produce parts of type 2. The first process consists of a fast intervention by machine 1 followed by a refinement performed by machine 2. The second process, performed entirely by machine 3, is much slower but produces finished parts. The duration of the refinement operation performed by machine 2 on parts processed by machine 1 is independent of the part type. Exactly three parts are in the system at any time, and finished parts are unloaded (instantaneously) from the system and immediately replaced by raw ones three at a time. Each machine processes one part at a time. For machines 1 and 2, parts of type 2 have nonpreemptive priority over parts of type 1. For each of machines 1, 2, and 3, parts of the same type are processed according to a first-come, first-served service discipline. A raw part loaded (instantaneously) into the system is of type 1 with probability $p \in (0, 1)$ and is of type 2 with probability $1 - p$. A part of type 2 goes to machine 1 with probability $q \in (0, 1)$ and to machine 3 with probability $1 - q$.

The successive times for machine 1 to process a part of type j are i.i.d. as a positive random variable $L_{1,j}$, the successive times for machine 2 to process a part are i.i.d. as a positive random variable L_2, and the successive times for machine 3 to process a part are i.i.d. as a positive random variable L_3. Each of these random variables has a continuous distribution function.

This system can be specified as an SPN with (marking-dependent) timed and immediate transitions; see Figure 2.13. Place d_1 contains one token if and only if machine 1 is processing a part; otherwise, place d_1 contains no tokens. Place d_2 (resp., place d_3) contains k (≥ 0) tokens if and only if k parts are awaiting processing or being processed by machine 2 (resp., machine 3). Places d_4, d_5, and d_6 each contain at most one token; there is a total of k (≥ 0) tokens in places d_4, d_5, and d_6 if and only if k finished parts are awaiting unloading. Place d_7 (resp., place d_8) contains k (≥ 0) tokens if and only if k raw parts of type 1 (resp., type 2) are awaiting processing by machine 1. Place d_9 contains one token if and only if machine 1 is idle; otherwise, place d_9 contains no tokens.

Transitions e_1, e_5, and e_6 are deterministic. Whenever transition $e_2 =$ "end of processing by machine 2" fires, it removes a token from place d_2 and deposits a token in one of places d_4, d_5, or d_6; the token is deposited in the lowest-numbered empty place. Formally, $p(s'; s, e_2) = 1$ when

$$s = (s_1, s_2, s_3, 0, 0, 0, s_7, s_8, s_9)$$
$$\text{and } s' = (s_1, s_2 - 1, s_3, 1, 0, 0, s_7, s_8, s_9),$$

when

$$s = (s_1, s_2, s_3, 1, 0, 0, s_7, s_8, s_9)$$
$$\text{and } s' = (s_1, s_2 - 1, s_3, 1, 1, 0, s_7, s_8, s_9),$$

and when

$$s = (s_1, s_2, s_3, 1, 1, 0, s_7, s_8, s_9)$$
$$\text{and } s' = (s_1, s_2 - 1, s_3, 1, 1, 1, s_7, s_8, s_9).$$

Similarly, transition $e_3 =$ "end of processing by machine 3" removes a token from place d_2 and deposits a token in one of places d_4, d_5, or d_6 whenever it fires. Whenever transition $e_4 =$ "unloading of finished parts and loading of raw parts" fires, it removes one token from each of places d_4, d_5, and d_6. Moreover, if n_1, n_2, and n_3 are nonnegative integers such that $n_1 + n_2 + n_3 = 3$, then with probability

$$p = 6(n_1! n_2! n_3!)^{-1} p^{n_1} q^{n_2} (1 - p)^{n_2 + n_3} (1 - q)^{n_3}$$

it deposits n_1 tokens in place d_7, n_2 tokens in place d_8, and n_3 tokens in place d_3. That is, a total of three tokens is assigned to places d_7, d_8, and d_3 according to a multinomial probability distribution with respective parameters p, $(1 - p)q$, and $(1 - p)(1 - q)$.

e_1 = end of processing by machine 1

e_2 = end of processing by machine 2

e_3 = end of processing by machine 3

e_4 = unloading of finished parts and loading of raw parts

e_5 = start of processing by machine 1 for part of type 1

e_6 = start of processing by machine 1 for part of type 2

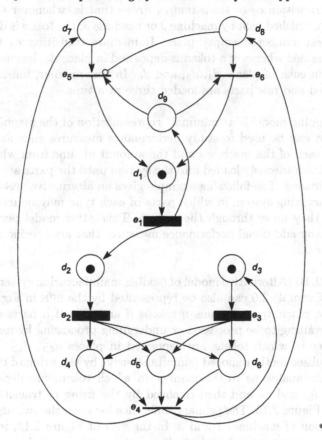

Figure 2.13. SPN representation of flexible manufacturing system.

The clock-setting distribution function for transition $e_1 = $ "end of processing by machine 1" depends explicitly on the current and new markings: if $s = (s_1, s_2, \ldots, s_9)$ and $s' = (s'_1, s'_2, \ldots, s'_9)$, then $F(x; s', e_1, s, e^*) = P\{L_{1,1} \leq x\}$ when $s'_7 = s_7 - 1$ and $F(x; s', e_1, s, e^*) = P\{L_{1,2} \leq x\}$ when $s'_8 = s_8 - 1$. The clock-setting distribution functions for the remaining timed transitions are defined in an obvious manner, and all speeds for enabled transitions are equal to 1.

Observe that the nonpreemptive priority of parts of type 2 over parts of type 1 for processing by machine 1 is modelled using inhibitor input places and immediate transitions in a manner similar to the SPN representation of the producer–consumer system in Example 2.1.

As mentioned above, a token is deposited in one of places d_4, d_5, or d_6 whenever transition e_2 or transition e_3 fires—that is, whenever there is a creation of a finished part by machine 2 or machine 3; the token is deposited in the lowest-numbered empty place. Immediate transition e_4 therefore becomes enabled whenever a token is deposited in place d_6, leaving exactly one token in each of places d_4, d_5, and d_6. In this manner, finished parts are unloaded and raw parts are loaded three at a time.

The foregoing model is a "minimal" representation of the manufacturing system that can be used to study performance measures such as the utilization of each of the machines and the amount of time from when three parts are simultaneously loaded into the system until the parts are simultaneously unloaded. The following example gives an alternative SPN model of the manufacturing system in which parts of each type may be more easily tracked as they move through the system. This latter model permits the study of many additional performance measures that are specific to a part of type 1 or 2.

EXAMPLE 2.10 (Alternative model of flexible manufacturing system). The system of Example 2.9 can also be represented by the SPN in Figure 2.14. In this SPN, place $d_{1,i,j}$ contains n tokens if and only if n parts of type i are either waiting to be processed or undergoing processing by machine j. The manner in which tokens are deposited in places $d_{4,1}$, $d_{4,2}$, and $d_{4,3}$ and then subsequently removed (simultaneously) by the firing of transition e_3 is exactly analogous to the manner in which tokens are deposited in places d_4, d_5, and d_6 and then removed by the firing of transition e_4 in the SPN of Figure 2.13. The primary difference between the two SPNs is the representation of machines 1 and 2. In the SPN of Figure 2.13, machine i ($i = 1, 2$) is represented by place d_i together with transition e_i. In the SPN of Figure 2.13, machine 1 is represented by the token that resides in one of places $d_{3,1}$ (when the machine is idle), $d_{2,1,1}$ (when the machine is processing a part of type 1), or $d_{2,2,1}$ (when the machine is processing a part of type 2); machine 2 is modelled similarly. This representation of each machine is similar to that of the channel in the producer–consumer

$e_{1,i,j}$ = start of processing by machine j for part of type i

$e_{2,i,j}$ = end of processing by machine j for part of type i

e_3 = unloading of finished parts and loading of raw parts

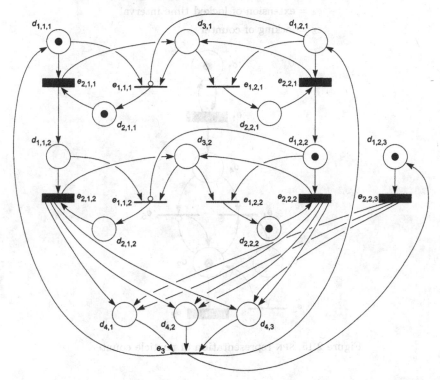

Figure 2.14. Alternative SPN representation of flexible manufacturing system.

models of Section 2.2.1 and—unlike the SPN in Figure 2.13—makes explicit the type of part that each machine is processing at each time point. Also unlike the SPN in Figure 2.13, the SPN in Figure 2.14 explicitly displays the nonpreemptive-priority mechanism for machine 2.

2.2.4 Resetting Clocks: Particle Counter

The clock for a transition $e \in E$ is not allowed to be reset when a transition $e^* \neq e$ triggers a marking change and transition e is enabled in both the old and the new marking. The following example illustrates a technique for getting around this restriction.

EXAMPLE 2.11 (Particle counter). Suppose that particles arrive, one at a time, at a counter. A particle arrives at time 0 and locks the counter for a time interval of fixed length T. If no further particles arrive in $(0, T]$, the

e_1 = arrival of particle

e_2 = end of locked time interval

e_3 = resetting of locked time interval

e_4 = extension of locked time interval

e_5 = locking of counter

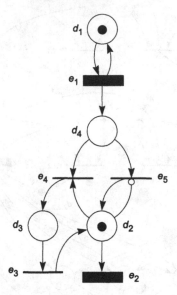

Figure 2.15. SPN representation of particle counter.

counter becomes unlocked at time T; the next particle gets registered and the counter is locked again for a time interval of length T. A particle that arrives when the counter is locked does not get registered but extends the locked interval so that the counter remains locked for an interval of length T after the arrival. The successive interarrival times for particles are i.i.d. as a random variable U with a continuous distribution function.

This system can be specified as an SPN with deterministic timed and immediate transitions; see Figure 2.15. Place d_1 contains exactly one token, reflecting the fact that the arrival process of particles is always active. Each of places d_2, d_3, and d_4 contains at most one token. Place d_2 contains a token if and only if the counter is locked, place d_3 contains a token if and only if the arrival of a particle extends the locked time interval, and place d_4 contains a token if and only if a particle has just arrived. All transitions are deterministic, and the clock-setting distribution functions for timed transitions are given by $F(x; s', e_1, s, e^*) = P\{U \leq x\}$ and $F(x; s', e_2, s, e^*) = 1_{[T, \infty)}(x)$. All speeds for enabled transitions are equal to 1.

Whenever the marking is $(1,1,0,0)$ and transition e_1 fires, so that a particle arrives while the counter is locked, immediate transition $e_4 = $ "extension of locked time interval" fires and timed transition $e_2 = $ "end of locked time interval" becomes disabled. Immediate transition $e_3 = $ "resetting of locked time interval" then fires, transition e_2 becomes enabled again, and the clock for transition e_2 is reset to the value T. In effect, the clock for transition e_2 is reset whenever the marking is $(1,1,0,0)$ and transition e_1 fires.

2.2.5 Compound Events: Slotted Ring

In many discrete-event systems, two or more events can occur simultaneously. As discussed in Section 2.3, the simultaneous occurrence of events can substantially complicate the specification of an SPN model. Sometimes these complications can be avoided by using a single transition to model multiple events that occur simultaneously in the system.

EXAMPLE 2.12 (Slotted ring). Consider a unidirectional ring network having a fixed number K of equal size slots and a fixed number of equally spaced ports, labelled $1, 2, \ldots, N$ in the direction of signal propagation; see Figure 2.16. At each port, constant-length message packets arrive according to a random process; the length equals the slot size. The propagation delay from one port to the next is a positive constant R. Assume that the number N of ports is a multiple of K and, so that there is no loss of utilization due to "unused bits," the time to transmit a message packet is NR/K. The lead "full/empty" (F/E) bit maintains the status of each slot. Subject to the restriction that no port may hold more than one slot simultaneously, a port that has a packet awaiting transmission and observes the status bit of an empty slot sets the bit to 1 (full) and starts transmission. Transmission ends when the slot contains the entire packet. When the status bit of the filled slot propagates back to the sending port, it resets the bit to 0 (empty) and releases the slot. The port releases the slot even if it has another packet awaiting transmission; this rule ensures that all ports have an opportunity to transmit. A port "holds" a slot from the time it sets the status bit to 1 until it releases the slot.

Assume that at most one packet awaits transmission at any time at any particular port; the successive times from end of transmission by port j until the arrival of the next packet for transmission by port j are i.i.d. as a positive random variable A_j with continuous distribution function.

This system can be specified as an SPN with timed transitions. For concreteness, suppose that there are $N = 4$ ports and $K = 2$ slots; see Figure 2.17. Set $k_1 = 3$, $k_2 = 4$, $k_3 = 1$, and $k_4 = 2$, and observe that the firing of transition $e_{2,j}$ $(1 \le j \le N)$ corresponds to the simultaneous observation of the slot 1 status bit by port j and the slot 2 status bit by port k_j.

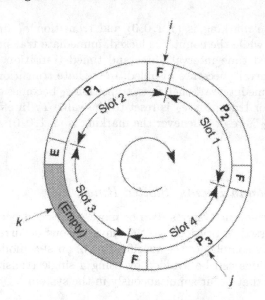

Figure 2.16. Slotted ring.

$e_{1,j}$ = arrival of packet for transmission by port j

$e_{2,j}$ = observation of slot 1 status bit by port j

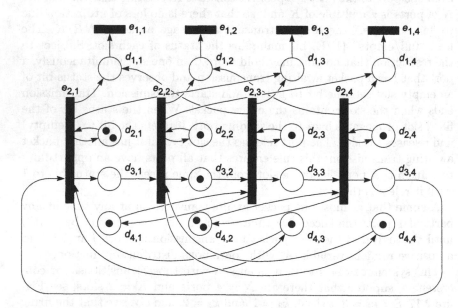

Figure 2.17. SPN representation of slotted ring (two slots, four ports).

Places $d_{1,j}$ and $d_{3,j}$ each contain at most one token. Place $d_{1,j}$ contains a token if and only if port j is not transmitting a packet and has no packet awaiting transmission. Place $d_{3,j}$ contains a token if and only if the status bit for slot 1 is propagating from port j to the next port. Places $d_{2,j}$ and $d_{4,j}$ each contain either one or two tokens. Place $d_{2,j}$ contains two tokens if and only if port j holds slot 1, and place $d_{4,j}$ contains two tokens if and only if port j holds slot 2. Because each of places $d_{2,j}$ and $d_{4,j}$ always contains at least one token, transition $e_{2,j}$ is always enabled when place $d_{3,j-1}$ contains a token.

Transition $e_{1,j}$ is deterministic for $1 \le j \le 4$. Whenever the marking is $s = (s_{1,1}, s_{2,1}, \dots, s_{4,4})$ and transition $e_{2,j} = $ "observation of slot 1 status bit by port j" fires, a token is removed from place $d_{3,j-1}$ and a token is deposited in place $d_{3,j}$, so that the slot 1 status bit starts to propagate to the next port. Moreover, if $s_{1,j} = 0$, $s_{4,j} = 1$, and $s_{2,l} = 1$ for $1 \le l \le 4$—so that port j has a packet waiting for transmission, port j does not hold slot 2, and no port holds slot 1—then a token also is deposited in place $d_{2,j}$ and port j starts transmission of a packet in slot 1. Similarly, if $s_{1,k_j} = 0$, $s_{2,k_j} = 1$, and $s_{4,l} = 1$ for $1 \le l \le 4$, then a token is deposited in place d_{4,k_j} and port k_j starts transmission of a packet in slot 2. Furthermore, if $s_{2,j} = 2$—so that port j has been holding slot 1—then a token is removed from place $d_{2,j}$ and port j releases slot 1. Similarly, if $s_{4,k_j} = 2$, then a token is removed from place d_{4,k_j} and port k_j releases slot 2. If $s_{4,j} = 2$—so that port j has just ended transmission of a packet in slot 2—then a token is deposited in place $d_{1,j}$ and port j starts to wait for the arrival of a packet. Similarly, if $s_{2,k_j} = 2$, then a token is deposited in place d_{1,k_j} and port k_j starts to wait for the arrival of a packet.

The clock-setting distribution functions are given by $F(x; s', e_{1,j}, s, e^*) = P\{A_j \le x\}$ and $F(x; s', e_{2,j}, s, e^*) = 1_{[R,\infty)}(x)$ for $1 \le j \le 4$. All speeds for timed transitions are equal to 1.

2.3 Concise Specification of New-Marking Probabilities

Because our formulation of the SPN model permits transitions to fire simultaneously, specification of new-marking probabilities potentially can be burdensome. Given a timed marking $s \in S$ and fixed marking $s' \in G$, for example, $2^{|E(s)|} - 1$ new-marking probabilities of the form $p(s'; s, E^*)$ must in principle be specified, one for each of the $2^{|E(s)|} - 1$ nonempty subsets $E^* \subseteq E(s)$. In this section we discuss several techniques for concise specification of new-marking probabilities.

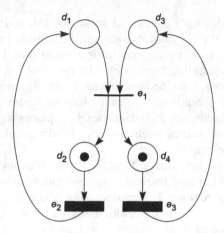

Figure 2.18. Example of a transition firing that never occurs.

2.3.1   Transition Firings That Never Occur

One elementary but useful technique for concise specification is to simply avoid specifying new-marking probabilities for transition firings that never occur. That is, a new-marking probability $p(s'; s, E^*)$ need not be specified explicitly if with probability 1 the transitions in E^* never fire simultaneously when the marking is s.

As an example, suppose that E^* contains both timed and immediate transitions. If $E^* \subseteq E(s)$ for some marking s, then s must be an immediate marking, and only the transitions in $E(s) \cap E'$ ($\neq E^*$) ever fire simultaneously when the marking is s. Hence probabilities of the form $p(\cdot; s, E^*)$ need not be specified.

As another example, suppose that each clock-setting distribution function is continuous and the marking s is timed. Then new-marking probabilities of the form $p(s'; s, E^*)$ with $|E^*| > 1$ need not be specified, because with probability 1 timed transitions never fire simultaneously. We have used this technique in all of the examples in Section 2.2.

As a final example, consider an SPN as in Figure 2.18 with marking set $G = \{s, s', s''\}$, where

$$s = (1, 0, 1, 0),$$
$$s' = (0, 1, 0, 1),$$

and

$$s'' = (1, 0, 0, 1).$$

Suppose that the initial marking is s, that all speeds for enabled transitions are equal to 1, and that each new clock reading for timed transition e_i

$(i = 2, 3)$ is uniformly distributed on an interval $[a_i, b_i]$. Also suppose that $b_2 < a_3$, so that new clock readings for transition e_2 are always smaller than new clock readings for transition e_3. Observe that the new-marking probabilities of the form $p(\,\cdot\,; s', e_3)$ need not be specified explicitly, because with probability 1 transition e_3 never fires when the marking is s'.

Remark 3.1. Suppose that we insist on specifying the new-marking probabilities of the form $p(\,\cdot\,; s', e_3)$. Observe that we must have $p(s'; s', e_3) = 1$ and $p(s; s', e_3) = p(s''; s', e_3) = 0$ if (1.3) is to be satisfied. If we also set $p(s; s'', e_3) = 1$, then transition e_3 does not behave as a deterministic transition when it fires and the marking is s', but does behave as a deterministic transition when it fires and the marking is s''. Because the former type of transition firing occurs with probability 0, we refer to e_3 (with a slight abuse of terminology) as a deterministic transition. In general, we refer to a transition as "deterministic" if it behaves as a deterministic transition except in scenarios that occur with probability 0.

2.3.2 Numerical Priorities

Many SPNs have the following property: whenever two or more transitions fire simultaneously, the net changes marking as if a subset of these transitions fire in succession. That is, there exists a representation of the form

$$p(s'; s, E^*) = p(s'; s, e_{j_1}, e_{j_2}, \dots, e_{j_l})$$

whenever $p(s'; s, E^*)$ is well defined, where $\{\, e_{j_1}, e_{j_2}, \dots, e_{j_l} \,\} \subseteq E^*$ and

$$p(s'; s, e_{j_1}, e_{j_2}, \dots, e_{j_l})$$
$$= \sum_{s_1, s_2, \dots, s_{l-1}} p(s_1; s, e_{j_1}) p(s_2; s_1, e_{j_2}) \cdots p(s'; s_{l-1}, e_{j_l}) \tag{3.2}$$

with the above sum taken over all sequences $s_1, s_2, \dots, s_{l-1}$ such that $e_{j_k} \in E(s_{k-1})$ for $2 \le k \le l$. [Thus $p(s'; s, e_{j_1}, e_{j_2}, \dots, e_{j_l})$ is the probability that the new marking is s' given that transitions $e_{j_1}, e_{j_2}, \dots, e_{j_l}$ successively trigger marking changes starting in marking s.] For such nets, it often suffices to explicitly specify only the "singleton" new-marking probabilities of the form $p(s'; s, e^*)$ and then give succinct rules for expressing a new-marking probability $p(s'; s, E^*)$ in terms of the singleton probabilities. These rules specify the elements of E^* that (in effect) successively fire and the order in which they fire. This approach is particularly effective when each transition is deterministic, so that specification of singleton probabilities is immediate. A simple and concise set of rules that suffices for all of the SPN models in this book can be based on the assignment of "priorities" to the transitions of the net. To simplify the exposition we restrict attention to SPNs in which all speeds are positive.

Before discussing priorities, we first introduce the notion of transitions in conflict.

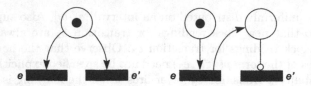

Figure 2.19. Two scenarios in which the firing of deterministic transition e causes transition e' to become disabled.

Definition 3.3. Two transitions e and e' are said to be in *conflict* if e and e' are both timed or both immediate and either

(i) $I(e) \cap I(e') \neq \varnothing$, or

(ii) $\big(J(e) \cap L(e')\big) \cup \big(J(e') \cap L(e)\big) \neq \varnothing$.

According to this definition, two timed transitions or two immediate transitions are in conflict if one of the transitions, when it fires, can potentially cause the other transition to become disabled. Such disabling occurs when e fires and either removes a token from a normal input place for e' (thereby decreasing the token count to 0) or deposits a token in an inhibitor input place for e'; see Figure 2.19. The transitive closure of the conflict relation is an equivalence relation on the set E and partitions E into mutually disjoint equivalence classes called *conflict sets*. Observe that, by definition, the transitions in a conflict set are either all timed or all immediate. Also observe that if two transitions—both timed or both immediate—are in different conflict sets, then the firing of one transition never causes the other transition to become disabled.

To concisely specify the behavior of the net when transitions fire simultaneously, we associate a *priority* (finite, nonnegative integer) with each transition of the net. In the graphical representation of an SPN, the priority of a transition is displayed in parentheses next to the transition; a transition for which no priority is explicitly displayed has priority 0. Denote by $\mathcal{P}(e)$ the priority of transition $e \in E$. We assume throughout that the priorities are such that $\mathcal{P}(e) \neq \mathcal{P}(e')$ whenever e and e' are in the same conflict set with $e \neq e'$. Heuristically, we define new-marking probabilities of the form $p(s'; s, E^*)$ in terms of the singleton probabilities and the priorities by applying the following two rules:

1. Whenever transitions within a conflict set fire simultaneously, the transition with the highest priority is selected to remove and deposit tokens in accordance with its associated singleton new-marking probabilities—that is, the net behaves as if the latter transition is the only one in the set that fires.

2. When transitions in different conflict sets fire simultaneously—and by the rule in (1) we can assume that, in effect, exactly one transition

fires in each set—the net behaves as if the transitions fire sequentially in order of decreasing priority.

Formally, suppose that all singleton new-marking probabilities have been specified, along with priorities $\{\,\mathcal{P}(e)\colon e \in E\,\}$. Denote by $Q_1, Q_2, \ldots, Q_k$ the conflict sets for the transitions. We specify a new-marking probability of the form $p(s'; s, E^*)$ as follows. Partition E^* into mutually disjoint nonempty subsets $E_1, E_2, \ldots, E_l$ such that each subset E_i is of the form $E^* \cap Q_j$ for some $j \in \{\,1, 2, \ldots, k\,\}$. Then for $1 \le i \le l$ denote by $\bar{e}_i$ the unique transition in E_i such that $\mathcal{P}(\bar{e}_i) = \max_{e \in E_i} \mathcal{P}(e)$. Finally, set

$$p(s'; s, E^*) = p(s'; s, \bar{e}_{\pi(1)}, \bar{e}_{\pi(2)}, \ldots, \bar{e}_{\pi(l)}), \qquad (3.4)$$

where $\bar{e}_{\pi(1)}, \bar{e}_{\pi(2)}, \ldots, \bar{e}_{\pi(l)}$ are the transitions $\bar{e}_1, \bar{e}_2, \ldots, \bar{e}_l$ ordered so that

$$\mathcal{P}(\bar{e}_{\pi(1)}) \ge \mathcal{P}(\bar{e}_{\pi(2)}) \ge \cdots \ge \mathcal{P}(\bar{e}_{\pi(l)}). \qquad (3.5)$$

In general, there may be more than one ordering such that (3.5) is satisfied. For the definition in (3.4) to make sense, we require that the right side of (3.4) have the same value for any two orderings. This requirement is satisfied by many SPNs encountered in practice, for example, SPNs with no marking-dependent transitions.

EXAMPLE 3.6 (Manufacturing cell with robots). Consider a manufacturing cell with two machines, two material-handling robots, two conveyors, a loading area for incoming raw parts, and an unloading area for outgoing finished parts. Robot 1 transfers raw parts, drawn as white squares in Figure 2.20, from the loading area to conveyor 1 and transfers finished parts, drawn as black squares, from conveyor 2 to the unloading area. Conveyor 1 moves raw parts to a designated position on the conveyor for transfer to a machine. Robot 2 transfers raw parts from conveyor 1 to the lowest-numbered available machine for processing and transfers finished parts from the machines to conveyor 2. Conveyor 2 moves finished parts to a designated position on the conveyor for transfer to the unloading area.

Raw parts are always available at the loading area. Each robot can handle only one part at a time. After a robot completes a transfer, the arm of the robot returns to a "null" position before starting another transfer. The arm of robot 1 does not leave its null position to transfer a raw part to conveyor 1 while a part is on the conveyor. The arm of robot 2 does not leave its null position to transfer a finished part to conveyor 2 while a part is on the conveyor and does not leave its null position to transfer a raw part to a machine while a part is at the machine. Thus, at any time there is at most one part on each conveyor and at most one part at each machine. Transfer of a finished part from conveyor 2 to the unloading area has (nonpreemptive) priority over transfer of a raw part from the loading area to conveyor 1. Transfer of a finished part from either machine to conveyor 2 has priority over transfer of a raw part from conveyor 1 to either machine, and transfer

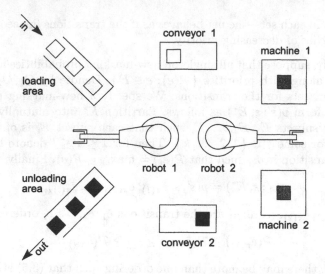

Figure 2.20. Manufacturing cell with robots.

of a finished part from machine 1 to conveyor 2 has priority over transfer of a finished part from machine 2.

The time for each of the actions performed by a robot is deterministic. The time for a conveyor to move a part is deterministic and may depend on the identity of the conveyor. The successive times for machine j to process a raw part are i.i.d. as a positive random variable L_j with continuous distribution function. We assume that the deterministic times for the actions performed by the robots and for the conveyors to move parts are such that with probability 1 no two events ever occur simultaneously.

This system can be specified as an SPN with deterministic timed and immediate transitions; see Figure 2.21. The interpretation of the transitions is given in Table 2.1. Each place contains at most one token; the interpretation of the tokens is given in Table 2.2. All transitions are deterministic, and all speeds for enabled transitions are equal to 1. The clock-setting distribution functions are defined in an obvious manner. Observe that the clock-setting distribution functions for transitions e_{17} and e_{20} explicitly depend on the current and new marking; no other clock-setting distribution functions exhibit such explicit dependence.

As can be seen from Figure 2.21, the priorities are given by $\mathcal{P}(e_{18}) = 1$, $\mathcal{P}(e_{19}) = 2$, $\mathcal{P}(e_{21}) = 2$, $\mathcal{P}(e_{22}) = 1$, $\mathcal{P}(e_{23}) = 4$, $\mathcal{P}(e_{24}) = 3$, and $\mathcal{P}(e) = 0$ otherwise. The relative values of $\mathcal{P}(e_{18})$, $\mathcal{P}(e_{19})$, and so forth reflect the relative priorities of the various operations performed by the robots. Observe that we can model different priority schemes for the robot operations without needing to change the bipartite graph of places and transitions.

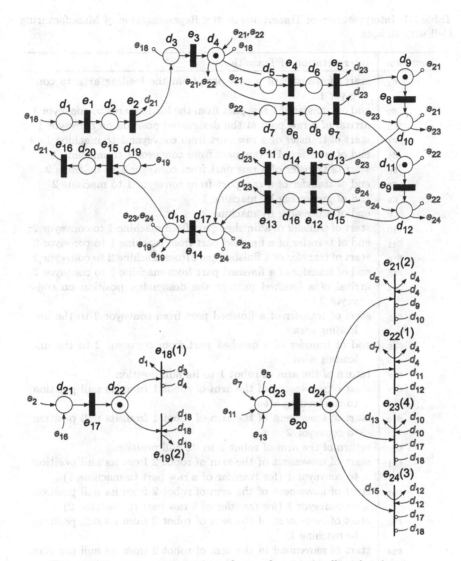

Figure 2.21. SPN representation of manufacturing cell with robots.

Table 2.1. Interpretation of Transitions in SPN Representation of Manufacturing Cell with Robots

Transition	Interpretation of Transition
e_1	start of transfer of a raw part from the loading area to conveyor 1
e_2	end of transfer of a raw part from the loading area to conveyor 1
e_3	arrival of a raw part at the designated position on conveyor 1
e_4	start of transfer of a raw part from conveyor 1 to machine 1
e_5	end of transfer of a raw part from conveyor 1 to machine 1
e_6	start of transfer of a raw part from conveyor 1 to machine 2
e_7	end of transfer of a raw part from conveyor 1 to machine 2
e_8	end of processing by machine 1
e_9	end of processing by machine 2
e_{10}	start of transfer of a finished part from machine 1 to conveyor 2
e_{11}	end of transfer of a finished part from machine 1 to conveyor 2
e_{12}	start of transfer of a finished part from machine 2 to conveyor 2
e_{13}	end of transfer of a finished part from machine 2 to conveyor 2
e_{14}	arrival of a finished part at the designated position on conveyor 2
e_{15}	start of transfer of a finished part from conveyor 2 to the unloading area
e_{16}	end of transfer of a finished part from conveyor 2 to the unloading area
e_{17}	return of the arm of robot 1 to its null position
e_{18}	start of movement of the arm of robot 1 from its null position to the loading area
e_{19}	start of movement of the arm of robot 1 from its null position to conveyor 2
e_{20}	return of the arm of robot 2 to its null position
e_{21}	start of movement of the arm of robot 2 from its null position to conveyor 1 (for transfer of a raw part to machine 1)
e_{22}	start of movement of the arm of robot 2 from its null position to conveyor 1 (for transfer of a raw part to machine 2)
e_{23}	start of movement of the arm of robot 2 from its null position to machine 1
e_{24}	start of movement of the arm of robot 2 from its null position to machine 2

Table 2.2. Interpretation of Places in SPN Representation of Manufacturing Cell with Robots

Place	Interpretation of Token in Place
d_1	the arm of robot 1 is moving from its null position to the loading area
d_2	robot 1 is transferring a raw part to conveyor 1
d_3	a raw part is being moved to the designated position on conveyor 1
d_4	a raw part is at the designated position on conveyor 1 awaiting transfer to a machine
d_5	the arm of robot 1 is moving from its null position to conveyor 1 (to transfer a raw part to machine 1)
d_6	robot 1 is transferring a raw part to machine 1
d_7	the arm of robot 1 is moving from its null position to conveyor 1 (to transfer a raw part to machine 2)
d_8	robot 1 is transferring a raw part to machine 2
d_9	machine 1 is processing a part
d_{10}	a finished part is at machine 1 awaiting transfer to conveyor 2
d_{11}	machine 2 is processing a part
d_{12}	a finished part is at machine 2 awaiting transfer to conveyor 2
d_{13}	the arm of robot 2 is moving from its null position to machine 1
d_{14}	robot 2 is transferring a finished part from machine 1 to conveyor 2
d_{15}	the arm of robot 2 is moving from its null position to machine 2
d_{16}	robot 2 is transferring a finished part from machine 2 to conveyor 2
d_{17}	a finished part is being moved to the designated position on conveyor 2
d_{18}	a raw part is at the designated position on conveyor 2 awaiting transfer to the unloading area
d_{19}	the arm of robot 1 is moving from its null position to conveyor 2
d_{20}	robot 1 is transferring a finished part from conveyor 2 to the unloading area
d_{21}	the arm of robot 1 is returning to its null position
d_{22}	the arm of robot 1 is in its null position
d_{23}	the arm of robot 2 is returning to its null position
d_{24}	the arm of robot 2 is in its null position

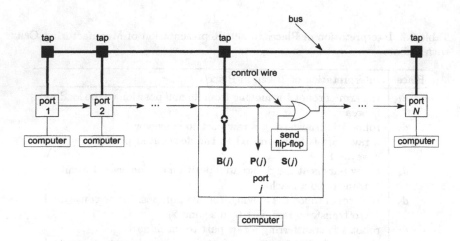

Figure 2.22. Collision-free bus network.

EXAMPLE 3.7 (Collision-free bus network). Consider a local area bus network with N ports, numbered $1, 2, \ldots, N$ from left to right; see Figure 2.22. Port j transmits and monitors message packets on the bidirectional bus at tap $\mathbf{B}(j)$. In addition to the bus, a unidirectional (left to right) logic control wire also links the ports. Associated with each port j is a flip-flop $\mathbf{S}(j)$ called the *send flip-flop*. Port j sets $\mathbf{S}(j)$ to 1 and resets $\mathbf{S}(j)$ to 0. The signal $\mathbf{P}(j)$, called the OR-signal, is tapped at the control wire input to port j and is the inclusive OR of the observed values of the send flip-flops of all ports to the left. Denote by T the propagation delay from end to end along the bus plus a small fixed quantity. Let $R(j)$ be the propagation delay along the control wire from port j to port N for $1 \le j \le N$; thus $R(1) > R(2) > \ldots > R(N) = 0$. Assume that signal propagation along the control wire is slower than along the bus in the sense that $R(1) > T$.

Distributed control scheme A1 is specified in terms of an algorithm for an individual port. When port j is not transmitting a packet and has no packets awaiting transmission, the arrival of a packet for transmission by port j initiates execution of the algorithm. If another packet is awaiting transmission by port j when this execution of the algorithm ends, the next execution begins immediately.

Algorithm A1

1. Set $\mathbf{S}(j)$ to 1.

2. Wait for a time interval $R(j) + T$.

3. Wait until the bus is observed at $\mathbf{B}(j)$ to be idle and $\mathbf{P}(j) = 0$; then start transmission of the packet, simultaneously resetting $\mathbf{S}(j)$ to 0.

Control scheme A1 is simple and asynchronous and provides collision-free communication among ports; that is, no two ports transmit signals that become electrically superimposed on the bus.

Assume that at most one packet awaits transmission at any time at any particular port; the successive times from end of transmission by port j until the arrival of the next packet for transmission by port j are i.i.d. as a positive random variable A_j with continuous distribution function. The successive times for port j to transmit a packet are i.i.d. as a positive random variable L_j with continuous distribution function. Transmission times are long in the sense that $P\{L_j > R(1) + T\} = 1$.

Denote the propagation delay along the bus between port i and port j by $T(i, j)$. Thus

$$T(i, j) = T(j, i) < T$$

and

$$T(i, j) + T(j, k) = T(i, k)$$

for $i < j < k$ or $i > j > k$.

This system can be specified as an SPN with deterministic timed and immediate transitions and a finite marking set; see Figure 2.23. (The figure displays the subnet that corresponds to a generic port j, where $1 < j < N$. The modifications required to obtain the subnet corresponding to port 1 or port N are straightforward.) The interpretation of the transitions is given in Table 2.3. Place $d_{6,j}$ contains at most $j - 1$ tokens for $2 \leq j \leq N$, and all other places contain at most one token. There is a token in place $d_{6,j}$ corresponding to each port k ($< j$) such that port j has observed the setting (to 1) of port k's flip-flop but has not yet observed the resetting (to 0) of this flip-flop. Thus place $d_{6,j}$ contains at least one token if and only if $\mathbf{P}(j) = 1$. The interpretation of the remaining places in the net is given in Table 2.4. The clock-setting distribution functions are defined in an obvious manner, and all speeds for enabled transitions are equal to 1. As can be seen from the figure, $\mathcal{P}(e_{6,j}) = \mathcal{P}(e_{8,j}) = 1$ and $\mathcal{P}(e_{9,j,k}) = 2$ for $1 \leq k < j \leq N$; the priorities for all other events are equal to 0.

Observe that, irrespective of propagation delays, transitions $e_{5,j}$ ($1 \leq j \leq N$) and $e_{6,j}$ can fire simultaneously, and similarly for transitions $e_{7,j}$ and $e_{8,j}$; that is, a port can observe an end of transmission and a start of transmission simultaneously. Indeed, transitions $e_{5,j}$ and $e_{6,j}$ fire simultaneously at time t whenever, at time $t - T(i, j)$ (with $i < j$), a packet awaits transmission by port i, the OR-signal $\mathbf{P}(i)$ is equal to 0, and port i observes

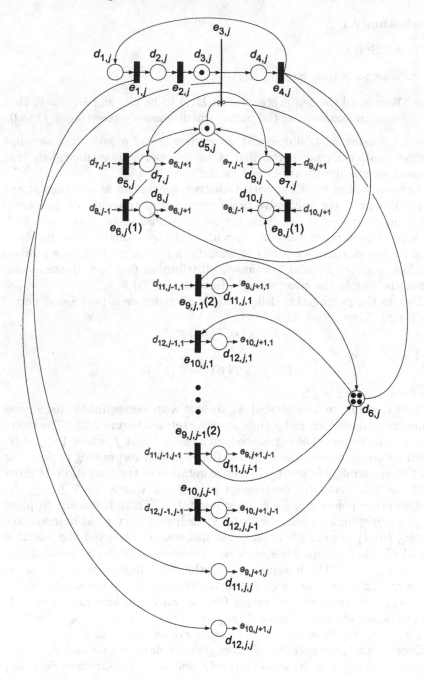

Port j

Figure 2.23. SPN representation of collision-free bus network.

Table 2.3. Interpretation of Transitions in SPN Representation of Collision-free Bus Network

Transition	Interpretation of Transition
$e_{1,j}$	setting (to 1) of flip-flop by port j
$e_{2,j}$	end of wait for $R(j) + T$
$e_{3,j}$	start of transmission by port j
$e_{4,j}$	end of transmission by port j
$e_{5,j}$	observation by port j of start of transmission by a port to the left
$e_{6,j}$	observation by port j of end of transmission by a port to the left
$e_{7,j}$	observation by port j of start of transmission by a port to the right
$e_{8,j}$	observation by port j of end of transmission by a port to the right
$e_{9,j,k}$	observation by port j of the setting (to 1) of flip-flop by port k
$e_{10,j,k}$	observation by port j of the resetting (to 0) of flip-flop by port k

Table 2.4. Interpretation of Places in SPN Representation of Collision-free Bus Network

Place	Interpretation of Token in Place
$d_{1,j}$	there is no packet awaiting transmission by port j and port j is not transmitting a packet
$d_{2,j}$	port j has set its flip-flop but has not yet completed the $R(j)+T$ wait
$d_{3,j}$	port j has completed the $R(j) + T$ wait but has not started transmission
$d_{4,j}$	port j is transmitting a packet
$d_{5,j}$	port j is observing transmission of a packet (by some port k with $k \neq j$) on the bus
$d_{7,j}$	the initial bit of a packet is propagating from port j to port $j+1$
$d_{8,j}$	the final bit of a packet is propagating from port j to port $j+1$
$d_{9,j}$	the initial bit of a packet is propagating from port j to port $j-1$
$d_{10,j}$	the final bit of a packet is propagating from port j to port $j-1$
$d_{11,j,k}$	the signal that port k has set its flip-flop (to 1) is propagating from port j to port $j+1$
$d_{12,j,k}$	the signal that port k has reset its flip-flop (to 0) is propagating from port j to port $j+1$

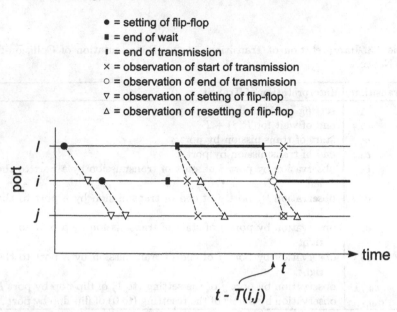

• = setting of flip-flop
▪ = end of wait
ı = end of transmission
× = observation of start of transmission
○ = observation of end of transmission
▽ = observation of setting of flip-flop
△ = observation of resetting of flip-flop

Figure 2.24. Timeline diagram for collision-free bus network.

an end of transmission by port l with $l < i$; see the timeline diagram in
Figure 2.24. (Roughly speaking, packets transmitted by ports i and l propagate "back to back" on the bus.) The assignment of priorities to transitions
ensures that the marking changes as if $e_{6,j}$ fires and then $e_{5,j}$ fires, that is,
as if port j first observes an end of transmission and then observes a start
of transmission. Also observe that transitions $e_{5,j}$ and $e_{6,j}$ need not fire
simultaneously, so that an attempt to model the simultaneous occurrence
of the corresponding events in the system by using a single transition as
in the slotted ring of Example 2.12 leads to a messy and complicated SPN
model.

Depending on the value of the propagation delays, other transitions may
also fire simultaneously. For example, transitions $e_{5,j}$ and $e_{7,l}$ fire simultaneously at time t if, for some $l < i < j$, transition $e_{3,i}$ fires at time
$t - T(i,j)$ and $T(i,j) = T(i,l)$. That is, port l and port j simultaneously
observe the start of transmission of a packet by port i if port l and port j
are equidistant from port i; see Figure 2.24. Whenever transitions $e_{5,j}$ and
$e_{7,l}$ fire simultaneously, the marking changes as if $e_{5,j}$ and then $e_{7,l}$ fires or,
equivalently, as if $e_{7,l}$ and then $e_{5,j}$ fires. The priorities for these two transitions are both equal to 0, reflecting the fact that the new marking does
not depend on the firing order. As another example, transitions $e_{2,j}$ and
$e_{9,j,i}$ can fire simultaneously; that is, port j can simultaneously complete
a wait of length $R(j) + T$ and observe the setting of a flip-flop by port i.
These events occur simultaneously if, for example, the packet interarrival-

time distributions have support on the positive integers and the constants $R(j)$, T, and so forth have integer values. Suppose that transitions $e_{2,j}$ and $e_{9,j,i}$ fire simultaneously and that

1. There is a packet awaiting transmission by port j,

2. $\mathbf{P}(j) = 0$, and

3. The bus is observed to be idle by port j

just before this transition firing. Because $\mathcal{P}(e_{9,j,i}) > \mathcal{P}(e_{2,j})$, the marking changes as if $e_{9,j,i}$ fires and then $e_{2,j}$ fires, and port j does not start transmission of a packet. Had the order of the priorities been reversed, port j would have started transmission of a packet.

The method of priorities can be generalized in various ways. We conclude our discussion by describing an extension in which two or more immediate transitions in a conflict set are allowed to have equal priorities. The idea is that, within each conflict set, the enabled immediate transitions having the highest priority are allowed to fire simultaneously, provided that the corresponding behavior of the SPN at such a firing is specified explicitly. Specifically, when the current (immediate) marking is s, denote by $E_j(s)$ ($1 \leq j \leq k$) the set of enabled immediate transitions within the jth conflict set that have the highest priority:

$$E_j(s) = \{\, e \in E' \cap E(s) \cap Q_j : \mathcal{P}(e) \geq \mathcal{P}(e') \text{ for all } e' \in E' \cap E(s) \cap Q_j \,\},$$

where, as before, $Q_1, Q_2, \ldots, Q_k$ are the conflict sets. In general, one or more of the sets $E_1(s), E_2(s), \ldots, E_k(s)$ may be empty; enumerate the nonempty subsets as $\bar{E}_1(s), \bar{E}_2(s), \ldots, \bar{E}_l(s)$, where $l = l(s) \leq k$. Then, for our extension, all new-marking probabilities of the form $p\big(s'; s, \bar{E}_i(s)\big)$ must be specified in addition to the singleton new-marking probabilities. The priorities of the transitions then determine the effective order in which the simultaneous transition firings for the different conflict sets occur. The details are as follows. We abuse notation slightly and denote by $\mathcal{P}\big(\bar{E}_i(s)\big)$ the common priority of the transitions in $\bar{E}_i(s)$. For arbitrary markings s and s' and transition sets $E_1, E_2, \ldots, E_l \subseteq E$, we can define quantities of the form $p(s'; s, E_1, E_2, \ldots, E_l)$ in analogy to (3.2); that is, $p(s'; s, E_1, E_2, \ldots, E_l)$ is the probability that the new marking is s' given that the sets of transitions $E_1, E_2, \ldots, E_l$ successively trigger marking changes starting in marking s. We then set

$$p(s'; s, E^*) = p\big(s'; s, \bar{E}_{\pi(1)}(s), \bar{E}_{\pi(2)}(s) \ldots, \bar{E}_{\pi(l)}(s)\big), \qquad (3.8)$$

where $\bar{E}_{\pi(1)}(s), \bar{E}_{\pi(2)}(s) \ldots, \bar{E}_{\pi(l)}(s)$ are the sets $\bar{E}_1(s), \bar{E}_2(s) \ldots, \bar{E}_l(s)$ ordered so that

$$\mathcal{P}\big(\bar{E}_{\pi(1)}(s)\big) \geq \mathcal{P}\big(\bar{E}_{\pi(2)}(s)\big) \geq \cdots \geq \mathcal{P}\big(\bar{E}_{\pi(l)}(s)\big). \qquad (3.9)$$

As before, we require that the right side of (3.8) have the same value for any two orderings that satisfy (3.9).

EXAMPLE 3.10 (Manufacturing cell with nondeterministic robots). Consider a manufacturing cell as in Example 3.6, except that if

(i) robot 2 is in its null position,

(ii) a raw part is on conveyor 1 awaiting transfer to a machine, and

(iii) there is no part either at machine 1 or machine 2,

then with fixed probability $q \in (0,1)$ robot 1 transfers the raw part to machine 1 and with probability $1-q$ transfers the part to machine 2. Similarly, whenever robot 2 is in its null position and there is a finished part at both machine 1 and machine 2 awaiting transfer to conveyor 2, with probability q robot 1 transfers the finished part at machine 1 to conveyor 2 and with probability $1-q$ transfers the finished part at machine 2 to conveyor 2. As in Example 3.6, transfer of a finished part from either machine to conveyor 2 has priority over transfer of a raw part from conveyor 1 to either machine.

This system can be specified as an SPN exactly as in Example 3.6, except that $\mathcal{P}(e_{21}) = \mathcal{P}(e_{22}) = 1$ and $\mathcal{P}(e_{23}) = \mathcal{P}(e_{24}) = 2$, and the new-marking probabilities are modified as follows. As before, all transitions are deterministic. For $s \in G(e_{21}) \cap G(e_{22})$ and $s' \in G$, set

$$p(s'; s, \{e_{21}, e_{22}\}) = qp(s'; s, e_{21}) + (1-q)p(s'; s, e_{22}).$$

Similarly, for $s \in G(e_{23}) \cap G(e_{24})$ and $s' \in G$, set

$$p(s'; s, \{e_{23}, e_{24}\}) = qp(s'; s, e_{23}) + (1-q)p(s'; s, e_{24}).$$

Then, for $s \in S'$, $s' \in G$, and $E^* = E(s) \cap E'$, the new-marking probability $p(s'; s, E^*)$ is defined as in (3.8).

2.4   Alternative Building Blocks

One drawback of our SPN formulation is that the marking set G must be specified—at least in principle—before specification of the new-marking probabilities, speeds, and clock-setting distributions. In this section we consider an alternative set of SPN building blocks that avoids this requirement. Denote by $\mathcal{Z}_+^L$ the set of all nonnegative, integer-valued vectors of length L. Then the building blocks consist of

- A finite set $D = \{d_1, d_2, \ldots, d_L\}$ of places

- A finite set $E = \{e_1, e_2, \ldots, e_M\}$ of transitions

- A (possibly empty) set $E' \subset E$ of immediate transitions

- Sets $I(e), L(e), J(e) \subseteq D$ of normal input places, inhibitor input places, and output places, respectively, for each $e \in E$

- A clock-setting distribution function $F(\,\cdot\,; e)$ for each $e \in E - E'$

- An initial marking $\bar{s}_0 \in \mathcal{Z}_+^L$

- A probability mass function $\tilde{p}(\,\cdot\,; E^*)$ on $\{-1, 0, 1\}^L$ for each $E^* \subseteq E$

There is no function $r(s, e)$; all clocks run down to 0 at unit rate. Moreover, the clock-setting distribution functions do not explicitly depend on the old marking, new marking, or set of transitions that trigger the marking change. The mechanism by which tokens are removed and deposited when the transitions in the set $E^* \subseteq E$ fire simultaneously also is independent of the old and new markings: when the marking is s and the transitions in E^* fire, the new marking is of the form $s + U(E^*)$, where the random variable $U(E^*)$ takes values in the set $\{-1, 0, 1\}^L$ and has probability mass function $\tilde{p}(\,\cdot\,; E^*)$. We assume that $\tilde{p}(u; E^*) = P\{U(E^*) = u\} > 0$ only if $u = (u_1, u_2, \ldots, u_L)$ satisfies the following two conditions.

1. $u_i = -1$ only if $d_i \in \bigcup_{e \in E^*} I(e)$.

2. $u_i = 1$ only if $d_i \in \bigcup_{e \in E^*} J(e)$.

We refer to SPNs that have the above building blocks as *restricted* SPNs.

A transition e of a restricted SPN is said to be *deterministic* if $\tilde{p}(u; \{e\}) = 1$, where $u = (u_1, u_2, \ldots, u_L)$ is given by

$$
u_i = \begin{cases}
-1 & \text{if } d_i \in I(e) - J(e); \\
1 & \text{if } d_i \in J(e) - I(e); \\
0 & \text{otherwise}
\end{cases}
$$

for $1 \leq i \leq L$. Thus with probability 1 the new marking is $s + u$ when the marking is s and a deterministic transition e fires.

The marking set G of a restricted SPN need not be specified explicitly. Rather, G can be defined in terms of the building blocks as follows. Write $s \to s'$ for $s, s' \in \mathcal{Z}_+^L$ if $P\{s + U(E^*) = s'\} > 0$ for some $E^* \subseteq E(s)$. We say that $s' \in \mathcal{Z}_+^L$ is *reachable* from $s \in \mathcal{Z}_+^L$ and write $s \rightsquigarrow s'$ if either $s \to s'$ or there exist markings $s^{(1)}, s^{(2)}, \ldots, s^{(n)} \in \mathcal{Z}_+^L$ ($n \geq 1$) such that $s \to s^{(1)} \to \cdots \to s^{(n)} \to s'$. Given these definitions, take

$$
G = \{s \in \mathcal{Z}_+^L : \bar{s}_0 \rightsquigarrow s\},
$$

the set of markings reachable from the initial marking $\bar{s}_0$. New-marking probabilities can be defined in terms of the building blocks by setting $p(s'; s, E^*) = \tilde{p}(s - s'; E^*)$ for $s', s \in \mathcal{Z}_+^L$ and $E^* \subseteq E(s)$.

Deterministic SPNs form an important subclass of restricted SPNs. An SPN is deterministic if every transition is deterministic and, whenever more than one immediate transition becomes enabled in a marking, the marking changes as if exactly one of the transitions—selected according to a probability distribution—fires. Thus, for $E^* = \{ e_{i_1}, e_{i_2}, \ldots, e_{i_l} \} \subseteq E'$, we have the representation

$$\tilde{p}(\,\cdot\,; E^*) = \sum_{k=1}^{l} a_k \tilde{p}(\,\cdot\,; e_{i_k}),$$

where $a_1, a_2, \ldots, a_l$ are probabilities that depend on E^* and sum to 1.

All our results for standard SPNs as defined in Section 2.1 automatically apply to restricted and deterministic SPNs. It is intuitively clear that restricted SPNs have less modelling power than standard SPNs. Nonetheless, restricted SPNs can model a usefully large class of discrete-event stochastic systems. It can be shown in particular that for any GSMP with finite state space, unit speeds, and a fixed initial state, there exists a restricted SPN with a marking process that behaves the same way—more precisely, the marking process "strongly mimics" the GSMP as defined in Chapter 4. If, with probability 1, events in the GSMP never occur simultaneously, then the GSMP can be strongly mimicked using a deterministic SPN. Moreover, for any SPN having unit speeds, a finite marking set, a fixed initial marking, and timed transitions that with probability 1 never fire simultaneously, there exists a deterministic SPN that behaves the same way; see Remarks 3.2 and 4.11 in Chapter 4.

In practice, it is often convenient to exploit the full generality of our original SPN formulation to obtain a concise representation of a specified system. Indeed, as shown by the queue with batch arrivals in Example 2.4, this generality sometimes is essential. On the other hand, if a system can be modelled as a deterministic SPN, then key properties such as k-boundedness, "liveness," and the existence of "invariants" can be determined using analysis techniques for ordinary Petri nets. An SPN is *live* if at least one transition is enabled in each reachable marking, and an *invariant* is a linear algebraic relation between the token counts in the places of the net that holds for every reachable marking.

Notes

The discussion of SPN building blocks in Section 2.1 follows Haas and Shedler (1989b). Both the notation and the formulation of the building blocks were originally motivated by the discussion of generalized semi-

Markov processes in Whitt (1980). As shown in Chapter 4, SPNs and GSMPs are closely related.

Kosaraju (1973) uses an untimed version of the producer–consumer system to illustrate the limited modelling power of ordinary Petri nets without inhibitor input places; see Section 7.1 in Peterson (1981). A discussion of PRI preemption can be found in Bobbio et al. (1995). The descriptions of ring and bus networks are based on work in Eswaran et al. (1978) and Loucks et al. (1982); see also Iglehart and Shedler (1983, 1984) and Haas and Shedler (1985a, 1985b). The flexible manufacturing system and the manufacturing cell with robots are presented in Ajmone Marsan et al. (1987) and Viswanadham and Narahari (1988), respectively; the current exposition of these models is based on the discussion in Haas and Shedler (1992). The particle-counter model of Example 2.11 corresponds to the "type II counter" described in Section 5.3 of Karlin and Taylor (1975); see also Haas and Shedler (1991).

The SPSIM prototype system for simulation of stochastic processes was developed by Jochens and Shedler (1989). SPSIM can be used to specify and simulate both GSMPs and SPNs. For details of the original SPSIM system and subsequent extensions, see Jochens and Shedler (1989), Bergman and Shedler (1993), and Shedler (1994).

Hack (1975) originally suggested the use of numerical priorities in ordinary Petri nets. Priority schemes of various types have since been incorporated into SPN formalisms; see, for example, Chapter 4 in Ajmone Marsan et al. (1995). The latter reference also discusses various notions of conflict between transitions. As indicated in Section 2.3, we view priorities not as a basic SPN building block, but rather as a convenient means for concise specification of the new-marking probabilities. For nets in which an enabled transition always remains enabled until it fires, Haas and Shedler (1987c) give conditions under which the value of the right side of (3.4) is independent of the ordering π.

The SPNs defined in Section 2.4 (especially the deterministic SPNs) are similar in spirit to many SPN formulations in the literature. For such nets, the set G is called the *reachability set* of the SPN. Determining the reachability set—or properties of the reachability set such as finiteness, k-boundedness, and liveness—is nontrivial. As mentioned previously, analysis methods for ordinary Petri nets are applicable when all transitions are deterministic; see Peterson (1981) and Reisig (1985) for an introduction to some of these methods, and see Jančar (2000) and Kosten and Tchoudaikina (1998) for recent discussions about the reachability problem.

3
The Marking Process

The marking process of an SPN records the marking as it evolves over continuous time. As discussed in Section 3.1, formal definition of the marking process is in terms of an underlying general state-space Markov chain that describes the net at successive marking changes. This definition leads to an algorithm for generating sample paths of the process.

Many performance measures such as long-run utilization, average revenue, availability, and throughput can be specified as time-average limits of the marking process or underlying chain—or as functions of such limits. In Section 3.2 we illustrate the specification of long-run performance measures through a variety of examples. In the process, we show how limit theorems in discrete time can be used to obtain limit theorems in continuous time. These results highlight the key role of the underlying chain in the analysis of long-run SPN behavior.

The "lifetime" of a marking process is the supremum of the successive times at which the marking changes. The lifetime must be almost surely (a.s.) infinite for time-average limits to be well defined. For some SPNs, however, infinitely many marking changes can occur in a finite time interval, so that the lifetime is finite. Such pathological behavior occurs if the process is absorbed into the set S' of immediate markings or if the marking changes occur ever more rapidly so that the sequence of occurrence times has an accumulation point. In the presence of nonexponential clock-setting distributions, this latter type of "explosion" can occur with probability 1 even when the expected time between successive marking changes increases linearly. In Section 3.3 we give conditions under which the lifetime is a.s. infinite. These conditions are mild and are satisfied by most SPNs encountered in practice. Our proof rests on a "geometric trials" recurrence criterion, which also is used in subsequent chapters to establish the regenerative property for both marking processes and sequences of delays.

When the marking process of an SPN is a continuous-time Markov chain (CTMC), the sequence of successive timed markings forms a discrete-time Markov chain and, given this sequence, the successive times between state transitions of the marking process are independent and exponentially distributed. This special structure makes it possible, in principle, to compute time-average limits either analytically or numerically. One might expect that the marking process of an SPN is a CTMC if each clock-setting distribution is exponential. This result is not quite true: the marking process can fail to have the Markov property when the clock-setting distribution function explicitly depends on the current and new marking. In the absence of such explicit dependence, however, the Markov property does indeed hold, as shown in Section 3.4. The proof of this result leads to explicit formulas for the elements of the infinitesimal generator matrix of the process. As a key step in establishing the Markov property, we determine the conditional distribution of the clock-reading vector, given the "partial history" of the underlying chain of the marking process. This conditional distribution plays a central role in the recurrence and regeneration results developed in subsequent chapters.

3.1 Definition of the Marking Process

In this section we define the marking process of an SPN in terms of a Markov chain that takes values in an uncountably infinite set. To prepare for this definition, we first give a brief introduction to general state-space Markov chains.

3.1.1 General State-Space Markov Chains

A Markov process is a stochastic process whose future evolution depends on the past and present only through the current state. Consider a Markov process that evolves in discrete time and takes values in an arbitrary state space Γ. If Γ is finite or countably infinite—the simplest and most familiar case—then the process is called a *discrete-time Markov chain* (DTMC); see Section A.2.4 for a discussion of DTMCs. A time-homogeneous DTMC can be characterized in terms of an initial distribution together with a "transition matrix." The (i,j)th entry of the matrix is the probability, starting in state i, that the chain next hits state j. When Γ is uncountably infinite, however, the probability that the chain hits a specified element of Γ typically is equal to 0, and the notion of a transition matrix is not useful.

The appropriate generalization of the transition matrix is the *transition kernel* P: the quantity $P(z, A)$ is the probability, starting in state z, that the chain next hits a state that is an element of the set A.

Definition 1.1. The discrete-time stochastic process $\{ Z_n : n \geq 0 \}$ defined on a probability space $(\Omega, \mathcal{F}, P_\mu)$ and taking values in Γ is a (time-homogeneous) *general state-space Markov chain* with initial distribution μ and transition kernel P if

$$P_\mu \{ Z_0 \in A \} = \mu(A) \tag{1.2}$$

and

$$P_\mu \{ Z_{n+1} \in A \mid Z_n, Z_{n-1}, \ldots, Z_0 \} = P(Z_n, A) \text{ a.s.} \tag{1.3}$$

for $n \geq 0$ and $A \subseteq \Gamma$.

We write P_μ for the probability law of the chain to emphasize the dependence on the initial distribution μ. We sometimes refer to a family of chains having a specified transition kernel P and indexed by the initial distribution μ somewhat loosely as "the chain with transition kernel P." Similarly, we sometimes say that a specified property holds for "the" chain "when the initial distribution is μ," meaning of course that the property holds for a specific member of the family.

Typically, μ and P are completely determined by the values $\{ \mu(A) : A \in \mathcal{A} \}$ and $\{ P(z, A) : z \in \Gamma \text{ and } A \in \mathcal{A} \}$, respectively, where $\mathcal{A}$ is a collection of subsets of Γ that have a relatively simple form. For example, when S is a finite or countably infinite set and $\Gamma \subseteq S \times \Re_+^K$ for some $K \geq 1$, we usually can take $\mathcal{A}$ to be the collection of all sets of the form

$$A = \{ s \} \times [0, a_1] \times [0, a_2] \times \cdots \times [0, a_K],$$

where $s \in S$ and $a_1, a_2, \ldots, a_K \geq 0$.

The finite-dimensional distributions of the chain can be computed using the relation

$$P_\mu \{ Z_0 \in A_0, Z_1 \in A_1, \ldots, Z_n \in A_n \}$$
$$= \int_{A_0} \mu(dz_0) \int_{A_1} P(z_0, dz_1) \cdots \int_{A_{n-1}} P(z_{n-2}, dz_{n-1}) P(z_{n-1}, A_n) \tag{1.4}$$

for $n \geq 0$ and $A_0, A_1, \ldots, A_n \subseteq \Gamma$. Denote by E_μ the expectation operator associated with P_μ. When the initial state is equal to $z \in \Gamma$ with probability 1, that is, $\mu(\{z\}) = 1$, we often write P_z for the probability law of the chain and E_z for the associated expectation. Define the n-step transition kernels for the chain by setting $P^n(z, A) = P_z \{ Z_n \in A \}$ for $n \geq 0$; observe that $P^0(z, A) = 1_A(z)$ and $P^1(z, A) = P(z, A)$. It follows from (1.4) that the kernels $\{ P^n : n \geq 0 \}$ satisfy the *Chapman–Kolmogorov* equations:

$$P^{n+m}(z, A) = \int_\Gamma P^n(z, dz') P^m(z', A) \tag{1.5}$$

for $z \in \Gamma$, $A \subseteq \Gamma$ and $m, n \geq 0$.

A chain can be defined by specifying an initial distribution μ and transition kernel P: for any choice of μ and P there exist a probability space $(\Omega, \mathcal{F}, P_\mu)$ and a stochastic process $\{Z_n : n \geq 0\}$ such that (1.2) and (1.3) hold. A standard construction of $\{Z_n : n \geq 0\}$ from μ and P uses Kolmogorov's existence theorem (Proposition 2.1 in the Appendix). In this construction, $\Omega = \Gamma^\infty$, so that each $\omega \in \Omega$ has the form $\omega = (\omega_0, \omega_1, \dots)$, where $\omega_n \in \Gamma$ for $n \geq 0$. The chain is then defined as the coordinate projection function on Γ^∞: $Z_n(\omega) = \omega_n$ for $n \geq 0$.

A general state-space Markov chain enjoys the *strong Markov property*, which asserts that the equality in (1.3) holds when the deterministic index n is replaced by a *stopping time* N:

$$P_\mu\{Z_{N+1} \in A \mid Z_N, Z_{N-1}, \dots, Z_0\} = P(Z_N, A) \text{ a.s.} \tag{1.6}$$

for $A \subseteq \Gamma$. Here N is a stopping time with respect to the chain $\{Z_n : n \geq 0\}$ if for each $n \geq 0$ the occurrence or nonoccurrence of the event $\{N = n\}$ is completely determined by $Z_0, Z_1, \dots, Z_n$; see Section A.1.5 for further discussion of stopping times.

3.1.2   Definition of the Continuous-Time Process

Formal definition of the marking process proceeds as follows. Recall that G is the set of markings of the SPN, S is the set of timed markings, and S' is the set of immediate markings. Similarly, E is the set of transitions and E' ($\subseteq E$) is the set of immediate transitions. Finally, recall that $E(s)$ is the set of enabled transitions and $r(s, e)$ is the speed at which the clock for enabled transition e runs down when the marking is s. Denote by $C(s)$ the set of possible *clock-reading vectors* when the marking is s:

$$C(s) = \{ c = (c_1, \dots, c_M) : c_i \geq 0$$
$$\text{and } c_i > 0 \text{ if and only if } e_i \in E(s) - E' \}.$$

Here the ith component of a clock-reading vector $c = (c_1, \dots, c_M)$ is the clock reading associated with transition e_i. Implicit in our definition is the convention that the reading on the clock for a disabled transition is 0. Beginning in marking s with clock-reading vector $c = (c_1, \dots, c_M) \in C(s)$, the time $t^*(s, c)$ to the next marking change is given by

$$t^*(s, c) = \min_{\{i : e_i \in E(s)\}} c_i/r(s, e_i), \tag{1.7}$$

where $c_i/r(s, e_i)$ is taken to be $+\infty$ when $r(s, e_i) = 0$. We sometimes refer to t^* as the *holding-time function* of the SPN. The set of transitions $E^*(s, c)$

that fire simultaneously and trigger the next marking change is given by

$$E^*(s, c) = \{ e_i \in E(s) \colon c_i - t^*(s,c)r(s,e_i) = 0 \}. \tag{1.8}$$

Observe that $E^*(s, c) = E' \cap E(s)$ whenever $s \in S'$ and $E^*(s, c) \subseteq E - E'$ whenever $s \in S$; in the former case, $t^*(s, c) = 0$.

Next consider a general state-space Markov chain $\{ (S_n, C_n) \colon n \geq 0 \}$ taking values in the set

$$\Sigma = \bigcup_{s \in G} (\{ s \} \times C(s)),$$

where $S_n = (S_{n,1}, S_{n,2}, \ldots, S_{n,L})$ represents the marking and $C_n = (C_{n,1}, C_{n,2}, \ldots, C_{n,M})$ represents the clock-reading vector just after the nth marking change. The transition kernel of the chain is given by

$$P((s, c), A) = p(s'; s, E^*) \prod_{e_i \in N} F(a_i; s', e_i, s, E^*) \prod_{e_i \in O} 1_{[0, a_i]}(c_i^*) \tag{1.9}$$

for all sets

$$A = \{ s' \} \times \{ (c_1', c_2', \ldots, c_M') \in C(s') \colon 0 \leq c_i' \leq a_i \text{ for } 1 \leq i \leq M \},$$

where $c_i^* = c_i - t^*(s,c)r(s,e_i)$, $E^* = E^*(s,c)$, $N = N(s'; s, E^*)$, and $O = O(s'; s, E^*)$. The right side of (1.9) is the probability, beginning with marking s and clock-reading vector c, that the SPN changes marking to s' with the reading c_i' on the clock associated with enabled transition $e_i \in E(s')$ set to a value in $[0, a_i]$. Specification of the transition kernel P for each set A of the above form is sufficient to uniquely determine P.

In more detail, the leftmost term on the right side of (1.9) is the probability that the new marking is s' when the current marking is s and the transitions in $E^* = E^*(s, c)$ fire simultaneously. Each remaining term represents the conditional probability that the clock for a transition e_i has a value in $[0, a_i]$ just after the marking change, given that the new marking is s'. The probabilities for the new transitions are multiplied together, since clocks for such transitions are set independently. For each old transition $e_i \in O(s'; s, E^*)$, the clock reading changes deterministically from c_i to $c_i^* = c_i - t^*(s, c)r(s, e_i)$. The probability that the clock reading for e_i has a value in $[0, a_i]$ just after the marking change is therefore equal to 0 or 1, depending on whether $c_i^* \in [0, a_i]$. Thus the joint probability that the clock for each old transition e_i has a value in $[0, a_i]$ is a product of indicator functions as in (1.9). For a transition $e_i \notin E(s')$, the associated clock reading is 0 by convention, so that $e_i \in [0, a_i]$ with probability 1 for any $a_i \geq 0$; the right side of (1.9) is therefore implicitly multiplied by a factor of 1 for each such e_i.

Denote by μ the initial distribution of the chain; that is, for any subset $B \subseteq \Sigma$, the quantity $\mu(B)$ represents the probability that $(S_0, C_0) \in B$. Denote by P_μ the probability law of the chain when the initial distribution is

μ. As discussed in Section 2.1, the initial marking s_0 is selected according to a (possibly degenerate) initial-marking distribution function ν_0 and then, for each enabled transition $e_i \in E(s_0)$, the corresponding clock reading $c_{0,i}$ is generated according to an initial clock-setting distribution function $F_0(\,\cdot\,; e_i, s_0)$. Thus the initial distribution μ is of the form

$$\mu(A) = \nu_0(s_0) \prod_{e \in E(s_0)} F_0(a_i; e, s_0) \tag{1.10}$$

for all sets

$$A = \{\, s_0 \,\} \times \{\, (c_{0,1}, \ldots, c_{0,M}) \in C(s_0) \colon 0 \le c_{0,i} \le a_i \text{ for } 1 \le i \le M \,\}.$$

Example 2.2 in the Appendix contains further details about the construction of the chain $\{\,(S_n, C_n) \colon n \ge 0\,\}$.

Finally, construct a continuous-time process $\{X(t) \colon t \ge 0\}$ from $\{\,(S_n, C_n) \colon n \ge 0\,\}$ in the following manner. Let ζ_n $(n \ge 0)$ be the (nonnegative, real-valued) time of the nth marking change: $\zeta_0 = 0$ and

$$\zeta_n = \sum_{k=0}^{n-1} t^*(S_k, C_k) \tag{1.11}$$

for $n \ge 1$. Let $\Delta \notin G$ and set

$$X(t) = \begin{cases} S_{N(t)} & \text{if } N(t) < \infty; \\ \Delta & \text{if } N(t) = \infty, \end{cases} \tag{1.12}$$

where

$$N(t) = \sup \{\, n \ge 0 \colon \zeta_n \le t \,\}. \tag{1.13}$$

The stochastic process $\{\, X(t) \colon t \ge 0 \,\}$ defined by (1.12) is the *marking process* of the SPN. By construction, the marking process takes values in the set $S \cup \{\Delta\}$ and has piecewise-constant, right-continuous sample paths. Observe that $X(t) = \Delta$ for at least one finite time point t if and only if the *lifetime* of the marking process, defined by

$$\tau_\Delta = \sup_{n \ge 0} \zeta_n,$$

is finite. As with Markov chains, we sometimes use loose terminology when referring to a family of marking processes that differ only in the initial distribution μ.

We often denote by $E_n^* = E^*(S_n, C_n)$ the random set of transitions that fire simultaneously and trigger the $(n+1)$st marking change $(n \ge 0)$ and by $t_n^* = t^*(S_n, C_n)$ the time between the nth and $(n+1)$st marking change. Let $\{\,\gamma(n) \colon n \ge 0\,\}$ be the indices of the successive marking changes at which the new marking is timed: $\gamma(-1) = -1$ and

$$\gamma(n) = \inf \{\, j > \gamma(n-1) \colon S_j \in S \,\} \tag{1.14}$$

for $n \geq 0$. Define the *embedded chain* $\{ (S_n^+, C_n^+) : n \geq 0 \}$ by setting

$$(S_n^+, C_n^+) = (S_{\gamma(n)}, C_{\gamma(n)}) \tag{1.15}$$

for $n \geq 0$. Suppose that $P_\mu \{ S_n \in S \text{ i.o.} \} = 1$, so that each random index $\gamma(n)$ is a.s. finite—sufficient conditions for this assumption to hold are given in Section 3.3.1. Because each random index $\gamma(n)$ is a stopping time with respect to the underlying chain $\{ (S_n, C_n) : n \geq 0 \}$, it follows from the strong Markov property for $\{ (S_n, C_n) : n \geq 0 \}$ that $\{ (S_n^+, C_n^+) : n \geq 0 \}$ is indeed a well-defined general state-space Markov chain. Denote by Σ^+ and μ^+ the state space and initial distribution, respectively, of the embedded chain:

$$\Sigma^+ = \{ (s, c) \in \Sigma : s \in S \}$$

and

$$\mu^+(A) = P_\mu \{ (S_0^+, C_0^+) \in A \}$$

for $A \subseteq \Sigma^+$.

3.1.3 Generation of Sample Paths

The form of the transition kernel in (1.9) leads to the following algorithm for generating sample paths of the underlying chain $\{ (S_n, C_n) : n \geq 0 \}$.

Algorithm 1.16 (Sample path generation for the underlying chain)

1. (Initialization) Set $\zeta = 0$. Select an initial marking $s \in G$ according to the probability mass function ν_0. For each enabled transition $e_i \in E(s)$, generate a corresponding clock reading c_i according to the clock-setting distribution function $F_0(\cdot\,; e_i, s)$. Set $c_i = 0$ for each $e_i \notin E(s)$.

2. Determine the set E^* of transitions that fire simultaneously and trigger the next marking change: $e_i \in E^*$ if and only if $c_i/r(s, e_i) \leq c_j/r(s, e_j)$ for all $j \neq i$. Also determine the time t^* to the next marking change as $t^* = c_{i^*}/r(s, e_{i^*})$, where i^* is any index such that $e_{i^*} \in E^*$.

3. Generate the new marking s' according to the probability mass function $p(\cdot\,; s, E^*)$.

4. For each transition $e_i \in N(s'; s, E^*) = E(s') - \big(E(s) - E^*\big)$, generate a new clock reading c_i' according to the distribution function $F(\cdot\,; s', e_i, s, E^*)$.

5. For each transition $e_i \in O(s'; s, E^*) = E(s') \cap \big(E(s) - E^*\big)$, set $c_i' = c_i - t^*(s, c)r(s, e_i)$.

6. For each transition $e_i \in \big(E(s) - E^*\big) - E(s')$, set $c_i' = 0$.

7. Go to step 2 and iterate with s' playing the role of s and c' the role of c.

At each marking change, the sets of transitions that become enabled and disabled must be determined. A naive approach to this task examines each transition $e \in E$; a better approach is as follows. Recall that $I(e)$, $L(e)$, and $J(e)$ are the sets of normal input places, inhibitor input places, and output places, respectively, for transition $e \in E$. Set

$$B_1(e^*) = \{ e \in E : I(e) \cap J(e^*) \neq \varnothing \text{ or } L(e) \cap I(e^*) \neq \varnothing \}$$

and

$$B_2(e^*) = \{ e \in E : I(e) \cap I(e^*) \neq \varnothing \text{ or } L(e) \cap J(e^*) \neq \varnothing \}.$$

The definition of the set $B_2(e^*)$ is closely related to the definition of conflict in Section 2.3.2: if $e \in B_2(e^*)$, then transition e^*, upon firing, can potentially remove a token from a normal input place for transition e or deposit a token in an inhibitor input place. The set $B_1(e^*)$ is defined in the opposite manner: if $e \in B_1(e^*)$, then transition e^*, upon firing, can potentially deposit a token in a normal input place for transition e or remove a token from an inhibitor input place. Observe that, at a marking change from s to s' triggered by the simultaneous firing of the transitions in E^*,

$$N(s'; s, E^*) \subseteq \bigcup_{e^* \in E^*} B_1(e^*) \tag{1.17}$$

and

$$(E(s) - E^*) - E(s') \subseteq \bigcup_{e^* \in E^*} B_2(e^*). \tag{1.18}$$

Typically, the sets $B_1(e^*)$ and $B_2(e^*)$ are small for each $e^* \in E^*$ and the set E^* is also small. Thus, even when the set E is large, relatively few transitions need be examined to update the set of enabled transitions from $E(s)$ to $E(s')$. Moreover, the sets $\{ B_1(e^*), B_2(e^*) : e^* \in E \}$ can be computed prior to generation of sample paths and then quickly accessed as needed.

A sample path of the marking process can be obtained from a sample path of the chain $\{ (S_n, C_n) : n \geq 0 \}$. As in (1.14), let $\{ \gamma(n) : n \geq 0 \}$ be the indices of the successive marking changes at which the new marking is timed. Also let ζ_n be the time of the nth marking change as defined in (1.11). A sample path of the marking process can be represented as a sequence $\{ (X_n, T_n) : n \geq 0 \}$, where $T_n = \zeta_{\gamma(n)}$ and $X_n = X(T_n)$. The following algorithm produces a realization of the sequence $\{ (X_n, T_n) : n \geq 0 \}$.

Algorithm 1.19 (Sample path generation for the marking process)

1. (Initialization) Set $k = -1$, $n = 0$, and $T_0 = 0$.

2. Increment k by 1.

3. If $t^*(S_k, C_k) = 0$, increment k by 1 repeatedly until $t^*(S_k, C_k) > 0$.

4. Set $X_n = S_k$ and $T_{n+1} = T_n + t^*(S_k, C_k)$.

5. Increment n by 1 and go to step 2.

3.2 Performance Measures

Long-run performance measures for an SPN are usually specified in terms of the marking process $\{ X(t) : t \geq 0 \}$ or underlying chain $\{ (S_n, C_n) : n \geq 0 \}$. In this section we give a brief survey of typical long-run performance measures and show that each such measure can be expressed as a function of time-average limits of the underlying chain. Thus an understanding of the long-run behavior of the underlying chain is essential when studying the long-run behavior of an SPN.

3.2.1 Simple Time-Average Limits and Ratios

Many performance measures of interest can be expressed as limits of the form

$$r(f) = \lim_{t \to \infty} \frac{1}{t} \int_0^t f\big(X(u)\big)\, du, \tag{2.1}$$

$$r(f_1, f_2) = \lim_{t \to \infty} \frac{\int_0^t f_1\big(X(u)\big)\, du}{\int_0^t f_2\big(X(u)\big)\, du}, \tag{2.2}$$

or

$$\tilde{r}(\tilde{f}_1, \tilde{f}_2) = \lim_{n \to \infty} \frac{\sum_{j=0}^n \tilde{f}_1(S_k, C_k)}{\sum_{k=0}^n \tilde{f}_2(S_k, C_k)}, \tag{2.3}$$

where f, f_1, and f_2 are real-valued functions defined on G, and $\tilde{f}_1$ and $\tilde{f}_2$ are real-valued functions defined on Σ.

EXAMPLE 2.4 (Producer–consumer system with nonpreemptive priority). For the system of Example 2.1 in Chapter 2, let r be the long-run fraction of time that the channel is busy; this quantity is often referred to as the *utilization* of the channel. Suppose that this system is modelled using the SPN in Figure 2.4. Then r can be specified as a limit of the form (2.1), where $f(s) = 1 - s_7$ for $s = (s_1, s_2, \ldots, s_7) \in G$. Suppose that the channel generates revenue at rate β_i whenever a transmission to consumer i is underway $(i = 1, 2)$. Then the system's long-run average revenue is of the form (2.1), where $f(s) = \beta_1 s_3 + \beta_2 s_6$ for $s = (s_1, s_2, \ldots, s_7) \in G$.

EXAMPLE 2.5 (System availability). Measures of long-run system availability often are of the form (2.1). Here $f(s) = 1$ if the marking s corresponds to a state in which the system is operational, and $f(s) = 0$ otherwise.

EXAMPLE 2.6 (Manufacturing cell with robots). For the system of Example 3.6 in Chapter 2, let r be the long-run utilization of robot 1 relative to robot 2. Suppose that this system is modelled using the SPN in Figure 2.21. Then r can be specified as a limit of the form (2.2), where $f_1(s) = 1 - s_{22}$ and $f_2(s) = 1 - s_{24}$ for $s = (s_1, s_2, \ldots, s_{24}) \in G$.

EXAMPLE 2.7 (Token ring). For the system of Example 2.6 in Chapter 2, let r be the long-run fraction of ring-token arrival times at port 1 at which there is a packet awaiting transmission. Suppose that this system is modelled using the SPN in Figure 2.10 and that, with probability 1, two or more events never occur simultaneously. Then r can be specified as a limit of the form (2.3), where

$$\tilde{f}_1(s, c) = \begin{cases} 1 & \text{if } E^*(s, c) = \{e_{3,1}\} \text{ and } s_{1,1} = 1; \\ 0 & \text{otherwise} \end{cases}$$

and

$$\tilde{f}_2(s, c) = \begin{cases} 1 & \text{if } E^*(s, c) = \{e_{3,1}\}; \\ 0 & \text{otherwise.} \end{cases}$$

3.2.2 Conversion of Limit Results to Continuous Time

This section is concerned with the problem of obtaining limit theorems for continuous-time performance measures—that is, performance measures expressed in terms of the marking process—from limit theorems for the underlying chain. Theorem 2.9 below, although elementary, provides a useful and general means of converting discrete-time results into limit theorems in continuous time. Let $\{X_n : n \geq 0\}$, $\{Y_n : n \geq 1\}$, $\{Y_n' : n \geq 1\}$, and $\{\Delta_n : n \geq 1\}$ be sequences of a.s. finite real-valued random variables with each Y_n' and Δ_n nonnegative, and let x, y, y', w, w', and δ be finite constants with $y' \geq 0$ and $\delta > 0$. Moreover, suppose that each Y_k and Y_k' can be represented in terms of a real-valued stochastic process $\{Z(t) : t \geq 0\}$ as

$$Y_k = Z(T_k) - Z(T_{k-1})$$

and

$$Y_k' = \sup_{T_{k-1} \leq t \leq T_k} \left| Z(t) - Z(T_{k-1}) \right|,$$

where $T_0 = 0$ and $T_k = \sum_{j=1}^{k} \Delta_j$ for $k \geq 1$. Theorem 2.9 is useful when we can establish limit results of the form

$$\lim_{n \to \infty} \frac{1}{n} \sum_{k=0}^{n-1} X_k = x \text{ a.s.,} \tag{2.8a}$$

$$\lim_{n \to \infty} \frac{1}{n} \sum_{k=1}^{n} \Delta_k = \delta \text{ a.s.,} \tag{2.8b}$$

$$\lim_{n \to \infty} \frac{1}{n} \sum_{k=1}^{n} X_{k-1} \Delta_k = w \text{ a.s.,} \tag{2.8c}$$

$$\lim_{n \to \infty} \frac{1}{n} \sum_{k=1}^{n} |X_{k-1}| \Delta_k = w' \text{ a.s.} \tag{2.8d}$$

$$\lim_{n \to \infty} \frac{1}{n} \sum_{k=1}^{n} Y_k = y \text{ a.s.,} \tag{2.8e}$$

or

$$\lim_{n \to \infty} \frac{1}{n} \sum_{k=1}^{n} Y_k' = y' \text{ a.s..} \tag{2.8f}$$

For $t \geq 0$, set $N(t) = \sup \{ n \geq 0 : T_n \leq t \}$ and $X(t) = X_{N(t)}$.

Theorem 2.9. *Let the sequences* $\{ X_n : n \geq 0 \}$, $\{ Y_n : n \geq 0 \}$, $\{ Y_n' : n \geq 0 \}$, *and* $\{ \Delta_n : n \geq 1 \}$ *be as above.*

(i) *If (2.8a) holds, then* $\lim_{n \to \infty} X_n/n = 0$ *a.s..*

(ii) *Without further conditions,* $\lim_{t \to \infty} N(t) = \infty$ *a.s.. If, moreover, (2.8b) holds, then* $\lim_{t \to \infty} N(t)/t = 1/\delta$ *a.s..*

(iii) *If (2.8a) and (2.8b) hold, then* $\lim_{t \to \infty} (1/t) \sum_{k=0}^{N(t)} X_n = x/\delta$ *a.s..*

(iv) *If (2.8b) and (2.8c) hold, and either (2.8d) holds or* $|X_{n-1}| \Delta_n / n \to 0$ *a.s., then* $\lim_{t \to \infty} (1/t) \int_0^t X(u) \, du = w/\delta$ *a.s..*

(v) *If (2.8b) and (2.8e) hold, and either (2.8f) holds or* $Y_n'/n \to 0$ *a.s., then* $\lim_{t \to \infty} Z(t)/t = y/\delta$ *a.s..*

Remark 2.10. Of course, if (2.8d) holds for some finite nonnegative w', then (2.8c) holds for some finite w. Similarly, (2.8e) holds whenever (2.8f) holds.

PROOF. The assertion in (i) follows from the fact that

$$\lim_{n\to\infty} \frac{X_n}{n} = \lim_{n\to\infty} \left[\frac{1}{n} \sum_{k=1}^{n} X_k - \left(\frac{n-1}{n} \right) \frac{1}{n-1} \sum_{k=1}^{n-1} X_k \right] = x - x = 0 \text{ a.s.}.$$

The first part of the assertion in (ii) follows because each Δ_n is a.s. finite by assumption: formally,

$$P\{ \lim_{t\to\infty} N(t) = \infty \} = P\{ \Delta_n < \infty \text{ for } n \geq 1 \}$$

$$\geq 1 - \sum_{n=1}^{\infty} P\{ \Delta_n = \infty \}$$

$$= 1,$$

where we have used Bonferroni's inequality [Proposition 1.1(vi) in the Appendix]. To prove the remaining part of the assertion in (ii), observe that $T_{N(t)} \leq t \leq T_{N(t)+1}$ for $t \geq 0$, so that

$$\frac{T_{N(t)}}{N(t)} \leq \frac{t}{N(t)} \leq \frac{T_{N(t)+1}}{N(t)}. \tag{2.11}$$

Thus, by (2.8b) and the fact that, as discussed above, $\lim_{t\to\infty} N(t) = \infty$ a.s., the outermost terms in (2.11) each converge to δ with probability 1, and the desired result follows. The assertion in (iii) follows from the assertions in (i) and (ii), because $N(t) \to \infty$ a.s. and

$$\lim_{t\to\infty} \frac{1}{t} \sum_{k=0}^{N(t)} X_n = \lim_{t\to\infty} \frac{N(t)}{t} \frac{1}{N(t)} \sum_{k=0}^{N(t)} X_n = \delta^{-1} x \text{ a.s.}.$$

The assertion in (iv) follows directly from the assertion in (v)—take $Z(t) = \int_0^t X(u)\, du$ and observe that $Y_n' \leq |X_{n-1}|\Delta_n$ for $n \geq 1$. To prove the assertion in (v), assume without loss of generality that $Z(0) = 0$ and write

$$\lim_{t\to\infty} \frac{Z(t)}{t} = \lim_{t\to\infty} \frac{\left(1/N(t)\right) \sum_{k=1}^{N(t)} Y_k + R_1(t)}{\left(1/N(t)\right) \sum_{k=1}^{N(t)} \Delta_k + R_2(t)},$$

where $R_1(t) = \left(Z(t) - Z(T_{N(t)}) \right)/N(t)$ and $R_2(t) = (t - T_{N(t)})/N(t)$. It suffices to show that the remainder terms $R_1(t)$ and $R_2(t)$ each converge to 0 a.s. as $t \to \infty$. To show that $\lim_{t\to\infty} R_1(t) = 0$ a.s., observe that

$$|R_1(t)| \leq \frac{Y_{N(t)+1}'}{N(t)}$$

for $t \geq 0$. Since $N(t) \to \infty$ a.s., the desired result follows immediately, provided that $Y_n'/n \to 0$ a.s.. If (2.8f) holds, then this latter convergence follows from the assertion in (i). An almost identical argument shows that $R_2(t) \to 0$ a.s., and the desired result follows.   □

EXAMPLE 2.12 (Time-average limits of the marking process). For an SPN with finite marking set G and underlying chain $\{(S_n, C_n): n \geq 0\}$, define the holding-time function t^* as in (1.7) and let f be a finite real-valued function defined on G. In later chapters we show that, under appropriate stability conditions,

$$\lim_{n \to \infty} \frac{1}{n} \sum_{k=0}^{n-1} t^*(S_k, C_k) = \delta \text{ a.s.,}$$

$$\lim_{n \to \infty} \frac{1}{n} \sum_{k=0}^{n-1} f(S_k) t^*(S_k, C_k) = w \text{ a.s.,}$$

and

$$\lim_{n \to \infty} \frac{1}{n} \sum_{k=0}^{n-1} |f(S_k)| t^*(S_k, C_k) = w' \text{ a.s.}$$

for finite constants δ, w, and w' with $\delta > 0$. It then follows from Theorem 2.9(iv) that a time-average limit of the form (2.1) can be expressed in the form (2.3), where $\tilde{f}_1(s, c) = f(s)t^*(s, c)$ and $\tilde{f}_2(s, c) = t^*(s, c)$. Similarly, a time-average limit of the form (2.2) can be expressed in the form (2.3), where $\tilde{f}_1(s, c) = f_1(s)t^*(s, c)$ and $\tilde{f}_2(s, c) = f_2(s)t^*(s, c)$.

3.2.3 Rewards and Throughput

Consider an SPN model in which rewards accrue continuously over time and also at an increasing sequence of random time points—the latter type of rewards are sometimes called *impulse rewards*. Specifically, suppose that

- Rewards accrue at finite rate $q(s)$ whenever the marking is equal to $s \in S$.

- Starting with marking s and clock-reading vector c just after a marking change, an impulse reward equal to $v(s, c)$ accrues at the next marking change.

For example, the function v might have the form

$$v(s, c) = \begin{cases} v_0 & \text{if } s = \tilde{s} \text{ and } E^*(s, c) = \{\tilde{e}\}; \\ 0 & \text{otherwise} \end{cases}$$

for some $\tilde{s} \in G$ and $\tilde{e} \in E(\tilde{s})$, so that an impulse reward of v_0 accrues whenever the current marking is equal to $\tilde{s}$ and transition $\tilde{e}$ fires. Denote by $R(t)$ the (random) total reward earned over the interval $[0, t]$. Formally, set $\tilde{h}(s, c) = q(s)t^*(s, c) + v(s, c)$ for $(s, c) \in \Sigma$ and set

$$R(t) = \sum_{k=0}^{N(t)} \tilde{h}(S_k, C_k) - D_1(t) - D_2(t), \qquad (2.13)$$

where $N(t)$ is the number of marking changes in the interval $(0, t]$, $D_1(t) = q(S_{N(t)})(\zeta_{N(t)+1} - t)$, and $D_2(t) = v(S_{N(t)}, C_{N(t)})$.

Theorem 2.14. *Suppose that*

$$\lim_{n \to \infty} \frac{1}{n} \sum_{k=0}^{n-1} t^*(S_k, C_k) = \delta \ a.s.$$

and

$$\lim_{n \to \infty} \frac{1}{n} \sum_{k=0}^{n-1} \tilde{h}(S_k, C_k) = x \ a.s.$$

for finite constants δ and x with $\delta > 0$. Also suppose that $\sup_{s \in S} |q(s)| < \infty$ and $\sup_{(s,c) \in G} |v(s,c)| < \infty$. Then

$$\lim_{t \to \infty} \frac{R(t)}{t} = \frac{x}{\delta} \ a.s..$$

PROOF. Set $\bar{v} = \sup_{(s,c) \in G} |v(s,c)|$ and $\bar{q} = \sup_{s \in S} |q(s)|$, and set

$$Y_n' = \sup_{\zeta_{n-1} \leq t \leq \zeta_n} |R(t) - R(\zeta_{n-1})|$$

for $n \geq 1$. Observe that $Y_n' \leq \bar{q}t^*(S_{n-1}, C_{n-1}) + \bar{v}$ for $n \geq 1$, so that $Y_n'/n \to 0$ a.s. by Theorem 2.9(i). The desired result now follows from Theorem 2.9(v)—take $Z(t) = R(t)$ and $\Delta_n = t^*(S_{n-1}, C_{n-1})$.   □

Remark 2.15. The assumption in Theorem 2.14 that $\sup_{s \in S} |q(s)| < \infty$ and $\sup_{(s,c) \in G} |v(s,c)| < \infty$ can be replaced by the assumption that

$$\lim_{n \to \infty} \frac{Y_n''}{n} = 0 \ a.s.,$$

where $Y_k'' = |q(S_k)|t^*(S_k, C_k) + |v(S_k, C_k)|$ for $k \geq 0$. Indeed, we have $\lim_{n \to \infty} Y_n'/n \leq \lim_{n \to \infty} Y_n''/n = 0$ a.s., so that the desired result follows from Theorem 2.9(v) as before. Of course, $\lim_{n \to \infty} Y''/n = 0$ a.s. whenever

$$\lim_{n \to \infty} \frac{1}{n} \sum_{k=0}^{n-1} Y_k'' < \infty \ a.s.,$$

by Theorem 2.9(i).

EXAMPLE 2.16 (Supply chain). Consider a simple "make-to-stock" supply chain for the manufacture and sale of finished items. The system consists of two suppliers (numbered 1 and 2), an original equipment manufacturer

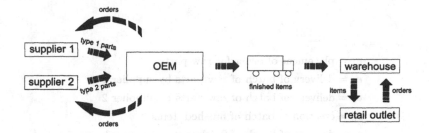

Figure 3.1. Supply chain.

(OEM), a truck, a warehouse, and a retail outlet; see Figure 3.1. The suppliers are located near the OEM and the warehouse is located near the retail outlet, but the OEM and warehouse are at some distance from each other. Supplier i ($i = 1, 2$) provides raw parts of type i, and the OEM produces finished items from these raw parts. Periodically—in expectation of demand for finished items—an order for one or more batches of parts of type 1 is sent to supplier 1 and, simultaneously, an order for the same number of batches of parts of type 2 is sent to supplier 2. Each supplier fills its respective order by delivering one batch at a time to the OEM. The OEM produces finished items one batch at a time—the manufacture of a batch of finished items requires one batch each of the two types of raw parts. The OEM is never idle when at least one batch of each type of raw part is available. The truck conveys finished items to the warehouse one batch at a time. To satisfy customer demands, the retail outlet periodically orders a batch of finished items from the warehouse. If at least one batch is available, then the order is immediately filled; if no batches are available, then the order is lost to the OEM, and the batch of finished items is provided by a competitor.

The time between successive placements of an order for raw parts is a positive constant. The number of batches of raw parts in an order is a positive integer constant that can depend (deterministically) on the state of the system just before the placement of the order—that is, on the number of batches of finished items on the truck and in the warehouse, the number of unfilled orders at each of the suppliers, and the current supply of raw parts at the OEM. The successive times for a supplier to deliver a batch of raw parts are i.i.d. as a positive random variable, as are the successive times to manufacture a batch of finished items, the successive times to convey a batch of finished items to the warehouse (and return the truck to the OEM), and the times between successive orders of finished items by the retail outlet.

This system can be specified as an SPN with timed and immediate transitions; see Figure 3.2. Each of places d_1 and d_9 always contains exactly one token, reflecting the fact that the placement of orders for both raw parts and finished items is always ongoing. Place d_4 (resp., d_5) contains n (≥ 0)

e_1 = placement of order for raw parts

e_4 = delivery of batch of raw parts by supplier 1

e_5 = delivery of batch of raw parts by supplier 2

e_6 = creation of batch of finished items

e_7 = delivery of batch of finished items to warehouse

e_8 = placement of order by retail outlet

e_9 = fulfillment of order for finished items

e_{10} = loss of order for finished items

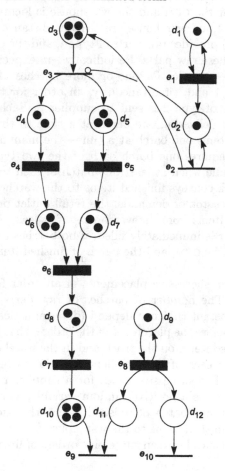

Figure 3.2. SPN representation of supply chain.

tokens if and only if supplier 1 (resp., supplier 2) has a backlog of n batches of raw parts that have been ordered but not yet delivered. Place d_6 (resp., d_7) contains n tokens if and only if there are n batches of parts of type 1 (resp., type 2) at the OEM. Place d_8 contains n tokens if and only if there are n batches of finished items either awaiting shipment or being conveyed to the warehouse. Place d_{10} contains n tokens if and only if there are n batches of finished items at the warehouse. Place d_{11} contains one token if an order from the retail outlet is being filled; otherwise, place d_{11} contains no tokens. Place d_{12} contains one token if an order from the retail outlet is about to be lost; otherwise, place d_{12} contains no tokens. We assume throughout that transitions never fire simultaneously.

All transitions except e_2 and e_8 are deterministic, and all speeds for enabled transitions are equal to 1. Whenever the marking is s and transition $e_1 = $ "placement of order for raw parts" fires, a token is deposited in place d_2 and transition e_2 becomes enabled. By means of a mechanism similar to that used for transition e_2 in the SPN model of the queue with batch arrivals—see Example 2.4 in Chapter 2—transition e_2 fires $m(s)$ times in succession before becoming disabled, thereby depositing $m(s)$ tokens in place e_3 and leaving place d_2 with zero tokens. Here $m(s)$ is a positive integer that depends in general on the marking s in which e_1 fires. Transition e_3 then fires $m(s)$ times in succession, depositing $m(s)$ tokens in each of places d_4 and d_5. In this manner, an order for $m(s)$ batches of raw parts is placed at each supplier. Whenever transition $e_8 = $ "placement of order by retail outlet" fires and place d_{10} contains at least one token, a token is deposited in place d_{11}; if place d_{10} contains no tokens, then a token is deposited in place d_{12}. Thus the order is filled if at least one batch of finished items is at the warehouse and is lost otherwise.

Denote by a_i the cost to the OEM of a batch of type i parts ($i = 1, 2$), and suppose that the OEM pays the supplier at the time of the order. Similarly, denote by b the cost to the retail outlet of a batch of finished items, and suppose that the retail outlet pays the OEM at the time of the order. Next, denote by h the cost to the OEM of conveying a batch of parts to the warehouse, and suppose that the OEM pays the trucker at the time of delivery. Finally, denote by u the inventory cost to the OEM per unit time for each batch of finished items stored at the warehouse, and denote by w the remaining costs to the OEM per unit time.

Define a reward structure as in (2.13) by setting $q(s) = w + u \cdot s_{10}$ and

$$
v(s, c) = \begin{cases} -(a_1 + a_2) & \text{if } E^*(s, c) = \{e_2\}; \\ -h & \text{if } E^*(s, c) = \{e_7\}; \\ b & \text{if } E^*(s, c) = \{e_9\}; \\ 0 & \text{otherwise} \end{cases}
$$

for $s = (s_1, s_2, \ldots, s_{12}) \in G$ and $c \in C(s)$. Then the long-run average reward coincides with the long-run average profit to the OEM.

By specializing the foregoing reward structure, we can formally specify a variety of throughput characteristics in discrete-event systems.

EXAMPLE 2.17 (Throughput of manufacturing cell with robots). For the SPN in Figure 2.21, define a reward structure as in (2.13) by setting $q(s) \equiv 0$ and

$$v(s,c) = \begin{cases} 1 & \text{if } E^*(s,c) = \{\tilde{e}\}; \\ 0 & \text{otherwise,} \end{cases}$$

where $\tilde{e} = e_{16} = $ "end of transfer of a finished part from conveyor 2 to the unloading area." Then the long-run average reward coincides with the long-run throughput of the manufacturing system.

3.2.4   General Functions of Time-Average Limits

As discussed above, many performance measures of interest can be expressed as ratios of time-average limits of the underlying chain.[1] Other performance measures can be expressed as more general functions of such time-average limits.

EXAMPLE 2.18 (Central moments). Let f be a real-valued function defined on S, and suppose that

$$\lim_{t \to \infty} \frac{1}{t} \int_0^t f(X(u)) \, du = r(f) \text{ a.s.}$$

for some finite constant $r(f)$. In this setting, long-run central moments may also be of interest, for example, the long-run variance $v(f)$ defined by

$$v(f) = \lim_{t \to \infty} \frac{1}{t} \int_0^t \left(f(X(u)) - r(f) \right)^2 du.$$

If

$$\lim_{t \to \infty} \frac{1}{t} \int_0^t f^2(X(u)) \, du = r(f^2) \text{ a.s.}$$

for some finite constant $r(f^2)$, then we can write $v(f) = r(f^2) - r^2(f)$. Set $\tilde{f}_1(s,c) = f(s)t^*(s,c)$, $\tilde{f}_2(s,c) = f^2(s)t^*(s,c)$, and $\tilde{f}_3(s,c) = t^*(s,c)$ for $(s,c) \in \Sigma$. Also set

$$\tilde{r}(\tilde{f}_i) = \lim_{n \to \infty} \frac{1}{n} \sum_{k=0}^{n-1} \tilde{f}_i(S_k, C_k)$$

---

[1]There has been no discussion so far of performance measures that pertain to system delays. Such performance measures are treated at length in Chapter 8.

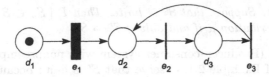

Figure 3.3. Absorption of the marking process into S'.

for $i = 1, 2, 3$. Provided that $\tilde{r}(\tilde{f}_1)$, $\tilde{r}(|\tilde{f}_1|)$, $\tilde{r}(\tilde{f}_2)$, and $\tilde{r}(\tilde{f}_3)$ are each well defined, an application of Theorem 2.9(iv) establishes the representation $v(f) = g(\tilde{r}(\tilde{f}_1), \tilde{r}(\tilde{f}_2), \tilde{r}(\tilde{f}_3))$, where $g(r_1, r_2, r_3) = (r_2/r_3) - (r_1/r_3)^2$. Analogous representations can be obtained for higher central moments.

3.3 The Lifetime of the Marking Process

Limits of the form $\lim_{t \to \infty} (1/t) \int_0^t f(X(u)) \, du$ are not well defined when the lifetime τ_Δ of the marking process is finite, because $f(X(t))$ is not defined for $t \geq \tau_\Delta$. In this section we show how this pathological situation can occur, and then we give mild conditions under which $\tau_\Delta = \infty$ a.s., so that the state space of the marking process can be restricted from $S \cup \{\Delta\}$ to S.

3.3.1 Absorption into the Set of Immediate Markings

The lifetime τ_Δ is finite if and only if an infinite number of marking changes occur in a finite time interval. This can occur if the sequence $\{S_n : n \geq 0\}$ is absorbed into the set S' of immediate markings. Indeed, write $\tau_\Delta = \sum_{n=0}^{\infty} t^*(S_n, C_n)$ and observe that the number of positive terms in the sum is finite unless $\{S_n : n \geq 0\}$ hits the set S of timed markings infinitely often.

EXAMPLE 3.1 (Absorption into S'). Consider an SPN with deterministic transitions as in Figure 3.3. The marking set is $G = \{(1, 0, 0), (0, 1, 0), (0, 0, 1)\}$ and the initial marking is $(1, 0, 0)$, as pictured in the figure. After leaving timed marking $(1, 0, 0)$, the marking process then alternates between the immediate markings $(0, 1, 0)$ and $(0, 0, 1)$, never returning to $(1, 0, 0)$.

Although in general it can be hard to determine whether $P_\mu\{S_n \in S \text{ i.o.}\} = 1$, the criterion given in Theorem 3.2 below often can be verified in practice. For $s \in S'$ and $s' \in G$, write $s \to s'$ if $p(s'; s, E(s) \cap E') > 0$. We write $S' \rightsquigarrow S$ if for each $s' \in S'$ there exists $s \in S$ such that either $s' \to s$ or there exist markings $s^{(1)}, s^{(2)}, \ldots, s^{(n)} \in S'$ $(n \geq 1)$ such that $s' \to s^{(1)} \to \cdots \to s^{(n)} \to s$.

Theorem 3.2. *Suppose that S' is finite. Then $P_\mu\{S_n \in S \text{ i.o.}\} = 1$ for any initial distribution μ if and only if $S' \rightsquigarrow S$.*

EXAMPLE 3.3 (Producer–consumer system with nonpreemptive priority). For the SPN of Example 2.1, observe that S' is finite because G is finite. It is trivial to verify that $S' \rightsquigarrow S$, and thus $P_\mu\{S_n \in S \text{ i.o.}\} = 1$ by Theorem 3.2.

Proving the necessity of the condition $S' \rightsquigarrow S$ in Theorem 3.2 is trivial. To prove sufficiency, we use the following generalization of the Borel–Cantelli lemma (Proposition 1.3 in the Appendix).

Lemma 3.4 (Geometric trials). *Let $\{Y_n : n \geq 0\}$ be a sequence of random variables defined on a probability space $(\Omega, \mathcal{F}, P)$ and taking values in a set S, and let A be a fixed subset of S. Suppose that there exists $\delta \in (0, 1]$ such that*

$$P\{Y_n \in A \mid Y_{n-1}, \ldots, Y_0\} \geq \delta \text{ a.s.} \tag{3.5}$$

for $n \geq 1$. Then $P\{Y_n \in A \text{ i.o.}\} = 1$.

PROOF. Define a sequence of random indices by $I_0 = 0$ and

$$I_k = \inf\{n > I_{k-1} : Y_n \in A\}$$

for $k \geq 1$. It suffices to show that $P\{I_k < \infty\} = 1$ for $k \geq 0$ because then, using Bonferroni's inequality,

$$P\{Y_n \in A \text{ i.o.}\} = P\{I_k < \infty \text{ for } k \geq 0\}$$
$$\geq 1 - \sum_{k=0}^{\infty} P\{I_k = \infty\}$$
$$= 1.$$

We use an inductive argument to show that each I_k is a.s. finite. Observe that I_0 is a.s. finite by definition and assume for induction that I_k is a.s. finite for some value of k. Using (3.5) it follows that

$$P\{I_{k+1} - I_k > n, I_k = j\}$$
$$= P\{Y_{j+n} \notin A, \ldots, Y_{j+1} \notin A, I_k = j\}$$
$$= E\Big[P\{Y_{j+n} \notin A, \ldots, Y_{j+1} \notin A, I_k = j \mid Y_{j+n-1}, \ldots, Y_0\}\Big]$$
$$= E\Big[1_{\{Y_{j+n-1}\notin A,\ldots,Y_{j+1}\notin A,I_k=j\}}P\{Y_{j+n} \notin A \mid Y_{j+n-1}, \ldots, Y_0\}\Big]$$
$$\leq E\Big[1_{\{Y_{j+n-1}\notin A,\ldots,Y_{j+1}\notin A,I_k=j\}}(1 - \delta)\Big]$$
$$= (1 - \delta)P\{I_{k+1} - I_k > n - 1, I_k = j\},$$

so that

$$P\{I_{k+1} - I_k > n, I_k = j\} \leq (1 - \delta)^n P\{I_k = j\} \tag{3.6}$$

for $n \geq 1$ and $j \geq 0$. Because $P\{I_k < \infty\} = 1$ by the induction hypothesis, we can sum (3.6) over j to obtain

$$P\{I_{k+1} - I_k > n\} \leq (1 - \delta)^n$$

for $n \geq 1$ so that, by Proposition 1.1(iv) in the Appendix, $I_{k+1} - I_k$ and hence I_{k+1} is a.s. finite. □

It follows from the proof of Lemma 3.4 that

$$P\{\tau_A > n\} \leq (1 - \delta)^n \tag{3.7}$$

for $n \geq 0$, where $\tau_A = \inf\{n \geq 1 : Y_n \in A\}$.

PROOF OF THEOREM 3.2. We prove sufficiency only. For each $s \in S'$, we can find an integer $k = k(s) \geq 1$ and a sequence of markings $s_1 \in S', s_2 \in S', \ldots, s_{k-1} \in S', s_k \in S$, depending on s, such that $s \to s_1 \to s_2 \to \cdots \to s_{k-1} \to s_k$; such a sequence exists because $S' \leadsto S$. There may in fact be many such sequences—fix one and set

$$\delta(s) = p(s_1; s, E' \cap E(s)) \prod_{j=2}^{k} p(s_j; s_{j-1}, E' \cap E(s_{j-1})).$$

Next, set $\delta = \min_{s \in S'} \delta(s) > 0$. Define an increasing sequence of random indices $\{\beta(n) : n \geq 0\}$ by setting $\beta(0) = 0$ and

$$\beta(n) = \begin{cases} \beta(n-1) + k(S_{\beta(n-1)}) & \text{if } S_{\beta(n-1)} \in S'; \\ \beta(n-1) + 1 & \text{if } S_{\beta(n-1)}, S_{\beta(n-1)+1} \in S; \\ \beta(n-1) + 1 + k(S_{\beta(n-1)+1}) & \text{if } S_{\beta(n-1)} \in S, S_{\beta(n-1)+1} \in S' \end{cases}$$

for $n \geq 1$. Also fix an initial distribution μ and set

$$Q_n(s) = P_\mu\{S_{\beta(n-1)+1} = s \mid S_{\beta(n-1)}, S_{\beta(n-2)}, \ldots, S_{\beta(0)}\}$$

for $n \geq 1$ and $s \in G$. Each $\beta(n)$ is an a.s. finite stopping time with respect to $\{(S_n, C_n) : n \geq 0\}$, and straightforward manipulations using the strong Markov property together with the form of the transition kernel in (1.9) show that

$$\begin{aligned} P_\mu\{S_{\beta(n)} \in S \mid S_{\beta(n-1)}, & S_{\beta(n-2)}, \ldots, S_{\beta(0)}\} \\ & \geq 1_{S'}(S_{\beta(n-1)}) \cdot \delta(S_{\beta(n-1)}) \\ & \quad + 1_S(S_{\beta(n-1)}) \cdot \left(\sum_{s \in S} 1 \cdot Q_n(s) + \sum_{s \in S'} \delta(s) \cdot Q_n(s) \right) \\ & \geq \delta \text{ a.s.} \end{aligned} \tag{3.8}$$

for $n \geq 1$. Lemma 3.4 now implies that $P_\mu\{S_{\beta(n)} \in S \text{ i.o.}\} = 1$, and hence $P_\mu\{S_n \in S \text{ i.o.}\} = 1$. □

Remark 3.9. Let $T_S = \inf\{n \geq 1 : S_n \in S\}$ be the first hitting time after 0 of the set of timed transitions and set $k = \sup_{s \in S'} k(s)$, where $k(\cdot)$ is defined as in the proof of Theorem 3.2. It follows from (3.7) and (3.8) that, under the conditions of the theorem, $k < \infty$ and

$$P_\mu\{T_S > l\} \leq (1 - \delta)^{\lfloor l/k \rfloor} \leq a\rho^l \tag{3.10}$$

for $l \geq 0$, where $a = (1 - \delta)^{-1}$, $\rho = (1 - \delta)^{1/k}$, and $\lfloor x \rfloor$ is the greatest integer less than or equal to x.

Remark 3.11. We can relax the requirement in Theorem 3.2 that S' be finite. In particular, the conclusion of the theorem holds provided that $S' \rightsquigarrow S$ and $\inf_{s \in S'} \delta(s) > 0$. When establishing the latter condition, we are free to make each $\delta(s)$ as large as possible by defining $\delta(s)$ in terms of the most likely path from s to the set S of timed markings.

3.3.2 Explosions

Even when $\{S_n : n \geq 0\}$ does not get absorbed into the set S', an infinite number of marking changes can occur in a finite time interval if the marking changes occur ever more rapidly so that the times $\{\zeta_n : n \geq 0\}$ have an accumulation point. We then say that an *explosion* has occurred at time $\tau_\Delta < \infty$.

EXAMPLE 3.12 (An explosive SPN). Consider an SPN with a single place d_1 and a single timed transition e_1 such that d_1 is both a normal input place and an output place for e_1. Whenever transition e_1 fires, it deposits a token in place d_1 (and does not remove a token from d_1). The initial marking is $s = (1)$, so that with probability 1 the sequence of successive markings is (1), (2), (3), and so forth. All speeds are equal to 1, and the clock-setting distribution functions are given by $F_0(x; e_1, (1)) = P\{A_1 \leq x\}$ and $F(x; (n), e_1, (n-1), e_1) = P\{A_n \leq x\}$ for $n \geq 2$, where $\{A_n : n \geq 1\}$ is a sequence of random variables such that $A_1 = 1$ with probability 1 and

$$A_n = \begin{cases} 1/n^2 & \text{with probability } (n^2 - 1)/n^2; \\ (n^5 - n^2 + 1)/n^2 & \text{with probability } 1/n^2 \end{cases}$$

for $n \geq 2$. Thus, starting from marking (n), the time until the next marking change is distributed as A_n and the expected time until this marking change is $E[A_n] = n$. Trivially, $P_\mu\{S_n \in S \text{ i.o.}\} = 1$. For this SPN, τ_Δ is distributed as $\sum_{n \geq 1} A_n$. It follows from the three-series theorem (Proposition 1.32 in the Appendix) that

$$P\{\tau_\Delta < \infty\} = P\left\{\sum_{n \geq 1} A_n < \infty\right\} = 1.$$

Thus, an infinite number of marking changes occur in a finite time interval with probability 1 even though the expected time between successive marking changes increases linearly.

3.3.3 Sufficient Conditions for Infinite Lifetimes

The following theorem gives conditions under which the lifetime of an SPN is a.s. infinite. The idea is to uniformly bound the speeds from above and impose a uniform bound—over all the clock-setting distribution functions for timed transitions—on the amount of probability mass that can be close to 0.

Theorem 3.13. *Suppose that*

(i) $P_\mu \{ S_n \in S \text{ i.o.} \} = 1$,

(ii) $\sup_{s,e} r(s,e) < \infty$, and

(iii) *there exists* $a > 0$ *such that*

$$\sup_{e' \in E - E'} \sup_{s',s,E^*} F(a; s', e', s, E^*) < 1.$$

Then $P_\mu \{ \tau_\Delta = \infty \} = 1$.

Observe that the conditions of Theorem 3.13 hold if either of the following conditions hold:

- $S' \rightsquigarrow S$ and the marking set G is finite.

- The condition in (i) holds and there are only finitely many distinct speeds and distinct clock-setting distribution functions.

PROOF. First suppose that the transitions $\{ e_1, e_2, \ldots, e_M \}$ are all timed. Set $r = \sup_{s,e} r(s,e)$ and $b = \sup_{e' \in E - E'} \sup_{s',s,E^*} F(a; s', e', s, E^*)$. Denote by N_n $(n \geq 1)$ the (random) set of new transitions just after the nth marking change: $N_n = N(S_n; S_{n-1}, E_{n-1}^*)$. Next, denote by I_k $(k \geq 0)$ the indicator variable that equals 1 if, at marking changes $kM, kM + 1, \ldots, (k+1)M$, each new clock reading exceeds the constant a:

$$I_k = \begin{cases} 1 & \text{if } C_{n,i} > a \text{ for } e_i \in N_n \text{ and } kM \leq n < (k+1)M; \\ 0 & \text{otherwise,} \end{cases}$$

where we take $N_0 = E(S_0)$. Because there are only M transitions, at least one transition must become enabled in the time interval $[\zeta_{kM}, \zeta_{(k+1)M}]$ and also fire in this interval. Because all speeds are bounded above by r, it follows that $\zeta_{(k+1)M} - \zeta_{kM} > a/r$ whenever $I_k = 1$. Using the hypothesis in (iii) and writing $\overline{F} = 1 - F$, we have

$$P_\mu \{ C_{kM,i} > a \text{ for } e_i \in N_{kM} \mid I_{k-1}, \ldots, I_0 \}$$

$$= E_\mu \Big[P_\mu \{ C_{kM,i} > a \text{ for } e_i \in N_{kM} \mid N_{kM}, S_{kM},$$

$$S_{kM-1}, E^*_{kM-1}, I_{k-1}, \ldots, I_0 \} \Big| I_{k-1}, \ldots, I_0 \Big]$$

$$= E_\mu \Bigg[\prod_{e_i \in N_{kM}} \overline{F}(a; S_{kM}, e_i, S_{kM-1}, E^*_{kM-1}) \Bigg| I_{k-1}, \ldots, I_0 \Bigg]$$

$$\geq E_\mu [(1-b)^{|N_{kM}|} \mid I_{k-1}, \ldots, I_0]$$

$$\geq (1-b)^M \text{ a.s.}$$

for $k \geq 0$. The above calculations can be repeated for sets N_{kM+1} through $N_{(k+1)M-1}$ to yield the inequality $P_\mu \{ I_k = 1 \mid I_{k-1}, \ldots, I_0 \} \geq (1-b)^{M^2}$ a.s. for $k \geq 0$. Using the geometric trials lemma, we find that

$$P_\mu \{ \tau_\Delta = \infty \} = P_\mu \Big\{ \sup_{n \geq 0} \zeta_n = \infty \Big\}$$

$$\geq P_\mu \{ \zeta_{(k+1)M} - \zeta_{kM} > a/r \text{ i.o.} \}$$

$$\geq P_\mu \{ I_k = 1 \text{ i.o.} \}$$

$$= 1,$$

and the desired result follows. Now suppose that there are one or more immediate transitions. Then the argument is almost the same as above, but we work with the embedded chain $\{ (S_n^+, C_n^+) : n \geq 0 \}$ defined at the end of Section 3.1.2. $\square$

EXAMPLE 3.14 (Producer–consumer system with nonpreemptive priority). As discussed previously, $S' \rightsquigarrow S$ for the SPN in Example 2.1. Because the marking set G is finite, it then follows that the lifetime of the marking process is a.s. infinite.

3.4   Markovian Marking Processes

In this section we show (Theorem 4.21) that the marking process of an SPN is a time-homogeneous CTMC provided that the clock associated with each transition is always set according to a fixed exponential distribution. Though intuitively plausible, this result is nontrivial to establish because the distribution of the clock-reading vector after a marking change, and hence the time between successive marking changes, is extremely complex for general clock-setting distributions. The proof of Theorem 4.21 rests on a representation (Lemma 4.10) of the conditional distribution of the clock-reading vector given the "partial history" of the underlying chain of the marking process. The proof also exploits the close connection between the definition of the marking process and the standard construction of a "minimal" CTMC.

3.4.1 Continuous-Time Markov Chains

Before proceeding with our main results we briefly review some basic facts about CTMCs. In the CTMC setting, the analog of the transition matrix of a DTMC—see Section A.2.4—is the *transition function* P^t. The quantity $P^t(s, s')$ is the probability, starting in state s, that the chain is in state s' exactly t time units later.

Definition 4.1. Let $\{X(t): t \geq 0\}$ be a stochastic process defined on a probability space $(\Omega, \mathcal{F}, P)$, taking values in a finite or countably infinite set S and having piecewise-constant, right-continuous sample paths. The process $\{X(t): t \geq 0\}$ is a (time-homogeneous) *continuous-time Markov chain* with initial distribution ν and transition function P^t if

$$P\{X(0) = s\} = \nu(s)$$

and

$$P\{X(t + u) = s \mid X(v): 0 \leq v \leq t\} = P^u(X(t), s) \text{ a.s.} \quad (4.2)$$

for $s \in S$ and $t, u \geq 0$.

Proposition 4.3 below characterizes the structure of a CTMC $\{X(t): t \geq 0\}$ prior to a possible "explosion" (as defined below). Let $\{\xi_n: n \geq 0\}$ be the sequence of successive state-transition times for the CTMC: $\xi_0 = 0$ and $\xi_n = \inf\{t > \xi_{n-1}: X(t) \neq X(\xi_{n-1})\}$. For $n \geq 0$, denote by $Y_n = X(\xi_n)$ the state hit by the chain at time ξ_n and by $T_n = \xi_{n+1} - \xi_n$ the holding time in state Y_n. If the chain is absorbed into state s, so that $X(t) = s$ for all $t \geq \xi_n$ and some $n \geq 0$, then we use the convention that $\xi_{n+1} = \xi_{n+2} = \cdots = \infty$ and $T_n = T_{n+1} = \cdots = \infty$. When $q = 0$, we take the exponential distribution with intensity q to be the improper distribution with unit probability mass at $+\infty$. If

$$\tau_\Delta \overset{\text{def}}{=} \sup_{n \geq 0} \xi_n < \infty,$$

then we say that an *explosion* has occurred at time τ_Δ; if τ_Δ is a.s. infinite, then we say that the CTMC is *nonexplosive*.

Proposition 4.3. *The stochastic process $\{Y_n: n \geq 0\}$ is a discrete-time Markov chain. Moreover, there exist nonnegative numbers $\{q(s): s \in S\}$ such that, given $\{Y_n: n \geq 0\}$, the random variables $\{T_n: n \geq 0\}$ are mutually independent and $P\{T_n \leq x\} = 1 - e^{-q(Y_n)x}$ for $x \geq 0$ and $n \geq 0$.*

We call $\{q(s): s \in S\}$ the *intensity vector* of the CTMC and $\{Y_n: n \geq 0\}$ the *embedded jump chain* of the CTMC. The transition matrix of the embedded jump chain is denoted by $W = \{W(s, s'): s, s' \in S\}$; observe that $W(s, s) = 0$ for $s \in S$.

Proposition 4.5 below is suggested by Proposition 4.3 and provides a means of constructing a CTMC from a vector $q_0 = \{q_0(s): s \in S\}$ of non-negative real numbers, a stochastic matrix $W_0 = \{W_0(s, s'): s', s \in S\}$, and a probability distribution $\nu = \{\nu(s): s \in S\}$. [We allow $W_0(s, s) > 0$ for one or more states $s \in S$.] To start the construction, define random variables $\{Y_n: n \geq 0\}$ and $\{T_n: n \geq 0\}$ on a probability space $(\Omega, \mathcal{F}, P)$ such that (1) the stochastic process $\{Y_n: n \geq 0\}$ is a DTMC with initial distribution ν and transition matrix W_0 and (2) given $\{Y_n: n \geq 0\}$, the random variables $\{T_n: n \geq 0\}$ are mutually independent and each T_n has an exponential distribution with intensity $q_0(Y_n)$. Kolmogorov's existence theorem ensures that such a definition is possible. Set $\zeta_0 = 0$ and $\zeta_n = \sum_{i=0}^{n-1} T_i$ for $n \geq 1$. Fix $\Delta \notin S$ and set

$$X(t) = \begin{cases} S_{N(t)} & \text{if } N(t) < \infty; \\ \Delta & \text{if } N(t) = \infty, \end{cases} \tag{4.4}$$

where $N(t) = \sup\{n \geq 0: \zeta_n \leq t\}$.

Proposition 4.5. *The stochastic process* $\{X(t): t \geq 0\}$ *defined by (4.4) is a time-homogeneous CTMC with initial distribution* ν. *The intensity vector* q *is given by* $q(s) = q_0(s)(1 - W_0(s, s))$ *for* $s \in S$, *and the transition matrix* W *for the embedded jump chain is given by* $W(s, s') = W_0(s, s')/(1 - W_0(s, s))$ *for* $s, s' \in S$ *with* $s \neq s'$.

When $P\{\tau_\Delta < \infty\} > 0$ there is, in general, more than one way to define the process after time τ_Δ so that it has piecewise-constant sample paths and satisfies the Markov property. All such processes behave identically up to time τ_Δ. Fix $s \in S$ and $u \geq 0$, and observe that for each such process $\{\bar{X}(t): t \geq 0\}$ we have

$$P\{\bar{X}(u) = s\} = P\{\bar{X}(u) = s, \ u < \tau_\Delta\} + P\{\bar{X}(u) = s, \ u \geq \tau_\Delta\}.$$

Moreover, the first term on the right side of the above equation is the same for each process. For the particular process $\{X(t): t \geq 0\}$ defined by (4.4), we have $X(u) = \Delta$ for $u \geq \tau_\Delta$, so that the second term on the right side is 0. Hence $P\{X(u) = s\} \leq P\{\bar{X}(u) = s\}$ for any process $\{\bar{X}(t): t \geq 0\}$ as above, and for this reason the process defined by (4.4) is called the *minimal* CTMC.

The special structure of a CTMC makes it possible (at least in principle) to compute time-average limits either analytically or numerically. Such computations are based on Proposition 4.6 below. Let $\{X(t): t \geq 0\}$ be a minimal CTMC with state space S, intensity vector q, and embedded jump chain $\{Y_n: n \geq 0\}$ having transition matrix W. Denote by W^n

$(n \geq 0)$ the nth power of the matrix W and set $\tau_s = \inf\{\, n > 0 \colon Y_n = s\,\}$. The chain $\{\, X(t) \colon t \geq 0 \,\}$ is *irreducible* if for each $s, s' \in S$ there exists $n = n(s, s') \in (0, \infty)$ such that $W^n(s, s') > 0$ and is *positive recurrent* if it is irreducible and $E_s[\tau_s] < \infty$ for $s \in S$; here E_s denotes the expectation when the CTMC starts in state s. Thus a CTMC is irreducible if the embedded jump chain is irreducible (as defined in Section A.2.4), and similarly for positive recurrence. It can be shown that an irreducible CTMC with a finite state space is necessarily positive recurrent. The *infinitesimal generator matrix* $Q = \{\, Q(s, s') \colon s, s' \in S \,\}$ of the CTMC is defined by setting

$$Q(s, s') = q(s)W(s, s')$$

for $s \neq s'$ and

$$Q(s, s) = -q(s).$$

The matrix Q is also known as the *intensity matrix* or *differential matrix* of the CTMC. Heuristically, starting in state s at time t, the probability that the chain jumps from s to s' during the interval $[t, t + \Delta t]$ is approximately equal to $Q(s, s')\Delta t + o(\Delta t)$ when Δt is small. Similarly, the probability that the chain jumps from s to some other state during the interval $[t, t + \Delta t]$ is approximately equal to $q(s)\Delta t + o(\Delta t)$. A probability distribution π on S is said to be a *stationary distribution* for $\{\, X(t) \colon t \geq 0 \,\}$ if and only if $\sum_{s \in S} \pi(s)P^t(s, s') = \pi(s')$ for $s' \in S$ and $t \geq 0$. Thus, if the initial state of the CTMC is selected according to π, then $X(t)$ is distributed according to π at each time $t > 0$.

Proposition 4.6. *Suppose that the CTMC $\{\, X(t) \colon t \geq 0 \,\}$ is nonexplosive, irreducible, and positive recurrent. Then there exists a unique stationary distribution π on the state space S of the chain. This distribution is determined as the normalized solution of the system of linear equations*

$$\pi Q = 0, \tag{4.7}$$

where π is interpreted as a row vector. Moreover, if f is a real-valued function such that $\sum_{s \in S} |f(s)|\pi(s) < \infty$, then

$$\lim_{t \to \infty} \frac{1}{t} \int_0^t f(X(u))\, du = \sum_{s \in S} f(s)\pi(s) \quad a.s.$$

for any initial distribution of the chain.

3.4.2 Conditional Distribution of Clock Readings

To establish the Markov property for a marking process, we need to determine the distribution of the clock-reading vector just after each marking change. Although the unconditional distribution of the clock-reading vector usually is complicated, it is possible to calculate certain conditional distributions. The key result in this direction is Lemma 4.10 below.

To prepare for Lemma 4.10, we first define the "partial history" of the underlying chain of an SPN. Let $\{X(t): t \geq 0\}$ be the marking process of an SPN and let $\{(S_n, C_n): n \geq 0\}$ be the underlying chain. Recall the definitions of t^* and E^* from (1.7) and (1.8), respectively, and set $t_n^* = t^*(S_n, C_n)$ and $E_n^* = E^*(S_n, C_n)$ for $n \geq 0$.

Definition 4.8. The *partial history* of the underlying chain up to the nth marking change $(n \geq 1)$ is the collection

$$\mathcal{F}_n = \{ S_0, E_0^*, t_0^*, S_1, E_1^*, t_1^*, \ldots, S_{n-1}, E_{n-1}^*, t_{n-1}^*, S_n \}. \qquad (4.9)$$

When $n = 0$, take $\mathcal{F}_0 = \{ S_0 \}$.

The partial history records the sequence of states, holding times, and sets of trigger events, but does not record detailed information about individual clock readings. Observe, however, that when a clock is set at time ζ_k and runs down to 0 at time ζ_l, triggering a marking change, detailed information about readings on the clock during $[\zeta_k, \zeta_l]$ can be inferred from $\mathcal{F}_n$ provided that $l \leq n$. If a transition is an old transition at time ζ_n, then one can infer from $\mathcal{F}_n$ the amount of time that has elapsed on the associated clock since the clock was most recently set; no other information about the clock reading is available.

A random variable γ taking values in the nonnegative integers is said to be a *stopping time* with respect to the increasing sequence $\{\mathcal{F}_n: n \geq 0\}$ if for each $n \geq 0$ the occurrence or nonoccurrence of the event $\{\gamma = n\}$ is completely determined by the values of the random variables in $\mathcal{F}_n$. For a stopping time γ we write

$$\mathcal{F}_\gamma = \{ \gamma, S_0, E_0^*, t_0^*, S_1, E_1^*, t_1^*, \ldots, S_{\gamma-1}, E_{\gamma-1}^*, t_{\gamma-1}^*, S_\gamma \}.$$

Recall the definition of the set of new transitions $N(s'; s, E^*)$ from Section 3.1.2, and let $\alpha(n, i)$ be the index (less than or equal to n) of the latest marking change at which the clock associated with enabled transition $e_i \in E(S_n)$ was set: $\alpha(0, i) = 0$ and

$$\alpha(n, i) = \max\{ k : 1 \leq k \leq n \text{ and } e_i \in N(S_k; S_{k-1}, E_{k-1}^*) \}$$

for $n \geq 1$. If the maximum is taken over an empty set, define $\alpha(n, i) = 0$; if $e_i \notin E(S_n)$, set $\alpha(n, i) = n$. Next, denote by $Z_{n,i}$ the amount of time that has elapsed on the clock associated with transition e_i between $\zeta_{\alpha(n,i)}$ and ζ_n: $Z_{n,i} = C_{\alpha(n,i),i} - C_{n,i}$.

We are now ready to state Lemma 4.10. The lemma asserts that the clock readings $\{ C_{\gamma,i}: e_i \in E - E' \}$ are conditionally independent, given the partial history up to a stopping time γ. If $e_i \in E(S_\gamma)$, then the conditional probability that the clock reading $C_{\gamma,i}$ exceeds x_i is computed as the probability that a sample from the clock-setting distribution for e_i exceeds $Z_{\gamma,i} + x_i$ given that the sample exceeds $Z_{\gamma,i}$. We use the convention $0/0 = 0$ throughout. For ease of exposition, we state our result for SPNs in which each timed transition is "simple" as in Definition 1.8 of Chapter 2.

Lemma 4.10. *Suppose that each timed transition is simple, and let γ be an a.s. finite stopping time with respect to $\{\mathcal{F}_n : n \geq 0\}$. Then*

$$P_\mu\{C_{\gamma,i} > x_i \text{ for } e_i \in H \mid \mathcal{F}_\gamma\}$$
$$= \begin{cases} \prod_{e_i \in H} \overline{F}(x_i + Z_{\gamma,i}; e_i)/\overline{F}(Z_{\gamma,i}; e_i) & \text{if } H \subseteq E(S_\gamma); \\ 0 & \text{otherwise} \end{cases} \quad (4.11)$$

with probability 1 for any subset $H \subseteq E - E'$ and nonnegative numbers $\{x_i : e_i \in H\}$.

PROOF. It suffices to prove the result when $\gamma \equiv k$ for an arbitrary but fixed constant $k \geq 0$ because then, for a general stopping time γ,

$$P_\mu\{C_{\gamma,i} > x_i \text{ for } e_i \in H \mid \mathcal{F}_\gamma\}$$
$$= \sum_{k=0}^{\infty} P_\mu\{C_{\gamma,i} > x_i \text{ for } e_i \in H, \gamma = k \mid \mathcal{F}_\gamma\}$$
$$= \sum_{k=0}^{\infty} 1_{\{\gamma=k\}} P_\mu\{C_{k,i} > x_i \text{ for } e_i \in H \mid \mathcal{F}_k\}$$
$$= \sum_{k=0}^{\infty} 1_{\{\gamma=k\}} 1_{\{H \subseteq E(S_k)\}} \prod_{e_i \in H} \frac{\overline{F}(x_i + Z_{k,i}; e_i)}{\overline{F}(Z_{k,i}; e_i)}$$
$$= \left(1_{\{H \subseteq E(S_\gamma)\}} \prod_{e_i \in H} \frac{\overline{F}(x_i + Z_{\gamma,i}; e_i)}{\overline{F}(Z_{\gamma,i}; e_i)}\right) \sum_{k=0}^{\infty} 1_{\{\gamma=k\}}$$
$$= 1_{\{H \subseteq E(S_\gamma)\}} \prod_{e_i \in H} \frac{\overline{F}(x_i + Z_{\gamma,i}; e_i)}{\overline{F}(Z_{\gamma,i}; e_i)} \quad \text{a.s.,}$$

and the desired result follows. To this end, fix H, $\{x_i : e_i \in H\}$, and $k \geq 0$. If $\gamma \equiv k = 0$, then (4.11) clearly holds, so suppose that $k > 0$. By standard properties of conditional probability—see (1.27) in the Appendix—it suffices to show that

$$P_\mu\{C_{k,i} > x_i \text{ for } e_i \in H, A\} = E_\mu\left[1_A \prod_{e_i \in H} \frac{\overline{F}(x_i + Z_{k,i}; e_i)}{\overline{F}(Z_{k,i}; e_i)}\right] \quad (4.12)$$

for all sets A of the form

$$A = \{\alpha(k,i) = l_i \text{ for } e_i \in E, \ S_m = s_m \text{ for } 0 \leq m \leq k,$$
$$E^*(S_m, C_m) = \tilde{E}_m \text{ for } 0 \leq m < k,$$
$$C_{m,i} \leq x_{m,i} \text{ for } 0 \leq m < l_i \text{ and } e_i \in E\},$$

where $0 \le l_i \le k$, $0 \le x_{m,i} < \infty$, $s_m \in G$, $\tilde{E}_m \subseteq E(s_m)$, and $E(s_k) \supseteq H$. Fix such a set A. Because both sides of (4.12) are trivially equal to zero if $P_\mu\{A\} = 0$, assume without loss of generality that A has positive P_μ-probability. Then A has the representation

$$A = \{\, S_m = s_m \text{ for } 0 \le m \le k, \ E^*(S_m, C_m) = \tilde{E}_m \text{ for } 0 \le m < k,$$
$$C_{m,i} \le x_{m,i} \text{ for } 0 \le m < l_i \text{ and } e_i \in E \,\};$$

that is, the random variables $\{\, \alpha(k, i) \colon e_i \in H \,\}$ do not appear explicitly in the representation of A because the values of these random variables are determined by the values of $S_0, S_1, \ldots, S_k$ and $E^*(S_0, C_0), E^*(S_1, C_1), \ldots, E^*(S_k, C_k)$. Thus there exist sets $A_0, A_1, \ldots, A_k \subseteq \Sigma$ such that

$$\{\, C_{k,i} > x_i \text{ for } e_i \in H, A \,\}$$
$$= \{\, (S_0, C_0) \in A_0, (S_1, C_1) \in A_1, \ldots, (S_k, C_k) \in A_k \,\}.$$

For example, if $m < \min\{\, l_i \colon e_i \in H \,\}$, then

$$A_m = \{\, s_m \,\} \times \{\, c = (c_1, \ldots, c_M) \in C(s_m) \colon$$
$$E^*(s_m, c) = \tilde{E}_m, \ c_i \le x_{m,i} \text{ for } e_i \in E \,\};$$

and if $m = k$, then

$$A_m = \{\, s_k \,\} \times \{\, c = (c_1, \ldots, c_M) \in C(s_k) \colon c_i > x_i \text{ for } e_i \in H \,\}.$$

Using (1.4), we then have

$$P_\mu\{\, C_{k,i} > x_i \text{ for } e_i \in H, A \,\}$$
$$= P_\mu\{\, (S_0, C_0) \in A_0, (S_1, C_1) \in A_1, \ldots, (S_k, C_k) \in A_k \,\}$$
$$= \int_{A_0} \mu\big(d(s_0, c_0)\big) \int_{A_1} P\big((s_0, c_0), d(s_1, c_1)\big) \qquad (4.13)$$
$$\cdots \int_{A_k} P\big((s_{k-1}, c_{k-1}), d(s_k, c_k)\big),$$

where μ and P are the initial distribution and transition kernel, respectively, of the underlying chain $\{\, (S_n, C_n) \colon n \ge 0 \,\}$.

The equality in (4.12) follows from (4.13) upon substituting the explicit expressions (1.10) and (1.9) for μ and P, respectively, into the multiple integral and using Fubini's theorem (Proposition 1.25 in the Appendix) to interchange the order of integration. Because these calculations are messy, we illustrate the basic ideas by giving the calculations for a simple specific SPN. Consider an SPN with four places and three (simple) deterministic transitions as in Figure 3.4. All speeds for enabled transitions are equal to 1. Set $s = (1, 1, 0, 0)$ and $s' = (0, 1, 1, 0)$, and suppose that the initial

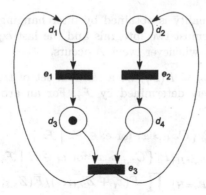

Figure 3.4. Example for proof of Theorem 4.10.

marking is equal to s with probability 1. We now establish (4.12) with $k = 1$, $H = \{e_2\}$, and

$$A = \{\, \alpha(1,1) = 1,\ \alpha(1,2) = 0,\ \alpha(1,3) = 1,$$
$$S_0 = s,\ S_1 = s',\ E^*(S_0, C_0) = \{e_1\},\ \text{and}\ C_{0,1} \leq x_{0,1} \,\}.$$

We can write

$$P_\mu \{\, C_{1,2} > x_2, A \,\} = P_\mu \{\, (S_0, C_0) \in A_0, (S_1, C_1) \in A_1 \,\},$$

where

$$A_0 = \{\, s \,\} \times \{\, (c_1, c_2, c_3) \in C(s) \colon c_1 < c_2 \text{ and } c_1 \leq x_{0,1} \,\}$$

and

$$A_1 = \{\, s' \,\} \times \{\, (c_1, c_2, c_3) \in C(s') \colon c_2 > x_2 \,\}.$$

Write $F_i(x) = F(x; e_i)$ for $i = 1, 2$ and observe that

$$P_\mu \{\, C_{1,2} > x_2, A \,\}$$

$$= \int_{A_0} \mu\big(d(s_0, c_0)\big)\, P\big((s_0, c_0), A_1\big)$$

$$= \int_0^{x_{0,1}} \int_{y_1}^\infty 1_{(x_2, \infty)}(y_2 - y_1)\, dF_2(y_2) dF_1(y_1)$$

$$= \int_0^{x_{0,1}} \int_0^\infty \int_0^\infty \frac{1_{(x_2, \infty)}(y_2 - y_1) 1_{(y_1, \infty)}(u)}{\overline{F}_2(y_1)}\, dF_2(u) dF_2(y_2) dF_1(y_1)$$

$$= \int_0^{x_{0,1}} \int_{y_1}^\infty \frac{\overline{F}_2(x_2 + y_1)}{\overline{F}_2(y_1)}\, dF_2(y_2) dF_1(y_1)$$

$$= E_\mu \left[1_A \frac{\overline{F}_2(x_2 + Z_{1,2})}{\overline{F}_2(Z_{1,2})} \right],$$

where the fourth equality is obtained by interchanging the order of integration for the innermost two integrals and the last equality exploits the fact that $Z_{1,2} = C_{0,1}$ whenever event A occurs.   □

Remark 4.14. Let $E_\gamma \subseteq E(S_\gamma)$ be a random set of transitions whose elements are completely determined by $\mathcal{F}_\gamma$. For an arbitrary fixed subset $H \subseteq E - E'$, we have

$$1_{\{E_\gamma=H\}} P_\mu \{ C_{\gamma,i} > x_i \text{ for } e_i \in E_\gamma \mid \mathcal{F}_\gamma \}$$
$$= 1_{\{E_\gamma=H\}} P_\mu \{ C_{\gamma,i} > x_i \text{ for } e_i \in H \mid \mathcal{F}_\gamma \}$$
$$= 1_{\{E_\gamma=H\}} \prod_{e_i \in H} \overline{F}(x_i + Z_{\gamma,i}; e_i)/\overline{F}(Z_{\gamma,i}; e_i)$$
$$= 1_{\{E_\gamma=H\}} \prod_{e_i \in E_\gamma} \overline{F}(x_i + Z_{\gamma,i}; e_i)/\overline{F}(Z_{\gamma,i}; e_i) \text{ a.s.,}$$

where the second equality follows from Lemma 4.10. Summing over all subsets $H \subseteq E - E'$, we find that

$$P_\mu \{ C_{\gamma,i} > x_i \text{ for } e_i \in E_\gamma \mid \mathcal{F}_\gamma \} = \prod_{e_i \in E_\gamma} \overline{F}(x_i + Z_{\gamma,i}; e_i)/\overline{F}(Z_{\gamma,i}; e_i) \text{ a.s..}$$

Remark 4.15. Lemma 4.10 can be generalized in a straightforward way to SPNs in which the timed transitions need not be simple. Set

$$U_n(x; e_i) = \begin{cases} F(x; S_{\alpha(n,i)}, e_i, S_{\alpha(n,i)-1}, E^*_{\alpha(n,i)-1}) & \text{if } \alpha(n,i) > 0; \\ F_0(x; e_i, S_0) & \text{if } \alpha(n,i) = 0 \end{cases}$$

and $\overline{U}_n = 1 - U_n$ for $n \geq 0$. The conditional distribution of the clock readings is then given by

$$P_\mu \{ C_{\gamma,i} > x_i \text{ for } e_i \in H \mid \mathcal{F}_\gamma \}$$
$$= \begin{cases} \prod_{e_i \in H} \overline{U}_\gamma(x_i + Z_{\gamma,i}; e_i)/\overline{U}_\gamma(Z_{\gamma,i}; e_i) & \text{if } H \subseteq E(S_\gamma); \quad (4.16) \\ 0 & \text{otherwise.} \end{cases}$$

The following result is an immediate consequence of Lemma 4.10 and gives a justification for "memoryless property" arguments in SPNs with exponential clock-setting distributions.

Corollary 4.17. *Suppose that γ is an a.s. finite stopping time with respect to $\{ \mathcal{F}_n : n \geq 0 \}$. Also suppose that each timed transition $e_i \in E - E'$ is simple with $F(x; e_i) = 1 - e^{-v(i)x}$ for some $v(i) \in (0, \infty)$. Then*

$$P_\mu \{ C_{\gamma,i} \leq x_i \text{ for } 1 \leq i \leq M \mid \mathcal{F}_\gamma \} = \prod_{e_i \in E(S_\gamma) \cap (E-E')} \left(1 - e^{-v(i)x_i}\right) \text{ a.s.}$$

for $x_1, x_2, \ldots, x_M \geq 0$.

The following variant of Lemma 4.10 is sometimes useful. Set $\tilde{\mathcal{F}}_n = \mathcal{F}_n - \{S_n\}$ for $n \geq 0$.

Corollary 4.18. *Suppose that each timed transition is simple, and let γ be an a.s. finite stopping time with respect to $\{\tilde{\mathcal{F}}_n : n \geq 0\}$. Then*

$$P_\mu\{S_\gamma = \bar{s} \text{ and } C_{\gamma,i} > x_i \text{ for } e_i \in H \mid \tilde{\mathcal{F}}_\gamma\}$$

$$= \begin{cases} p(\bar{s}; S_{\gamma-1}, E^*_{\gamma-1}) \prod_{e_i \in H} \overline{F}(x_i + Z_{\gamma,i}; e_i)/\overline{F}(Z_{\gamma,i}; e_i) & \text{if } H \subseteq E(\bar{s}); \\ 0 & \text{otherwise} \end{cases}$$

with probability 1 for any marking $\bar{s} \in G$, subset $H \subseteq E - E'$, and nonnegative numbers $\{x_i : e_i \in H\}$.

PROOF. Fix $\bar{s}$, H, and $\{x_i : e_i \in H\}$. We give the proof for the case $H \subseteq E(\bar{s})$; the proof for the case $H \not\subseteq E(\bar{s})$ is similar. Set $g(s) = 1_{\{\bar{s}\}}(s)$ and $h(c) = \prod_{e_i \in H} 1_{(x_i, \infty)}(c_i)$ for $s \in G$ and $c = (c_1, c_2, \ldots, c_M) \in C(s)$. Also, for $s' \in G$ and $u = (s, E^*, z, t^*)$ with $s \in G$, $E^* \subset E(s)$, $z = (z_1, z_2, \ldots, z_M) \in \Re_+^M$, and $t^* \geq 0$, set

$$w(s', u) = \prod_{e_i \in H \cap N(s'; s, E^*)} \overline{F}(x_i) \prod_{e_i \in H \cap O(s'; s, E^*)} \frac{\overline{F}(x_i + z_i + t^* r(s, e_i); e_i)}{\overline{F}(z_i + t^* r(s, e_i); e_i)}.$$

With this notation, the assertion of the corollary can be written as

$$E_\mu[g(S_\gamma)h(C_\gamma) \mid \tilde{\mathcal{F}}_\gamma] = p(\bar{s}; S_{\gamma-1}, E^*_{\gamma-1})w(\bar{s}, U_{\gamma-1}),$$

where $U_{\gamma-1} = (S_{\gamma-1}, E^*_{\gamma-1}, Z_{\gamma-1}, t^*_{\gamma-1})$ and $Z_{\gamma-1} = (Z_{\gamma-1,1}, \ldots, Z_{\gamma-1,M})$.

Using Lemma 4.10 and the fact that $U_{\gamma-1}$ is determined by $\tilde{\mathcal{F}}_\gamma$, we find that

$$\begin{aligned} E_\mu[g(S_\gamma)h(C_\gamma) \mid \tilde{\mathcal{F}}_\gamma] &= E_\mu[E_\mu[g(S_\gamma)h(C_\gamma) \mid \mathcal{F}_\gamma] \mid \tilde{\mathcal{F}}_\gamma] \\ &= E_\mu[g(S_\gamma) E_\mu[h(C_\gamma) \mid \mathcal{F}_\gamma] \mid \tilde{\mathcal{F}}_\gamma] \\ &= E_\mu[g(S_\gamma)w(S_\gamma, U_{\gamma-1}) \mid \tilde{\mathcal{F}}_\gamma] \\ &= w(\bar{s}, U_{\gamma-1})E_\mu[g(S_\gamma) \mid \tilde{\mathcal{F}}_\gamma] \text{ a.s..} \end{aligned}$$

Set $\mathcal{G}_n = \{(S_0, C_0), (S_1, C_1), \ldots, (S_n, C_n)\}$ for $n \geq 0$, and observe that $\tilde{\mathcal{F}}_n \subseteq \mathcal{G}_{n-1}$ for each n. Using the strong Markov property for the underlying chain and the specific form of the transition kernel, we have

$$\begin{aligned} E_\mu[g(S_\gamma) \mid \tilde{\mathcal{F}}_\gamma] &= E_\mu[E_\mu[g(S_\gamma) \mid \mathcal{G}_{\gamma-1}] \mid \tilde{\mathcal{F}}_\gamma] \\ &= E_\mu[p(\bar{s}; S_{\gamma-1}, E^*_{\gamma-1}) \mid \tilde{\mathcal{F}}_\gamma] \\ &= p(\bar{s}; S_{\gamma-1}, E^*_{\gamma-1}) \text{ a.s.,} \end{aligned}$$

and the desired result follows. $\square$

Lemma 4.19 is similar to Lemma 4.10 and is used in subsequent chapters. Two clock readings $C_{\gamma,i}$ and $C_{\gamma',i'}$ observed at random marking changes γ and γ', respectively, are said to be *disjoint* if either (1) $i \neq i'$ or (2) $i = i'$ and, with probability 1, transition e_i fires or becomes disabled between marking changes γ and γ'. Lemma 4.19 asserts that the clock readings in a collection are conditionally mutually independent, given the partial history up to a stopping time γ, if the clock readings are pairwise disjoint and each clock reading observed after γ corresponds to a new clock setting.

Lemma 4.19. *Let* $\gamma, \gamma_1, \gamma_2, \ldots, \gamma_{n+m}$ $(m, n \geq 0)$ *be a.s. finite stopping times with respect to* $\{\mathcal{F}_n : n \geq 0\}$, *and let* $C_{\gamma_1, i_1}, \ldots, C_{\gamma_n, i_n}, C_{\gamma_{n+1}, i_{n+1}},$ $\ldots, C_{\gamma_{n+m}, i_{n+m}}$ *be pairwise disjoint clock readings. Suppose that each timed transition is simple and, with probability 1,*

(i) $\max_{1 \leq l \leq n} \gamma_l \leq \gamma \leq \min_{n+1 \leq l \leq n+m} \gamma_l$, *and*

(ii) $e_{i_l} \in N(S_{\gamma_l}; S_{\gamma_l - 1}, E^*_{\gamma_l - 1})$ *for* $n + 1 \leq l \leq n + m$.

Then

$$P_\mu \{ C_{\gamma_l, i_l} > x_l \text{ for } 1 \leq l \leq n + m \mid \mathcal{F}_\gamma \}$$
$$= \prod_{l=1}^{n} P_\mu \{ C_{\gamma_l, i_l} > x_l \mid \mathcal{F}_\gamma \} \prod_{l=n+1}^{n+m} \overline{F}(x_l; e_{i_l})$$

with probability 1 for all $x_1, x_2, \ldots, x_{n+m} \geq 0$.

The intuition behind the proof of Lemma 4.19 is as follows. If $\gamma_l > \gamma$, then $\mathcal{F}_\gamma$ contains no information that will "distort" the conditional distribution of the new clock reading C_{γ_l, i_l} to be anything other than that of an independent sample from $F(\cdot; e_{i_l})$. This assertion follows because γ is a stopping time. If $\gamma_l < \gamma$ and the transition enabled just after the γ_lth marking change fires before the γth marking change, then the information in $\mathcal{F}_\gamma$ completely determines the value of C_{γ_l, i_l}. It follows that the conditional probability of the event $\{ C_{\gamma_l, i_l} > x_l \}$ factors out of the joint conditional probability expression—see Proposition 1.29 in the Appendix. If $\gamma_l < \gamma$ and the transition enabled just after the γ_lth marking change has not fired before the γth marking change, then the event $\{ C_{\gamma_l, i_l} > x_l \}$ can be reexpressed as an event of the form $\{ C_{\gamma, i_l} > x'_l \}$, and Lemma 4.10 applies.

3.4.3 The Markov Property

The following example shows that even when all clock-setting distributions are exponential, the marking process may not be a CTMC if the intensities depend on the current marking, new marking, and set of transitions that trigger the marking change.

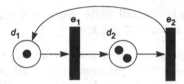

Figure 3.5. Non-Markovian SPN with exponential clock-setting distributions.

EXAMPLE 4.20 (Non-Markovian SPN with exponential clock-setting distributions).   Consider an SPN with two places and two deterministic timed transitions as in Figure 3.5. Fix $N > 1$, and suppose that the initial marking for this SPN is $(N, 0)$ with probability 1, so that the two places always contain a total of N tokens—in the figure, $N = 3$. All speeds for enabled transitions are equal to 1. Transition e_1 is simple, with $F(x; e_1) = 1 - e^{-v(0)x}$. The clock-setting distribution function for transition e_2 is given by

$$F\big(x; (s_1 + 1, s_2 - 1), e_2, (s_1, s_2), e_2\big) = 1 - e^{-v(1)x}$$

for $(s_1, s_2) \in G$ with $s_2 \geq 1$, and

$$F\big(x; (N - 1, 1), e_2, (N, 0), e_1\big) = 1 - e^{-v(2)x},$$

where $v(0)$, $v(1)$, and $v(2)$ are positive numbers with $v(1) \neq v(2)$. This SPN corresponds to a finite-capacity single-server queue in which the service-time distribution for the job that initiates a busy period differs from the service-time distribution for the other jobs that arrive during the busy period. Using (4.16), it can be shown that

$$P_\mu \{ C_{k,2} > x \mid S_k = (N - 1, 1), S_{k-1} = (N, 0) \} = e^{-v(2)x},$$

but

$$P_\mu \{ C_{k,2} > x \mid S_k = (N - 1, 1), S_{k-1} = (N - 2, 2) \} = e^{-v(1)x}$$

for $x \geq 0$. Thus, given the sequence of markings $\{ S_n : n \geq 0 \}$, the holding time in state $S_k = (N - 1, 1)$ is exponentially distributed but the intensity depends on more than just S_k. The marking process cannot possibly be a CTMC, as this would contradict Proposition 4.3.

Theorem 4.21 asserts that the marking process is a CTMC provided that each timed transition is simple and has an exponential clock-setting distribution. Recall from (1.14) that the random indices $\{ \gamma(n) : n \geq 0 \}$ correspond to the successive marking changes at which the new marking is timed. For timed markings $s, s' \in S$, let $p^+(s'; s, E^*)$ be the probability that the next timed marking is s' when the current marking is s and the transitions in E^* trigger a marking change:

$$p^+(s'; s, E^*) = \sum \left(p(s_1; s, E^*) \prod_{j=2}^{k} p\big(s_j; s_{j-1}, E' \cap E(s_{j-1})\big) \right),$$

where the summation is over all finite sequences $s_1, \ldots, s_k$ $(k \geq 1)$ such that $s_k = s'$ and $s_j \in S'$ for $1 \leq j < k$.

Theorem 4.21. *Suppose that each timed transition $e_i \in E - E'$ is simple with $F(x; e_i) = 1 - e^{-v(i)x}$ for some $v(i) \in (0, \infty)$. Also suppose that $P_\mu \{ S_n \in S \ i.o. \} = 1$. Then the marking process $\{ X(t) : t \geq 0 \}$ is a time-homogeneous* CTMC. *The initial distribution is given by*

$$\nu(s) = P_\mu \{ S_{\gamma(0)} = s \}$$

for $s \in S$, the intensity vector is given by

$$q(s) = \sum_{e_i \in E(s)} \left(1 - p^+(s; s, e_i) \right) r(s, e_i) v(i)$$

for $s \in S$, and the transition matrix for the embedded jump chain is given by

$$W(s, s') = \begin{cases} \sum_{e_i \in E(s)} \dfrac{r(s, e_i) v(i)}{q(s)} p^+(s'; s, e_i) & \text{if } s' \neq s; \\ 0 & \text{if } s' = s \end{cases}$$

for $s', s \in S$. If $\sup_{s,e} r(s, e) < \infty$, then the chain is nonexplosive.

To prove Theorem 4.21, we need the following result, which specifies the conditional joint distribution of the clock-reading vector $C_{\gamma(n)}$ and marking $S_{\gamma(n+1)}$, given the partial history $\mathcal{F}_{\gamma(n)}$. For $n \geq 0$ and $1 \leq i \leq M$, define a (random) distribution function $U_{n,i}$ on $[0, \infty)$ by setting

$$U_{n,i}(x) = \begin{cases} 1 - e^{-v(i)x} & \text{if } e_i \in E(S_{\gamma(n)}); \\ 1_{[0,\infty)}(x) & \text{if } e_i \notin E(S_{\gamma(n)}) \end{cases}$$

for $x \geq 0$. Set

$$U_n(x) = \prod_{i=1}^{M} U_{n,i}(x_i)$$

for $n \geq 0$ and $x = (x_1, x_2, \ldots, x_M) \in [0, \infty)^M$.

Lemma 4.22. *Suppose that each timed transition $e_i \in E - E'$ is simple with $F(x; e_i) = 1 - e^{-v(i)x}$ for some $v(i) \in (0, \infty)$. Also suppose that $P_\mu \{ S_n \in S \ i.o. \} = 1$. Then*

$$P_\mu \{ C_{\gamma(n),i} \leq x_i \text{ for } 1 \leq i \leq M, S_{\gamma(n+1)} = s \mid \mathcal{F}_{\gamma(n)} \}$$
$$= \int_{[0,x_1] \times \cdots \times [0,x_M]} p^+ \left(s; S_{\gamma(n)}, E^*(S_{\gamma(n)}, c) \right) dU_n(c) \ a.s. \tag{4.23}$$

for any $n \geq 0$, $s \in S$, and $x_1, x_2, \ldots, x_M \geq 0$.

PROOF. Fix $n \geq 0$, $s \in S$, and $x_1, x_2, \ldots, x_M \geq 0$. Observe that $\gamma(n)$, which is a.s. finite by hypothesis, is also a stopping time with respect to $\{\mathcal{F}_n : n \geq 0\}$. Because $S_{\gamma(n)}$ is determined by the values of the random variables in $\mathcal{F}_{\gamma(n)}$, it follows from Corollary 4.17 that

$$P_\mu \left\{ S_{\gamma(n)} = s, C_{\gamma(n),i} \leq x_i \text{ for } 1 \leq i \leq M \mid \mathcal{F}_{\gamma(n)} \right\}$$
$$= 1_{\{s\}}(S_{\gamma(n)}) \prod_{i=1}^{M} U_{n,i}(x_i) \text{ a.s.} \tag{4.24}$$

for $s \in S$. Moreover, using (1.9) and the strong Markov property, it is straightforward to show that

$$P_\mu \left\{ S_{\gamma(n+1)} = s \mid S_{\gamma(n)}, C_{\gamma(n)} \right\}$$
$$= p^+\left(s; S_{\gamma(n)}, E^*(S_{\gamma(n)}, C_{\gamma(n)})\right) \text{ a.s.} \tag{4.25}$$

for $s \in S$ and $n \geq 0$. Finally, we have

$$P_\mu \left\{ C_{\gamma(n),i} \leq x_i \text{ for } 1 \leq i \leq M, S_{\gamma(n+1)} = s \mid \mathcal{F}_{\gamma(n)} \right\}$$
$$= E_\mu \left[P_\mu\{ C_{\gamma(n),i} \leq x_i \text{ for } 1 \leq i \leq M, S_{\gamma(n+1)} = s \mid \mathcal{F}_{\gamma(n)}, C_{\gamma(n)} \} \right.$$
$$\left. \Big| \mathcal{F}_{\gamma(n)} \right]$$
$$= E_\mu \left[P_\mu \left\{ S_{\gamma(n+1)} = s \mid \mathcal{F}_{\gamma(n)}, C_{\gamma(n)} \right\} \prod_{i=1}^{M} 1_{\{C_{\gamma(n),i} \leq x_i\}} \Big| \mathcal{F}_{\gamma(n)} \right]$$
$$= E_\mu \left[p^+\left(s; S_{\gamma(n)}, E^*(S_{\gamma(n)}, C_{\gamma(n)})\right) \prod_{i=1}^{M} 1_{\{C_{\gamma(n),i} \leq x_i\}} \Big| \mathcal{F}_{\gamma(n)} \right] \text{ a.s.,}$$
$$\tag{4.26}$$

where the third equality follows from the strong Markov property and (4.25). It follows directly from (4.24) that the rightmost expression in (4.26) is equal to the right side of (4.23). $\qquad\square$

PROOF OF THEOREM 4.21. Set $T_n = t^*(S_{\gamma(n)}, C_{\gamma(n)})$ and $Y_n = S_{\gamma(n)}$ for $n \geq 0$, where the sequence of random indices $\{\gamma(n) : n \geq 0\}$ is defined by (1.14). Also set $q_0(s) = \sum_{e_i \in E(s)} r(s, e_i) v(i)$ for $s \in S$ and

$$W_0(s, s') = \sum_{e_i \in E(s)} \frac{r(s, e_i) v(i)}{q_0(s)} p^+(s'; s, e_i)$$

for $s, s' \in S$. Comparing the definition of the process $\{X(t) : t \geq 0\}$ to that of a minimal CTMC, we see that if

(i) $\{Y_n : n \geq 0\}$ is a DTMC with transition matrix W_0, and

(ii) given $\{Y_n : n \geq 0\}$, the random variables $\{T_n : n \geq 0\}$ are mutually independent and each T_n is exponentially distributed with intensity $q_0(Y_n)$,

then the first two assertions of the theorem follow. The conditions in (i) and (ii) hold if and only if

$$P_\mu\{Y_0 = s_0, T_0 > t_0, \ldots, Y_n = s_n, T_n > t_n, Y_{n+1} = s_{n+1}\}$$
$$= P_\mu\{Y_0 = s_0\} \prod_{k=0}^{n} e^{-q_0(s_k)t_k} W_0(s_k, s_{k+1}) \tag{4.27}$$

for $n \geq 0$, $s_0, \ldots, s_{n+1} \in S$, and $t_0, \ldots, t_n \geq 0$. To establish (4.27), observe that

$$P_\mu\{T_n > t, Y_{n+1} = s \mid \mathcal{F}_{\gamma(n)}\}$$

$$= P_\mu\Big\{\min_{e_i \in E(S_{\gamma(n)})} r^{-1}(S_{\gamma(n)}, e_i)C_{\gamma(n),i} > t, S_{\gamma(n+1)} = s \mid \mathcal{F}_{\gamma(n)}\Big\}$$

$$= \sum_{e_i \in E(S_{\gamma(n)})} P_\mu\Big\{E^*_{\gamma(n)} = \{e_i\}, \, r^{-1}(S_{\gamma(n)}, e_i)C_{\gamma(n),i} > t,$$
$$S_{\gamma(n+1)} = s \mid \mathcal{F}_{\gamma(n)}\Big\}$$

$$= \sum_{e_i \in E(S_{\gamma(n)})} \frac{r(S_{\gamma(n)}, e_i)v(i)}{\sum_{e_k \in E(S_{\gamma(n)})} r(S_{\gamma(n)}, e_k)v(k)}$$
$$p^+(s; S_{\gamma(n)}, e_i) \exp\Big(-\sum_{e_k \in E(S_{\gamma(n)})} r(S_{\gamma(n)}, e_k)v(k)t\Big)$$

$$= \sum_{e_i \in E(S_{\gamma(n)})} \frac{r(S_{\gamma(n)}, e_i)v(i)}{q(S_{\gamma(n)})} p^+(s; S_{\gamma(n)}, e_i)e^{-q(S_{\gamma(n)})t}$$

$$= W_0(Y_n, s)e^{-q_0(Y_n)t} \text{ a.s.}$$

for $n \geq 0$, $t \geq 0$, and $s \in S$. Here the third equality follows from Lemma 4.22 and the well-known fact that if $X_1, X_2, \ldots, X_n$ are mutually independent exponential random variables with respective intensities $q_1, q_2, \ldots, q_n$, then, setting $M_n = \min(X_1, X_2, \ldots, X_n)$ and $q^* = q_1 + q_2 + \cdots + q_n$,

$$P\{M_n = X_i \text{ and } M_n > x\} = \Big(\frac{q_i}{q^*}\Big)e^{-q^* x}$$

for $1 \leq i \leq n$ and $x \geq 0$. Thus,

$$P_\mu\{T_n > t, Y_{n+1} = s \mid Y_0, \ldots, Y_n, T_0, \ldots, T_{n-1}\}$$
$$= E_\mu\Big[P_\mu\{T_n > t, Y_{n+1} = s \mid \mathcal{F}_{\gamma(n)}\} \mid Y_0, \ldots, Y_n, T_0, \ldots, T_{n-1}\Big]$$

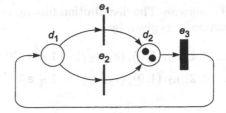

Figure 3.6. Markovian SPN with no simple timed transitions.

$$= E_\mu\left[W_0(Y_n, s)e^{-q(Y_n)t} \mid Y_0, \ldots, Y_n, T_0, \ldots, T_{n-1}\right]$$
$$= W_0(Y_n, s)e^{-q(Y_n)t} \text{ a.s.} \tag{4.28}$$

for $n \geq 0$, $t \geq 0$, and $s \in S$. A simple inductive argument using (4.28) yields (4.27), and the first two assertions of the theorem follow. To prove the final assertion, observe that if $\sup_{s,e} r(s, e) < \infty$, then the conditions of Theorem 3.13 are satisfied and the lifetime of the marking process is infinite; equivalently, the chain is nonexplosive.　　□

The conditions in Theorem 4.21 are sufficient but not necessary for the marking process to be a time-homogeneous CTMC. The following example shows that the marking process may be a CTMC even when one or more timed transitions are not simple.

EXAMPLE 4.29 (Markovian SPN with no simple timed transitions). Consider an SPN with two places and three transitions as in Figure 3.6. The marking set is $G = \{(1, 0), (0, 1), (0, 2)\}$. Whenever place d_1 contains a token and transitions e_1 and e_2 fire simultaneously, the token is removed from place d_1. Moreover, either one token is deposited in place d_2 or two tokens are deposited, each scenario occurring with probability $1/2$. Whenever place d_2 contains exactly one token and transition e_3 fires, the token is removed from place d_2 and a token is deposited in place d_1. Whenever place d_2 contains two tokens and transition e_3 fires, a token is removed from place d_2 (and no tokens are deposited in place d_1). Thus the new-marking probabilities are given by

$$p((0, 1); (1, 0), \{e_1, e_2\}) = 1/2,$$
$$p((0, 2); (1, 0), \{e_1, e_2\}) = 1/2,$$
$$p((1, 0); (0, 1), e_3) = 1,$$
$$p((0, 1); (0, 2), e_3) = 1,$$

and $p(s'; s, E^*) = 0$ otherwise. The distribution function of new clock readings for timed transition e_3 is given by

$$F\big(x; (0,1), e_3, (1,0), \{e_1, e_2\}\big) = 1 - e^{-v(1)x},$$
$$F\big(x; (0,2), e_3, (1,0), \{e_1, e_2\}\big) = 1 - e^{-v(2)x},$$

and

$$F\big(x; (0,1), e_3, (0,2), e_3\big) = 1 - e^{-v(1)x},$$

where $v(1), v(2) > 0$ and $v(1) \neq v(2)$. With probability 1, the initial marking is equal to $(1,0)$. Using arguments similar to the proof of Theorem 4.21, it can be shown that the marking process is a CTMC with state space $S = \{(0,1), (0,2)\}$. The intensity vector is given by

$$q(s) = \begin{cases} v(1)/2 & \text{if } s = (0,1); \\ v(2) & \text{if } s = (0,2), \end{cases}$$

and the transition matrix for the embedded jump chain is given by

$$W(s, s') = \begin{cases} 1 & \text{if } s' \neq s; \\ 0 & \text{if } s' = s. \end{cases}$$

Notes

Our definition of the marking process follows Haas and Shedler (1989b). As with the SPN building blocks, this definition was originally motivated by the discussion of generalized semi-Markov processes in Whitt (1980). A comprehensive treatment of general state-space Markov chains can be found in Meyn and Tweedie (1993a); see also Asmussen (1987a, Section I.6), Chung (1967, Section 9.2), and Durrett (1991, Section 5.6).

Chiola (1991) first proposed efficient methods, based essentially on the relations in (1.17) and (1.18), for updating the set of currently enabled transitions when generating sample paths of the marking process. Techniques for efficient generation of sample paths on parallel computers have been studied by Ferscha and Richter (1997), among others.

The assertion in Theorem 2.9(ii) is often presented in the context of renewal theory, in which the starting assumption is that the sequence $\{\Delta_n : n \geq 1\}$ consists of i.i.d. random variables; see, for example, p. 58 of Ross (1983). The result in Theorem 2.9(iv) appears as Proposition 2 in Glynn and Iglehart (1988).

For some recent discussions about simulation of supply chains, see, for example, the papers of Archibald et al. (1999), Ingalls and Kasales (1999), and Viswanadham and Raghavan (2000). With a concomitant increase in complexity, the model in Example 2.16 can be extended so that, for example, the order size for raw parts also depends on explicit forecasts of customer demand.

The sufficient conditions given in Theorem 3.13 for the lifetime of a marking process to be a.s. infinite can be viewed as an extension of the sufficient condition for CTMCs given in Theorem 3.23 of Çinlar (1975, Chapter 8). This latter condition requires that $\sup_s q(s) < \infty$. Other conditions that rule out explosions in CTMCs can be found in Section 8.3 in Çinlar and in Sections II.2 and II.3 in Asmussen (1987a). The geometric trials lemma (Lemma 3.4) used in the proof of Theorem 3.13 can be derived from the martingale convergence theorem; see Hall and Heyde (1980, Corollary 2.3).

Our treatment of CTMCs follows Asmussen (1987a). Alternative characterizations of recurrence and irreducibility in CTMCs, as well as other aspects of the fundamental theory of continuous-time chains, can be found in Asmussen's book, as well as in the books of Chung (1967), Çinlar (1975), Karlin and Taylor (1975), and Kohlas (1982).

Much of the literature on SPNs concerns nets in which the marking process is Markovian. In this setting, the marking process is typically defined directly as a CTMC, essentially by specifying an infinitesimal generator matrix in terms of the SPN building blocks. A typical goal is to compute the stationary probability distribution of the marking process by solving the system of equations in (4.7). This task can be nontrivial, especially when the size of the state space S is very large. Consequently, much effort has been expended in developing efficient solution techniques. One class of techniques tries to exploit symmetries in the model; in the CTMC setting these techniques sometimes are referred to as "lumping" methods. SPN-type frameworks have proven to be convenient for specifying model symmetries and for using these symmetries to facilitate computation of stationary probabilities; see Chiola et al. (1988, 1993). A number of authors such as Boucherie (1994) and Coleman (1993) have studied SPNs for which the system of equations in (4.7) has a "product-form" solution that is amenable to efficient computation. Techniques for obtaining bounds and approximations to time-average limits have been investigated by Campos et al. (1994) and others. Recently, attention has focused on numerical methods for SPNs in which the marking process contains an embedded semi-Markov process (Choi et al., 1994) and on SPNs in which the clock-setting distributions are either deterministic or exponential (Lindemann and Shedler, 1996; Puliafito et al., 1998). When a Markovian marking process is sufficiently complex so that simulation is an attractive alternative, the Markov property can be exploited to increase simulation efficiency—see Hordijk et al. (1976) and, for an extension of the idea to semi-Markovian marking processes, Fox and Glynn (1985).

A markedly different approach to both the analysis and simulation of certain SPNs is to focus not on the stochastic processes associated with the net, but rather on a set of recursive equations that directly describes the sequence of transition firing times. See, for example, Baccelli (1992), Baccelli and Canales (1993), and Baccelli et al. (1993, 1996).

4
Modelling Power

The examples in Chapter 2 show how the SPN building blocks can be used to formally specify a variety of discrete-event stochastic systems. The question then arises as to exactly how large a class of discrete-event systems can be modelled within the SPN framework. Although this question cannot be answered precisely, the modelling power of SPNs can usefully be compared with that of *generalized semi-Markov processes* (GSMPs).

The GSMP is the traditional model for the underlying stochastic process of a discrete-event system, and a wide range of computer, communication, manufacturing, and transportation systems have been modelled as GSMPs. Thus GSMPs are a good benchmark for assessment of modelling power. Moreover, the methodology that we develop for comparing the SPN and GSMP formalisms can be used to investigate a variety of other modelling power questions that arise in the study of discrete-event stochastic systems. For example, it may be of interest to determine whether inhibitor input places actually increase the modelling power of SPNs.

Although GSMPs are similar to SPNs, the two formal systems differ in the event-scheduling mechanism, the state-transition mechanism, and the form of the state space. A GSMP is a continuous-time stochastic process that makes a state transition when one or more "events" associated with the occupied state occur. Unlike an SPN state, which is a vector of token counts, a GSMP state can be an element of an arbitrary finite or countably infinite set. Moreover, the set of "active" (i.e., scheduled) events associated with a GSMP state is explicitly specified by the modeller—and can be an arbitrary subset of the set of all events—whereas the set of enabled transitions associated with the marking of an SPN is determined by the SPN graph. Events associated with a state compete to trigger the next state transition, and each set of trigger events has its own probability distribution for determining the new state. In contrast to the new-marking probabilities of an SPN, there are no constraints on the state-transition probabilities of a GSMP. At each state transition, new events may be scheduled. For each of these new events, a clock indicating the time until the event is scheduled to occur is set according to a probability distribution that depends on the old state, the new state, and the set of events that triggers the state transition. Clock readings for new events are always positive with probability 1, so that there is no analog of an immediate transition. If a scheduled event is not in the set of events that triggers a state transition but is associated with the new state, then its clock continues to run down (at a state-dependent speed); if such an event is not associated with the new state, it is cancelled and the corresponding clock reading is discarded. As with the marking process of an SPN, a GSMP is defined in terms of a general state-space Markov chain that describes the state and clock-reading vector at successive state-transition times. Further details of the GSMP formalism are given in Section 4.1.

As can be seen from the foregoing description, GSMPs have a more general state-transition mechanism, event-scheduling mechanism, and form of the state space than SPNs. This greater degree of generality means, however, that it can be hard to come up with the "right" state definition and set of events from scratch when modelling a complex system as a GSMP. Also, GSMPs are not particularly amenable to top-down or bottom-up modelling. For these reasons the SPN building blocks often are easier to use than the GSMP building blocks. Because of their more specialized structure, however, it might be conjectured that SPNs have less modelling power than GSMPs.

In Section 4.3 we show that, on the contrary, SPNs have at least the modelling power of GSMPs; this result establishes SPNs as an attractive general framework for performance analysis of discrete-event stochastic systems. Specifically, for any GSMP there exists an SPN with a marking process such that the two processes (and their underlying chains) have the same finite-dimensional distributions under an appropriate mapping between the state spaces. This notion of "strong mimicry" is discussed in Section 4.2.

To establish the modelling power result, we use the building blocks of the given GSMP to construct a "canonical" SPN. We then display a mapping from the state space of the underlying chain of the canonical SPN to the state space of the underlying chain of the GSMP that preserves the initial distribution, transition kernel, and holding-time function. In general, the canonical SPN has random inputs and outputs as well as timed and immediate transitions, and the number of tokens in a place is unbounded. When the state space of the given GSMP is finite, there exists a 2-bounded canonical SPN; if no scheduled events of the GSMP can be cancelled, only timed transitions are required. When the state space of the GSMP is finite and the current state and trigger event uniquely determine the next state, there exists a 1-bounded canonical SPN in which all transitions are deterministic. No inhibitor input places are needed in any of the canonical SPNs.

Is the modelling power of SPNs strictly greater than that of GSMPs? In light of the above modelling power results, such an assertion might appear plausible because an SPN can have one or more immediate transitions but a GSMP does not have "immediate events." Indeed, one can easily construct an SPN that is not a "special case" of a GSMP in that the embedded chain is not the underlying chain of any GSMP and the marking process does not coincide with any GSMP—see Example 4.1 below, as well as the adjacent discussion of the particle-counter model. In Section 4.4, however, we show (Theorem 4.6) that for any SPN with timed and immediate transitions there exists a GSMP that strongly mimics the marking process of the SPN. The state of the canonical GSMP consists essentially of a timed marking along with a representation of how the clock associated with each timed transition was set since the last timed marking. The events of the GSMP correspond to the timed transitions. In combination with the results of Section 4.3, Theorem 4.6 shows that SPNs have the same modelling power as GSMPs. Also, as shown in Chapter 5, Theorem 4.6 is useful when establishing recurrence properties for SPNs—the theorem provides a means of avoiding complications caused by the presence of immediate transitions.

4.1   Generalized Semi-Markov Processes

The basic components of a GSMP model are

- A finite or countably infinite set $\underline{S}$ of *states*

- A finite set $\underline{E} = \{\, \underline{e}_1, \underline{e}_2, \ldots, \underline{e}_M \,\}$ of *events*

- A mapping $\underline{s} \mapsto \underline{E}(\underline{s})$ from $\underline{S}$ to the nonempty subsets of $\underline{E}$

- State-transition probabilities of the form $\underline{p}(\underline{s}'; \underline{s}, \underline{E}^*)$

- Finite nonnegative *speeds* of the form $\underline{r}(\underline{s}, \underline{e})$

- Clock-setting distribution functions of the form $\underline{F}(\,\cdot\,; \underline{s}', \underline{e}', \underline{s}, \underline{E}^*)$

The set $\underline{E}(\underline{s})$ is the set of *active* events in state $\underline{s}$, that is, the set of all events that can possibly occur in state $\underline{s}$. Observe that $\underline{E}(\underline{s})$ is a GSMP building block that is explicitly specified by the modeller. In contrast, a set $E(s)$ in an SPN is specified indirectly by means of the normal input and inhibitor input functions. Similarly to a new-marking probability in an SPN, the state-transition probability $\underline{p}(\underline{s}'; \underline{s}, \underline{E}^*)$ is the probability that the new state is $\underline{s}'$ given that the events in $\underline{E}^*$ occur simultaneously in state $\underline{s}$. As in an SPN, a *clock* is associated with each event $\underline{e} \in \underline{E}$. The clock for an active event records the remaining time until the event is scheduled to occur; $\underline{r}(\underline{s}, \underline{e})$ is the speed at which the clock associated with event $\underline{e}$ runs down in state $\underline{s}$. At a transition from $\underline{s}$ to $\underline{s}'$ triggered by the simultaneous occurrence of the events in the set $\underline{E}^*$, a clock reading is generated for each *new event* $\underline{e}' \in \underline{N}(\underline{s}'; \underline{s}, \underline{E}^*) = \underline{E}(\underline{s}') - (\underline{E}(\underline{s}) - \underline{E}^*)$ according to $\underline{F}(\,\cdot\,; \underline{s}', \underline{e}', \underline{s}, \underline{E}^*)$. We assume that $\underline{F}(0; \underline{s}', \underline{e}', \underline{s}, \underline{E}^*) = 0$ for $\underline{e}' \in E$, so that an event never occurs at the instant that it becomes active. (Thus a GSMP has no analog of an immediate transition.) For each *old event* $\underline{e}' \in \underline{O}(\underline{s}'; \underline{s}, \underline{E}^*) = \underline{E}(\underline{s}') \cap (\underline{E}(\underline{s}) - \underline{E}^*)$, the old clock reading is kept after the state transition. For $\underline{e}' \in (\underline{E}(\underline{s}) - \underline{E}^*) - \underline{E}(\underline{s}')$, event $\underline{e}'$ (that was active before the events in $\underline{E}^*$ occurred) is cancelled after the state transition and the clock reading is discarded. When the state is $\underline{s}$ and the set $\underline{E}^*$ of events that simultaneously trigger a state transition is $\underline{E}^* = \{\, \underline{e}^* \,\}$, we often write $\underline{p}(\underline{s}'; \underline{s}, \underline{e}^*)$ for $\underline{p}(\underline{s}'; \underline{s}, \{\underline{e}^*\})$, and so forth.

The GSMP is the stochastic process that records the state of the system as it evolves over continuous time. Similarly to the marking process of an SPN, the formal definition of a GSMP is in terms of a general state-space Markov chain $\{\, (\underline{S}_n, \underline{C}_n) \colon n \geq 0 \,\}$, where $\underline{S}_n$ represents the state and $\underline{C}_n = (\underline{C}_{n,1}, \underline{C}_{n,2}, \ldots, \underline{C}_{n,M})$ represents the clock-reading vector just after the nth state transition. The state space of the chain is $\underline{\Sigma} = \bigcup_{\underline{s} \in \underline{S}}(\{\, \underline{s} \,\} \times \underline{C}(\underline{s}))$, where $\underline{C}(\underline{s})$ is the set of possible clock-reading vectors in state $\underline{s}$:

$$\underline{C}(\underline{s}) = \{\, \underline{c} = (\underline{c}_1, \ldots, \underline{c}_M) \colon \underline{c}_i \geq 0 \text{ and } \underline{c}_i > 0 \text{ if and only if } \underline{e}_i \in \underline{E}(\underline{s}) \,\}.$$

As with SPNs, the initial state $\underline{s}_0$ is selected according to an *initial-state distribution* ν_0 defined on S. Then, for each active event $\underline{e}_i \in \underline{E}(\underline{s}_0)$, an initial clock reading is generated according to an *initial clock-setting distribution function* $\underline{F}_0(\,\cdot\,; \underline{e}_i, \underline{s}_0)$. Thus the initial distribution $\underline{\mu}$ of the underlying chain is of the form

$$\underline{\mu}(A) = \underline{\nu}_0(\underline{s}_0) \prod_{\underline{e} \in \underline{E}(\underline{s}_0)} \underline{F}_0(a_i; \underline{e}, \underline{s}_0)$$

for all sets

$$A = \{\, \underline{s}_0 \,\} \times \{\, (\underline{c}_{0,1}, \ldots, \underline{c}_{0,M}) \in \underline{C}(\underline{s}_0) \colon 0 \leq \underline{c}_{0,i} \leq a_i \text{ for } 1 \leq i \leq M \,\}.$$

The transition kernel of the chain is specified in terms of the GSMP building blocks by a formula identical to (1.9) in Chapter 3. In this specification, we define the following quantities identically to their SPN counterparts:

$$\underline{t}^*(\underline{s}, \underline{c}) = \min_{\{\, i \colon \underline{e}_i \in \underline{E}(\underline{s}) \,\}} \{\, \underline{c}_i \underline{r}^{-1}(\underline{s}, \underline{e}_i) \,\},$$

$$\underline{c}_i^*(\underline{s}, \underline{c}) = \underline{c}_i - \underline{t}^*(\underline{s}, \underline{c})\underline{r}(\underline{s}, \underline{e}_i),$$

and

$$\underline{E}^*(\underline{s}, \underline{c}) = \{\, \underline{e}_i \in \underline{E}(\underline{s}) \colon \underline{c}_i^*(\underline{s}, \underline{c}) = 0 \,\}$$

for $\underline{s} \in \underline{S}$, $\underline{c} = (\underline{c}_1, \underline{c}_2, \ldots, \underline{c}_M) \in \underline{C}(s)$, and $\underline{e}_i \in \underline{E}(\underline{s})$. In the preceding definition of the *holding-time function* $\underline{t}^*$, we take $\underline{c}_i \underline{r}^{-1}(\underline{s}, \underline{e}_i)$ to be $+\infty$ when $\underline{r}(\underline{s}, \underline{e}_i) = 0$. Beginning in state $\underline{s}$ with clock-reading vector $\underline{c}$, the quantity $\underline{t}^*(\underline{s}, \underline{c})$ is the time to the next state transition and $\underline{E}^*(\underline{s}, \underline{c})$ is the *trigger event set*; that is, the set of events that trigger the state transition.

The GSMP is the stochastic process $\{\underline{X}(t): t \geq 0\}$, where $\underline{X}(t)$ is the state of the system at time $t \geq 0$. Formal specification of $\{\underline{X}(t): t \geq 0\}$ in terms of the chain $\{(\underline{S}_n, \underline{C}_n): n \geq 0\}$ proceeds exactly as in (1.11)–(1.13) in Chapter 3. As with the marking process of an SPN, the GSMP takes values in the set $\underline{S} \cup \{\underline{\Delta}\}$ and has piecewise-constant, right-continuous sample paths. Here $\underline{\Delta}$ corresponds to the state of the system after a possible explosion; such explosions are ruled out whenever

1. $\sup_{\underline{s}, \underline{e}} \underline{r}(\underline{s}, \underline{e}) < \infty$.

2. There exists $a > 0$ such that $\sup_{\underline{s}', \underline{e}', \underline{s}, \underline{E}^*} \underline{F}(a; \underline{s}', \underline{e}', \underline{s}, \underline{E}^*) < 1$.

The proof of this assertion is almost identical to that of Theorem 3.13 in Chapter 3.

EXAMPLE 1.1 (Cyclic queues). Consider a closed network of queues with two single-server service centers and K (≥ 2) jobs. A job that completes service at center 1 moves to center 2; a job that completes service at center 2 moves to center 1. Both queueing disciplines are first-come, first-served. Successive service times at center i ($i = 1, 2$) are i.i.d. as a positive random variable L_i. Initially, all jobs are at center 2 and a job is just starting service. Let $\underline{X}(t)$ be the number of jobs waiting or in service at center 2 at time t.

Formal specification of the process $\{\underline{X}(t): t \geq 0\}$ is as a GSMP with state space $\underline{S} = \{0, 1, \ldots, K\}$ and event set $\underline{E} = \{\underline{e}_1, \underline{e}_2\}$, where $\underline{e}_i = $ "service completion at center i." For $\underline{s} \in \underline{S}$, event $\underline{e}_1 \in \underline{E}(\underline{s})$ if and only if $\underline{s} < K$ and $\underline{e}_2 \in \underline{E}(s)$ if and only if $\underline{s} > 0$. The state-transition probabilities are given by $\underline{p}(\underline{s}+1; \underline{s}, \underline{e}_1) = 1$ for $0 \leq \underline{s} < K$, $\underline{p}(\underline{s}-1; \underline{s}, \underline{e}_2) = 1$ for $0 < \underline{s} \leq K$, $\underline{p}(\underline{s}; \underline{s}, \{\underline{e}_1, \underline{e}_2\}) = 1$ for $0 < \underline{s} < K$, and $\underline{p}(\underline{s}'; \underline{s}, \underline{E}^*) = 0$ otherwise.

The clock-setting distribution functions are given by $\underline{F}(x; \underline{s}', \underline{e}_i, \underline{s}, \underline{E}^*) = P\{L_i \leq x\}$ for $i = 1, 2$, and all speeds are equal to 1. The initial-state distribution is given by $\underline{\nu}_0(K) = 1$, and the initial clock-setting distribution function for e_2 is $\underline{F}_0(x; e_2, K) = P\{L_2 \leq x\}$.

Observe that the sets of new events are given by $\underline{N}(1; 0, \underline{e}_1) = \{\underline{e}_1, \underline{e}_2\}$, $\underline{N}(\underline{s}+1; \underline{s}, \underline{e}_1) = \{\underline{e}_1\}$ for $0 < \underline{s} < K-1$, $\underline{N}(K; K-1, \underline{e}_1) = \varnothing$, $\underline{N}(0; 1, \underline{e}_2) = \varnothing$, $\underline{N}(\underline{s} - 1; \underline{s}, \underline{e}_2) = \{\underline{e}_2\}$ for $1 < \underline{s} < K$, $\underline{N}(K - 1; K, \underline{e}_2) = \{\underline{e}_1, \underline{e}_2\}$, and $\underline{N}(\underline{s}; \underline{s}, \{\underline{e}_1, \underline{e}_2\}) = \{\underline{e}_1, \underline{e}_2\}$ for $0 < \underline{s} < K$. The sets of old events are given by $\underline{Q}(1; 0, \underline{e}_1) = \varnothing$, $\underline{Q}(\underline{s} + 1; \underline{s}, \underline{e}_1) = \{\underline{e}_2\}$ for $0 < \underline{s} < K$, $\underline{Q}(\underline{s} - 1; \underline{s}, \underline{e}_2) = \{\underline{e}_1\}$ for $1 \leq \underline{s} < K$, $\underline{Q}(K - 1; K, \underline{e}_2) = \varnothing$, and $\underline{Q}(\underline{s}; \underline{s}, \{\underline{e}_1, \underline{e}_2\}) = \varnothing$ for $0 < \underline{s} < K$. The set $(\underline{E}(\underline{s}) - \underline{E}^*) - \underline{E}(\underline{s}')$ of cancelled events equals $\varnothing$ for $\underline{s}, \underline{s}' \in \underline{S}$ and $\underline{E}^* \subseteq \underline{E}(\underline{s})$.

In analogy with SPNs—see (4.9) in Chapter 3—we can define the partial history of the underlying chain $\{(\underline{S}_n, \underline{C}_n): n \geq 0\}$ of a GSMP. Set $\underline{t}_n^* = \underline{t}^*(\underline{S}_n, \underline{C}_n)$ and $\underline{E}_n^* = \underline{E}^*(\underline{S}_n, \underline{C}_n)$ for $n \geq 0$. Then the partial history $\underline{\mathcal{F}}_n$ of the underlying chain up to the nth state transition $(n \geq 1)$ is defined by

$$\underline{\mathcal{F}}_n = \{\, \underline{S}_0, \underline{E}_0^*, \underline{t}_0^*, \underline{S}_1, \underline{E}_1^*, \underline{t}_1^*, \ldots, \underline{S}_{n-1}, \underline{E}_{n-1}^*, \underline{t}_{n-1}^*, \underline{S}_n \,\}.$$

[Take $\underline{\mathcal{F}}_0 = \{\, \underline{S}_0 \,\}$.] The following result can be established using an argument similar to the proof of Lemma 4.10 in Chapter 3.

Lemma 1.2. *Let γ be an a.s. finite stopping time with respect to $\{\underline{\mathcal{F}}_n : n \geq 0\}$. Then, with probability 1,*

$$P_{\underline{\mu}}\{\, \underline{C}_{\gamma, i} > x_i \text{ for } \underline{e}_i \in \underline{H} \mid \mathcal{F}_\gamma \,\}$$
$$= \begin{cases} \prod_{\underline{e}_i \in \underline{H}} P_{\underline{\mu}}\{\, \underline{C}_{\gamma, i} > x_i \mid \mathcal{F}_\gamma \,\} & \text{if } \underline{H} \subseteq \underline{E}(\underline{S}_\gamma); \\ 0 & \text{otherwise} \end{cases}$$

for any subset $\underline{H} \subseteq \underline{E}$ and nonnegative numbers $\{\, x_i : \underline{e}_i \in \underline{H} \,\}$.

Lemma 1.2 asserts that the clock readings of a GSMP, observed at a stopping time γ, are conditionally independent given the partial history up to the γth state transition.

4.2   Mimicry and Strong Mimicry

In this section we formalize (in Definitions 2.1 and 2.7) two senses in which the marking process of an SPN can mimic a GSMP. We then give sufficient conditions (Theorem 2.10) for "strong" mimicry.

4.2.1   Definitions

Let $\{\, \underline{X}(t): t \geq 0 \,\}$ be a GSMP with state space $\underline{S}$, holding-time function $\underline{t}^*$, and underlying chain $\{\, (\underline{S}_n, \underline{C}_n): n \geq 0 \,\}$ having initial distribution $\underline{\mu}$. Also let $\{\, X(t): t \geq 0 \,\}$ be the marking process of an SPN with timed marking set S, holding-time function t^*, and underlying chain $\{\, (S_n, C_n): n \geq 0 \,\}$ having initial distribution μ.

Definition 2.1. The marking process $\{\, X(t): t \geq 0 \,\}$ is said to *mimic* the GSMP $\{\, \underline{X}(t): t \geq 0 \,\}$ if there exists a mapping λ from S onto $\underline{S}$ such that $\{\, \underline{X}(t): t \geq 0 \,\}$ and $\{\, \lambda X(t): t \geq 0 \,\}$ have the same finite-dimensional distributions; that is,

$$P_{\underline{\mu}}\{\, \underline{X}(t_1) = \underline{s}_1, \ldots, \underline{X}(t_m) = \underline{s}_m \,\}$$
$$= P_\mu\{\, \lambda X(t_1) = \underline{s}_1, \ldots, \lambda X(t_m) = \underline{s}_m \,\}$$

for $m \geq 1$, $0 \leq t_1 < t_2 < \cdots < t_m$, and $\underline{s}_1, \underline{s}_2, \ldots, \underline{s}_m \in \underline{S}$.

Because both $\{\underline{X}(t)\colon t \geq 0\}$ and $\{X(t)\colon t \geq 0\}$ have piecewise-constant sample paths, the finite-dimensional distributions of these processes completely determine their continuous-time properties. For example, if the process $\{X(t)\colon t \geq 0\}$ mimics $\{\underline{X}(t)\colon t \geq 0\}$ and

$$P_{\underline{\mu}}\left\{ \lim_{t\to\infty} \frac{1}{t} \int_0^t f(\underline{X}(u))\,du = r(f) \right\} = 1$$

as $t \to \infty$ for a real-valued function f and constant $r(f)$, then it can be shown that

$$P_\mu\left\{ \lim_{t\to\infty} \frac{1}{t} \int_0^t f(\lambda X(u))\,du = r(f) \right\} = 1.$$

The following example shows, however, that even if the marking process of an SPN mimics a GSMP, the behavior of the SPN and GSMP may appear different when the two models are observed at successive marking changes (resp., state transitions).

EXAMPLE 2.2 (Cyclic queues with feedback). Consider the network of queues of Example 1.4 in Chapter 2. Recall that the system consists of two single-server service centers and N (≥ 2) jobs. With fixed probability $p \in (0,1)$, a job that completes service at center 1 moves to center 2 and with probability $1-p$ joins the tail of the queue at center 1. A job that completes service at service center 2 moves to center 1. The queueing discipline at each center is first-come, first-served. Suppose that successive service times at center i ($i = 1, 2$) are independent and exponentially distributed with mean $1/q_i$. Also suppose that initially all jobs are at center 2 and a job starts service. Let $\underline{X}(t)$ be the number of jobs waiting or in service at service center 2 at time t. Formal specification of the process $\{\underline{X}(t)\colon t \geq 0\}$ is as a GSMP with state space $\underline{S} = \{0, 1, \ldots, N\}$ and event set $\underline{E} = \{\underline{e}_1, \underline{e}_2\}$, where $\underline{e}_i =$ "service completion at center i." To model the feedback, we set $\underline{p}(\underline{s}; \underline{s}, \underline{e}_1) = 1 - p$ and $\underline{p}(\underline{s} + 1; \underline{s}, \underline{e}_1) = p$ for $0 \leq \underline{s} < N$. The clock-setting distributions are given by $\underline{F}(x; \underline{s}', \underline{e}_1, \underline{s}, \underline{E}^*) = 1 - \exp(-q_1 x)$ and $\underline{F}(x; \underline{s}', \underline{e}_2, \underline{s}, \underline{E}^*) = 1 - \exp(-q_2 x)$. The remaining details of the specification are left to the reader.

An argument similar to the proof of Theorem 4.21 in Chapter 3 shows that the process $\{\underline{X}(t)\colon t \geq 0\}$ is a CTMC. The intensity vector q is

$$q = (pq_1, pq_1 + q_2, pq_1 + q_2, \ldots, pq_1 + q_2, q_2), \tag{2.3}$$

the transition matrix W is

$$W = \begin{pmatrix} 0 & 1 & 0 & 0 & \cdots & 0 & 0 & 0 \\ b & 0 & a & 0 & \cdots & 0 & 0 & 0 \\ 0 & b & 0 & a & \cdots & 0 & 0 & 0 \\ \vdots & \vdots & \vdots & \vdots & \ddots & \vdots & \vdots & \vdots \\ 0 & 0 & 0 & 0 & \cdots & b & 0 & a \\ 0 & 0 & 0 & 0 & \cdots & 0 & 1 & 0 \end{pmatrix}, \tag{2.4}$$

$$e_1 = \text{service completion at center 1}$$
$$e_2 = \text{service completion at center 2}$$

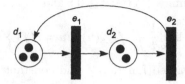

Figure 4.1. An SPN that mimics cyclic queues with feedback.

where $a = pq_1/(pq_1 + q_2)$ and $b = q_2/(pq_1 + q_2)$, and the initial distribution v is

$$v = (0, 0, \ldots, 0, 1). \tag{2.5}$$

Next, consider an SPN with two timed deterministic transitions as in Figure 4.1; in every marking, the two places contain a combined total of exactly N tokens. The clock-setting distribution functions are $F(x; s', e_1, s, e) = 1 - \exp(-pq_1 x)$ and $F(x; s', e_2, s, e) = 1 - \exp(-q_2 x)$. All speeds for enabled transitions are equal to 1. The initial-marking distribution is given by $\nu_0\big((0, N)\big) = 1$ and the initial clock-setting distribution function for e_2 is $F_0\big(x; e_2, (0, N)\big) = 1 - \exp(-q_2 x)$.

Denote the marking process of the SPN by $\{X(t): t \geq 0\}$ and define a mapping $\lambda\colon S \mapsto \underline{S}$ by $\lambda s = s_2$ for $s = (s_1, s_2) \in S$. An application of Theorem 4.21 in Chapter 3 shows that the process $\{\lambda X(t): t \geq 0\}$ is a CTMC with intensity vector q, transition matrix W, and initial distribution v given by (2.3)–(2.5), respectively. Because the intensity vector, transition matrix, and initial distribution are the same for $\{\lambda X(t): t \geq 0\}$ and $\{\underline{X}(t): t \geq 0\}$, these two processes have the same finite-dimensional distributions. Thus the marking process of the SPN mimics the GSMP. Observe, however, that the SPN model does not exhibit the feedback behavior that occurs in the GSMP model. In this sense the SPN model does not behave identically to the GSMP model even though the marking process of the SPN mimics the GSMP.

The following example illustrates a stronger notion of mimicry that more effectively captures the notion of identical stochastic behavior.

EXAMPLE 2.6 (Cyclic queues with feedback). Modify the SPN of Example 2.2 so that $F(x; s', e_1, s, e) = 1 - \exp(-q_1 x)$. Also modify the new-marking probabilities so that $p(s; s, e_1) = 1 - p$ and $p\big((s_1 - 1, s_2 + 1); s, e_1\big) = p$ for $s = (s_1, s_2) \in S$. This SPN is similar to the SPN given in Example 1.4 of Chapter 2. The marking process of this SPN mimics $\{\underline{X}(t): t \geq 0\}$ in the sense of Definition 2.1 (under the mapping λ of Example 2.2). The marking process also mimics $\{\underline{X}(t): t \geq 0\}$ in the following, stronger sense. Denote the underlying chain of $\{\underline{X}(t): t \geq 0\}$ by $\{(\underline{S}_n, \underline{C}_n): n \geq 0\}$ and the state space of $\{(\underline{S}_n, \underline{C}_n): n \geq 0\}$ by $\underline{\Sigma}$. Similarly, let $\{(S_n, C_n): n \geq 0\}$ be the underlying chain of the marking process and Σ be the state space of $\{(S_n, C_n): n \geq 0\}$. Define the mapping $\phi: \Sigma \mapsto \underline{\Sigma}$ by $\phi(s, c) = (\lambda s, c)$. It can be shown that $\{\phi(S_n, C_n): n \geq 0\}$ and $\{(\underline{S}_n, \underline{C}_n): n \geq 0\}$ have the same finite-dimensional distributions. Thus marking changes with "feedback" of a token mimic the feedback-type state transitions of the GSMP.

Motivated by the above discussion, we give the following definition. Recall from Definition 1.15 in Chapter 3 that the embedded chain $\{(S_n^+, C_n^+): n \geq 0\}$ of the marking process records the marking and clock-reading vector at each marking change for which the new marking is timed. As before, we denote the state space of the embedded chain by Σ^+ and the initial distribution by μ^+.

Definition 2.7. The marking process $\{X(t): t \geq 0\}$ is said to *strongly mimic* the GSMP $\{\underline{X}(t): t \geq 0\}$ if

(i) there exists a mapping λ from S onto $\underline{S}$ such that the processes $\{\underline{X}(t): t \geq 0\}$ and $\{\lambda X(t): t \geq 0\}$ have the same finite-dimensional distributions; and

(ii) there exists a mapping ϕ from Σ^+ onto $\underline{\Sigma}$ of the form $\phi(s, c) = (\lambda s, \eta(s, c))$ such that the discrete-time processes $\{(\underline{S}_n, \underline{C}_n): n \geq 0\}$ and $\{\phi(S_n^+, C_n^+): n \geq 0\}$ have the same finite-dimensional distributions.

Clearly, strong mimicry implies mimicry by definition. On the other hand, Example 2.2 shows that mimicry need not imply strong mimicry; that is, condition (i) of Definition 2.7 can hold while condition (ii) fails to hold. The following example shows that, conversely, there can exist a mapping $\phi = (\lambda, \eta)$ such that $\{(\underline{S}_n, \underline{C}_n): n \geq 0\}$ and $\{\phi(S_n^+, C_n^+): n \geq 0\}$ have the same finite-dimensional distributions but $\{\underline{X}(t): t \geq 0\}$ and $\{\lambda X(t): t \geq 0\}$ do not. Thus condition (i) in Definition 2.7 is not redundant.

EXAMPLE 2.8 (Alternating renewal process with constant holding times). Consider a GSMP with state space $\underline{S} = \{1, 2\}$ and event set $\underline{E} = \{\underline{e}\}$. The state-transition probabilities are $\underline{p}(2; 1, \underline{e}) = \underline{p}(1; 2, \underline{e}) = 1$ and the clock-setting distribution functions are

$$\underline{F}(x; 1, \underline{e}, 2, \underline{e}) = 1_{[1,\infty)}(x) \quad \text{and} \quad \underline{F}(x; 2, \underline{e}, 1, \underline{e}) = 1_{[2,\infty)}(x).$$

All speeds $r(s, e)$ for active events are equal to 1. The GSMP is initially in state 1 and the initial clock reading is equal to 1. Thus the GSMP visits state 1 for one time unit, then visits state 2 for two time units, then visits state 1 for one time unit, and so forth.

Next, consider an SPN with two timed deterministic transitions as in Figure 4.1, except that in every marking the two places contain a combined total of exactly one token, so that the marking set G $(= S)$ is

$G = \{(0,1),(1,0)\}$. The clock-setting distribution functions are

$$F\big(x;(1,0),e_1,(0,1),e_2\big) = 1_{[2,\infty)}(x)$$

$$\text{and } F\big(x;(0,1),e_2,(1,0),e_1\big) = 1_{[1,\infty)}(x).$$

All speeds for enabled transitions are equal to 1. The initial marking of the SPN is $(1,0)$ and the initial clock reading for transition e_1 is equal to 2. Thus the marking process visits state $(1,0)$ for two time units, then visits state $(0,1)$ for one time unit, then visits state $(1,0)$ for two time units, and so forth. The embedded chain $\{(S_n^+, C_n^+) : n \geq 0\}$ coincides with the underlying chain $\{(S_n, C_n) : n \geq 0\}$ because there are no immediate transitions.

Set $\lambda(1,0) = 1$, $\lambda(0,1) = 2$, $\eta((1,0),(2,0)) = 1$, and $\eta((0,1),(0,1)) = 2$. With probability 1, the successive states of $\{(\underline{S}_n, \underline{C}_n) : n \geq 0\}$ are $(1,1)$, $(2,2)$, $(1,1)$, $(2,2)$, ... and the successive states of $\{(S_n^+, C_n^+) : n \geq 0\}$ are $\big((1,0),(2,0)\big)$, $\big((0,1),(0,1)\big)$, $\big((1,0),(2,0)\big)$, $\big((0,1),(0,1)\big)$, ... , so that condition (ii) of Definition 2.7 holds. Condition (i) of Definition 2.7 fails to hold with λ defined as above: for example,

$$P_\mu\{\underline{X}(1.5) = 1\} = 0 \neq 1 = P_\mu\{\lambda X(1.5) = 1\}.$$

4.2.2 Sufficient Conditions for Strong Mimicry

Theorem 2.10 below gives sufficient conditions for strong mimicry and hence for mimicry. This result asserts that the marking process of an SPN strongly mimics a GSMP if there exists a mapping ϕ that preserves the initial distribution and transition kernel of the embedded chain and also preserves the holding-time function. The conditions of the theorem ensure that

$$P_{\underline{\mu}}\{(\underline{S}_0, \underline{C}_0) \in \underline{A}\} = P_\mu\{\phi(S_0^+, C_0^+) \in \underline{A}\},$$

$$P_{\underline{\mu}}\{(\underline{S}_{n+1}, \underline{C}_{n+1}) \in \underline{A} \mid (\underline{S}_n, \underline{C}_n) = (\underline{s}, \underline{c})\}$$
$$= P_\mu\{\phi(S_{n+1}^+, C_{n+1}^+) \in \underline{A} \mid \phi(S_n^+, C_n^+) = (\underline{s}, \underline{c})\},$$

and

$$\underline{t}^*\big(\phi(S_n^+, C_n^+)\big) = t^*(S_n^+, C_n^+)$$

for $\underline{A} \subseteq \underline{\Sigma}$, $(\underline{s}, \underline{c}) \in \underline{\Sigma}$, $t \geq 0$, and $n \geq 0$.

We use the following notation throughout. If ϕ is a mapping from a set Σ to another set $\underline{\Sigma}$ and $\underline{A}$ is a subset of $\underline{\Sigma}$, then $\phi^{-1}\underline{A}$ denotes the set $\{x \in \Sigma : \phi x \in \underline{A}\}$ and ϕA (where $A \subseteq \Sigma$) denotes the set $\{\phi x : x \in A\}$. With a slight abuse of notation, we also denote by ϕ the mapping from Σ^∞ to $\underline{\Sigma}^\infty$ given by

$$\phi(x_0, x_1, \ldots) = (\phi x_0, \phi x_1, \ldots)$$

for $x_0, x_1, \ldots \in \Sigma$. Similarly, if D (resp., $\underline{D}$) is a set of functions from $[0, \infty)$ to Σ (resp., $\underline{\Sigma}$), we denote by ϕ the mapping from D to $\underline{D}$ defined by setting $\phi x = \underline{x}$, where $\underline{x}(t) = \phi(x(t))$ for $t \geq 0$.

Theorem 2.10 requires the existence of a mapping ϕ—from the state space of the embedded chain of the SPN to the state space of the underlying chain of the GSMP—that preserves initial distributions and transition kernels. In applications it is convenient to ignore this preservation requirement when dealing with zero-probability events. To this end, we introduce the notion of an "inaccessible" set.

Definition 2.9. Let ϕ be a mapping from Σ^+ onto $\underline{\Sigma}$. A set $\underline{H} \subseteq \underline{\Sigma}$ is said to be *inaccessible* with respect to ϕ if

$$P_{\underline{\mu}}\{ (\underline{S}_n, \underline{C}_n) \in \underline{H} \text{ for some } n \geq 0 \}$$
$$= P_{\mu}\{ \phi(S_n^+, C_n^+) \in \underline{H} \text{ for some } n \geq 0 \} = 0.$$

Theorem 2.10. *Suppose that there exist a mapping ϕ from Σ^+ onto $\underline{\Sigma}$ of the form $\phi(s,c) = (\lambda s, \eta(s,c))$ and a set $\underline{H}$ inaccessible with respect to ϕ such that*

(i) $\underline{t}^*(\phi(s,c)) = t^*(s,c)$ *for all* $(s,c) \in \Sigma^+$,

(ii) $\underline{\mu}(\underline{A}) = \mu^+(\phi^{-1}\underline{A})$ *for all* $\underline{A} \subseteq \underline{\Sigma} - \underline{H}$, *and*

(iii) $\underline{P}(\phi(s,c), \underline{A}) = P^+((s,c), \phi^{-1}\underline{A})$ *for all* $(s,c) \in \Sigma^+ - \phi^{-1}\underline{H}$ *and* $\underline{A} \subseteq \underline{\Sigma} - \underline{H}$.

Then $\{ X(t) : t \geq 0 \}$ *strongly mimics* $\{ \underline{X}(t) : t \geq 0 \}$.

PROOF. We first show that $\{ (\underline{S}_n, \underline{C}_n) : n \geq 0 \}$ and $\{ \phi(S_n^+, C_n^+) : n \geq 0 \}$ have the same finite-dimensional distributions. Set $\underline{P}^1(\underline{x}, \underline{A}) = \underline{P}(\underline{x}, \underline{A})$ for $\underline{A} \subseteq \underline{\Sigma}$ and $\underline{x} \in \underline{\Sigma}$ and recursively define

$$\underline{P}^n(\underline{x}, \underline{A}_1, \ldots, \underline{A}_n) = \int_{\underline{A}_1} \underline{P}^{n-1}(\underline{y}, \underline{A}_2, \ldots, \underline{A}_n) \, \underline{P}(\underline{x}, d\underline{y})$$

for $n \geq 2$, $\underline{A}_1, \underline{A}_2, \ldots, \underline{A}_n \subseteq \underline{\Sigma}$, and $\underline{x} \in \underline{\Sigma}$. Similarly, define probabilities $P^n(x, A_1, \ldots, A_n)$ for $n \geq 1$, $A_1, A_2, \ldots, A_n \subseteq \Sigma^+$, and $x \in \Sigma^+$ in terms of P^+. It follows from (1.4) in Chapter 3 that

$$P_{\underline{\mu}}\{ \underline{A} \} = P_{\underline{\mu}}\{ (\underline{S}_0, \underline{C}_0) \in \underline{A}_0, (\underline{S}_1, \underline{C}_1) \in \underline{A}_1, \ldots, (\underline{S}_n, \underline{C}_n) \in \underline{A}_n \}$$
$$= \int_{\underline{A}_0} \underline{\mu}(d\underline{z}_0) \int_{\underline{A}_1} \underline{P}(d\underline{z}_1, \underline{z}_0) \cdots \int_{\underline{A}_n} \underline{P}(d\underline{z}_n, \underline{z}_{n-1}) \tag{2.11}$$
$$= \int_{\underline{A}_0} \underline{P}^n(\underline{x}, \underline{A}_1, \ldots, \underline{A}_n) \, \underline{\mu}(d\underline{x})$$

for every $n \geq 0$ and set

$$\underline{A} = \underline{A}_0 \times \underline{A}_1 \times \cdots \times \underline{A}_n \times \underline{\Sigma} \times \underline{\Sigma} \times \cdots \subseteq \underline{\Sigma}^\infty. \tag{2.12}$$

A corresponding result holds for P_μ. We now show that the probabilities $\underline{P}^n$ and P^n satisfy

$$\underline{P}^n(\phi x, \underline{A}_1, \ldots, \underline{A}_n) = P^n(x, \phi^{-1}\underline{A}_1, \ldots, \phi^{-1}\underline{A}_n) \qquad (2.13)$$

for all $n \geq 1$, $x \in \Sigma^+ - \phi^{-1}\underline{H}$, and sets $\underline{A}_1, \underline{A}_2, \ldots, \underline{A}_n \subseteq \Sigma - \underline{H}$. First, observe that the assertion (2.13) reduces to condition (iii) when $n = 1$. Assume for induction that (2.13) holds for a fixed value of n and observe that

$$
\begin{aligned}
\underline{P}^{n+1}(\phi x, \underline{A}_1, \ldots, \underline{A}_{n+1}) &= \int_{\underline{A}_1} \underline{P}^n(\underline{y}, \underline{A}_2, \ldots, \underline{A}_{n+1})\, \underline{P}(\phi x, d\underline{y}) \\
&= \int_{\underline{A}_1} \underline{P}^n(\underline{y}, \underline{A}_2, \ldots, \underline{A}_{n+1})\, P^+(x, \phi^{-1}d\underline{y}) \\
&= \int_{\phi^{-1}\underline{A}_1} \underline{P}^n(\phi \underline{y}, \underline{A}_2, \ldots, \underline{A}_{n+1})\, P^+(x, dy) \\
&= \int_{\phi^{-1}\underline{A}_1} P^n(y, \phi^{-1}\underline{A}_2, \ldots, \phi^{-1}\underline{A}_{n+1})\, P^+(x, dy) \\
&= P^{n+1}(x, \phi^{-1}\underline{A}_1, \ldots, \phi^{-1}\underline{A}_{n+1}),
\end{aligned}
$$

where the second equality follows from condition (iii), the third equality follows from a "change-of-variable" result (Proposition 1.24 in the Appendix), and the fourth equality follows from the induction hypothesis. Using (2.11), condition (ii), and (2.13), we find that

$$
\begin{aligned}
P_\mu\{\underline{A}\} &= \int_{\underline{A}_0} \underline{P}^n(\underline{x}, \underline{A}_1, \ldots, \underline{A}_n)\, \mu(d\underline{x}) \\
&= \int_{\underline{A}_0} \underline{P}^n(\underline{x}, \underline{A}_1, \ldots, \underline{A}_n)\, \mu^+(\phi^{-1}d\underline{x}) \\
&= \int_{\phi^{-1}\underline{A}_0} \underline{P}^n(\phi \underline{x}, \underline{A}_1, \ldots, \underline{A}_n)\, \mu^+(d\underline{x}) \\
&= \int_{\phi^{-1}\underline{A}_0} P^n(x, \phi^{-1}\underline{A}_1, \ldots, \phi^{-1}\underline{A}_n)\, \mu^+(d\underline{x}) \\
&= P_\mu\{\phi^{-1}\underline{A}\}
\end{aligned}
$$

for any set $\underline{A} \subseteq \Sigma^\infty$ of the form (2.12) with each $\underline{A}_i$ a subset of $\Sigma - \underline{H}$. For a general set $\underline{A}$ of the form (2.12), the above argument and the inaccessibility assumption on $\underline{H}$ together imply that

$$P_\mu\{\underline{A}\} = P_\mu\{\underline{A} \cap \underline{B}\} = P_\mu\{\phi^{-1}(\underline{A} \cap \underline{B})\} = P_\mu\{\phi^{-1}\underline{A}\},$$

where $\underline{B} = (\Sigma - \underline{H})^n \times \Sigma^\infty$. Thus the processes $\{(\underline{S}_n, \underline{C}_n): n \geq 0\}$ and $\{\phi(S_n^+, C_n^+): n \geq 0\}$ have the same finite-dimensional distributions.

It remains to show that the processes $\{\underline{X}(t) \colon t \geq 0\}$ and $\{\lambda X(t) \colon t \geq 0\}$ have the same finite-dimensional distributions. For ease of exposition, suppose that $\{(\underline{S}_n, \underline{C}_n) \colon n \geq 0\}$ and $\{(S_n^+, C_n^+) \colon n \geq 0\}$ each has been defined using the standard construction for general state-space Markov chains discussed at the end of Section 3.1.1. Thus the underlying sample spaces of the chains are $\underline{\Sigma}^\infty$ and Σ^∞, respectively.[1] Set $\underline{\Gamma} = \underline{S} \times \Re_+$ and let $\underline{\Psi}$ be the mapping from $\underline{\Sigma}$ to $\underline{\Gamma}$ defined by $\underline{\Psi}(\underline{s}, \underline{c}) = (\underline{s}, \underline{t}^*(\underline{s}, \underline{c}))$. Also let $\underline{D}(\underline{S})$ be the set of possible sample paths of the process $\{\underline{X}(t) \colon t \geq 0\}$; that is, $\underline{D}(\underline{S})$ is the set of right-continuous piecewise-constant functions from $[0, \infty)$ to $\underline{S} \cup \{\underline{\Delta}\}$. Next define a mapping $\underline{\Phi}$ from $\underline{\Gamma}^\infty$ to $\underline{D}(\underline{S})$ as follows: for $\underline{g} = ((\underline{s}_0, \underline{t}_0), (\underline{s}_1, \underline{t}_1), \dots) \in \underline{\Gamma}^\infty$ and $t \geq 0$, set

$$n(\underline{g}, t) = \inf \{ n \geq 0 \colon t_0 + t_1 + \cdots + t_n > t \},$$

and then set $\underline{\Phi}\underline{g} = \underline{x}$, where $\underline{x}$ is the unique element of $\underline{D}(\underline{S})$ that satisfies

$$\underline{x}(t) = \begin{cases} \underline{s}_{n(\underline{g},t)} & \text{if } n(\underline{g}, t) < \infty; \\ \underline{\Delta} & \text{if } n(\underline{g}, t) = \infty. \end{cases}$$

It follows from these definitions that[2] $\underline{X}(t, \underline{\omega}) = (\underline{\Phi}\underline{\Psi}\underline{\omega})(t)$ for $\underline{\omega} \in \underline{\Sigma}^\infty$ and $t \geq 0$. Define sets $D(S)$ and Γ and mappings Ψ and Φ in a similar manner, and let θ be the mapping from Σ^∞ to $(\Sigma^+)^\infty$ defined by

$$\theta((s_0, c_0), (s_1, c_1), \dots) = ((s_{\gamma(0)}, c_{\gamma(0)}), (s_{\gamma(1)}, c_{\gamma(1)}), \dots),$$

where $\gamma(\cdot)$ is defined by (1.14) in Chapter 3. Observe that $X(t, \omega) = (\Phi\Psi\theta\omega)(t)$ for $\omega \in \Sigma^\infty$ and $t \geq 0$. To establish mimicry, it therefore suffices to show that

$$P_\mu \{ \underline{\Psi}^{-1}\underline{\Phi}^{-1}\underline{A} \} = P_\mu \{ \theta^{-1}\Psi^{-1}\Phi^{-1}\lambda^{-1}\underline{A} \} \tag{2.14}$$

for $\underline{A} \subseteq \underline{D}(\underline{S})$. To this end, set $\Lambda(s, t) = (\lambda s, t)$ for $(s, t) \in \Gamma$. Observe that by condition (i)

$$\underline{\Psi}\phi x = \Lambda\Psi x \tag{2.15}$$

for $x \in \Sigma$. Also observe that by definition

$$\Lambda^{-1}\underline{\Phi}^{-1}\underline{A} = \Phi^{-1}\lambda^{-1}\underline{A} \tag{2.16}$$

for $\underline{A} \subseteq \underline{D}(\underline{S})$. Since $\{(\underline{S}_n, \underline{C}_n) \colon n \geq 0\}$ and $\{\phi(S_n^+, C_n^+) \colon n \geq 0\}$ have the same finite-dimensional distributions, it also follows that

$$P_\mu \{ \underline{B} \} = P_\mu \{ \theta^{-1}\phi^{-1}\underline{B} \} \tag{2.17}$$

---

[1] When the foregoing chains are each defined on some probability space other than the standard one, the proof goes through almost exactly as described, except that an additional mapping comes into play for each chain, namely the mapping from an element of the sample space to the corresponding sample path of the chain.

[2] Recall here the notational conventions introduced just before Definition 2.9.

for $\underline{B} \subseteq \underline{\Sigma}^{\infty}$. Thus, for a set $\underline{A} \subseteq \underline{D}(\underline{S})$,

$$
\begin{aligned}
P_{\underline{\mu}} \left\{ \underline{\Psi}^{-1} \underline{\Phi}^{-1} \underline{A} \right\} &= P_{\underline{\mu}} \left\{ \theta^{-1} \phi^{-1} \underline{\Psi}^{-1} \underline{\Phi}^{-1} \underline{A} \right\} \\
&= P_{\underline{\mu}} \left\{ \theta^{-1} \Psi^{-1} \Lambda^{-1} \underline{\Phi}^{-1} \underline{A} \right\} \\
&= P_{\underline{\mu}} \left\{ \theta^{-1} \Psi^{-1} \Phi^{-1} \lambda^{-1} \underline{A} \right\}
\end{aligned}
$$

by (2.15)–(2.17), and (2.14) follows.   □

The quantities μ^+ and P^+ often can be computed in a straightforward manner from μ and P when verifying the conditions of Theorem 2.10. Moreover, it suffices to examine only sets $\underline{A}$ of the form

$$
\underline{A} = \{\underline{s}\} \times \{ (\underline{c}_1, \dots, \underline{c}_M) : 0 \le \underline{c}_i \le a_i \text{ for } 1 \le i \le M \}. \tag{2.18}
$$

EXAMPLE 2.19 (Producer–consumer system with nonpreemptive priority). As discussed in Example 2.1 in Chapter 2, the producer–consumer system with nonpreemptive priority can be modelled as an SPN—see Figure 2.4. This system can also be modelled within the GSMP framework. Specifically, set $\underline{X}(t) = (U_1(t), U_2(t), V(t))$, where $U_i(t)$ denotes the number of items awaiting transmission in buffer i at time t and

$$
V(t) = \begin{cases} i & \text{if transmission of an item to consumer } i \\ & \quad \text{is underway at time } t; \\ 0 & \text{if no transmission is underway at time } t. \end{cases}
$$

Formal specification of the process $\{\underline{X}(t) : t \ge 0\}$ is as a GSMP. The state space S is the set of all elements

$$
(u_1, u_2, v) \in \{0, 1, \dots, B_1\} \times \{0, 1, \dots, B_2\} \times \{0, 1, 2\}
$$

such that

1. $v > 0$ whenever $u_1 + u_2 > 0$.

2. $u_1 + 1_{\{1\}}(v) \le B_1$.

3. $u_2 + 1_{\{2\}}(v) \le B_2$.

The event set is $\underline{E} = \{\underline{e}_1, \underline{e}_2, \underline{e}_3\}$, where $\underline{e}_i =$ "creation of item by producer i" $(i = 1, 2)$ and $\underline{e}_3 =$ "end of transmission." For $\underline{s} = (u_1, u_2, v)$, $\underline{e}_i \in \underline{E}(\underline{s})$ $(i = 1, 2)$ if and only if $u_i + 1_{\{i\}}(v) < B_i$, and $\underline{e}_3 \in \underline{E}(\underline{s})$ if and only if $v > 0$.

The state-transition probabilities are as follows. If $\underline{e} = \underline{e}_1 =$ "creation of item by producer 1," then $\underline{p}(\underline{s}'; \underline{s}, \underline{e}) = 1$ when

$$
\underline{s} = (u_1, u_2, v) \text{ with } v > 0 \qquad \text{and} \qquad \underline{s}' = (u_1 + 1, u_2, v)
$$

and when

$$
\underline{s} = (0, 0, 0) \qquad \text{and} \qquad \underline{s}' = (0, 0, 1).
$$

If $\underline{e} = \underline{e}_2 = $ "creation of item by producer 2," then $\underline{p}(\underline{s}'; \underline{s}, \underline{e}) = 1$ when

$$\underline{s} = (u_1, u_2, v) \text{ with } v > 0 \qquad \text{and} \qquad \underline{s}' = (u_1, u_2 + 1, v)$$

and when

$$\underline{s} = (0, 0, 0) \qquad \text{and} \qquad \underline{s}' = (0, 0, 2).$$

If $\underline{e} = \underline{e}_3 = $ "end of transmission," then $\underline{p}(\underline{s}'; \underline{s}, \underline{e}) = 1$ when

$$\underline{s} = (u_1, u_2, v) \text{ with } u_1 > 0 \qquad \text{and} \qquad \underline{s}' = (u_1 - 1, u_2, 1),$$

when

$$\underline{s} = (0, u_2, v) \text{ with } u_2 > 0 \qquad \text{and} \qquad \underline{s}' = (0, u_2 - 1, 2),$$

and when

$$\underline{s} = (0, 0, v) \qquad \text{and} \qquad \underline{s}' = (0, 0, 0).$$

All other state-transition probabilities $\underline{p}(\underline{s}'; \underline{s}, \underline{e})$ are equal to 0. For $\underline{s}, \underline{s}' = (u_1', u_2', v') \in \underline{S}$ and $\underline{e} \in \underline{E}(\underline{s})$, the clock-setting distribution functions are $\underline{F}(x; \underline{s}', \underline{e}_1, \underline{s}, \underline{e}) = P\{A_1 \leq x\}$, $\underline{F}(x; \underline{s}', \underline{e}_2, \underline{s}, \underline{e}) = P\{A_2 \leq x\}$, and $\underline{F}(x; \underline{s}', \underline{e}_3, \underline{s}, \underline{e}) = P\{L_{v'} \leq x\}$. All speeds for active events are equal to 1.

We now establish conditions (i)–(iii) of Theorem 2.10 with

$$\underline{H} = \{(\underline{s}, \underline{c}) \in \underline{\Sigma} : |\underline{E}^*(\underline{s}, \underline{c})| > 1\}.$$

For $s = (s_1, \ldots, s_7) \in S$ and $c = (c_1, c_2, \ldots, c_6) \in C(s)$, set $\eta(s, c) = (\underline{c}_1, \underline{c}_2, \underline{c}_3)$, where $\underline{c}_1 = c_1$, $\underline{c}_2 = c_4$, and

$$\underline{c}_3 = \begin{cases} c_3 & \text{if } s_3 = 1; \\ c_6 & \text{if } s_6 = 1. \end{cases}$$

Also set $\lambda s = (u_1, u_2, v)$, where $u_1 = s_2$, $u_2 = s_5$ and

$$v = \begin{cases} 1 & \text{if } s_3 = 1; \\ 2 & \text{if } s_6 = 1; \\ 0 & \text{if } s_7 = 1. \end{cases}$$

Finally, set $\phi(s, c) = (\lambda s, \eta(s, c))$. Denote the initial distribution of the GSMP by $\underline{\mu}$ and set

$$\mu(A) = \underline{\mu}(\phi(A \cap \Sigma^+))$$

for $A \subseteq \Sigma$. To see that condition (i) holds, fix

$$s = (s_1, s_2, 1, s_4, s_5, 0, 0) \qquad \text{and} \qquad c = (c_1, 0, c_3, c_4, 0, 0), \qquad (2.20)$$

where $s_1 + s_2 = B_1 - 1$ and $s_4 + s_5 = B_2$. Then $\lambda s = (s_2, s_5, 1)$, $\eta(s, c) = (c_1, c_4, c_3)$, and

$$\underline{t}^*(\phi(s, c)) = \min(c_1, c_3, c_4) = t^*(s, c).$$

Similar computations establish condition (i) for every other pair $(s, c) \in \Sigma^+$. Now fix $\underline{A} \subseteq \underline{\Sigma}$. Observe that ϕ is a one-to-one mapping, so that there exists a unique subset $B \subseteq \Sigma^+$ such that $\underline{A} = \phi B$. It follows from the definition of μ that $\mu^+(B) = \mu(B) = \underline{\mu}(\phi B)$. Formal substitution of $B = \phi^{-1}\underline{A}$ into the latter expression yields $\underline{\mu}(\underline{A}) = \mu^+(\phi^{-1}\underline{A})$. Since $\underline{A}$ is arbitrary, this establishes condition (ii). Finally, fix s and c as in (2.20) with $s_2 \geq 1$ and $c_3 < \min(c_1, c_4)$, and observe that $(s, c) \in \Sigma^+ - \phi^{-1}\underline{H}$. Set

$$\underline{A} = \{ (s_2 - 1, s_5, 1) \} \times \{ (\underline{c}_1', \underline{c}_2', \underline{c}_3') : 0 \leq \underline{c}_i' \leq a_i \text{ for } 1 \leq i \leq 3 \}.$$

Then

$$\underline{P}(\phi(s, c), \underline{A}) = 1_{[0, a_1]}(c_1 - c_3) 1_{[0, a_2]}(c_4 - c_3) P\{ L_1 \leq a_3 \}.$$

On the other hand,

$$\phi^{-1}\underline{A} = \{ (s_1 + 1, s_2 - 1, 1, s_4, s_5, 0, 0) \}$$
$$\times \{ (c_1', 0, c_3', c_4', 0, 0) : 0 \leq c_1' \leq a_1, \ 0 \leq c_3' \leq a_3, \text{ and } 0 \leq c_4' \leq a_2 \}$$

and

$$P^+\big((s, c), \phi^{-1}\underline{A}\big) = 1_{[0, a_1]}(c_1 - c_3) 1_{[0, a_2]}(c_4 - c_3) P\{ L_1 \leq a_3 \}$$
$$= \underline{P}(\phi(s, c), A).$$

Similar computations establish condition (iii) for every other pair $(s, c) \in \Sigma^+ - \phi^{-1}\underline{H}$ and set $\underline{A} \subset \underline{\Sigma}$ of the form (2.18). Thus the marking process of the SPN strongly mimics the GSMP.

We conclude this section by giving a corollary to Theorem 2.10 that is applicable to SPNs with no immediate transitions. Although the scope of this result is somewhat limited, the conditions on the building blocks of the SPN and GSMP are relatively easy to check.[3]

Corollary 2.21. *Suppose that $E' = \varnothing$. Also suppose that there exist a mapping λ from S to $\underline{S}$ and a one-to-one mapping ψ from E to $\underline{E}$ such that*

(i) $\underline{E}(\lambda s) = \psi E(s)$,

(ii) $\underline{p}(\lambda s'; \lambda s, \psi E^*) = p(s'; s, E^*)$,

(iii) $\underline{F}(\cdot ; \lambda s', \psi e', \lambda s, \psi E^*) = F(\cdot ; s', e', s, E^*)$,

---

[3]In the presence of immediate transitions, it appears difficult to state simple corollaries to Theorem 2.10 that involve direct conditions on the building blocks. This difficulty arises from the fact that clocks can be set at marking changes for which either the old or new marking is immediate.

(iv) $\underline{r}(\lambda s, \psi e) = r(s, e),$

(v) $\underline{F}_0(\,\cdot\,; \psi e, \lambda s) = F_0(\,\cdot\,; e, s),$ *and*

(vi) $\underline{\nu}_0(\lambda s) = \nu_0(s)$

for all s, s', e, e', and E^*. Then $\{\,X(t): t \geq 0\,\}$ strongly mimics $\{\,\underline{X}(t): t \geq 0\,\}$.

PROOF. Write $E = \{\,e_1, e_2, \ldots, e_M\,\}$ and $\underline{E} = \{\,\underline{e}_1, \underline{e}_2, \ldots, \underline{e}_M\,\}$, and assume without loss of generality that $\psi e_i = \underline{e}_i$ for $1 \leq i \leq M$. For $s \in S$ and $c = (c_1, c_2, \ldots, c_M) \in C(s)$, set $\eta(s, c) \equiv \eta(c) = c$ and $\phi(s, c) = (\lambda s, \eta(s, c))$. Then, for example, we have

$$
\begin{aligned}
t^*(s, c) &= \min_{\{\,i:\, e_i \in E(s)\,\}} c_i / r(s, e_i) \\
&= \min_{\{\,i:\, e_i \in E(s)\,\}} c_i / \underline{r}(\lambda s, \psi e_i) \\
&= \min_{\{\,i:\, \underline{e}_i \in \underline{E}(\lambda s)\,\}} c_i / \underline{r}(\lambda s, \underline{e}_i) \\
&= \underline{t}^*(\phi(s, c))
\end{aligned}
$$

for $(s, c) \in \Sigma^+$, and condition (i) of Theorem 2.10 holds. Similar arguments then establish the remaining conditions of Theorem 2.10. □

Remark 2.22. Observe that if with probability 1 the transitions in a set E^* never fire simultaneously when the marking is s and similarly for the set ψE^* and state λs, then the conclusion of Corollary 2.21 holds even if conditions (ii) and (iii) fail to hold for s and E^*. Similarly, if $\nu_0(s) = 0$, then the conclusion of Corollary 2.21 holds even if condition (v) fails to hold for marking s and a transition $e \in E(s)$. Such "strengthenings" of Corollary 2.21 are directly analogous to the use of the inaccessible set $\underline{H}$ in Theorem 2.10 and are applied throughout without further comment.

4.3 Mimicry Theorems for Marking Processes

In this section we show that SPNs have at least the modelling power of GSMPs. We start by providing some modelling-power results for several restricted classes of GSMPs. As might be expected, each of these classes can be mimicked by a correspondingly restricted class of SPNs. Our first result concerns GSMPs with a finite state space in which the current state and trigger event set uniquely determine the next state. Any such GSMP can be mimicked by the marking process of a 1-bounded SPN with deterministic timed and immediate transitions. Our next result asserts that for any GSMP with a finite state space there exists a 2-bounded SPN having a marking process that strongly mimics the GSMP; if events are never cancelled,

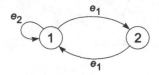

Figure 4.2. State-transition diagram for two-state GSMP with $E(1) = \{e_1, e_2\}$ and $E(2) = \{e_1\}$.

no immediate transitions are required. Finally, we show in Theorem 3.4 that for any GSMP having a countably infinite state space there exists a 2-bounded SPN having a marking process that strongly mimics the GSMP. Each of these results is proved in the same way: we use the building blocks of the GSMP to construct a canonical SPN and then display mappings that satisfy the conditions of Theorem 2.10 or Corollary 2.21.

4.3.1   Finite-State Processes

Theorem 3.1. *Suppose that all nonzero state-transition probabilities of a* GSMP *with finite state space are equal to 1. Then there exists a 1-bounded* SPN *with deterministic transitions having a marking process that strongly mimics the* GSMP.

PROOF. Without loss of generality, suppose that the state space of the GSMP is $\underline{S} = \{1, 2, \ldots, K\}$ and the event set is $\underline{E} = \{\underline{e}_1, \underline{e}_2, \ldots, \underline{e}_M\}$. Denote the state-transition probabilities by $\underline{p}(\underline{s}'; \underline{s}, \underline{E}^*)$, the set of active events in state $\underline{s}$ by $\underline{E}(\underline{s})$, and so forth.

As mentioned above, the idea is to construct a canonical SPN and then show that the marking process of this SPN mimics the GSMP. We illustrate the basic ideas that underlie the canonical SPN construction by means of a simple example. Consider a GSMP with state space $\underline{S} = \{1, 2\}$, event set $\underline{E} = \{\underline{e}_1, \underline{e}_2\}$, and active event sets given by $\underline{E}(1) = \{e_1, e_2\}$ and $\underline{E}(2) = \{e_1\}$. The state-transition probabilities are given by $\underline{p}(2; 1, \underline{e}_1) = \underline{p}(1; 1, \underline{e}_2) = \underline{p}(1; 2, \underline{e}_1) = 1$; see Figure 4.2. Each event is "simple" in that the clock-setting distribution for the event does not depend explicitly on the old state, new state, or set of trigger events. All speeds for active events are equal to 1. The canonical SPN for this GSMP is displayed in Figure 4.3. Place d_j $(j = 1, 2)$ contains a token if and only if the current state of the GSMP is j, and place $d_{1,i}$ contains a token $(i = 1, 2)$ if and only if event $\underline{e}_i$ (of the GSMP) is currently active. There is a token in place $d_{2,i}$ if and only if event $\underline{e}_i$ has just occurred, and there is a token in place $d_{3,i}$ if and only if active event $\underline{e}_i$ is to be cancelled. All the transitions are deterministic. When there is a token in place d_1, for example, and transition $e_{1,1}$ fires (i.e., the GSMP is in state 1 and event $\underline{e}_1$ occurs), a token is deposited in place $d_{2,1}$, and exactly one of the immediate transitions of the form $e_{i,j,k}$ becomes enabled, namely $e_{1,1,2}$. When $e_{1,1,2}$ fires, it removes a token from

$e_{1,i}$ = event $\underline{e}_i$ (of the GSMP) triggers a state transition

$e_{2,i}$ = event $\underline{e}_i$ (of the GSMP) is cancelled

$e_{i,j,k}$ = the GSMP makes a transition from state j to k when event $\underline{e}_i$ occurs

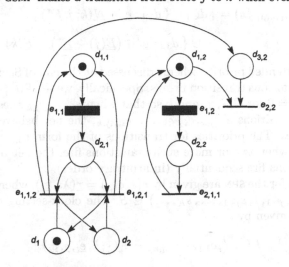

Figure 4.3. SPN representation of two-state GSMP.

place d_1 and deposits a token in place d_2; such a firing corresponds to a transition of the GSMP from state 1 to state 2. Moreover, $e_{1,1,2}$ deposits tokens in places $d_{1,1}$ and $d_{3,2}$ when it fires, so that

1. Transition $e_{1,1}$—which corresponds to event $\underline{e}_1$ of the GSMP—becomes enabled.

2. Immediate transition $e_{2,2}$ becomes enabled and fires, causing transition $e_{1,2}$—which corresponds to event $\underline{e}_2$ of the GSMP—to become disabled.

Thus the transitions become enabled or disabled in accordance with the event-scheduling mechanism of the GSMP. The clock-setting distribution and speeds for transition $e_{1,i}$ ($i = 1, 2$) are the same as the clock-setting distribution and speeds for event $\underline{e}_i$ in the GSMP.

For a general GSMP, the canonical SPN is constructed along similar lines. The SPN has a place d_j for each state j of the GSMP and a transition $e_{1,i}$ for each event $\underline{e}_i$. If the GSMP makes a transition from state j to k when event $\underline{e}_i$ occurs, then the canonical SPN contains a deterministic transition $e_{i,j,k}$. If the events in a set $\underline{E}^* = \{ \underline{e}_{i_1}, \underline{e}_{i_2}, \ldots, \underline{e}_{i_l} \}$ can occur simultaneously in the GSMP and trigger a transition from state j to k, then the SPN contains an immediate transition denoted $e_{i_1,\ldots,i_l,j,k}$. The set of normal input places is

$$I(e_{i_1,\ldots,i_l,j,k}) = \{\, d_j \,\} \cup \{\, d_{2,i_1}, d_{2,i_2}, \ldots, d_{2,i_l} \,\},$$

and the set of output places is

$$J(e_{i_1,\ldots,i_l,j,k}) = \{\, d_k \,\} \cup \{\, d_{1,l} \colon \underline{e}_l \in \underline{N}(k; j, \underline{E}^*) \,\}$$

$$\cup \{\, d_{3,l} \colon \underline{e}_l \in \big(\underline{E}(j) - \underline{E}^*\big) - \underline{E}(k) \,\}.$$

We use the extended priority scheme discussed at the end of Section 2.3.2 to handle simultaneous transition firings. Specifically, we set $\mathcal{P}(e_{i_1,\ldots,i_l,j,k}) = 1$ and $\mathcal{P}(e_{i,j,k}) = 0$ for all i, j, and k so that when $e_{i_1,\ldots,i_l,j,k}$ fires simultaneously with transitions $e_{i_1,j,k}, e_{i_2,j,k}, \ldots, e_{i_m,j,k}$, the net behaves as if only $e_{i_1,\ldots,i_l,j,k}$ fires. The priorities for transitions of the form $e_{2,i}$ are all equal to 0, so that when two or more such transitions fire, the net behaves as if these transitions fire sequentially (in arbitrary order).

The speeds for the SPN are given by $r(s, e_{1,i}) = \underline{r}(\lambda s, \underline{e}_i)$, where $\lambda s = j$ for $s = (s_1, \ldots, s_{j-1}, 1, s_{j+1}, \ldots, s_K, \ldots) \in S$. The clock-setting distribution functions are given by

$$F(\,\cdot\,; s', e_{1,l}, s, e_{i,j,k}) = \underline{F}(\,\cdot\,; k, \underline{e}_l, j, \underline{e}_i).$$

Set $\phi(s, c) = \big(\lambda s, \eta(s, c)\big)$ for $(s, c) \in \Sigma^+$, where $\eta(s, c) = (c_{1,1}, c_{2,1}, \ldots, c_{M,1})$ for $s \in S$ and $c = (c_{1,1}, c_{1,2}, \ldots, c_{1,M}, \ldots) \in C(s)$. Finally, set $\mu(A) = \underline{\mu}\big(\phi(A \cap \Sigma^+)\big)$ for $A \subseteq \Sigma$. Tedious but straightforward calculations show that the mapping ϕ satisfies the conditions of Theorem 2.10. Thus the marking process of the above SPN strongly mimics the GSMP. □

Remark 3.2. The canonical SPN constructed in the proof of Theorem 3.1 can be used with relatively minor modifications to prove the assertion in Section 2.4 that for any GSMP with a finite state space, unit speeds, and a fixed initial state, there exists a restricted SPN with a marking process that strongly mimics the GSMP. The primary changes in the canonical SPN are that

- For each event in the GSMP there are, in general, several corresponding timed transitions in the SPN, one for each of the distinct distribution functions used in the GSMP to set the clock for the event.

- The SPN contains a deterministic immediate transition of the form $e_{i,j,k}$ for each i, j, and k such that $\underline{p}(k; j, \underline{e}_i) > 0$, and similarly for transitions of the form $e_{i_1,\ldots,i_l,j,k}$.

If, with probability 1, events in the GSMP never occur simultaneously, then the canonical SPN is deterministic in the sense of Section 2.4.

Theorem 3.3 concerns GSMPs with finite state space and arbitrary state-transition probabilities.

$e_i =$ event $\underline{e}_i$ (of the GSMP) triggers a state transition

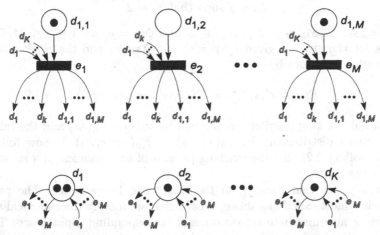

Figure 4.4. SPN representation of GSMP with finite state space and no cancelled events.

Theorem 3.3. *For any* GSMP *with finite state space there exists a 2-bounded* SPN *with random inputs and outputs having a marking process that strongly mimics the* GSMP. *If active events are never cancelled, no immediate transitions are required.*

PROOF. Consider an arbitrary but fixed GSMP and, as in the proof of Theorem 3.1, suppose that the state space of the GSMP is of the form $\underline{S} = \{\,1, 2, \ldots, K\,\}$ and the event set is $\underline{E} = \{\,\underline{e}_1, \underline{e}_2, \ldots, \underline{e}_M\,\}$. First suppose that $\big(\underline{E}(\underline{s}) - \underline{E}^*\big) - \underline{E}(\underline{s}') = \varnothing$ for all $\underline{s}'$, $\underline{s}$, and $\underline{E}^*$, so that active events are never cancelled. Construct a canonical SPN with finite state space as in Figure 4.4. Place d_j contains two tokens if and only if the GSMP is in state j; otherwise, place d_j contains one token. Place $d_{1,j}$ contains one token if and only if event $\underline{e}_j$ is active; otherwise, place $d_{1,j}$ contains no tokens. Whenever place d_j contains two tokens and transition e_i fires, one token is removed from each of places d_j and $d_{1,i}$. Moreover, one token is deposited in exactly one of places $d_1, d_2, \ldots, d_K$; the probability that the token is deposited in place d_k $(1 \le k \le K)$ is $\underline{p}(k; j, \underline{e}_i)$. Finally, given that a token is deposited in place d_k, tokens are deposited in places $d_{1,i_1}, d_{1,i_2}, \ldots, d_{1,i_l}$, where the indices $i_1, i_2, \ldots, i_l$ are such that[4] $\underline{N}(k; j, \underline{e}_i) = \{\,\underline{e}_{i_1}, \underline{e}_{i_2}, \ldots, \underline{e}_{i_l}\,\}$. Similar marking changes occur when two or more transitions fire simultaneously. Formally,

$$p(s'; s, E^*) = \underline{p}(\lambda s'; \lambda s, \psi E^*),$$

---

[4]Recall that $\underline{N}(k; j, \underline{e}_i)$ is the set of new events for the GSMP when $\underline{e}_i$ triggers a transition from state j to k.

where

$$\lambda s = j \text{ such that } s_j = 2$$

for $s = (s_1, \ldots, s_K, s_{1,1}, \ldots, s_{M,1}) \in S$ and $\psi e_i = \underline{e}_i$ for $1 \leq i \leq M$. The speeds for the SPN are given by $r(s, e) = \underline{r}(\lambda s, \psi e)$ and the clock-setting distribution functions by

$$F(\cdot; s', e', s, E^*) = \underline{F}(\cdot; \lambda s', \psi e', \lambda s, \psi E^*).$$

The initial-marking distribution is given by $\nu_0(s) = \underline{\nu}_0(\lambda s)$ and the initial clock-setting distributions by $F_0(\cdot; e, s) = \underline{F}_0(\cdot; \psi e, \lambda s)$. It now follows from Corollary 2.21 that the marking process of the canonical SPN strongly mimics the GSMP.

Now suppose that event $\underline{e}_i$ of the GSMP can be cancelled. The proof proceeds almost exactly as above, except that we modify the canonical SPN by adding an immediate transition and corresponding input place. This new transition and new place are used to mimic the cancellation of events in the same manner as transition $e_{2,2}$ and place $d_{3,2}$ are used in Figure 4.3.

□

4.3.2 Countable-State Processes

We now give a mimicry result for GSMPs with a countably infinite state space.

Theorem 3.4. *For any GSMP with a countably infinite state space there exists an SPN with random inputs and outputs, timed transitions, and immediate transitions having a marking process that strongly mimics the GSMP. No inhibitor input places are required.*

PROOF. Consider a GSMP with state space $\underline{S} = \{1, 2, \ldots\}$ and event set $\underline{E} = \{\underline{e}_1, \underline{e}_2, \ldots, \underline{e}_M\}$. First suppose that, with probability 1, events in the GSMP never occur simultaneously. Construct a canonical SPN consisting of a place d_0 and M identical subnets—one subnet for each event in the GSMP. Figure 4.5 displays place d_0 and the subnet corresponding to a generic GSMP event $\underline{e}_i$. For ease of exposition, we first display a canonical SPN that has inhibitor input places and then show how to modify the SPN to contain only normal input places.

Place d_0 contains $\underline{s}$ tokens if and only if the GSMP is in state $\underline{s}$. Place $d_{0,i}$ contains one token if and only if event $\underline{e}_i$ of the GSMP is active; otherwise, place $d_{0,i}$ contains no tokens.

Suppose that place d_0 contains $\underline{s}$ tokens and transition $e_{0,i}$ fires; this scenario corresponds to the occurrence of event $\underline{e}_i$ in state $\underline{s}$. Then either transition $e_{3,i}$ fires a random number of times in succession before becoming disabled—resulting in a random number of tokens being deposited in

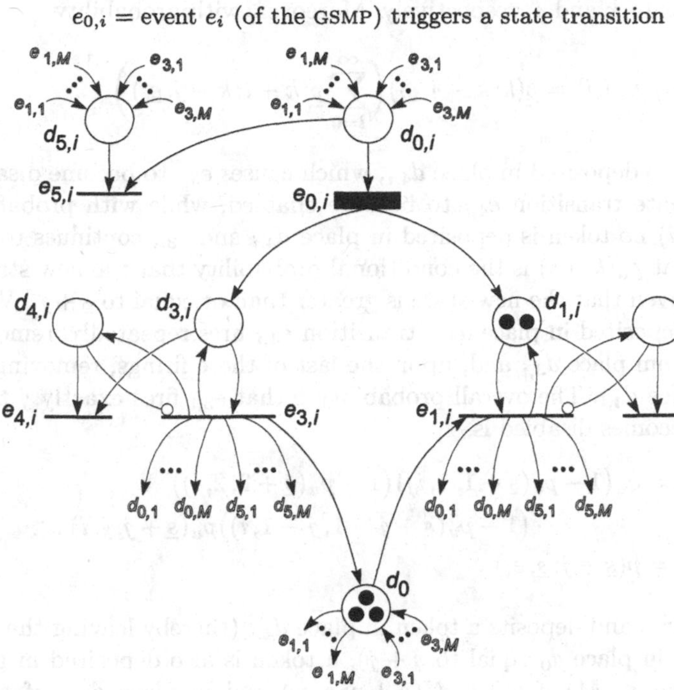

$e_{0,i}$ = event e_i (of the GSMP) triggers a state transition

Figure 4.5. SPN representation of GSMP with countably infinite state space.

place d_0—or transition $e_{1,i}$ fires a random number of times in succession—resulting in a random number of tokens being removed from place d_0. The mechanism by which either transition $e_{3,i}$ or $e_{1,i}$ fires is essentially the same as in the SPN model of a queue with batch arrivals given in Section 2.2.2; the probability that place d_0 contains $\underline{s} + l$ tokens after the assorted immediate transitions stop firing is $\underline{p}(\underline{s} + l; \underline{s}, \underline{e}_i)$, where $-(\underline{s} - 1) \leq l < \infty$. Moreover, similarly to the previous canonical SPNs, tokens are deposited in places of the form $d_{0,j}$ or $d_{5,j}$ so that transitions of the form $e_{0,j}$ become enabled or disabled in accordance with the event-scheduling mechanism of the GSMP.

In more detail, transition $e_{0,i}$ deposits a token in either place $d_{3,i}$ or place $d_{1,i}$ when it fires; the token is deposited in place $d_{3,i}$ with probability $q_u = \sum_{j=1}^{\infty} \underline{p}(\underline{s} + j; \underline{s}, \underline{e}_i)$ and in place $d_{1,i}$ with probability $1 - q_u$. (Here q_u is the probability that the new state $\underline{s}'$ satisfies $\underline{s}' > \underline{s}$.) If the token is deposited in place $d_{3,i}$, then immediate transition $e_{3,i}$ fires a random number of times, in the following manner. Whenever $e_{3,i}$ fires with $k - 1$ tokens in place d_0 and $j - 1$ tokens in place $d_{3,i}$ ($k \geq \underline{s} + 1$ and $j \geq 1$), one token is deposited in each of places d_0 and $d_{3,i}$, bringing the respective token counts to k and j, respectively. Moreover, with probability

$$p_u(k,j,i) = \underline{p}(k; k-j, \underline{e}_i)\left(\sum_{l=0}^{\infty} \underline{p}(k+l; k-j, \underline{e}_i)\right)^{-1},$$

a token is also deposited in place $d_{4,i}$, which causes $e_{3,i}$ to become disabled and immediate transition $e_{4,i}$ to become enabled, while with probability $1 - p_u(k,j,i)$ no token is deposited in place $d_{4,i}$ and $e_{3,i}$ continues to fire. [Observe that $p_u(k,j,i)$ is the conditional probability that the new state is $k = \underline{s} + j$, given that the new state is greater than or equal to $\underline{s} + j$.] When a token is deposited in place $d_{4,i}$, transition $e_{4,i}$ fires repeatedly, removing all tokens from place $d_{3,i}$ and, upon the last of these firings, removing the token in place $d_{4,i}$. The overall probability p that $e_{3,i}$ fires exactly j times and then becomes disabled is

$$\begin{aligned} p &= q_u\big(1 - p_u(\underline{s}+1, 1, i)\big)\big(1 - p_u(\underline{s}+2, 2, i)\big) \\ &\quad \cdots \big(1 - p_u(\underline{s}+j-1, j-1, i)\big)p_u(\underline{s}+j, j, i) \\ &= \underline{p}(\underline{s}+j; \underline{s}, e_i). \end{aligned}$$

When $e_{3,i}$ fires and deposits a token in place $d_{4,i}$ (thereby leaving the final token count in place d_0 equal to $\underline{s}+j$), a token is also deposited in place $d_{0,m}$ $(1 \le m \le M)$ if $\underline{e}_m \in \underline{N}(\underline{s}+j; \underline{s}, \underline{e}_i)$ and in place $d_{5,m}$ if $\underline{e}_m \in \big(\underline{E}(\underline{s}) - \{\underline{e}_i\}\big) - \underline{E}(\underline{s}+j)$. The clock for each newly enabled transition $e_{0,m}$ is set according to the distribution function $\underline{F}(\cdot; \underline{s}+j, \underline{e}_m, \underline{s}, \underline{e}_i)$. Thus the SPN emulates the event-scheduling mechanism of the GSMP at a transition from state $\underline{s}$ to $\underline{s}+j$. Observe that the sole purpose of place $d_{3,i}$ is to keep count of the number of times that transition $e_{3,i}$ has fired, allowing the SPN to "remember" that the initial token count in place d_0 was $\underline{s}$. Transitions $e_{1,i}$ and $e_{2,i}$ fire in an analogous manner, changing the token count in place d_0 from $\underline{s}$ to $\underline{s} - j$ with probability $\underline{p}(\underline{s}-j; \underline{s}, \underline{e}_i)$.

The speeds for the canonical SPN are given by $r(s,e) = \underline{r}(\lambda s, \psi e)$, where

$$\lambda s = s_0$$

for $s = (s_0, s_{0,1}, \ldots, s_{5,1}, \ldots, s_{0,M} \cdots, s_{5,M}) \in S$ and $\psi e_{0,i} = \underline{e}_i$ for $1 \le i \le M$. For $s \in S$ and $c = (c_{0,1}, \ldots, c_{5,1}, \ldots, c_{0,M}, \ldots, c_{5,M}) \in C(s)$, set

$$\eta(s,c) = (c_{0,1}, c_{0,2}, \ldots, c_{0,M}).$$

Finally, the initial distribution is given by $\mu(A) = \underline{\mu}(\phi A)$ for $A \subseteq \Sigma$, where $\phi(s,c) = \big(\lambda s, \eta(s,c)\big)$ for $(s,c) \in \Sigma$. A straightforward argument shows that the conditions of Theorem 2.10 hold, so that the marking process of the canonical SPN strongly mimics the GSMP.

Now suppose that two or more events of the GSMP can occur simultaneously. The proof proceeds almost exactly as above, except that we modify

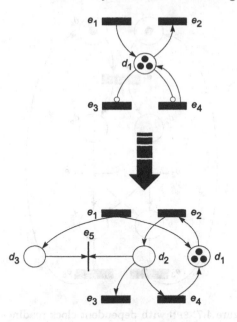

Figure 4.6. SPN representation with no inhibitor inputs.

the canonical SPN by adding additional subnets, each of which corresponds to a set $\underline{E}^*$ of events that can occur simultaneously.

The inhibitor input places used in the construction of the canonical SPN are convenient, but not essential. An SPN always can be modified so that (1) the modified SPN has no inhibitor input places and (2) the marking process of the modified SPN strongly mimics the marking process of the original SPN in a sense analogous to Definition 2.7. This modification depends critically on the use of random outputs and immediate transitions and is illustrated using the subnet in the top portion of Figure 4.6. This subnet captures the various possible relationships between a place and a transition. To eliminate the need for inhibitor input places, modify the subnet by adding two places d_2 and d_3 and a deterministic immediate transition e_5 as in the bottom portion of Figure 4.6. The idea is to modify the subnet so that place d_2 contains one token if and only if place d_1 contains no tokens and contains no tokens only if place d_1 contains at least one token. To this end, we change the new-marking probabilities so that the transitions behave as follows. Whenever place d_1 contains only one token and transition e_2 fires and removes this token, e_2 also deposits a token in place d_2. Whenever there is one token in place d_2, no tokens in place d_1, and transition e_4 fires and deposits a token in d_1, transition e_4 also removes the token in d_2. If, rather than e_4, transition e_1 fires and deposits a token in d_1, then e_1 also deposits a token in place d_3, which causes immediate transition e_5 to fire

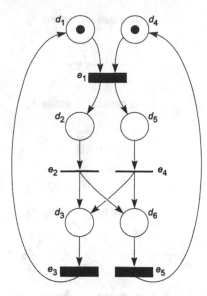

Figure 4.7. SPN with dependent clock readings.

and remove the token in d_2. Otherwise, transitions e_1, e_2, e_3, and e_4 remove
and deposit tokens as in the original SPN.   □

4.4   Converse Results

Because SPNs may have immediate transitions, the marking process of an
SPN need not behave like a GSMP. Consider, for example, the SPN model of
the particle counter from Example 2.11 in Chapter 2—see Figure 2.15. Recall that when the marking process makes a state transition from $(1,1,0,0)$
to $(1,1,0,0)$ triggered by the firing of transition e_1, the clock for transition
e_2 appears to be reset. Such resetting is not allowed in the GSMP framework. There also exist SPNs in which the clock readings just after a specified
marking change are conditionally dependent given the partial history of the
embedded chain up to the marking change.[5] As shown by Lemma 1.2, such
dependence cannot occur in GSMPs.

EXAMPLE 4.1 (SPN with dependent clock readings).   Consider the SPN
displayed in Figure 4.7. The set of immediate markings is $S' = \{(0,1,0,0,0, 0), (0,0,0,0,1,0)\}$ and the set of timed markings is the set of all elements

---

[5]In analogy to the partial history $\mathcal{F}_n$ of the underlying chain—see Section 3.4.2—we
define the partial history $\mathcal{F}_n^+$ of the embedded chain by setting $\mathcal{F}_0^+ = \{S_0^+\}$ and $\mathcal{F}_n^+ = \{S_0^+, E_0^+, t_0^+, S_1^+, E_1^+, t_1^+, \ldots, S_{n-1}^+, E_{n-1}^+, t_{n-1}^+, S_n^+\}$ for $n \geq 1$, where $t_n^+ = t^*(S_n^+, C_n^+)$
and $E_n^+ = E^*(S_n^+, C_n^+)$ for $n \geq 0$.

$(s_1, s_2, \ldots, s_6) \in \{0, 1\}^6$ such that $s_2 = s_5 = 0$, $s_1 + s_3 = 1$, and $s_4 + s_6 = 1$. All transitions except e_1 are deterministic. Whenever transition e_1 fires, it removes one token from each of places d_1 and d_4 and deposits one token in either place d_2 or d_5; the token is deposited in place d_2 with probability $1/2$ and in place d_5 with probability $1/2$. The clock-setting distributions for transitions e_3 and e_5 are given by

$$F(x; s', e_3, s, e^*) \equiv F(x; e_3, e^*) = \begin{cases} 1_{[1,\infty)}(x) & \text{if } e^* = e_2; \\ 1_{[2,\infty)}(x) & \text{if } e^* = e_4 \end{cases}$$

and $F(x; s', e_5, s, e^*) \equiv F(x; e_5, e^*) = F(x; e_3, e^*)$. All speeds for enabled transitions are equal to 1. Suppose that the initial marking is $(1, 0, 0, 1, 0, 0)$ and let γ be the random index of the first marking change at which the new marking is $(0, 0, 1, 0, 0, 1)$ Observe that, for example,

$$P\{ C_{\gamma,3} = 2, C_{\gamma,5} = 2 \mid \mathcal{F}_\gamma^+ \} = 1/2$$

but

$$P\{ C_{\gamma,3} = 2 \mid \mathcal{F}_\gamma^+ \} P\{ C_{\gamma,5} = 2 \mid \mathcal{F}_\gamma^+ \} = 1/4.$$

That is, the clock readings for transitions e_3 and e_5 just after the γth marking change are not conditionally independent given $\mathcal{F}_\gamma^+$. It follows that the marking process cannot be a GSMP, as this would violate Lemma 1.2.

In light of the foregoing examples, one might conjecture that there exist SPNs that cannot be mimicked by GSMPs (in a sense analogous to mimicry of GSMPs by SPNs). In this section we show that, to the contrary, for any SPN with timed and immediate transitions, there exists a GSMP that strongly mimics the marking process of the SPN. It then follows from this result and the results in Section 4.3 that SPNs and GSMPs have the same modelling power.

The definition of strong mimicry by a GSMP of the marking process of an SPN is analogous to Definition 2.7. As before, let $\{\underline{X}(t): t \geq 0\}$ be a GSMP with state space $\underline{S}$ and underlying chain $\{(\underline{S}_n, \underline{C}_n): n \geq 0\}$, and let $\{X(t): t \geq 0\}$ be a marking process of an SPN with timed marking set S and underlying chain $\{(S_n, C_n): n \geq 0\}$.

Definition 4.2. The GSMP $\{\underline{X}(t): t \geq 0\}$ is said to *strongly mimic* the marking process $\{X(t): t \geq 0\}$ if

(i) there exists a mapping λ from $\underline{S}$ onto S such that $\{X(t): t \geq 0\}$ and $\{\lambda \underline{X}(t): t \geq 0\}$ have the same finite-dimensional distributions, and

(ii) there exists a mapping ϕ from $\underline{\Sigma}$ onto Σ^+ of the form $\phi(\underline{s}, \underline{c}) = (\lambda \underline{s}, \eta(\underline{s}, \underline{c}))$ such that the discrete-time processes $\{(S_n^+, C_n^+): n \geq 0\}$ and $\{\phi(\underline{S}_n, \underline{C}_n): n \geq 0\}$ have the same finite-dimensional distributions.

To prove our main result, we use the building blocks of the SPN to construct a canonical GSMP that strongly mimics the marking process. The state of the GSMP consists essentially of a timed marking along with a representation of how the clock associated with each timed transition was set since the last timed marking. The events of the GSMP correspond to the timed transitions. If, moreover, enabled transitions of the SPN can become disabled and then enabled again during a sojourn in the set of immediate markings (resulting in an apparent "resetting" of the corresponding clocks), then the canonical GSMP requires additional events and further augmentation of the state space. The following examples illustrate these ideas and motivate our general construction of the canonical GSMP.

EXAMPLE 4.3 (Particle counter). Using the building blocks of the SPN shown in Figure 2.15, construct a GSMP with state space

$$\underline{S} = \{\, (1,0,0,0,0), (1,1,0,0,1), (1,1,0,0,2) \,\}$$

and event set

$$\underline{E} = \{\, \underline{e}_1, \underline{e}_{2,1}, \underline{e}_{2,2} \,\}.$$

Observe that each state is of the form $\underline{s} = (s, u)$, where s is a timed marking of the SPN and $u \in \{\, 0, 1, 2 \,\}$. The idea is that events $\underline{e}_{2,1}$ and $\underline{e}_{2,2}$ correspond to transition e_2 and at most one of these events is active at any time. Whenever the clock for transition e_2 is "reset," event $\underline{e}_{2,i}$ is cancelled and event $\underline{e}_{2,3-i}$ becomes active, where $i = 1$ or 2. The state of the GSMP consists of the marking s of the SPN along with a component u that keeps track of whether $\underline{e}_{2,1}$ or $\underline{e}_{2,2}$ is currently active. Some details of the construction are as follows.

For $\underline{s} = (s, u) \in \underline{S}$,

$$\underline{e}_{2,1} \in \underline{E}(\underline{s}) \text{ if and only if } e_2 \in E(s) \text{ and } u = 1$$

and

$$\underline{e}_{2,2} \in \underline{E}(\underline{s}) \text{ if and only if } e_2 \in E(s) \text{ and } u = 2.$$

All speeds $\underline{r}(\underline{s}, \underline{e})$ for active events are equal to 1.

If $\underline{e}^* = \underline{e}_1$, then the state-transition probability $\underline{p}(\underline{s}'; \underline{s}, \underline{e}^*) = 1$ when

$$\underline{s} = (1,0,0,0,0) \quad \text{and} \quad \underline{s}' = (1,1,0,0,1),$$

when

$$\underline{s} = (1,1,0,0,1) \quad \text{and} \quad \underline{s}' = (1,1,0,0,2),$$

and when

$$\underline{s} = (1,1,0,0,2) \quad \text{and} \quad \underline{s}' = (1,1,0,0,1).$$

If $\underline{e}^* = \underline{e}_{2,1}$, then $\underline{p}(\underline{s}'; \underline{s}, \underline{e}^*) = 1$ when

$$\underline{s} = (1,1,0,0,1) \quad \text{and} \quad \underline{s}' = (1,0,0,0,0).$$

If $\underline{e}^* = \underline{e}_{2,2}$, then $\underline{p}(\underline{s}'; \underline{s}, \underline{e}^*) = 1$ when

$$\underline{s} = (1, 1, 0, 0, 2) \qquad \text{and} \qquad \underline{s}' = (1, 0, 0, 0, 0).$$

All other state-transition probabilities $\underline{p}(\underline{s}'; \underline{s}, \underline{e})$ are equal to 0. The clock-setting distribution functions are given by $\underline{F}(x; \underline{s}', \underline{e}_1, \underline{s}, \underline{e}^*) = P\{U \leq x\}$ and $\underline{F}(x; \underline{s}', \underline{e}_{2,1}, \underline{s}, \underline{e}^*) = \underline{F}(x; \underline{s}', \underline{e}_{2,2}, \underline{s}, \underline{e}^*) = 1_{[T,\infty)}(x)$. This GSMP strongly mimics the marking process of the SPN.

EXAMPLE 4.4 (SPN with dependent clock readings). Consider the SPN of Example 4.1. Using the building blocks of the SPN, construct a GSMP with event set $\underline{E} = \{\underline{e}_1, \underline{e}_3, \underline{e}_5\}$ and state space $\underline{S}$ consisting of all elements $(s_1, s_2, \ldots, s_6, v) \in S \times \{0, 2, 4\}$ such that $v = 0$ whenever $\min(s_3, s_6) = 0$.

The idea is that whenever the SPN changes marking from $(1, 0, 0, 1, 0, 0)$ to $(0, 1, 0, 0, 0, 0)$ to $(0, 0, 1, 0, 0, 1)$—so that the clocks for transitions e_3 and e_5 are set according to $F(\cdot; e_3, e_2)$ and $F(\cdot; e_5, e_2)$—the GSMP makes a transition from state $(1, 0, 0, 1, 0, 0, 0)$ to state $(0, 0, 1, 0, 0, 1, 2)$. Similarly, whenever the SPN changes marking from $(1, 0, 0, 1, 0, 0)$ to $(0, 0, 0, 0, 1, 0)$ to $(0, 0, 1, 0, 0, 1)$—so that the clocks for transitions e_3 and e_5 are set according to $F(\cdot; e_3, e_4)$ and $F(\cdot; e_5, e_4)$—the GSMP makes a transition from state $(1, 0, 0, 1, 0, 0, 0)$ to state $(0, 0, 1, 0, 0, 1, 4)$. Thus the last component of the GSMP state is used to keep track of the distribution function used to set the clocks for transitions e_3 and e_5. Formally, we set $\underline{p}(\underline{s}'; \underline{s}, \underline{e}_1) = 1/2$ when $\underline{s} = (1, 0, 0, 1, 0, 0, 0)$ and $\underline{s}' = (0, 0, 1, 0, 0, 1, 2)$, and when $\underline{s} = (1, 0, 0, 1, 0, 0, 0)$ and $\underline{s}' = (0, 0, 1, 0, 0, 1, 4)$. Moreover, for $\underline{s}, \underline{s}' = (s_1', \ldots, s_6', v') \in \underline{S}$ and $i = 3, 5$, we set $\underline{F}(\cdot; \underline{s}', \underline{e}_i, \underline{s}, \underline{e}_1) = F(\cdot; e_i, e_{v'})$. The remaining building blocks are defined in an obvious way. For example, the speeds are given by $\underline{r}(\underline{s}, \underline{e}) = r(\lambda\underline{s}, \psi\underline{e})$, where $\lambda(s_1, \ldots, s_6, v) = (s_1, \ldots, s_6)$ for $\underline{s} = (s_1, \ldots, s_6) \in \underline{S}$ and $\psi\underline{e}_i = e_i$ for $i = 1, 3, 5$. This GSMP strongly mimics the marking process of the SPN.

Theorem 4.5 is analogous to Theorem 2.10 and gives sufficient conditions under which a GSMP strongly mimics the marking process of an SPN.

Theorem 4.5. *Suppose that there exists a mapping ϕ from $\underline{\Sigma}$ onto Σ^+ of the form $\phi(\underline{s}, \underline{c}) = (\lambda\underline{s}, \eta(\underline{s}, \underline{c}))$ such that*

(i) $t^*(\phi(\underline{s}, \underline{c})) = \underline{t}^*(\underline{s}, \underline{c})$ *for all* $(\underline{s}, \underline{c}) \in \underline{\Sigma}$,

(ii) $\mu^+(A) = \underline{\mu}(\phi^{-1}A)$ *for all* $A \subseteq \Sigma^+$, *and*

(iii) $P^+(\phi(\underline{s}, \underline{c}), A) = \underline{P}((\underline{s}, \underline{c}), \phi^{-1}A)$ *for all* $(\underline{s}, \underline{c}) \in \underline{\Sigma}$ *and* $A \subseteq \Sigma^+$.

Then $\{\underline{X}(t): t \geq 0\}$ *strongly mimics* $\{X(t): t \geq 0\}$.

Theorem 4.6. *For any SPN with timed and immediate transitions, there exists a GSMP that strongly mimics the marking process of the SPN.*

PROOF. Consider a fixed but arbitrary SPN, and assume without loss of generality that the set of timed transitions is $E - E' = \{e_1, e_2, \ldots, e_m\}$ and the set of immediate transitions is $E' = \{e_{m+1}, e_{m+2}, \ldots, e_M\}$. We construct a canonical GSMP as follows. Whenever the SPN changes marking to (timed) marking s, the GSMP makes a state transition to state $\underline{s} = (s, w, u)$. The component

$$w = \big(\bar{s}(1), s(1), v(1), \bar{s}(2), s(2), v(2), \ldots, \bar{s}(m), s(m), v(m)\big)$$

records how each clock was set since the last timed marking. The quantities $s(i)$ and $\bar{s}(i)$ are the old and new markings when the clock for timed transition e_i was set. The vector $v(i) = (v_1(i), v_2(i), \ldots, v_M(i))$ encodes the set $E^*(i)$ of transitions that fired simultaneously and triggered the marking change from $s(i)$ to $\bar{s}(i)$: $v_j(i) = 1$ if $e_j \in E^*(i)$ and $v_j(i) = 0$ if $e_j \notin E^*(i)$. If the clock for transition e_i was not set since the last timed marking, then $\big(\bar{s}(i), s(i), v(i)\big) = (\mathbf{0}^L, \mathbf{0}^L, \mathbf{0}^M)$, where $\mathbf{0}^n$ denotes a 0-vector of length n. As suggested by Example 4.3, the GSMP must have—in general—two events $\underline{e}_{i,1}$ and $\underline{e}_{i,2}$ that correspond to timed transition e_i $(1 \leq i \leq m)$; at most one of these events is active at any time. The component $u = (u_1, u_2, \ldots, u_m)$ keeps track of which events are active: u_i equals 2 if event $\underline{e}_{i,2}$ is active, equals 1 if event $\underline{e}_{i,1}$ is active, and equals 0 if neither $\underline{e}_{i,1}$ nor $\underline{e}_{i,2}$ is active. Thus, for $\underline{s} = (s, w, u) \in \underline{S}$ and $1 \leq i \leq m$,

$$\underline{e}_{i,1} \in \underline{E}(\underline{s}) \text{ if and only if } e_i \in E(s) \text{ and } u_i = 1$$

and

$$\underline{e}_{i,2} \in \underline{E}(\underline{s}) \text{ if and only if } e_i \in E(s) \text{ and } u_i = 2.$$

For definiteness, we always enable $\underline{e}_{i,1}$ in preference to $\underline{e}_{i,2}$; e.g., if $\underline{E}(\underline{s}) \cap \{\underline{e}_{i,1}, \underline{e}_{i,2}\} = \varnothing$ and the GSMP makes a transition to a state $\underline{s}' = (s', w', u')$ such that $e_i \in E(s')$, then $\underline{e}_{i,1} \in \underline{E}(\underline{s}')$. The speeds of the GSMP are defined by setting $\underline{r}(\underline{s}, \underline{e}_{i,j}) = r(s, e_i)$ for $\underline{s} = (s, w, u) \in \underline{S}$ and $\underline{e}_{i,j} \in \underline{E}(\underline{s})$.

For $\underline{s} = (s, w, u) \in \underline{S}$, $\underline{E}^* = \{\underline{e}_{i_1,j_1}, \underline{e}_{i_2,j_2}, \ldots, \underline{e}_{i_l,j_l}\} \subseteq \underline{E}(\underline{s})$, and $\underline{s}' = (s', w', u') \in \underline{S}$ with $w' = \big(\bar{s}'(1), s'(1), v'(1), \ldots, \bar{s}'(m), s'(m), v'(m)\big)$, the state-transition probability $\underline{p}(\underline{s}'; \underline{s}, \underline{E}^*)$ is of the form

$$\underline{p}(\underline{s}'; \underline{s}, \underline{E}^*) = \sum_{s^{(0)}, \ldots, s^{(k)}} p\big(s^{(1)}; s^{(0)}, E^*\big) p\big(s^{(2)}; s^{(1)}, E(s^{(1)}) \cap E'\big)$$
$$\cdots p\big(s^{(k)}; s^{(k-1)}, E(s^{(k-1)}) \cap E'\big),$$

where $E^* = \{e_{i_1}, e_{i_2}, \ldots, e_{i_l}\}$. Here the sum is over all sequences $s = s^{(0)}, s^{(1)}, \ldots, s^{(k-1)}, s^{(k)} = s'$ with $s^{(j)} \in S'$ for $0 < j < k$ that are consistent with the values of u, u', w, and w'. For example, if $u_3 = 1$ and $u_3' = 2$—indicating that the clock for e_3 was reset at least once—then, to be consistent, a sequence must contain at least one $s^{(j)}$ for which $e_3 \notin E(s^{(j)})$.

At a state transition for which $\underline{p}(\underline{s}'; \underline{s}, \underline{E}^*) > 0$, the clock-setting distribution function for a new event $\underline{e}_{i,j}$ is given by

$$\underline{F}(\,\cdot\,; \underline{s}', \underline{e}_{i,j}, \underline{s}, \underline{E}^*) = F(\,\cdot\,; \bar{s}'(i), e_i, s'(i), E^*(i)),$$

where $E^*(i) = \{\, e_j \colon v_j'(i) = 1 \,\}$.

Define the initial distribution μ of the GSMP as follows. For each $s \in S$ select $w(s)$ and $u(s)$ such that $(s, w(s), u(s)) \in \underline{S}$ and write $\theta_1(s) = (s, w(s), u(s))$; thus, θ_1 is a one-to-one mapping from S to a proper subset of $\underline{S}$. For $s \in S$ and $c = (c_1, c_2, \ldots, c_m, 0, 0, \ldots, 0) \in C(s)$, set $\theta_2(s, c) = (c_{1,1}, c_{1,2}, \ldots, c_{m,1}, c_{m,2})$, where

$$(c_{i,1}, c_{i,2}) = \begin{cases} (0,0) & \text{if } u_i(s) = 0; \\ (c_i, 0) & \text{if } u_i(s) = 1; \\ (0, c_i) & \text{if } u_i(s) = 2 \end{cases}$$

for $1 \le i \le m$. Finally, set

$$\mu(\underline{A}) = \mu^+(\theta^{-1}\underline{A})$$

for $\underline{A} \subseteq \underline{\Sigma}$, where

$$\theta(s, c) = \big(\theta_1(s), \theta_2(s, c)\big)$$

for $(s, c) \in \bigcup_{s \in S}\big(\{\, s \,\} \times C(s)\big)$.

For $\underline{s} = (s, w, u) \in \underline{S}$ and $\underline{c} = (c_{1,1}, c_{1,2}, \ldots, c_{m,1}, c_{m,2}) \in \underline{C}(\underline{s})$, set $\lambda \underline{s} = s$ and $\eta(\underline{s}, \underline{c}) = (c_1, c_2, \ldots, c_M)$, where

$$c_i = \begin{cases} 0 & \text{if } u_i = 0; \\ c_{i,1} & \text{if } u_i = 1; \\ c_{i,2} & \text{if } u_i = 2r \end{cases}$$

for $1 \le i \le m$ and $c_i = 0$ for $m < i \le M$. Define the mapping $\phi \colon \underline{\Sigma} \mapsto \Sigma$ by $\phi(\underline{s}, \underline{c}) = (\lambda \underline{s}, \eta(\underline{s}, \underline{c}))$ for $(\underline{s}, \underline{c}) \in \underline{\Sigma}$. Straightforward calculations show that the mapping ϕ satisfies the conditions of Theorem 4.5, so that $\{\, \underline{X}(t) \colon t \ge 0 \,\}$ strongly mimics $\{\, X(t) \colon t \ge 0 \,\}$. $\square$

Remark 4.7. Observe that if the SPN has a finite marking set, the GSMP constructed in the proof of Theorem 4.6 has a finite state space. Moreover, if (with probability 1) no timed transitions of the SPN fire simultaneously, then (with probability 1) no events of the GSMP occur simultaneously. Also observe that if all enabled timed transitions remain enabled when there is a marking change and the new marking is immediate, it suffices for the events of the GSMP to be in one-to-one correspondence with the transitions of the SPN and for the state of the GSMP to be of the form $\underline{s} = (s, w)$.

Remark 4.8. It follows directly from Theorems 3.4 and 4.6 that SPNs and GSMPs have the same modelling power.

We conclude this chapter by showing that an "irreducible" SPN with finite state space can always be mimicked by an "irreducible" GSMP. Recall from Section 3.3.1 that, for $s \in S'$ and $s' \in G$, we write $s \to s'$ if $p\big(s'; s, E(s) \cap E'\big) > 0$. Extend this notation to the case where $s \in S$ and $s' \in G$ by writing $s \to s'$ if $p(s'; s, e)r(s, e) > 0$ for some $e \in E(s)$. Next, write $s \rightsquigarrow s'$ if either $s \to s'$ or there exist markings $s^{(1)}, s^{(2)}, \ldots, s^{(n)} \in G$ $(n \geq 1)$ such that $s \to s^{(1)} \to \cdots \to s^{(n)} \to s'$. Clearly, the relation $\rightsquigarrow$ is transitive.

Definition 4.9. An SPN with marking set G is said to be *irreducible* if $s \rightsquigarrow s'$ for each $s, s' \in G$.

We can define the relation $\rightsquigarrow$ for a GSMP in a completely analogous manner and say that a GSMP is irreducible if $\underline{s} \rightsquigarrow \underline{s}'$ for all $\underline{s}, \underline{s}' \in \underline{S}$.

In general, the canonical GSMP constructed in the proof of Theorem 4.6 need not be irreducible even if the marking process of the SPN is irreducible. The construction can be modified, however, to obtain an irreducible GSMP that strongly mimics the marking process of the SPN when the marking set is finite.

Corollary 4.10. *For any irreducible SPN with a finite marking set, there exists an irreducible GSMP with a finite state space that strongly mimics the marking process of the SPN.*

The idea of the proof is as follows. Consider the GSMP constructed in Theorem 4.6 with (finite) state space $\underline{S}$ and event set $\underline{E}$. For the GSMP, write $\underline{s} \longleftrightarrow \underline{s}'$ if $\underline{s} \rightsquigarrow \underline{s}'$ and $\underline{s}' \rightsquigarrow \underline{s}$. Observe that the relation $\longleftrightarrow$ is an equivalence relation on $\underline{S}$ and, since $\underline{S}$ is finite, induces a finite number of equivalence classes on $\underline{S}$. At least one of these equivalence classes, say $\underline{S}^0 \subseteq \underline{S}$, must be closed; that is, $\underline{s}' \in \underline{S}^0$ whenever $\underline{s} \in \underline{S}^0$ and $\underline{s} \rightsquigarrow \underline{s}'$. (Otherwise, there exist two states $\underline{s}$ and $\underline{s}'$ that belong to different equivalence classes but $\underline{s} \longleftrightarrow \underline{s}'$, a contradiction.) It follows from the irreducibility of the SPN that for each $s \in S$ there exists at least one pair (w, u) such that $(s, w, u) \in \underline{S}^0$. Now consider the GSMP with state space $\underline{S}^0$ and event set $\underline{E}^0 = \underline{E}$ such that $\underline{E}^0(\underline{s})$, $p^0(\underline{s}'; \underline{s}, \underline{E}^*)$, $r^0(\underline{s}, \underline{e})$, and $\underline{F}^0(\,\cdot\,; \underline{s}', \underline{e}', \underline{s}, \underline{E}^*)$ coincide with the quantities $\underline{E}(\underline{s})$, $p(\underline{s}'; \underline{s}, \underline{E}^*)$, $r(\underline{s}, \underline{e})$, and $\underline{F}(\,\cdot\,; \underline{s}', \underline{e}', \underline{s}, \underline{E}^*)$ defined in Theorem 4.6 for $\underline{s}, \underline{s}' \in \underline{S}^0$. Define the initial distribution μ^0 analogously to μ in Theorem 4.6, but define the mapping θ so that μ^0 is concentrated on $\Sigma^0 = \bigcup_{\underline{s} \in \underline{S}^0}\big(\{\,\underline{s}\,\} \times \underline{C}(\underline{s})\big)$. This GSMP is irreducible and the mapping ϕ (as in Theorem 4.6) satisfies the conditions of Theorem 4.5.

Remark 4.11. The foregoing results can be used to establish the assertion given in Section 2.4 that for any SPN having unit speeds, a finite marking set, a fixed initial marking, and timed transitions that with probability 1 never fire simultaneously, there exists a "deterministic SPN" that behaves the same way. The idea is that, as shown in this section, the marking process of the former SPN can be strongly mimicked by a GSMP having a finite state space, unit speeds, a fixed initial state, and events that with probability 1 never occur simultaneously. This GSMP can in turn be strongly mimicked by a deterministic SPN; see Remark 3.2.

Notes

Our discussion of modelling power follows Haas and Shedler (1988, 1989a, 1991); these references give further details of the canonical SPN and GSMP constructions. In the literature for ordinary (untimed, deterministic) Petri nets, modelling power is defined in terms of the possible sequences of markings of the net; there is no notion either of the probability that a given sequence is realized or of marking changes occurring at continuous time points. For example, a Petri net is said to mimic a Turing machine—see Motwani and Raghavan (1995, p. 16)—if, for any initial state of the machine, the net generates the same sequence of states as the machine under an appropriate mapping between the state spaces. It is well known that inhibitor input places are needed for Petri nets to have the same modelling power as Turing machines in the sense that for any Turing machine there exists a Petri net that mimics the machine; see Peterson (1981, Sec. 7.3). This result is in contrast to the theorems in Section 4.3, which show that permitting inhibitor input places does not increase the modelling power of the SPN formalism.

The GSMP model originated in the work of Matthes (1962) and König et al. (1967, 1974). Our formulation follows the treatment in Whitt (1980), modified as in Shedler (1993, Ch. 6) to permit simultaneous occurrence of events. Interesting discussions of the role of GSMPs in the study of discrete-event systems can be found in Glynn (1989b), Glasserman (1991), and Glasserman and Yao (1994). There is also a large literature dealing with conditions under which the steady-state distribution of a GSMP depends on the clock-setting distribution functions only through their means; see, for example, Miyazawa (1993), Coyle and Taylor (1995), and references therein.

5
Recurrence

The marking process of an SPN must be stable for time-average limits to be well defined and for simulation-based estimation techniques to be applicable. Although nontrivial, establishing stability properties for a specified SPN is therefore a key step in a methodologically sound simulation study.

Stability of the marking process typically follows from stability of the underlying general state-space Markov chain used to define the marking process. Perhaps the most basic notion of stability for such a chain is "Harris recurrence." A Harris recurrent chain has the property that any "dense enough" set of states is hit infinitely often with probability 1. Thus a Harris recurrent chain is stable in that it does not systematically drift off toward the outer reaches of the state space—fix a dense set of states that is compact, and observe that the chain repeatedly returns to this set. We require that each target set be dense because an individual state typically is hit with probability 0 when the state space of the chain is uncountably infinite.

As discussed in Section 5.1, one means for establishing Harris recurrence is to show that

1. The chain is "ϕ-irreducible" in that any (dense enough) set of states can be reached with positive probability from any initial state.

2. The chain "drifts" toward a specified "petite" subset of the state space whenever the chain lies outside of this subset.

We consider irreducible finite-state SPNs with positive speeds and give "positive density" and moment conditions on the clock-setting distributions under which a drift condition holds.

In the context of regenerative simulation—see Chapter 6—it usually suffices to show that the chain hits a specified set of states infinitely often with probability 1. The successive times at which the chain hits the set typically correspond to "regeneration points" at which the chain probabilistically restarts. The foregoing drift approach can be specialized to establish the desired recurrence property for the specified set. Alternatively, the geometric trials technique described in Section 5.2 can be used to establish recurrence. This technique, which is based on Lemma 3.4 in Chapter 3, exploits the detailed structure of the SPN model and avoids the somewhat restrictive positive density assumptions used in the drift approach.

5.1   Drift Criteria

In this section, we formally define ϕ-irreducibility and Harris recurrence and present a drift criterion for recurrence (Theorem 1.13). We then give conditions (Theorem 1.22) on the building blocks of an SPN under which the drift criterion is satisfied.

5.1.1   Harris Recurrence and Drift

Just as irreducibility and (positive) recurrence play a key role in the theory of Markov chains with a finite or countably infinite state space, ϕ-irreducibility and (positive) Harris recurrence, defined below, are central to the study of general state-space chains. Consider such a chain $\{\, Z_n : n \geq 0 \,\}$ with state space Γ, along with a nontrivial measure ϕ—see Section A.1.2—on subsets of Γ.

Definition 1.1. The chain $\{\, Z_n : n \geq 0 \,\}$ is ϕ-*irreducible* if for each $z \in \Gamma$ and $A \subseteq \Gamma$ with $\phi(A) > 0$, there exists $n > 0$ (possibly depending on both z and A) such that $P^n(z, A) > 0$.

Thus a chain is ϕ-irreducible if any "dense enough" set of states (as measured by ϕ) can be reached from any initial state after a finite number of steps with positive probability. Not surprisingly, ϕ-irreducibility can also be characterized in terms of "hitting times" to sufficiently dense sets. Specifically, denote by τ_A the hitting time of a set $A \subseteq \Gamma$: $\tau_A = \inf\{\, n \geq 1 : Z_n \in A \,\}$. Then $\{\, Z_n : n \geq 0 \,\}$ is ϕ-irreducible if and only if $P_z\{\, \tau_A < \infty \,\} > 0$ for all $z \in \Gamma$ and $A \subseteq \Gamma$ with $\phi(A) > 0$.

EXAMPLE 1.2 (Random walk on the real line).   Define a discrete-time process $\{\, Z_n : n \geq 0 \,\}$ by setting $Z_0 = 0$ and $Z_n = Z_{n-1} + X_n$, where $\{\, X_n : n \geq 1 \,\}$ is a sequence of i.i.d. real-valued random variables. Then $\{\, Z_n : n \geq 0 \,\}$ is a Markov chain with transition kernel $P(z, A) = P\{\, X_1 \in A - z \,\}$, where $A - z = \{\, x - z : x \in A \,\}$ is the set A translated by z. Suppose that X_1 has a density function f that is positive on the real line. Fix a set $A \subseteq \Re$ such that $\mu^{\mathrm{Leb}}(A) > 0$, where μ^{Leb} denotes Lebesgue measure—see Section A.1.2 for a discussion of μ^{Leb}. Observe that $\mu^{\mathrm{Leb}}(A -$

$z) = \mu^{\text{Leb}}(A) > 0$ for $z \in \Gamma$ because Lebesgue measure is invariant under translation. It follows that $P(z, A) = \int_{A-z} f(x)\,dx > 0$ because the integral of a positive function over a set of positive Lebesgue measure is always positive—see Lemma 1.23 in the Appendix. Thus $P^n(z, A) > 0$ for $A \subseteq \Gamma$, $z \in \Gamma$, and $n = 1$, and the chain is ϕ-irreducible with $\phi = \mu^{\text{Leb}}$.

In applications the measure ϕ often is a modification of (possibly multi-dimensional) Lebesgue measure.

Definition 1.3. The chain $\{Z_n : n \geq 0\}$ is *Harris recurrent with recurrence measure ϕ* if it is ϕ-irreducible and $P_z\{Z_n \in A \text{ i.o.}\} = 1$ for all $z \in \Gamma$ and $A \subseteq \Gamma$ with $\phi(A) > 0$.

Harris recurrence can be viewed as a strengthening of ϕ-irreducibility: from any initial state, every dense enough set of states not only can be reached with positive probability, but also is hit infinitely often with probability 1.

A Harris recurrent chain admits an *invariant measure*, that is, a measure π_0 on subsets of Γ that satisfies

$$\int P(z, A)\,\pi_0(dz) = \pi_0(A) \tag{1.4}$$

for $A \subseteq \Gamma$. The measure π_0 is unique to within a multiplicative constant. If $\pi_0(\Gamma) < \infty$, then $\pi(\,\cdot\,) = \pi_0(\,\cdot\,)/\pi_0(\Gamma)$ is an invariant *probability* measure, and (1.4) can be rewritten as $P_\pi\{Z_1 \in A\} = \pi(A)$ for $A \subseteq \Gamma$. That is, if the initial state of the chain Z_0 is distributed according to π, then Z_1 is also distributed according to π. (It then follows from the Markov property that Z_k is distributed according to π for $k \geq 0$ and that the chain is "stationary" as defined in Section A.2.2.)

Definition 1.5. The chain $\{Z_n : n \geq 0\}$ is *positive Harris recurrent with recurrence measure ϕ* if it is Harris recurrent with recurrence measure ϕ and admits an invariant probability measure.

Given a positive Harris recurrent chain with invariant probability measure π and a real-valued function f defined on Γ, we often write

$$\pi(f) = \int f(z)\,\pi(dz) = E_\pi\left[f(Z_0)\right]$$

for the expected value of a function f with respect to π, and write

$$\pi(|f|) = \int |f(z)|\,\pi(dz).$$

The quantity $\pi(f)$ is well defined and finite whenever $\pi(|f|) < \infty$.

As with chains on a finite or countably infinite state space, chains on a general state space can exhibit "periodic" or "aperiodic" behavior. To makes these concepts precise, we first define the notion of a "d-cycle."

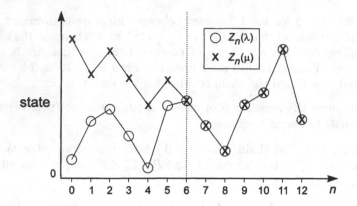

Figure 5.1. Coupling of two Markov chains (coupling epoch $N = 6$).

Definition 1.6. A *d-cycle* of a ϕ-irreducible chain $\{Z_n : n \geq 0\}$ is a finite collection $\{\Gamma_1, \Gamma_2, \ldots, \Gamma_d\}$ of disjoint subsets of Γ such that $\phi(\Gamma - \bigcup_{i=1}^d \Gamma_i) = 0$ and $P(x, \Gamma_{i+1}) = 1$ for $x \in \Gamma_i$ and $1 \leq i \leq d$. (Take $\Gamma_{i+1} = \Gamma_1$ when $i = d$.)

Thus if the initial state of the chain is an element of, say, Γ_1, then with probability 1 the chain will next hit the set Γ_2, and so forth, according to the pattern $\Gamma_1 \to \Gamma_2 \to \cdots \to \Gamma_d \to \Gamma_1 \to \cdots$ *ad infinitum*. The set of states that do not belong to any Γ_i is "negligible" in that the ϕ-measure of this set is 0. It can be shown that at least one d-cycle always exists for a ϕ-irreducible chain.

Definition 1.7. The *period* of a ϕ-irreducible chain $\{Z_n : n \geq 0\}$ is the largest d for which a d-cycle exists; the chain is called *aperiodic* if $d = 1$ and *periodic* if $d > 1$.

Closely tied to the aperiodicity property is the notion of a "Harris ergodic" chain.

Definition 1.8. The chain $\{Z_n : n \geq 0\}$ is *Harris ergodic* if it is positive Harris recurrent and aperiodic.

Our primary interest in Harris ergodic chains stems from the fact that they are amenable to "coupling" arguments.

Definition 1.9. The chain $\{Z_n : n \geq 0\}$ *admits coupling* if for any two initial distributions μ and λ there exist on a common probability space versions $\{Z_n(\mu) : n \geq 0\}$ and $\{Z_n(\lambda) : n \geq 0\}$ of the chain—having respective initial distributions μ and λ—along with an a.s. finite random index N such that $Z_n(\mu) = Z_n(\lambda)$ for $n \geq N$.

Thus, with probability 1 the two sample paths merge into a single path after a finite number of state transitions; see Figure 5.1.

Proposition 1.10. *A chain $\{Z_n : n \geq 0\}$ having a stationary distribution admits coupling if and only if it is Harris ergodic.*

By choosing the initial distribution λ in Definition 1.9 to be the invariant distribution π, Proposition 1.10 often can be used to extend results for a stationary Harris ergodic chain to a nonstationary version of the chain having some arbitrary initial distribution $\mu \neq \pi$. In Chapter 7 we use this approach to establish the validity of certain "consistent estimation" methods for SPNs.

Proposition 1.13 below gives conditions under which a chain $\{Z_n : n \geq 0\}$ is positive Harris recurrent. A key hypothesis of Proposition 1.13 is that the chain drift toward a specified "petite" subset of the state space whenever the chain lies outside this subset.

Definition 1.11. A subset $B \subseteq \Gamma$ is *petite* with respect to the chain $\{Z_n : n \geq 0\}$ if there exist a probability distribution q on the nonnegative integers and a nontrivial measure ψ such that

$$\inf_{z \in B} \sum_{n=0}^{\infty} q(n) P^n(z, A) \geq \psi(A)$$

for all $A \subseteq \Gamma$.

Equivalently, the subset B is petite if there exists a nonnegative integer-valued random variable N, independent of $\{Z_n : n \geq 0\}$, such that

$$\inf_{z \in B} P_z \{Z_N \in A\} \geq \psi(A)$$

for all $A \subseteq \Gamma$. A trivial example of a petite set is given by $B = \{\bar{z}\}$, where $\bar{z} \in \Gamma$; for this set, the above inequality holds with $N \equiv 1$ and $\psi(\cdot) = P(\bar{z}, \cdot)$. It can be shown that there exists at least one petite set of positive ϕ-measure for a ϕ-irreducible chain. In applications, compact (i.e., closed and bounded) sets often serve as petite sets. The following result gives a useful characterization of petiteness.

Proposition 1.12. *Suppose that the chain $\{Z_n : n \geq 0\}$ is ϕ-irreducible. A set $B \subseteq \Gamma$ is petite with respect to $\{Z_n : n \geq 0\}$ if for each set $A \subseteq \Gamma$ with $\phi(A) > 0$ there exists a finite positive integer $n = n(A)$ such that*

$$\inf_{z \in B} P_z \{\tau_A \leq n\} > 0.$$

For real-valued functions f and g, both defined on Γ, write $f = O(g)$ if $\sup_{x \in \Gamma} |f(x)|/|g(x)| < \infty$. (Here we take $0/0 = 0$.)

Proposition 1.13. *Suppose that the chain $\{Z_n : n \geq 0\}$ is ϕ-irreducible. Also suppose that there exist a petite set B, an integer $m \geq 1$, a function $v : \Gamma \mapsto [1, \infty)$, and a real number $\beta \in (0, 1)$ such that*

$$E_z [v(Z_m) - v(Z_0)] \leq -\beta v(z) \tag{1.14}$$

for all $z \in \Gamma - B$, and

$$\sup_{z \in B} E_z \left[v(Z_m) - v(Z_0) \right] < \infty. \tag{1.15}$$

Then $\{ Z_n : n \geq 0 \}$ is positive Harris recurrent with recurrence measure ϕ and hence admits an invariant probability measure π. Moreover, $\pi(|f|) < \infty$ for any function f such that $f = O(v)$.

For $z \notin B$, the quantity $v(z)$ can be viewed as the "distance" between state z and the set B. The quantity $E_z \left[v(Z_m) - v(Z_0) \right]$ in (1.14) and (1.15) is called the m-step expected *drift* of the chain. Thus the condition in (1.14) asserts that the m-step expected drift is strictly negative whenever the chain lies outside B; the exact "rate of drift" is specified by the function βv. The condition in (1.14) is usually called a "geometric" drift criterion: whenever the chain lies outside B, the distance function v is required to decrease in expectation not merely by some positive amount but by a factor[1] of β.

5.1.2 The Positive Density Condition

In this section we give conditions—encapsulated in the "positive density assumption" PD given below—under which the embedded chain of the marking process of an SPN is ϕ-irreducible and satisfies the drift criteria for stability in (1.14) and (1.15). As usual, we assume that the initial distribution of the underlying chain is of the form given by (1.10) in Chapter 3.

Denote by $\mathcal{G}^+$ the set of distribution functions on $[0, \infty)$ that have a convergent LaPlace–Stieltjes transform in a neighborhood of the origin. That is, $F \in \mathcal{G}^+$ if and only if there exists $a_F > 0$ such that $\int_0^\infty e^{ux} \, dF(x) < \infty$ for $u \in [0, a_F]$. Observe that each distribution function $F \in \mathcal{G}^+$ has finite moments of all orders. Many common distribution functions belong to $\mathcal{G}^+$, for example, the uniform, exponential, gamma, beta, and truncated normal distributions.

A nonnegative function G is a *component* of a distribution function F if G is not identically equal to 0 and $G \leq F$. If G is a component of F and G is absolutely continuous—see Section A.1.3—so that G has a density function g, then we say that g is a *density component* of F. For example, let X be a random variable such that $X = 2$ with probability 0.5 and X takes on a value randomly and uniformly distributed between 0 and 1 with probability 0.5. The distribution function F of X can be written as

---

[1]A more general form of drift criterion is obtained by replacing βv by some arbitrary function $g \colon \Gamma \mapsto [1, \infty)$. When $g(z) \equiv c$ for some $c > 0$, the drift criterion reduces to a general state-space version of *Foster's criterion* (Proposition 2.18 in the Appendix) for positive recurrence in chains with a countable state space.

$F = 0.5F_1 + 0.5F_2$, where $F_1(x) = 1_{[2,\infty)}(x)$ for $x \geq 0$ and

$$F_2(x) = \begin{cases} 0 & \text{if } x < 0; \\ x & \text{if } 0 \leq x \leq 1; \\ 1 & \text{if } x > 1. \end{cases}$$

The function $G(x) = 0.5F_2(x)$ is a component of F and $g(x) = 0.5 \cdot 1_{[0,1]}(x)$ is a density component. Observe that in this example F has a density component even though F is not absolutely continuous. In general, if F is the distribution function of a random variable X and F has a density component g, then $P\{a \leq X \leq b\} \geq \int_a^b g(x)\,dx$ for $-\infty < a \leq b < \infty$. If F is absolutely continuous with density function f, then f is trivially a density component of F.

Definition 1.16. *Assumption PD is said to hold for a specified* SPN *if*

(i) the marking set G is finite,

(ii) the SPN is irreducible as in Definition 4.9 of Chapter 4,

(iii) all speeds are positive, and

(iv) there exists $0 < \bar{x} < \infty$ such that each clock-setting distribution function $F(\,\cdot\,; s', e', s, e^*)$ and $F_0(\,\cdot\,; e', s)$ with $e' \in E - E'$ belongs to $\mathcal{G}^+$ and has a density component that is positive and continuous on $(0, \bar{x})$.

If Assumption PD holds and each clock-setting distribution $F(\,\cdot\,; s', e', s, e^*)$ and $F_0(\,\cdot\,; e', s)$ is absolutely continuous with corresponding density function $f(\,\cdot\,; s', e', s, e^*)$ and $f_0(\,\cdot\,; e', s)$, then we always take the "density components" to be f and f_0 by convention.

As usual, denote by Σ and Σ^+ the state spaces of the underlying chain $\{(S_n, C_n) \colon n \geq 0\}$ and embedded chain $\{(S_n^+, C_n^+) \colon n \geq 0\}$, respectively. Whenever Assumption PD holds, we define $\bar{\phi}$ be the unique measure on subsets of Σ^+ such that

$$\bar{\phi}(\{s\} \times [0, x_1] \times [0, x_2] \times \cdots \times [0, x_M]) = \prod_{\{i \colon e_i \in E(s)\}} \min(x_i, \bar{x}) \quad (1.17)$$

for all $s \in S$ and $x_1, x_2, \ldots, x_M \geq 0$. If, for example, a set $B \subseteq \Sigma^+$ is of the form $B = \{s\} \times A$ with $E(s) = E$, then $\bar{\phi}(B)$ is equal to the Lebesgue measure of the set $A \cap [0, \bar{x}]^M$.

Remark 1.18. Observe that if Assumption PD holds, then there exists a real number $q > 0$ such that

$$\int_0^\infty e^{qx}\,dF(x; s', e', s, e^*) < \infty \quad (1.19)$$

and

$$\int_0^\infty e^{qx}\, dF_0(x; e', s) < \infty \tag{1.20}$$

for all s', s, e', and e^*.

Now consider an SPN with marking set G, timed marking set S, transition set E, and underlying and embedded chains with respective state spaces Σ and Σ^+. For $b > 0$, denote by H_b the set of all states $(s, c) \in \Sigma^+$ such that each clock reading is bounded above by b:

$$H_b = \Big\{ (s, c) \in \Sigma^+ : \max_{1 \leq i \leq M} c_i \leq b \Big\}. \tag{1.21}$$

Finally, set

$$h_q(s, c) = \exp\Big(q \max_{1 \leq i \leq M} c_i \Big)$$

for $q \geq 0$, $s \in S$, and $c = (c_1, c_2, \ldots, c_M) \in C(s)$.

Theorem 1.22. *If Assumption PD holds, then*

(i) *the embedded chain* $\{ (S_n^+, C_n^+) : n \geq 0 \}$ *is* $\bar{\phi}$-*irreducible, where* $\bar{\phi}$ *is defined by* (1.17)*, and*

(ii) *for each* $b > 0$ *the set* H_b *defined by* (1.21) *is petite with respect to* $\{ (S_n^+, C_n^+) : n \geq 0 \}$.

Moreover, for some $m \geq 1$*, all* q *satisfying* (1.19) *and* (1.20)*, and all sufficiently large* b,

(iii) $\sup_{(s,c) \in H_b} E_{(s,c)} \big[h_q(S_m^+, C_m^+) - h_q(S_0^+, C_0^+) \big] < \infty$*, and*

(iv) *there exists* $\beta \in (0, 1)$ *such that*

$$E_{(s,c)} \big[h_q(S_m^+, C_m^+) - h_q(S_0^+, C_0^+) \big] \leq -\beta h_q(s, c)$$

for $(s, c) \in \Sigma^+ - H_b$.

The proof of Theorem 1.22 is rather long and is given in the next subsection.

Remark 1.23. The irreducibility Assumption PD requires is a structural property of the net and does not by itself imply irreducibility for the underlying chain, embedded chain, or marking process. Indeed, in the absence of constraints on the clock-setting distributions there can exist markings $s, s' \in S$ such that s is hit with positive probability and $s \rightsquigarrow s'$, but

$$P_\mu \{ S_n = s \text{ and } S_{n+k} = s' \text{ for some } n, k \geq 0 \} = 0. \tag{1.24}$$

Such a situation is illustrated in Example 1.25 below. Theorem 1.22 shows, however, that such anomalous behavior is ruled out by the remaining conditions in Definition 1.16.

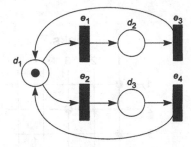

Figure 5.2. An irreducible SPN with a marking that is never hit.

EXAMPLE 1.25 (Irreducible SPN with a marking that is never hit).   Consider an SPN with three places and four timed transitions as in Figure 5.2. The state space of the SPN is $G = S = \{(1,0,0),(0,1,0),(0,0,1)\}$. Suppose that each timed transition e_i is deterministic and simple, with each successive new clock reading for e_i uniformly distributed on a specified interval $[a_i, b_i]$. Also suppose that $P_\mu \{S_0 = (1,0,0)\} = 1$. Observe that this SPN is irreducible; in particular, $s \to s'$, where $s = (1,0,0)$ and $s' = (0,0,1)$. If $b_1 < a_2$, however, then with probability 1 transition e_1 always fires before transition e_2, so that (1.24) holds. Moreover, setting $A = s' \times C(s')$, we see that $P_\mu \{(S_n, C_n) \in A \text{ i.o.}\} = 0$ for any initial distribution μ—we emphasize that the probability of hitting A infinitely often is 0 even though $\bar\phi(A) > 0$ for any choice of $\bar x > 0$, where $\bar\phi$ is defined by (1.17). Of course, this SPN does not satisfy Assumption PD since the clock-setting distribution function for transition e_2 does not have a density component that is positive on an interval of the form $(0, \bar x)$.

The following result is an immediate consequence of Proposition 1.13 and Theorem 1.22.

Corollary 1.26. *Suppose that Assumption PD holds for an* SPN. *Then the embedded chain of the marking process is positive Harris recurrent with recurrence measure $\bar\phi$ given by (1.17) and hence admits a stationary distribution π. Moreover, if q satisfies (1.19) and (1.20), then $\pi(|f|) < \infty$ for any function f such that $f = O(h_q)$.*

EXAMPLE 1.27 (Telephone system).   Consider a telephone system with N telephones connected to a switchboard by lines numbered $1, 2, \ldots, N$. The switchboard has K links numbered $1, 2, \ldots, K$, each of which can connect any two lines, subject to the restriction that only one connection at a time can be made to each line; see Figure 5.3. If more than one link is available and the called line is not in use, a placed call is connected (instantaneously) on the lowest-numbered available link. The system is a lost-call system in the sense that any call is immediately lost if no connection can be made when it is placed. A call is lost if at least one link is available but the called

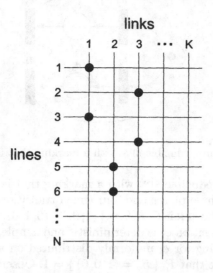

Figure 5.3. Telephone system.

× = call placed at line

I = completion of call on link

∇ = busy call

□ = blocked call

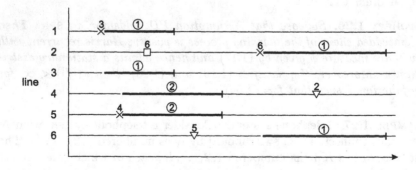

Figure 5.4. Timeline diagram for telephone system (six lines, two links). A circled number represents the link on which a call is connected, and a number displayed above an ×, ∇, or □ represents the destination of a call (or attempted call).

$e_{1,i}$ = call placed at line i

$e_{2,m}$ = end of call connected on link m

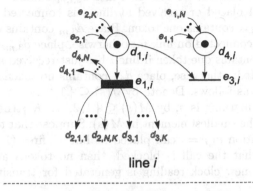

line i

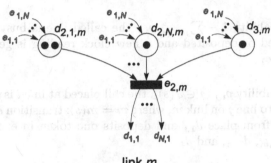

link m

Figure 5.5. SPN representation of telephone system.

line is in use (a *busy call*) and a call is lost if no link is available (a *blocked call*). Figure 5.4 shows a timeline diagram for the telephone system with $N = 6$ lines and $K = 2$ links. The initial call (placed at line 1 to line 3) is connected on link 1 and the next call (placed at line 5 to line 3) is connected on link 2. The third call (placed at line 2 to line 6) is a blocked call and the fourth call (placed at line 6 to line 5) is a busy call.

Successive durations of calls placed at line i are i.i.d. as a positive random variable L_i, and the successive times from the end of a call placed or received at line i to the next call placed at line i are i.i.d. as a positive random variable A_i. After a lost call placed at line i, the time to the next call placed at line i is also distributed as A_i. Whenever a call is placed at line i, the called line is line j with (independent) probability p_{ij}. A line cannot place a call to itself, and thus $p_{ii} = 0$ for $1 \leq i \leq N$.

This system can be specified as a 2-bounded SPN with unit speeds, $N+K$ timed transitions, and N deterministic immediate transitions. The SPN consists of N subnets corresponding to the N lines and K subnets corresponding to the K links; Figure 5.5 displays subnets for a generic line i and a generic link m. Place $d_{1,i}$ contains one token if and only if line i is idle; otherwise, place $d_{1,i}$ contains no tokens. Place $d_{2,i,m}$ contains two tokens if and only if a call placed or received at line i is connected on link m; otherwise, place $d_{2,i,m}$ contains one token. Place $d_{3,m}$ contains one token if and only if a call is connected on link m; otherwise, place $d_{3,m}$ contains no tokens. Place $d_{4,i}$ contains one token if line i has just received a call and is about to be connected; otherwise, place $d_{4,i}$ contains no tokens.

The SPN behaves as follows. Denote by $J(s) \subseteq \{1, 2, \ldots, N\}$ the set of idle lines when the marking is s, by $M(s) \subseteq \{1, 2, \ldots, K\}$ the set of idle links, and by $m(s)$ the smallest element in $M(s)$. Suppose that the marking is $s \in S$ and transition $e_{1,i} = $ "call placed at line i" fires $(1 \leq i \leq N)$. If $M(s) = \varnothing$, so that the call is blocked, then no tokens are removed or deposited and a new clock reading is generated for transition $e_{1,i}$. If $M(s) \neq \varnothing$, then

1. With probability $1 - \sum_{j \in J(s)} p_{i,j}$ the called line is busy: no tokens are removed or deposited and a new clock reading is generated for transition $e_{1,i}$.

2. With probability $p_{i,j}$ $(j \in J(s))$, the call placed at line i is successfully connected to line j on link m, where $m = m(s)$: transition $e_{1,i}$ removes one token from place $d_{1,i}$ and deposits one token in each of places $d_{2,i,m}$, $d_{2,j,m}$, $d_{3,m}$, and $d_{4,j}$.

Observe that when a token is deposited in place $d_{4,j}$ as in (2) above, immediate transition $e_{3,j}$ fires and removes the token in place $d_{1,j}$, thereby causing transition $e_{1,j}$ to become disabled. Now suppose that transition $e_{2,m} = $ "end of call connected on link m" fires $(1 \leq m \leq K)$ and each of places $d_{2,i,m}$ and $d_{2,j,m}$ contains two tokens for some i and j. Then one token is removed from each of places $d_{2,i,m}$, $d_{2,j,m}$, and $d_{3,m}$, and one token is deposited in each of places $d_{1,i}$ and $d_{1,j}$, so that link m, line i, and line j each become idle.

Suppose that for some $a > 0$ the random variables $L_1, L_2, \ldots, L_N$ each are distributed according to a uniform distribution on $[0, a]$ and $A_1, A_2, \ldots,$ A_N are each distributed according to an exponential distribution function with intensity q for some $q > 0$. Also suppose that we wish to show that $P\{S_n = \tilde{s} \text{ i.o.}\} = 1$, where $\tilde{s}$ is the unique timed marking in which all links are idle. Equivalently, we wish to show that $P\{(S_n, C_n) \in A \text{ i.o.}\} = 1$, where $A = \{(s, c) \in \Sigma^+ : s = \tilde{s}\}$. It is not hard to show that $s \rightsquigarrow \tilde{s}$ and $\tilde{s} \rightsquigarrow s'$ for all $s, s' \in G$, so that the SPN is irreducible. Thus Assumption PD holds with $\bar{x} = a$ and the embedded chain $\{(S_n^+, C_n^+) : n \geq 0\}$ is Harris recurrent with recurrence measure $\bar{\phi}$. Because $\bar{\phi}(A) = a^N > 0$, the desired result follows from Corollary 1.26.

The following example shows how Corollary 1.26 can be used in an indirect way to show that a specified subset of $\Sigma - \Sigma^+$ is hit infinitely often with probability 1 by the underlying chain.

EXAMPLE 1.28 (Flexible manufacturing system). For the SPN of Example 2.9 in Chapter 2, recall that the firing of transition e_4 corresponds to the unloading of finished parts and the loading of raw parts. Suppose we wish to show that $P\{(S_n, C_n) \in A \text{ i.o.}\} = 1$, where

$$A = \{(s, c) \in \Sigma : E^*(s, c) = \{e_4\}\} \subset \Sigma - \Sigma^+$$

and E^* is given by (1.8) in Chapter 3. Also suppose that there exists $0 < \bar{x} \leq \infty$ such that each of the distribution functions for the processing-time random variables $L_{1,1}$, $L_{1,2}$, L_2, and L_3 belongs to $\mathcal{G}^+$ and has a density component that is positive and continuous on $(0, \bar{x})$. Then Assumption PD holds because the SPN is irreducible with finite marking set and positive speeds. Set $A^+ = \{(s, c) : s = \bar{s}\}$, where $\bar{s} = (0, 0, 1, 1, 1, 0, 0, 0, 0)$. Whenever the marking is $\bar{s}$, there are two finished parts in the system and machine 3 is processing a part. Observe that $A^+ \subset \Sigma^+$ and $\bar{\phi}(A^+) = \bar{x} > 0$, so that $P\{(S_n^+, C_n^+) \in A^+ \text{ i.o.}\} = 1$ by Corollary 1.26 and hence $P\{(S_n, C_n) \in A^+ \text{ i.o.}\} = 1$. The desired result now follows because $(S_{n+1}, C_{n+1}) \in A$ whenever $(S_n, C_n) \in A^+$.

5.1.3 Proof of Theorem 1.22

For ease of exposition, we assume throughout that all speeds are equal to 1 and that all transitions are simple as in Definition 1.8 of Chapter 3. We assume initially that all transitions are timed, so that the embedded chain coincides with the underlying chain; we then show how to extend the proof to handle immediate transitions.

Irreducibility and Petite Sets

Suppose that Assumption PD holds and that all transitions are timed. Thus there exists $0 < \bar{x} \leq \infty$ such that each clock-setting distribution function has a density component that is positive and continuous on $(0, \bar{x})$. We establish both the $\bar{\phi}$-irreducibility of $\{(S_n, C_n) : n \geq 0\}$ and the petiteness of H_b for $b \geq 0$ through a sequence of lemmas.

Lemma 1.29. *Let $A \subseteq \Sigma$ satisfy $\bar{\phi}(A) > 0$. Then for each $\bar{s} \in S$ there exist a set $\bar{B} = \bar{B}(\bar{s}, A) \subseteq C(\bar{s}) \cap [0, \bar{x}]^M$, an integer $n = n(\bar{s}, A) \leq |S|$, and a real number $\delta = \delta(\bar{s}, A) > 0$ such that*

(i) $\bar{\phi}(\{\bar{s}\} \times \bar{B}) > 0$, and

(ii) $P^n((s, c), A) \geq \delta$ for all $(s, c) \in \{\bar{s}\} \times \bar{B}$.

PROOF. For $s, s' \in S$ with $s \neq s$, let $d(s, s')$ be the smallest integer k such that $s \to s_1 \to \cdots \to s_k = s'$ for some $s_1, s_2, \ldots, s_k \in S$. Because the SPN is irreducible, the "distance measure" d is well defined with $d \leq |S|$. For $n \geq 1$, denote by μ_n^{Leb} Lebesgue measure on $\mathfrak{R}^n$.

It suffices to prove the lemma for a set A of the form $\{\bar{s}'\} \times \bar{A}$, where $\bar{A} \subseteq [0, \bar{x}]^M \cap C(\bar{s}')$. For this choice of A, we show that the conclusion of the lemma holds with $n(\bar{s}, A) = d(\bar{s}, \bar{s}')$. Suppose at first that $d(\bar{s}, \bar{s}') = 1$, so that $p(\bar{s}'; \bar{s}, \bar{e}) > 0$ for some $\bar{e} \in E(\bar{s})$. We construct the desired set $\bar{B}$ when $O(\bar{s}'; \bar{s}, \bar{e}) \neq \varnothing$ and $E(\bar{s}') = E$; the construction for each other possible scenario is similar. Under our assumptions, $\bar{\phi}(A) = \mu_M^{\mathrm{Leb}}(\bar{A}) > 0$. Assume without loss of generality that $O(\bar{s}'; \bar{s}, \bar{e}) = \{e_1, e_2, \ldots, e_k\}$ for some $1 \leq k < M$ and that $\bar{e} = e_M$. Thus $E(\bar{s}) = \{e_1, \ldots, e_k, e_M\}$ and $N(\bar{s}'; \bar{s}, \bar{e}) = \{e_{k+1}, e_{k+2}, \ldots, e_M\}$. Set $\bar{A}_\epsilon = \bar{A} \cap [\epsilon, \bar{x} - \epsilon]^M$, where $\epsilon \in (0, \bar{x}/2)$ is chosen small enough so that $\mu_M^{\mathrm{Leb}}(\bar{A}_\epsilon) > 0$. For $v = (v_1, v_2, \ldots, v_k) \in [\epsilon, \bar{x} - \epsilon]^k$, set

$$\bar{A}_\epsilon(v) = \Big\{ (a_1, a_2, \ldots, a_{M-k}) \in [\epsilon, \bar{x} - \epsilon]^{M-k} :$$
$$(v_1, \ldots, v_k, a_1, \ldots, a_{M-k}) \in \bar{A}_\epsilon \Big\}.$$

Because $\mu_M^{\mathrm{Leb}}(\bar{A}_\epsilon) > 0$ and, by Fubini's theorem (Proposition 1.25 in the Appendix),

$$\mu_M^{\mathrm{Leb}}(\bar{A}_\epsilon) = \int_{[\epsilon, \bar{x} - \epsilon]^k} \mu_{M-k}^{\mathrm{Leb}}(\bar{A}_\epsilon(v)) \, \mu_k^{\mathrm{Leb}}(dv),$$

there exist a set $Q \subseteq [\epsilon, \bar{x} - \epsilon]^k$ and a real number $\gamma > 0$ such that $\mu_k^{\mathrm{Leb}}(Q) > 0$ and $\mu_{M-k}^{\mathrm{Leb}}(\bar{A}_\epsilon(v)) > \gamma$ for $v \in Q$—see Lemma 1.22 in the Appendix. We now show that the desired set $\bar{B}$ is given by

$$\bar{B} = \Big\{ c = (c_1, c_2, \ldots, c_M) \in C(\bar{s}) :$$
$$0 < c_M < \epsilon \text{ and } (c_1 - c_M, c_2 - c_M, \ldots, c_k - c_M) \in Q \Big\}.$$

We see by inspection that $\bar{B} \subseteq [0, \bar{x}]^M$. Moreover, it follows from Fubini's theorem and the invariance of Lebesgue measure under translation that $\bar{\phi}(\{\bar{s}\} \times \bar{B}) = \epsilon \mu_k^{\mathrm{Leb}}(Q) > 0$. For $1 \leq i \leq M - k$ let $f(\cdot; e_{k+i})$ be a density component of $F(\cdot; e_{k+i})$ as in Assumption PD, and for $y = (y_1, y_2, \ldots, y_{M-k}) \in \mathfrak{R}^{M-k}$ set $w(y) = \prod_{i=1}^{M-k} f(y_i; e_{k+i})$. By the continuity and positivity assumptions on the density components, it follows that

$$w^* \stackrel{\mathrm{def}}{=} \inf_{y \in [\epsilon, \bar{x} - \epsilon]^{M-k}} w(y) > 0.$$

Observe that $c_i \geq c_M + \epsilon$ for $1 \leq i \leq k$ whenever $c = (c_1, c_2, \ldots, c_M) \in \bar{B}$, so that $E^*(\bar{s}, c) = \{ e_M \}$ and

$$P\big((\bar{s}, c), \{ \bar{s}' \} \times \bar{A}\big) \geq P\big((\bar{s}, c), \{ \bar{s}' \} \times \bar{A}_\epsilon\big)$$

$$\geq p(\bar{s}'; \bar{s}, e_M) \int_{\bar{A}_\epsilon(\tilde{c})} w(y) \mu^{\mathrm{Leb}}_{M-k}(dy)$$

$$\geq \delta,$$

where $\tilde{c} = (c_1 - c_M, c_2 - c_M, \ldots, c_k - c_M) \in Q$ and $\delta = p(\bar{s}'; \bar{s}, e_M) w^* \gamma > 0$. This establishes the lemma when $d(\bar{s}, \bar{s}') = 1$. The general result follows in a straightforward manner by induction on $d(\bar{s}, \bar{s}')$, using the above argument together with the Chapman–Kolmogorov equations—see (1.5) in Chapter 3. $\qquad\square$

We now partition Σ into a finite collection $\mathcal{Q}$ of mutually disjoint subsets. Elements $(s, c) = (s, c_1, c_2, \ldots, c_M)$ and $(s', c') = (s', c'_1, c'_2, \ldots, c'_M)$ belong to the same subset $Q \in \mathcal{Q}$ if and only if $s = s'$ and the clock readings are in the same relative order, that is,

$$c_i \begin{Bmatrix} < \\ = \\ > \end{Bmatrix} c_j \quad \text{if and only if} \quad c'_i \begin{Bmatrix} < \\ = \\ > \end{Bmatrix} c'_j$$

for all $1 \leq i, j \leq M$. For each $Q \in \mathcal{Q}$ and $\epsilon > 0$ set

$$Q_\epsilon = \big\{ (s, c) \in Q : c \in [0, \epsilon]^M \big\}.$$

Lemma 1.30. *Let $A \subseteq \Sigma$ satisfy $\bar{\phi}(A) > 0$. Then for each $Q \in \mathcal{Q}$ there exist real numbers $\epsilon = \epsilon(Q, A) > 0$ and $\delta = \delta(Q, A) > 0$ together with an integer $n = n(Q, A) \leq |S| + M$ such that $P^n\big((s, c), A\big) \geq \delta$ for $(s, c) \in Q_\epsilon$.*

PROOF. For ease of exposition, we prove the lemma under the assumption that $E(s) = E$ for all $s \in S$; extending the proof to handle arbitrary sets of active events is straightforward. We also fix $\bar{s} \in S$ and give the proof for the set $Q = \{ (\bar{s}, c_1, c_2, \ldots, c_M) \in \Sigma : c_1 < c_2 < \cdots < c_M \}$, the proof for each other set in $\mathcal{Q}$ being similar. Let $s_1, s_2, \ldots, s_M \in S$ be such that

$$p_0 \overset{\mathrm{def}}{=} p(s_1; \bar{s}, e_1) p(s_2; s_1, e_2) \cdots p(s_M; s_{M-1}, e_M) > 0.$$

By Lemma 1.29 there exist a set $\bar{B} = \bar{B}(s_M, A) \subseteq C(s_M) \cap [0, \bar{x}]^M$, a real number $\delta_0 = \delta_0(s_M, A) > 0$, and an integer $l = l(s_M, A) \leq |S|$ such that $\bar{\phi}(\{ s_M \} \times \bar{B}) = \mu^{\mathrm{Leb}}_M(\bar{B}) > 0$ and $P^l((s, c), A) \geq \delta_0$ for all $(s, c) \in \{ s_M \} \times \bar{B}$. Set $\bar{B}_\epsilon = \bar{B} \cap [\epsilon, \bar{x} - \epsilon]^M$, where $\epsilon \in (0, \bar{x}/2)$ is chosen small enough so that $\mu^{\mathrm{Leb}}_M(\bar{B}_\epsilon) > 0$. Fix $\bar{c} = (\bar{c}_1, \bar{c}_2, \ldots, \bar{c}_M) \in C(\bar{s})$ such that $0 < \bar{c}_1 < \bar{c}_2 < \cdots < \bar{c}_M < \epsilon$. It suffices to show that

$$P^{M+l}\big((\bar{s}, \bar{c}), A\big) \geq \delta > 0, \tag{1.31}$$

where δ does not depend explicitly on $\bar{c}$. For $y = (y_1, y_2, \ldots, y_M) \in \Re_+^M$, set $w(y) = \prod_{i=1}^M f(y_i; e_i)$, where $f(\cdot; e)$ is a density component of $F(\cdot; e)$ as in Assumption PD. Also set $v = v(\bar{c}) = (\bar{c}_M - \bar{c}_1, \bar{c}_M - \bar{c}_2, \ldots, \bar{c}_M - \bar{c}_{M-1}, 0)$ and denote by $\bar{B}_\epsilon + v$ the set $\bar{B}_\epsilon$ translated by the vector v. Observe that

$$P^M\big((\bar{s}, \bar{c}), \{\bar{s}\} \times \bar{B}_\epsilon\big)$$
$$\geq P_{(\bar{s}, \bar{c})}\{ S_1 = s_1, S_2 = s_2, \ldots, S_M = s_M,$$
$$(C_{1,1}, C_{2,2}, \ldots, C_{M,M}) \in \bar{B}_\epsilon + v \} \qquad (1.32)$$
$$\geq p_0 \int_{\bar{B}_\epsilon + v} w(y)\, \mu_M^{\text{Leb}}(dy).$$

By construction, $v \leq (\epsilon, \epsilon, \ldots, \epsilon)$, so that $\bar{B}_\epsilon + v \subseteq [\epsilon, \bar{x}]^M$ and hence $w(y) > 0$ for all $y \in \bar{B}_\epsilon + v$. Since, in addition, $\mu_M^{\text{Leb}}(\bar{B}_\epsilon + v) = \mu_M^{\text{Leb}}(\bar{B}_\epsilon) > 0$, it follows that the rightmost term in (1.32) is positive. This term can be viewed as a (continuous) function of v. Denote by v^* the value of v that minimizes this function over the compact set $[0, \epsilon]^M$. It follows from the Chapman–Kolmogorov equations that (1.31) holds with

$$\delta = \delta_0 p_0 \int_{\bar{B}_\epsilon + v^*} w(y)\, \mu_M^{\text{Leb}}(dy) > 0. \qquad \square$$

Lemma 1.33. *The chain* $\{(S_n, C_n) : n \geq 0\}$ *is* $\bar{\phi}$*-irreducible, where* $\bar{\phi}$ *is defined by (1.17). Moreover, the set* H_b *defined by (1.21) is petite with respect to* $\{(S_n, C_n) : n \geq 0\}$ *for each* $b > 0$.

PROOF. Fix a set $A \subseteq \Sigma$ with $\bar{\phi}(A) > 0$. Using notation as in Lemma 1.30, set $n = n(A) = \max_{Q \in \mathcal{Q}} n(Q, A) \leq |S| + M$, $\epsilon = \min_{Q \in \mathcal{Q}} \epsilon(Q, A) > 0$, and $\delta = \min_{Q \in \mathcal{Q}} \delta(Q, A) > 0$. It follows from Lemma 1.30 that

$$P_{(s,c)}\{\tau_A \leq n\} \geq \delta > 0 \qquad (1.34)$$

for all $(s, c) \in \Sigma_\epsilon$, where $\Sigma_\epsilon = \{(s, c) \in \Sigma : c \in [0, \epsilon]^M\}$. We now derive an analogous result for the hitting time of the set Σ_ϵ, starting from an arbitrary state $(\bar{s}, \bar{c}) \in \Sigma$.

For $k \geq 0$, set $W_k = 1$ if $\epsilon/2 < C_{n,i} < \epsilon$ for $e_i \in N(S_n; S_{n-1}, E_{n-1}^*)$ and $kM \leq n < (k+1)M$; otherwise, set $W_k = 0$. Thus W_k is the indicator of the event in which, at marking changes $kM, kM+1, \ldots, (k+1)M-1$, each new clock reading lies in the interval $(\epsilon/2, \epsilon)$. Observe that[2] $\zeta_{(k+1)M} - \zeta_{kM} > \epsilon/2$ whenever $W_k = 1$. Denote by $\lceil x \rceil$ the smallest integer greater than or equal to x. Setting

$$k^*(\bar{c}) = \left\lceil 2 \max_{1 \leq i \leq M} \bar{c}_i / \epsilon \right\rceil$$

---

[2]Recall from (1.11) in Chapter 3 that ζ_n is the time of the nth marking change.

and $\gamma(\bar{c}) = \prod_{i=1}^{M} \left(F(\epsilon; e_i) - F(\epsilon/2; e_i) \right)^{Mk^*(\bar{c})}$, we find that

$$P^{Mk^*(\bar{c})}\left((\bar{s}, \bar{c}), \Sigma_\epsilon \right) \geq P_{(\bar{s}, \bar{c})} \left\{ W_1 = \cdots = W_{k^*(\bar{c})} = 1 \right\} \geq \gamma(\bar{c}) > 0. \quad (1.35)$$

By (1.34) and (1.35),

$$P_{(\bar{s}, \bar{c})} \left\{ \tau_A \leq Mk^*(\bar{c}) + n(A) \right\} \geq \gamma(\bar{c})\delta. \quad (1.36)$$

The desired results follow immediately from (1.36) and Proposition 1.12.
□

Expected Drift

We now establish the assertions in (iii) and (iv) of Theorem 1.22 with m equal to M (the total number of transitions). This result completes the proof of the theorem under the assumption that there are no immediate transitions. For ease of exposition, we suppose that $E(s) = E$ for all $s \in S$; the argument is similar when $E(s) \subset E$ for one or more markings $s \in S$. (Indeed, the disabling of transitions can only accelerate the drift toward a set H_b.) We frequently write $E[X; B] = E[X1_B]$, where $1_B = 1$ if event B occurs and $1_B = 0$ otherwise. Denote by $x \vee y$ the maximum of x and y.

To establish Theorem 1.22(iii), we actually prove the stronger result that

$$\sup_{(s,c) \in \Sigma} E_{(s,c)} \left[h_q(S_M, C_M) - h_q(S_0, C_0) \right] < \infty.$$

It suffices to show that

$$\sup_{(s,c) \in Q} E_{(s,c)} \left[h_q(S_M, C_M) - h_q(S_0, C_0) \right] < \infty \quad (1.37)$$

for $Q \in \mathcal{Q}$, where $\mathcal{Q}$ is a finite partition of Σ as in Lemma 1.30. We give the argument for a subset $Q \in \mathcal{Q}$ such that $(s, c) = (s, c_1, c_2, \ldots, c_M) \in Q$ only if $c_M > c_i$ for $1 \leq i < M$; the argument for each other element of $\mathcal{Q}$ is similar. For $1 \leq i \leq M$ and $j \geq 1$, denote by $A_{i,j}$ the jth successive new clock reading generated for transition e_i. Thus $\{ A_{i,j} : 1 \leq i \leq M, j \geq 1 \}$ is a collection of mutually independent random variables with $P_\mu \{ A_{i,j} \leq x \} = F(x; e_i)$ for all i and j. Set

$$A' = \min_{1 \leq i,j \leq M} A_{i,j}$$

and

$$A'' = \max_{1 \leq i,j \leq M} A_{i,j}.$$

Denote by B the event in which $C_{M,M} > C_{i,M}$ for $1 \leq i < M$ and $e_M \in O(S_{n+1}; S_n, E_n^*)$ for $0 \leq n < M$. Thus event B occurs if and only if transition e_M does not fire during the first M marking changes and, just after the Mth marking change, the clock reading for transition e_M is greater than the clock readings for the other transitions. Fix a state $(s, c) \in Q$, and observe that

$$
\begin{aligned}
E_{(s,c)} & [h_q(S_M, C_M) - h_q(S_0, C_0); B] \\
& = E_{(s,c)}[e^{q(c_M - \zeta_M)} - e^{qc_M}; B] \\
& \leq 0.
\end{aligned}
$$

Next, denote by B^c the complement of event B. Observe that if the initial state is an element of Q and event B^c occurs, then the clock with the largest reading just after the Mth marking change was set sometime during the first M marking changes. It follows that

$$
\begin{aligned}
E_{(s,c)} [h_q(S_M, C_M) - h_q(S_0, C_0); B^c] & \leq E_{(s,c)} [h_q(S_M, C_M); B^c] \\
& \leq E_{(s,c)} [e^{qA''}; B^c] \\
& \leq \sum_{i=1}^{M} \sum_{j=1}^{M} E_{(s,c)} [e^{qA_{i,j}}] \qquad (1.38) \\
& = M \sum_{i=1}^{M} \gamma_q(i),
\end{aligned}
$$

where $\gamma_q(i) = \int_0^\infty e^{qx} \, dF(x; e_i) < \infty$. Thus

$$
\begin{aligned}
E_{(s,c)} & [h_q(S_M, C_M) - h_q(S_0, C_0)] \\
& = E_{(s,c)} [h_q(S_M, C_M) - h_q(S_0, C_0); B] \\
& \quad + E_{(s,c)} [h_q(S_M, C_M) - h_q(S_0, C_0); B^c] \\
& \leq M \sum_{i=1}^{M} \gamma_q(i) \\
& < \infty.
\end{aligned}
$$

Because (s, c) is an arbitrary element of Q, (1.37) holds.

To establish Theorem 1.22(iv), fix $b > 0$ and $(s, c) \in (\Sigma - H_b) \cap Q$, where Q is as before. Thus $c_M > c_i$ for $1 \leq i < M$ and $c_M > b$. Suppose that event B occurs, so that transition e_M does not fire during the first M marking changes. If follows that, during the first M marking changes, the clock for at least one transition in $\{ e_1, e_2, \ldots, e_{M-1} \}$ is set and then runs down to 0. Of these transitions, select the one with the smallest index. Denote by A^* the length of the interval from the first time during $[0, \zeta_M]$ that the clock for this distinguished transition is set until the clock runs down to 0. Thus A^* is a (randomly determined) element of the set $\{ A_{i,j} : 1 \leq i, j \leq M \}$ and $\zeta_M \geq A^*$. Using the mean-value theorem we find that, for some random variable W with $0 \leq W \leq A^*/c_M$,

$$E_{(s,c)} \left[h_q(S_M, C_M) - h_q(S_0, C_0); B \right]$$
$$= E_{(s,c)} \left[e^{q(c_M - \zeta_M)} - e^{qc_M}; B \right]$$
$$\leq E_{(s,c)} \left[e^{qc_M(1 - A^*/c_M)} - e^{qc_M}; B \right]$$
$$= E_{(s,c)} \left[-qA^* e^{qc_M(1-W)}; B \right] \tag{1.39}$$
$$\leq -qe^{qc_M} E_{(s,c)} \left[A^* e^{-qA^*}; B \right]$$
$$\leq -qe^{qc_M} \theta,$$

where $\theta = E_{(s,c)} \left[A' e^{-qA''} \right]$. Observe that θ does not depend on (s, c) and that $\theta < \infty$ under our distributional assumptions. It follows from (1.38) and (1.39) that

$$E_{(s,c)} \left[h_q(S_2, C_2) - h_q(S_0, C_0) \right] \leq g(b) h_q(s, c),$$

where $g(b) = Me^{-qb} \sum_{i=1}^{M} \gamma_q(i) - q\theta$. Fix $\epsilon \in (0, 1)$ small enough so that

$$\beta \stackrel{\text{def}}{=} \epsilon q\theta < 1.$$

Clearly, $g(b) \to -q\theta$ as $b \to \infty$, so that if b is sufficiently large, then $g(b) \leq -\beta$ and Theorem 1.22(iv) holds for $(s, c) \in (\Sigma - H_b) \cap Q$. Similar arguments apply to each other element of Q, and the desired result follows.

Immediate Transitions

We have established Theorem 1.22 under the assumption that all transitions are timed. We now extend this result to SPNs with one or more immediate transitions. Because it appears hard to modify the foregoing proof to handle this general case, we apply an indirect approach.

By Corollary 4.10 in Chapter 4, there exists an irreducible GSMP with a finite state space that strongly mimics the marking process of the SPN. Let $\{ (\underline{S}_n, \underline{C}_n) : n \geq 0 \}$ be the underlying chain of this GSMP. Denote by $\underline{\Sigma}$ the state space of the underlying chain and by μ the initial distribution. Also let ψ be the mapping from $\underline{\Sigma}$ onto Σ^+ such that $\{ (S_n^+, C_n^+) : n \geq 0 \}$ and $\{ \psi(\underline{S}_n, \underline{C}_n) : n \geq 0 \}$ have the same finite-dimensional distributions. Define a function $\underline{h}_q$ on $\underline{\Sigma}$ analogously to the function h_q defined on Σ^+. Similarly, for $b > 0$, define a set $\underline{H}_b \subseteq \underline{\Sigma}$ analogously to the set $H_b \subseteq \Sigma^+$. It follows from the specific definition of $\underline{\Sigma}$ given in the proof of Corollary 4.10 in Chapter 4 that

$$h_q(\psi(\underline{s}, \underline{c})) = \underline{h}_q(\underline{s}, \underline{c}) \tag{1.40}$$

for $(\underline{s}, \underline{c}) \in \underline{\Sigma}$ and $q \geq 0$. Moreover, for $b > 0$,

$$\psi(\underline{\Sigma} - \underline{H}_b) = \Sigma^+ - H_b. \tag{1.41}$$

Observe that the proof thus far can be applied essentially without change to establish the assertions of Theorem 1.22 for the underlying chain of the mimicking GSMP. We can therefore pick $b > 0$ large enough so that

$$E_{(\underline{s},\underline{c})}\left[\underline{h}_q(\underline{S}_m, \underline{C}_m) - \underline{h}_q(\underline{S}_0, \underline{C}_0)\right] \leq -\beta \underline{h}_q(\underline{s}, \underline{c})$$

for some $\beta \in (0,1)$ and all $(\underline{s}, \underline{c}) \in \underline{\Sigma} - \underline{H}_b$, where m is the number of events in the GSMP. Now fix $(s, c) \in \Sigma^+ - H_b$. By (1.41), there exists $(\underline{s}, \underline{c}) \in \underline{\Sigma} - \underline{H}_b$ such that $\psi(\underline{s}, \underline{c}) = (s, c)$. We then have

$$\begin{aligned}
E_{(s,c)} &\left[h_q(S_m^+, C_m^+) - h_q(S_0^+, C_0^+)\right] \\
&= E_{(\underline{s},\underline{c})}\left[h_q(\psi(\underline{S}_m, \underline{C}_m)) - h_q(\psi(\underline{S}_0, \underline{C}_0))\right] \\
&= E_{(\underline{s},\underline{c})}\left[\underline{h}_q(\underline{S}_m, \underline{C}_m) - \underline{h}_q(\underline{S}_0, \underline{C}_0)\right] \\
&\leq -\beta \underline{h}_q(\underline{s}, \underline{c}) \\
&= -\beta h_q(s, c),
\end{aligned}$$

where the first equality follows from Corollary 4.10 in Chapter 4 and the remaining two equalities follow from (1.40). Thus we have established Theorem 1.22(iv). The remaining assertions of Theorem 1.22 are proved in a similar manner.

5.2   The Geometric Trials Technique

The results in the previous section give conditions under which the embedded chain $\{(S_n^+, C_n^+): n \geq 0\}$ hits any dense enough set of states infinitely often with probability 1. As discussed earlier, it sometimes suffices to show that the embedded or underlying chain hits one particular set of states infinitely often with probability 1, that is,

$$P\{(S_n, C_n) \in A \text{ i.o.}\} = 1 \tag{2.1}$$

for some specified set $A \subset \Sigma$. Such a set is said to be *recurrent* with respect to $\{(S_n, C_n): n \geq 0\}$. If Assumption PD holds and $A \subseteq \Sigma^+$ with $\bar{\phi}(A) > 0$, then (2.1) follows immediately from Corollary 1.26.

In this section, we give methods for establishing recurrence that do not require positive density assumptions on the clock-setting distribution functions. Such methods are useful because many SPN models have one or more clock-setting distribution functions that have support on some finite or countably infinite set of points or on an interval not of the form $[0, u]$. In SPN models of computer networks, for example, propagation delays often are modelled as deterministic constants, leading to degenerate clock-setting distribution functions that put all of the probability mass on a single point; see Examples 2.6, 2.7, 2.12, and 3.7 in Chapter 2. Similarly, in SPN models of manufacturing systems, the time required for a robot to execute a movement or for a conveyor to transport a part often is modelled as a deterministic constant or as a random variable that is bounded away from 0; see Example 3.6 in Chapter 2.

Sometimes the detailed structure of a specified SPN model can be exploited in a direct way to establish recurrence, as illustrated by the following example.

EXAMPLE 2.2 (Flexible manufacturing system). As in Example 1.28, suppose we wish to show that (2.1) holds with

$$A = \{\, (s,c) \in \Sigma : E^*(s,c) = \{\, e_4 \,\} \,\}.$$

We can establish (2.1) without imposing the positive density assumptions on the clock-setting distributions that are used in Example 1.28. The only requirement is that $L_{1,1}$, $L_{1,2}$, L_2, and L_3 each be a.s. finite. Denote by $\theta(n)$ the random index of the nth marking change at which the underlying chain hits the set A. By considering the possible sample paths of $\{\, (S_n, C_n) : n \geq 0 \,\}$, it can be seen that $\theta(0) \leq 9$ for any choice of initial state and, moreover, $\theta(n) - \theta(n-1) \leq 9$ for $n \geq 1$. Thus each $\theta(n)$ is a.s. finite and (2.1) holds.

Although brute-force recurrence arguments as in Example 2.2 do not require positive density assumptions on the clock readings, they are applicable only to extremely simple SPN models. In the remainder of this section we therefore focus on a geometric trials technique that avoids the positive density assumptions of Corollary 1.26 and can be used to establish recurrence even in very complex SPN models.

5.2.1 A Geometric Trials Criterion

It can often be difficult to show directly that $P\{\, (S_n, C_n) \in A \text{ i.o.} \,\} = 1$ for a specified set A. In such cases the following two-step approach can be useful. First, find a set $B \supset A$ for which it is easy to show that $P\{\, (S_n, C_n) \in B \text{ i.o.} \,\} = 1$. Equivalently, find a set B for which it is easy to show that $\beta(n)$ is a.s. finite for $n \geq 1$, where $\beta(n)$ is the random index of the nth marking change at which the underlying chain hits the set B. Next, show that

$$P_\mu \{\, (S_{\beta(n)}, C_{\beta(n)}) \in A \text{ i.o.} \,\} = 1.$$

Throughout, we restrict attention to sets of the form $A = \{\, (s,c) \in \Sigma : s \in \bar{G} \,\}$, where $\bar{G} \subseteq G$. Thus the goal is to show that

$$P_\mu \{\, S_{\beta(n)} \in \bar{G} \text{ i.o.} \,\} = 1. \tag{2.3}$$

In this case, the set $\bar{G}$ is said to be recurrent; if $\bar{G} = \{\, \bar{s} \,\}$ for some $\bar{s} \in G$, then $\bar{s}$ is said to be recurrent. The primary tool for establishing (2.3) is the geometric trials lemma—Lemma 3.4 in Chapter 3—which we now recast as Lemma 2.4. In the lemma $\{\mathcal{F}_n : n \geq 0\}$ denotes the increasing sequence of partial histories of the underlying chain $\{(S_n, C_n) : n \geq 0\}$; see Section 3.4.2.

Lemma 2.4. *Let $\{\beta(n) : n \geq 1\}$ and $\{\alpha(n) : n \geq 1\}$ be increasing sequences of a.s. finite random indices such that each $\alpha(n)$ and each $\beta(n)$ is a stopping time with respect to $\{\mathcal{F}_n : n \geq 0\}$ and, moreover, $\beta(n-1) \leq \alpha(n) < \beta(n)$ for $n \geq 1$. [Take $\beta(0) = 0$.] Suppose that*

$$P_\mu \left\{ S_{\beta(n)} \in \bar{G} \mid \mathcal{F}_{\alpha(n)} \right\} \geq \delta \ \text{a.s.} \tag{2.5}$$

for some $\delta > 0$ and all $n \geq 1$. Then $P_\mu \left\{ S_{\beta(n)} \in \bar{G} \ i.o. \right\} = 1$.

PROOF. Fix $n \geq 1$ and set

$$Z_n = \begin{cases} 1 & \text{if } S_{\beta(n)} \in \bar{G}; \\ 0 & \text{otherwise.} \end{cases}$$

Observe that the values of $Z_1, Z_2, \ldots, Z_{n-1}$ are completely determined by $\mathcal{F}_{\beta(n-1)}$, and hence by $\mathcal{F}_{\alpha(n)}$, so that

$$\begin{aligned}
P_\mu \left\{ Z_n = 1 \mid Z_{n-1}, \ldots, Z_1 \right\} \\
&= E_\mu \left[P_\mu \left\{ Z_n = 1 \mid \mathcal{F}_{\alpha(n)} \right\} \mid Z_{n-1}, \ldots, Z_1 \right] \\
&= E_\mu \left[P_\mu \left\{ S_{\beta(n)} \in \bar{G} \mid \mathcal{F}_{\alpha(n)} \right\} \mid Z_{n-1}, \ldots, Z_1 \right] \\
&\geq E_\mu \left[\delta \mid Z_{n-1}, \ldots, Z_1 \right] \\
&= \delta \ \text{a.s.,}
\end{aligned}$$

and the desired result follows from the geometric trials lemma.   □

The random times $\{\alpha(n) : n \geq 0\}$ are chosen for convenience; as discussed in the following subsections, (2.5) can be more easily established for some random times than for others.

5.2.2 GNBU Distributions

When establishing recurrence using Corollary 1.26, we require that the SPN be irreducible and each clock-setting distribution function have a density component that is positive and continuous on an interval of the form $(0, \bar{x}]$. Use of Lemma 2.4, on the other hand, leads to conditions on the SPN building blocks that depend on the particular SPN of interest. A typical requirement is that certain of the new clock readings be generated according to "GNBU" distribution functions. The class of GNBU distribution functions generalizes the "new better than used" distribution functions that arise in the statistical theory of reliability.

Definition 2.6. A distribution function F with support on $[0, \infty)$ is *new better than used* (NBU) if and only if

$$\overline{F}(x + y) \leq \overline{F}(x)\overline{F}(y)$$

for $x, y \geq 0$, where $\overline{F} = 1 - F$.

Suppose, for example, that F is the distribution function for the random lifetime L of a machine and that $P\{L > y\} > 0$ for some $y > 0$. If F is NBU, then

$$P\{L - y > x \mid L > y\} \leq P\{L > x\}$$

for $x \geq 0$. That is, the survival probability for a machine of age y is less than the corresponding survival probability for a new machine. Equivalently,

$$P\{L - y \leq x \mid L > y\} \geq P\{L \leq x\}$$

for $x \geq 0$, so that the residual lifetime of a machine of age y is stochastically smaller—see Definition 1.7 in the Appendix—than the lifetime of a new machine.

NBU distributions arise frequently in applications. For example, the distribution function of a random variable L is NBU if L is a.s. equal to a fixed constant. Moreover, an absolutely continuous distribution function F with density function f is NBU if the *failure rate* $r(t) = f(t)/\overline{F}(t)$ is nondecreasing in t. Examples of such distributions include the exponential distribution (which has a constant failure rate), the Weibull distribution with shape parameter greater than 1, the gamma distribution with shape parameter greater than 1, and the truncated normal distribution.

If a distribution function F is NBU, then for sufficiently large x the ratio $\overline{F}(x + y)/\overline{F}(y)$ is bounded away from 1 as a function of y. The generalized NBU (GNBU) distribution functions are characterized by this boundedness property.

Definition 2.7. A distribution function F with support on $[0, \infty)$ is GNBU *with lower bound* x^* if and only if

$$\sup_{y \geq 0} \frac{\overline{F}(x + y)}{\overline{F}(y)} < 1 \qquad (2.8)$$

for $x > x^*$, where we take $0/0 = 0$.

Observe that if (2.8) holds for $x = x_0$, then (2.8) holds for any $x \geq x_0$. Lemma 2.9 gives some conditions under which a distribution function is GNBU. Recall that the *essential supremum* of a distribution function F, written ess sup F, is defined as $\sup\{x: F(x) < 1\}$. Similarly, the *essential infimum* of F, written ess inf F, is defined as $\inf\{x: F(x) > 0\}$.

Lemma 2.9. *Suppose that F is the distribution function of a nonnegative random variable.*

(i) *If F is* NBU, *then F is* GNBU *with lower bound $x^* = \operatorname{ess\,inf} F$.*

(ii) *If F is absolutely continuous with a density function f that is positive on* $(\operatorname{ess\,inf} F, \infty)$ *and satisfies*

$$\lim_{y \to \infty} \frac{f(x^* + y)}{f(y)} < 1 \tag{2.10}$$

for some $x^* > 0$, then F is GNBU *with lower bound* $\max(x^*, \operatorname{ess\,inf} F)$.

(iii) *If F is absolutely continuous with a density function f that is positive on a finite interval $[a, b]$ and equal to 0 elsewhere, then F is* GNBU *with lower bound $x^* = a$.*

(iv) *If there exist a continuous* NBU *distribution function G and a constant $c \in (0, \infty)$ such that*

$$\lim_{x \to \infty} \frac{\overline{F}(x)}{\overline{G}(x)} = c,$$

then F is GNBU.

PROOF. If F is NBU and $x > \operatorname{ess\,inf} F$, then

$$\sup_{y \geq 0} \frac{\overline{F}(x + y)}{\overline{F}(y)} \leq \overline{F}(x) < 1.$$

To prove the assertion in (ii), pick $x > \max(x^*, \operatorname{ess\,inf} F)$ and observe that

$$\lim_{y \to \infty} \frac{\overline{F}(x + y)}{\overline{F}(y)} \leq \lim_{y \to \infty} \frac{\overline{F}(x^* + y)}{\overline{F}(y)} = \lim_{y \to \infty} \frac{f(x^* + y)}{f(y)} < 1$$

where the equality follows from l'Hopital's rule. Pick $b > 0$ and $v < 1$ such that $\overline{F}(x + y)/\overline{F}(y) \leq v$ for $y > b$. Because $x > \operatorname{ess\,inf} F$ and f is positive on $(\operatorname{ess\,inf} F, \infty)$, the continuous function $g(y) = \overline{F}(x + y)/\overline{F}(y)$ is strictly less than 1 for all $y \in [0, b]$. Set $u = \sup_{0 \leq y \leq b} g(y)$ and observe that $u < 1$ because a continuous function attains its maximum value over a compact set. The desired result now follows because

$$\sup_{y \geq 0} \frac{\overline{F}(x + y)}{\overline{F}(y)} = \max\left(\sup_{0 \leq y \leq b} \frac{\overline{F}(x + y)}{\overline{F}(y)}, \sup_{y > b} \frac{\overline{F}(x + y)}{\overline{F}(y)} \right)$$

$$\leq \max(u, v)$$

$$< 1.$$

To prove the assertion in (iii), it suffices to show that (2.8) holds for every $x \in (a, b)$. Pick such an x and observe that, under our assumptions,

$\overline{F}$ is strictly decreasing on $[a, b]$. Also observe that $\overline{F}(x+y)/\overline{F}(y) < 1$ for $y = 0$ and

$$\frac{\overline{F}(x+y)}{\overline{F}(y)} < \frac{\overline{F}(x+y)}{\overline{F}(x)} < 1$$

for $0 < y < x$. For $y \geq x$, we have $\overline{F}(y) \geq \overline{F}(y+x)$, with $\overline{F}(y) = \overline{F}(y+x)$ only if $y > b$, in which case $\overline{F}(x+y)/\overline{F}(y) = 0$.

To prove the assertion in (iv), pick $x, \epsilon > 0$ such that

$$\frac{\overline{F}(x+y)}{\overline{G}(x+y)} \leq c + \epsilon$$

for $y \geq 0$. It follows from the NBU property of G that

$$\sup_{y \geq 0} \frac{\overline{F}(x+y)}{\overline{F}(y)} \leq (c+\epsilon)\overline{G}(x)\left(\inf_{y \geq 0} \frac{\overline{F}(y)}{\overline{G}(y)}\right)^{-1}. \tag{2.11}$$

It suffices to show that

$$\inf_{y \geq 0} \frac{\overline{F}(y)}{\overline{G}(y)} > 0,$$

since then the term on the right side of (2.11) is less than 1 for sufficiently large x. The above inequality follows by an argument similar to the proof of the assertion in (ii)—use the fact that, since G is continuous, the function $h(y) = \overline{F}(y)/\overline{G}(y)$ is lower semicontinuous and hence attains its infimum over any interval of the form $[0, b]$. $\qquad\square$

Many distribution functions are GNBU but not NBU. For example, if F is any non-NBU distribution function such that $F(u) = 1$ for some $u < \infty$, then F is GNBU with lower bound u. Other examples include mixtures of exponential distributions and gamma distributions with shape parameter less than 1. To establish the GNBU property for these distributions, apply Lemma 2.9(ii); alternatively, Lemma 2.9(iv) can be used to show that mixtures of exponential distributions are GNBU—take $G(x) = 1 - \exp(-bx)$ for an appropriate constant b. The foregoing gamma distribution functions, far from being NBU, are *new worse than used* (NWU) in that $\overline{F}(x+y) \geq \overline{F}(x)\overline{F}(y)$ for $x, y \geq 0$ (with strict inequality for at least one value of x and y).

As shown by the following result, a GNBU distribution has finite moments of all orders.

Lemma 2.12. *If F is GNBU, then $\int_0^\infty x^r \, dF(x) < \infty$ for $r \geq 0$.*

PROOF. Let x^* be the GNBU lower bound for F. Fix $x > x^*$ and set

$$\gamma = \gamma(x) = \sup_{y \geq 0} \frac{\overline{F}(x+y)}{\overline{F}(y)} < 1.$$

An easy inductive argument shows that $\overline{F}(kx + y) \le \gamma^k \overline{F}(y)$ for $y \ge 0$ and $k \in \{0, 1, 2, \ldots\}$. In particular, $\overline{F}(kx) \le \gamma^k$. Fix $r > 1$ and use a standard identity—see (1.13) in the Appendix—to obtain

$$
\int_0^\infty y^r \, dF(y) = \int_0^\infty r y^{r-1} \overline{F}(y) \, dy
$$

$$
= \sum_{k=0}^\infty \int_{kx}^{(k+1)x} r y^{r-1} \overline{F}(y) \, dy
$$

$$
\le r x^r \sum_{k=0}^\infty (k+1)^{r-1} \gamma^k
$$

$$
< \infty. \qquad \square
$$

We conclude this section by establishing some additional properties of GNBU distributions that are useful when verifying the geometric trials recurrence criterion in (2.5).

Lemma 2.13. *Let* $A_1, A_2, \ldots, A_m$ *be mutually independent random variables with distribution functions* $F_1, F_2, \ldots, F_m$, *and suppose that each* F_i *is* GNBU *with lower bound* x_i^*. *Then*

$$
\sup_{y_1, \ldots, y_m \ge 0} P\left\{ \sum_{i=1}^m (A_i - y_i) > x \;\middle|\; A_i > y_i \text{ for } 1 \le i \le m \right\} < 1 \qquad (2.14)
$$

for $x > x_1^* + x_2^* + \cdots + x_m^*$.

PROOF. The proof is by induction on m. For $m = 1$ the desired result (2.14) reduces to (2.8). Assume for induction that (2.14) holds for some $m \ge 1$. Fix $\epsilon > 0$, $x > x_1^* + x_2^* + \cdots + x_{m+1}^* + \epsilon$, and $y_1, y_2, \ldots, y_{m+1} \ge 0$. Define events G, H_m, and H_{m+1} by setting

$$
G = \left\{ \sum_{i=1}^m (A_i - y_i) \le x - x_{m+1}^* - \epsilon \right\},
$$

$$
H_m = \{ A_i > y_i \text{ for } 1 \le i \le m \},
$$

and

$$
H_{m+1} = \{ A_i > y_i \text{ for } 1 \le i \le m+1 \}.
$$

Also set

$$
\gamma_{m+1} = \sup_{y \ge 0} \frac{\overline{F}_{m+1}(x_{m+1}^* + \epsilon + y)}{\overline{F}_{m+1}(y)}.
$$

Recall that 1_H denotes the random variable that equals 1 if event H occurs and equals 0 otherwise and that H^c denotes the complement of event H. Setting

$$
\theta = \sup_{y_1, \ldots, y_m \ge 0} P\{ G^c \mid H_m \},
$$

we find that

$$P\left\{ \sum_{i=1}^{m+1}(A_i - y_i) > x \,\middle|\, H_{m+1} \right\}$$

$$= E\left[P\left\{ \sum_{i=1}^{m+1}(A_i - y_i) > x \,\middle|\, H_{m+1}, A_1, \ldots, A_m \right\} \,\middle|\, H_{m+1} \right]$$

$$= E\left[1_{H_m}\left(\frac{\overline{F}_{m+1}\big(y_{m+1} + x - \sum_{i=1}^m (A_i - y_i)\big)}{\overline{F}_{m+1}(y_{m+1})} \right) \,\middle|\, H_{m+1} \right]$$

$$\leq E\left[1_{H_m \cap G}\left(\frac{\overline{F}_{m+1}\big(y_{m+1} + x - \sum_{i=1}^m (A_i - y_i)\big)}{\overline{F}_{m+1}(y_{m+1})} \right) + 1_{H_m \cap G^c} \,\middle|\, H_{m+1} \right]$$

$$\leq E\left[1_{H_m \cap G}\left(\frac{\overline{F}_{m+1}\big(y_{m+1} + x_{m+1}^* + \epsilon\big)}{\overline{F}_{m+1}(y_{m+1})} \right) + 1_{H_m \cap G^c} \,\middle|\, H_{m+1} \right]$$

$$\leq \gamma_{m+1} P\{ G \mid H_m \} + P\{ G^c \mid H_m \}$$

$$\leq \gamma_{m+1}(1 - \theta) + \theta.$$

Since $\theta < 1$ by the induction hypothesis, $\gamma_{m+1} < 1$ by (2.8), and $y_1, y_2, \ldots,$ y_{m+1} are arbitrary, the desired result follows.   □

An immediate consequence of (2.14) is that

$$\inf_{y_1, \ldots, y_m \geq 0} P\left\{ \sum_{i=1}^m (A_i - y_i) \leq x \,\middle|\, A_i > y_i \text{ for } 1 \leq i \leq m \right\} > 0 \qquad (2.15)$$

for $x > x_1^* + x_2^* + \cdots + x_m^*$. This latter inequality can be generalized as follows.

Lemma 2.16. *For some $m \geq 1$, let $A_1, A_2, \ldots, A_m, B, Q$ be nonnegative random variables with respective distribution functions $F_1, F_2, \ldots, F_m, G,$ H. Suppose that $A_1, \ldots, A_m$ are mutually independent and independent of both B and Q, and that each F_i is GNBU with lower bound x_i^*. Also suppose that $x_1^* + \cdots + x_m^* + b < q$, where $b = \text{ess inf } G$ and $q = \text{ess sup } H$. Then*

$$\inf_{y_1, \ldots, y_m \geq 0} P\left\{ \sum_{i=1}^m (A_i - y_i) + B \leq Q \,\middle|\, A_i > y_i \text{ for } 1 \leq i \leq m \right\} > 0.$$

The result in Lemma 2.16 follows directly from (2.15) after conditioning on B and Q.

5.2.3 A Simple Recurrence Argument

GNBU distributional assumptions often can be combined with a "sample path condition" and a "positivity condition" to establish the geometric trials recurrence criterion in (2.5). We illustrate our general approach by means of a simple example.

EXAMPLE 2.17 (Token ring). For the system of Example 2.6 in Chapter 2, suppose that the distribution function F_j of each interarrival-time random variable A_j is NBU. Recall that R_j is the time for the ring token to propagate from port j to the next port, and suppose that

$$\operatorname{ess\,inf} F_j < R_N \qquad (2.18)$$

for $1 \le j \le N$.

Consider the SPN representation of the token ring given in Figure 2.10, and denote by $\beta(n) + 1$ the random index of the nth marking change at which transition $e_{3,1} = $ "observation of ring token by port 1" fires—thus $E^*(S_{\beta(n)}, C_{\beta(n)}) = \{e_{3,1}\}$ and $S_{\beta(n)}$ is the marking just before the firing of $e_{3,1}$. Suppose we wish to show that $P_\mu\{S_{\beta(n)} = \bar{s} \text{ i.o.}\} = 1$, where $\bar{s} = (1, 0, 0, 0, \ldots, 1, 0, 0, 0, 1, 0, 0, 1)$. Observe that the marking is $\bar{s}$ if and only if all ports have a packet awaiting transmission and the ring token is propagating from port N to port 1. Let $\alpha(n)$ be the index of the nth marking change at which transition $e_{3,1}$ becomes enabled—that is, at which the ring token begins to propagate from port N to port 1—and suppose that $\alpha(1) = 0$. Observe that there can be at most $2N$ packet arrivals, N observations of the ring token by a port, and N packet transmissions in the time interval $[\zeta_{\beta(n)+1}, \zeta_{\beta(n+1)+1}]$. It follows that

$$\beta(n+1) - \beta(n) = \big(\beta(n+1) + 1\big) - \big(\beta(n) + 1\big) \le 4N.$$

Similarly, $\beta(1) \le 4N$. Thus each $\beta(n)$, and hence each $\alpha(n)$, is a.s. finite.

Set $\bar{G} = \{\bar{s}\}$. Fix $n \ge 1$ and denote by I_n the random set of indices of the ports having no packet awaiting transmission at time $\zeta_{\alpha(n)}$. Clearly, $S_{\beta(n)} \in \bar{G}$ [that is, $S_{\beta(n)} = \bar{s}$] if for each $j \in I_n$ there is an arrival in the interval $[\zeta_{\alpha(n)}, \zeta_{\beta(n)+1})$ of a packet for transmission by port j. Thus

$$P_\mu\{S_{\beta(n)} \in \bar{G} \mid \mathcal{F}_{\alpha(n)}\}$$
$$\ge P_\mu\{C_{\alpha(n),1,j} \le R_N \text{ for } j \in I_n \mid \mathcal{F}_{\alpha(n)}\} \text{ a.s..}$$

As in Section 3.4.2, define $Z_{n,1,j}$ to be the amount of time that has elapsed on the clock for transition $e_{1,j}$ between the most recent clock-setting time prior to ζ_n and time ζ_n itself. Since each F_j is NBU, it follows from Remark 4.14 in Chapter 3 that

$$P_\mu\big\{\, C_{\alpha(n),1,j} \leq R_N \text{ for } j \in I_n \mid \mathcal{F}_{\alpha(n)} \,\big\}$$

$$= \prod_{j \in I_n} \left(1 - \frac{\overline{F}_j(R_N + Z_{\alpha(n),1,j})}{\overline{F}_j(Z_{\alpha(n),1,j})} \right)$$

$$\geq \prod_{j \in I_n} F_j(R_N)$$

$$\geq \prod_{j=1}^N F_j(R_N) \text{ a.s..}$$

Each quantity $F_j(R_N)$ is positive by (2.18), so that (2.5) holds with $\delta = \prod_{j=1}^N F_j(R_N)$. The desired result now follows from Lemma 2.4. Observe that Corollary 1.26 cannot be used to establish recurrence: the clock-setting distribution functions for transitions $e_{3,1}, e_{3,2}, \ldots, e_{3,N}$ are degenerate and therefore do not satisfy the positive density condition in Assumption PD.

The key steps of the recurrence argument in the foregoing example are as follows:

1. Show that $S_{\beta(n)} \in \bar{G}$ if the clock readings for the enabled events in a specified set $\tilde{E}$ are "small enough" just after the $\alpha(n)$th marking change. This implication constitutes the "sample path condition." In Example 2.17, $\tilde{E} = \{\, e_{1,1}, \ldots, e_{1,N} \,\}$ and "small enough" means that each clock reading is less than R_N.

2. Require that each event in $\tilde{E}$ has an NBU clock-setting distribution function. Then the probability that the clock readings at time $\zeta_{\alpha(n)}$ for the enabled events in $\tilde{E}$ are small enough is bounded below by the probability that fresh samples from the clock-setting distributions are small enough. This step in the argument rests on an appropriate representation of conditional clock-reading distributions; in Example 2.17, we use the representation given by Lemma 4.10 in Chapter 3.

3. Impose a "positivity condition" on the clock-setting distribution functions which ensures that the latter probability in (2) is positive. This positive probability value serves as the constant δ in (2.5), and the desired result follows. In Example 2.17, the positivity condition is given by (2.18).

It is easy to weaken the NBU assumption in the foregoing argument and require only that each F_j be GNBU with lower bound x_j^* satisfying

$$x_j^* < R_N. \tag{2.19}$$

Set $\gamma_j(x) = \sup_{y \geq 0} \overline{F}_j(x+y)/\overline{F}_j(y)$ for $1 \leq j \leq N$. Then

$$P_\mu\left\{ C_{\alpha(n),1,j} \leq R_N \text{ for } j \in I_n \mid \mathcal{F}_{\alpha(n)} \right\}$$

$$= \prod_{j \in I_n} \left(1 - \frac{\overline{F}_j(R_N + Z_{\alpha(n),1,j})}{\overline{F}_j(Z_{\alpha(n),1,j})} \right)$$

$$\geq \prod_{j \in I_n} \left(1 - \gamma_j(R_N) \right)$$

$$\geq \prod_{j=1}^{N} \left(1 - \gamma_j(R_N) \right) \text{ a.s..}$$

It follows from (2.19) and the definition of the GNBU property that $\left(1 - \gamma_j(R_N) \right) > 0$ for each j, so that (2.5) holds with $\delta = \prod_{j=1}^{N}\left(1 - \gamma_j(R_N) \right)$.

In the remainder of the chapter, we show how arguments such as those given above can be extended and applied to a variety of SPN models. In each of our examples, one or more of the clock-setting distribution functions fails to satisfy the positive density condition in Assumption PD, so that Corollary 1.26 is not applicable.

5.2.4   Recurrence Theorems

We can extend the argument in Example 2.17 not only by replacing the NBU distributional assumptions with weaker GNBU assumptions, but also by using more elaborate sample path and positivity conditions. Theorem 2.21 below is a general result in this direction and is applicable to a variety of models encountered in practice. In the theorem the sequences $\{\beta(n): n \geq 1\}$ and $\{\alpha(n): n \geq 0\}$ are as in Lemma 2.4, and we define G_α to be the state space of the process $\{S_{\alpha(n)}: n \geq 1\}$. In addition, $\{k(i,j,s): s \in G_\alpha, \ 1 \leq i,j \leq M\}$ is a collection of finite nonnegative integers such that

$$k(i,j) \stackrel{\text{def}}{=} \sup_{s \in G_\alpha} k(i,j,s) < \infty \tag{2.20}$$

for each i and j. Finally, denote by $\alpha(n,j,l)$ $(n \geq 1, 1 \leq j \leq M$, and $l \geq 1)$ the random index of the lth marking change after $\alpha(n)$ at which transition e_j becomes enabled and by $A_{n,j,l} = C_{\alpha(n,j,l),j}$ the value of the corresponding new clock reading for e_j. For ease of exposition, we suppose that all transitions are simple and all speeds are equal to 1; extending the results in this section to the general case is straightforward.

Theorem 2.21. *Let $\tilde{E} \subseteq E - E'$, $e_q \in E - E'$, and $\bar{G} \subseteq G$; and let $\{x_i^*: e_i \in \tilde{E}\}$ be a collection of nonnegative numbers. Also let $\{\beta(n): n \geq 1\}$ and $\{\alpha(n): n \geq 0\}$ be as in Lemma 2.4 and $\{k(i,j,s): s \in G_\alpha, 1 \leq i, j \leq M\}$ be nonnegative integers satisfying (2.20). Set $\tilde{E}_n = \tilde{E} \cap E(S_{\alpha(n)})$ and $K_n(i,j) = k(i,j,S_{\alpha(n)})$, and suppose that*

(i) *for each $e_i \in \tilde{E}$ the clock-setting distribution function $F(\cdot\,; e_i)$ is* GNBU *with lower bound x_i^*,*

(ii) *$e_q \in N(S_{\alpha(n)}; S_{\alpha(n)-1}, E_{\alpha(n)-1}^*)$ and*

$$P_\mu\left\{ S_{\beta(n)} \in \bar{G} \mid \mathcal{F}_{\alpha(n)} \right\}$$
$$\geq P_\mu\left\{ C_{\alpha(n),i} + \sum_{j=1}^{M} \sum_{l=1}^{K_n(i,j)} A_{n,j,l} < C_{\alpha(n),q}, \; e_i \in \tilde{E}_n \mid \mathcal{F}_{\alpha(n)} \right\} \; a.s.$$
(2.22)

for $n \geq 0$, and

(iii) *the positivity condition*

$$x_i^* + \sum_{j=1}^{M} k(i,j)y_j < z \qquad \text{for } e_i \in \tilde{E}$$
(2.23)

holds, where $z = \operatorname{ess\,sup} F(\cdot\,; e_q)$ and $y_j = \operatorname{ess\,inf} F(\cdot\,; e_j)$ for $1 \leq j \leq M$.

Then $P_\mu\{ S_{\beta(n)} \in \bar{G} \; i.o. \} = 1$.

PROOF. Fix $n \geq 1$. For $e_i \in \tilde{E}$, write $F_i(\cdot) = F(\cdot\,; e_i)$ and set $\gamma_i(x) = \sup_{y \geq 0} \overline{F}_i(x+y)/\overline{F}_i(y)$. Also write

$$U_{n,i} = C_{\alpha(n),q} - \sum_{j=1}^{M} \sum_{l=1}^{K_n(i,j)} A_{n,j,l}$$

and set $\mathcal{G}_n = \{U_{n,i}: e_i \in \tilde{E}_n\}$. Next, set

$$\tilde{U}_i = B_q - \sum_{j=1}^{M} \sum_{l=1}^{k(i,j)} A_{j,l},$$

where B_q is an independent sample from $F(\cdot\,; e_q)$ and each $A_{j,l}$ is an independent sample from $F(\cdot\,; e_j)$. Observe that $K_n(i,j) \leq k(i,j)$ a.s. for each i and j, so that $\tilde{U}_i$ is stochastically smaller than $U_{n,i}$ for each i. As before, denote by $Z_{n,i}$ the amount of time that has elapsed on the clock for transition e_i between the most recent clock-setting time prior to ζ_n and time ζ_n itself. We then have

$$P_\mu \left\{ S_{\beta(n)} \in \bar{G} \mid \mathcal{F}_{\alpha(n)} \right\}$$

$$\geq P_\mu \left\{ C_{\alpha(n),i} \leq U_{n,i} \text{ for } e_i \in \tilde{E}_n \mid \mathcal{F}_{\alpha(n)} \right\}$$

$$= E_\mu \left[P\left\{ C_{\alpha(n),i} \leq U_{n,i} \text{ for } e_i \in \tilde{E}_n \mid \mathcal{F}_{\alpha(n)}, \mathcal{G}_n \right\} \mid \mathcal{F}_{\alpha(n)} \right]$$

$$= E_\mu \left[\prod_{e_i \in \tilde{E}_n} \left(1 - \frac{\overline{F}_i(U_{n,i} + Z_{\alpha(n),i})}{\overline{F}_i(Z_{\alpha(n),i})} \right) \Bigg| \mathcal{F}_{\alpha(n)} \right]$$

$$\geq E \left[\prod_{e_i \in \tilde{E}} \left(1 - \gamma_i(\tilde{U}_i) \right) \right] \text{ a.s.,}$$

where the first inequality follows from condition (ii) of the theorem and the second equality follows from Lemmas 4.10 and 4.19 in Chapter 3. To complete the proof, let w_i ($i \in \tilde{E}$) be the essential supremum of the distribution of $\tilde{U}_i$ and observe that $w_i = z - \sum_{j=1}^M k(i,j)y_j$. Next, write $\tilde{E} = \{ e_{i_1}, e_{i_2}, \ldots, e_{i_r} \}$ and, for $u = (u_1, u_2, \ldots, u_r) \in \Re_+^r$, set

$$g(u) = \prod_{m=1}^r \left(1 - \gamma_{i_m}(u_m) \right).$$

Denote by H the distribution function of the random vector $(\tilde{U}_{i_1}, \ldots, \tilde{U}_{i_r})$, and set $R = [x_{i_1}^*, w_{i_1}] \times \cdots \times [x_{i_r}^*, w_{i_r}]$. Observe that

$$E \left[\prod_{e_i \in \tilde{E}} \left(1 - \gamma_i(\tilde{U}_i) \right) \right] \geq \delta,$$

where $\delta = \int_R g \, dH$. Condition (i) of the theorem implies that g is positive on the set R, and condition (iii) implies that $\int_R dH > 0$. Thus $\delta > 0$—see Lemma 1.23 in the Appendix—and the desired result follows by Lemma 2.4.
□

EXAMPLE 2.24 (Cyclic queues). Consider a closed network of queues with two single-server service centers and N (≥ 2) jobs. A job that completes service at center 1 moves to center 2; a job that completes service at center 2 moves to center 1. Both queueing disciplines are first-come, first-served. Successive service times at center i ($i = 1, 2$) are i.i.d. as a positive random variable L_i. The random variable L_1 is uniformly distributed on the interval $[a, b]$ for some $0 < a < b$, and the random variable L_2 is uniformly

e_1 = service completion at center 1

e_2 = service completion at center 2

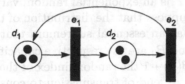

Figure 5.6. SPN representation of cyclic queues (five jobs).

distributed on the interval $[0, (N-1)a + \epsilon]$ for some $\epsilon > 0$. This system can be specified as an SPN with two timed deterministic transitions as in Figure 5.6.

Denote by $\beta(n)+1$ the random index of the nth marking change at which transition e_2 fires. Suppose we wish to show that $P_\mu\{ S_{\beta(n)} = \bar{s} \text{ i.o.} \} = 1$, where $\bar{s} = (0, N)$. Let $\alpha(n)$ be the random index of the nth marking change at which transition e_2 becomes enabled. Clearly, every $\beta(n)$ and $\alpha(n)$ is a.s. finite. Observe that $S_{\beta(n)} = \bar{s}$ if all the jobs at center 1 at time $\zeta_{\alpha(n)}$ complete service and move to center 2 during the interval $[\zeta_{\alpha(n)}, \zeta_{\beta(n)+1})$, so that

$$P_\mu\big\{ S_{\beta(n)} = \bar{s} \mid \mathcal{F}_{\alpha(n)} \big\}$$
$$\geq P_\mu\big\{ C_{\alpha(n),1} + A_{n,1,1} + \cdots + A_{n,1,J} < C_{\alpha(n),2} \mid \mathcal{F}_{\alpha(n)} \big\} \text{ a.s.,}$$

where $J = S_{\alpha(n),1} - 1$ and $A_{n,1,1}, A_{n,1,2}, \ldots$ are the successive center 1 service times that start after $\zeta_{\alpha'(n)}$. That is, (2.22) holds with

- $\bar{G} = \{ \bar{s} \}$,

- $e_q = e_2$,

- $\tilde{E} = \{ e_1 \}$,

- $k(i, j, s) = s_1 - 1$ for $i = 1$, $j = 1$ and $s = (s_1, s_2) \in G_\alpha$, and
  $k(i, j, s) = 0$ otherwise.

By Lemma 2.9(iii), the distribution function of L_1 is GNBU with lower bound $x_1^* = a$. Moreover, $G_\alpha = \{ (s_1, s_2) \in S : s_1 \leq N - 1 \}$. Thus the positivity condition (2.23) holds with

$$x_1^* + \sum_{j=1}^{M} k(1, j) y_1 = x_1^* + k(1, 1) y_1 = a + (N-2)a = (N-1)a$$

and $z = (N-1)a + \epsilon$. The desired result now follows from Theorem 2.21.

EXAMPLE 2.25 (Producer–consumer system with nonpreemptive priority).
For the system of Example 2.1 in Chapter 2, suppose that the creation-time
random variables A_1 and A_2 are each distributed as $Y + a$, where a is a
positive constant and Y is an exponential random variable with intensity q
for some $q > 0$. Also suppose that the distribution of the transmission-time
random variable L_1 has an essential supremum that exceeds $\max((B_1 -
1)a, B_2 a)$, where B_1 and B_2 are the respective capacities of buffers 1 and 2
as before. Denote by $\beta(n) + 1$ the random index of the nth marking change
at which transition $e_3 =$ "end of transmission to consumer 1" fires.

Suppose we wish to show that $P_\mu\{ S_{\beta(n)} = \bar{s} \text{ i.o.} \} = 1$, where $\bar{s} =
(0, B_1 - 1, 1, 0, B_2, 0, 0)$. The marking is $\bar{s}$ if and only if there are B_1 items in
buffer 1—one of which is being transmitted to consumer 1—and B_2 items
in buffer 2.

Denote by $\alpha(n)$ the random index of the nth marking change at which
transition $e_2 =$ "start of transmission to consumer 1" fires. Using the fact
that producer–consumer pair 1 has nonpreemptive priority over producer–
consumer pair 2 for use of the channel, it is straightforward to show that
each $\beta(n)$, and hence each $\alpha(n)$, is a.s. finite.

Fix $n \geq 1$ and observe that $S_{\beta(n)} = \bar{s}$ if producers 1 and 2 create $S_{\alpha(n),1}$
and $S_{\alpha(n),4}$ items, respectively, in the interval $[\zeta_{\alpha(n)}, \zeta_{\beta(n)+1})$. Suppose that
at time $\zeta_{\alpha(n)}$ both producer 1 and producer 2 are creating an item. Then the
foregoing event certainly will occur if, starting at time $\zeta_{\alpha(n)}$, the residual
creation time $C_{\alpha(n),1}$ plus the sum of the next $S_{\alpha(n),1} - 1$ creation times
for producer 1 is less than $\zeta_{\beta(n)+1} - \zeta_{\alpha(n)}$, and similarly for $C_{\alpha(n),4}$ plus
the sum of the next $S_{\alpha(n),4} - 1$ creation times for producer 2. A similar
analysis holds for other possible scenarios at time $\zeta_{\alpha(n)}$, and it follows that
(2.22) holds with

- $\bar{G} = \{\bar{s}\}$,

- $e_q = e_3$,

- $\tilde{E} = \{e_1, e_4\}$,

- $k(i, i, s) = s_i - 1$ for $i = 1, 4$ and $s = (s_1, s_2, \ldots, s_7) \in G_\alpha$, and
  $k(i, j, s) = 0$ otherwise.

By Lemma 2.9(ii), the common distribution of A_1 and A_2 is GNBU with
lower bound a, and the positivity condition (2.23) holds with

$$x_1^* + \sum_{j=1}^{M} k(1, j)y_j = x_1^* + k(1, 1)y_1 = a + (B_1 - 2)a = (B_1 - 1)a$$

and, similarly,

$$x_4^* + \sum_{j=1}^{M} k(4, j)y_j = x_4^* + k(4, 4)y_4 = a + (B_2 - 1)a = B_2 a.$$

The desired result now follows from Theorem 2.21.

EXAMPLE 2.26 (Collision-free bus network). For the system of Example 3.7 in Chapter 2, suppose that the interarrival-time random variables $A_2, A_3, \ldots, A_N$ have GNBU distribution functions with respective lower bounds $x_2^*, x_3^*, \ldots, x_N^*$. Also suppose that

$$x_j^* + R(j) + T < z \tag{2.27}$$

for $2 \leq j \leq N$, where z is the essential supremum of the distribution of the transmission-time random variable L_1. Denote by $\beta(n) + 1$ the random index of the nth marking change at which transition $e_{4,1} = $ "end of transmission by port 1" fires.

Suppose we wish to show that $P_\mu\{ S_{\beta(n)} = \bar{s} \text{ i.o.} \} = 1$, where $\bar{s}$ is the unique marking such that $\bar{s}_{3,j} = 1$ for $2 \leq j \leq N$ and $\bar{s}_{4,1} = 1$. The marking is $\bar{s}$ if and only if a transmission by port 1 is underway and ports 2 through N each have a packet awaiting transmission, have completed the $R(j) + T$ wait, and have observed the setting (to 1) of the flip-flop by all ports to the left.

Denote by $\alpha(n)$ the random index of the nth marking change at which transition $e_{3,1} = $ "start of transmission by port 1" fires. Using the fact that the OR-signal for port 1 is always equal to 0 (since there are no ports to the left), it can be shown that each $\beta(n)$, and hence each $\alpha(n)$, is a.s. finite.

Fix $n \geq 1$ and suppose that at time $\zeta_{\alpha(n)}$ no port has a packet awaiting transmission. Observe that $S_{\beta(n)} = \bar{s}$ if each port j ($2 \leq j \leq N$) receives a packet for transmission and completes the $R(j) + T$ wait in the interval $[\zeta_{\alpha(n)}, \zeta_{\beta(n)+1})$. A similar analysis holds for other possible scenarios at time $\zeta_{\alpha(n)}$, and it follows that (2.22) holds with[3]

- $\bar{G} = \{ \bar{s} \}$,

- $e_q = e_{4,1}$,

- $\tilde{E} = \{ e_{l,j} : l = 1, 2 \text{ and } 2 \leq j \leq N \}$,

- $k(\{1,j\}, \{2,j\}, s) = 1$ for $2 \leq j \leq N$ and $s \in G_\alpha$ such that $s_{1,j} = 1$, and $k(\cdot, \cdot, s) = 0$ otherwise.

Each clock-setting distribution function $F(\cdot; e_{1,j})$ is GNBU by assumption. Moreover, it follows from our previous discussion that each (degenerate) distribution function $F(\cdot; e_{2,j})$ is NBU and hence GNBU with lower bound $x_{2,j}^* = R(j) + T$ by Lemma 2.9. Observe that each inequality in the positivity condition (2.23) is of the form $x_j^* + R(j) + T < z$ or $x_{2,j}^* < z$. Because $x_{2,j}^* = R(j) + T$, it follows from (2.27) that the positivity condition in (2.23) holds. The desired result now follows from Theorem 2.21.

---

[3]For this SPN model, each transition is doubly or triply subscripted, and the notation in (2.22) is modified accordingly.

The next result is a variant of Theorem 2.21 in which the sample path condition consists of a single inequality, but this inequality involves a sum of residual clock readings. The proof is similar to that of Theorem 2.21 and uses Lemma 2.16. In the theorem, the sequences $\{\beta(n): n \geq 1\}$ and $\{\alpha(n): n \geq 0\}$ are as in Lemma 2.4 and, as before, G_α is the state space of the process $\{S_{\alpha(n)}: n \geq 1\}$. In addition, $\{k(j,s): s \in G_\alpha, 1 \leq j \leq M\}$ is a collection of finite nonnegative integers such that

$$k(j) \stackrel{\text{def}}{=} \sup_{s \in G_\alpha} k(j,s) < \infty \qquad (2.28)$$

for each j. As before, the quantity $\alpha(n, j, l)$ is the random index of the lth marking change after $\alpha(n)$ at which transition e_j becomes enabled, and $A_{n,j,l} = C_{\alpha(n,j,l),j}$.

Theorem 2.29. *Let $\tilde{E} \subseteq E - E'$, $e_q \in E - E'$, and $\bar{G} \subseteq G$; and let $\{x_i^*: e_i \in \tilde{E}\}$ be a collection of nonnegative numbers. Also let $\{\beta(n): n \geq 1\}$ and $\{\alpha(n): n \geq 0\}$ be as in Lemma 2.4 and $\{k(j,s): s \in G_\alpha, 1 \leq j \leq M\}$ be nonnegative integers satisfying (2.28). Set $\tilde{E}_n = \tilde{E} \cap E(S_{\alpha(n)})$ and $K_n(j) = k(j, S_{\alpha(n)})$, and suppose that*

(i) for each $e_i \in \tilde{E}$ the clock-setting distribution function $F(\cdot\,; e_i)$ is GNBU with lower bound x_i^*,

(ii) $e_q \in N(S_{\alpha(n)}; S_{\alpha(n)-1}, E_{\alpha(n)-1}^*)$ and

$$P_\mu \left\{ S_{\beta(n)} \in \bar{G} \mid \mathcal{F}_{\alpha(n)} \right\}$$
$$\geq P_\mu \left\{ \sum_{e_i \in \tilde{E}_n} C_{\alpha(n),i} + \sum_{j=1}^{M} \sum_{l=1}^{K_n(j)} A_{n,j,l} < C_{\alpha(n),q} \;\middle|\; \mathcal{F}_{\alpha(n)} \right\} \quad a.s.$$
$$(2.30)$$

for $n \geq 0$, and

(iii) the positivity condition

$$\sum_{e_i \in \tilde{E}} x_i^* + \sum_{j=1}^{M} k(j) y_j < z \qquad (2.31)$$

holds, where $z = \operatorname{ess\,sup} F(\cdot\,; e_q)$ and $y_j = \operatorname{ess\,inf} F(\cdot\,; e_j)$ for $1 \leq j \leq M$.

Then $P_\mu \{ S_{\beta(n)} \in \bar{G} \text{ i.o.} \} = 1$.

EXAMPLE 2.32 (Cyclic queues). Consider a closed network of queues with three single-server service centers and N (≥ 2) jobs. A job that completes service at center i ($i = 1, 2$) moves to center $i + 1$; a job that completes

e_1 = service completion at center 1

e_2 = service completion at center 2

e_3 = service completion at center 3

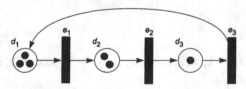

Figure 5.7. SPN representation of cyclic queues (three tandem servers and six jobs).

service at center 3 moves to center 1. All queueing disciplines are first-come, first-served. Successive service times at center i ($i = 1, 2, 3$) are i.i.d. as a positive random variable L_i. Both L_1 and L_2 have a truncated normal distribution with density $f(x) = (2/\pi)^{-1/2}\exp(-x^2/2)$ for $x \geq 0$. L_3 is uniformly distributed on $[1, 5]$. This system can be specified as an SPN with three timed deterministic transitions as in Figure 5.7.

Denote by $\beta(n)+1$ the random index of the nth marking change at which transition e_3 = "service completion at center 3" fires. Suppose we wish to show that $P_\mu\{S_{\beta(n)} = \bar{s} \text{ i.o.}\} = 1$, where $\bar{s} = (0, 0, N)$. Let $\alpha(n)$ be the random index of the nth marking change at which transition e_3 becomes enabled. Clearly, every $\beta(n)$ and $\alpha(n)$ is a.s. finite. Fix $n \geq 1$ and observe that $S_{\beta(n)} = \bar{s}$ if each of the jobs at centers 1 and 2 at time $\zeta_{\alpha(n)}$ moves to center 3 during the interval $[\zeta_{\alpha(n)}, \zeta_{\beta(n)+1})$. Suppose there are at least two jobs at center 1 and at center 2 at time $\zeta_{\alpha(n)}$. Then, for a job waiting in queue at center 1, the time for the job to move to center 3 is the sum of the job's residual waiting time (in queue) at center 1, the job's next service time at center 1, the job's next waiting time at center 2, and the job's next service time at center 2. An upper bound U_n on this total time is obtained by summing

1. The residual service time of the job in service at center 1 (at time $\zeta_{\alpha(n)}$)

2. The next center 1 service time for each job in queue at center 1

3. The next center 2 service time for each job in queue at center 1

4. The residual service time of the job in service at center 2

5. The next center 2 service time for each job in queue at center 2

Indeed, U_n is an upper bound on the time for *any* job at center 1 or 2 to move to center 3. Thus $S_{\beta(n)} = \bar{s}$ if U_n does not exceed $\zeta_{\beta(n)+1} - \zeta_{\alpha(n)} = $

$C_{\alpha(n),3}$. A similar analysis applies to each other possible scenario at time $\zeta_{\alpha(n)}$, and (2.30) holds with

- $\bar{G} = \{\bar{s}\}$,

- $e_q = e_3$,

- $\tilde{E} = \{e_1, e_2\}$,

- $k(1, s) = s_1 - 1$, $k(2, s) = s_1 + s_2 - 1$, and $k(3, s) = 0$ for $s = (s_1, s_2, s_3) \in G_\alpha$.

As mentioned previously, the truncated normal distribution is NBU and hence GNBU with lower bound 0. Because the essential infimum of this distribution also is equal to 0, the positivity condition (2.31) holds trivially. The desired result now follows from Theorem 2.21.

5.2.5 Some Ad-Hoc Recurrence Arguments

The foregoing recurrence theorems, though applicable to a variety of SPN models, certainly do not cover all possible SPNs of interest. We conclude the present chapter by showing how Lemmas 4.10 and 4.19 in Chapter 3, Lemma 2.16 in the current chapter, and extensions of these results can be used to establish recurrence directly for some specific SPN models. For each model, the idea is to show that there exists a collection of positive constants $\{\delta(s^+): s^+ \in G_\alpha\}$ such that

$$P_\mu\left\{S_{\beta(n)} \in \bar{G} \mid \mathcal{F}_{\alpha(n)}\right\} \geq \delta(S_{\alpha(n)}) \text{ a.s.} \tag{2.33}$$

for $n \geq 0$; here $\{\beta(n): n \geq 1\}$ and $\{\alpha(n): n \geq 1\}$ are as in Lemma 2.4 and G_α is the state space of $\{S_{\alpha(n)}: n \geq 1\}$. Provided that the set G_α is finite, the inequality in (2.5) holds because $\delta(S_{\alpha(n)}) \geq \delta$ a.s., where $\delta = \min_{s^+ \in G_\alpha} \delta(s^+) > 0$. The recurrence of A then follows from Lemma 2.4.

EXAMPLE 2.34 (Manufacturing cell with robots). For the system of Example 3.6 in Chapter 2, denote by R_1 the (constant) time for robot 1 to return to its null position after transfer of a part to conveyor 1; we assume that this time is greater than the time for robot 1 to return to its null position after transfer of a part to the unloading area. Similarly, denote by R_2 the (constant) time for robot 2 to return to its null position after transfer of a part to machine 1. Suppose that the machine 2 processing-time random variable L_2 has an exponential distribution with intensity q for some $q > 0$. Also suppose that the distribution function of the machine 1 processing-time random variable L_1 has an infinite essential supremum. Denote by $\beta(n) + 1$ the random index of the nth marking change at which transition $e_8 = $ "end of processing by machine 1" fires.

Suppose we wish to show that $P_\mu\{S_{\beta(n)} = \bar{s} \text{ i.o.}\} = 1$, where $\bar{s}$ is the unique marking such that $\bar{s}_4 = \bar{s}_9 = \bar{s}_{11} = \bar{s}_{22} = \bar{s}_{24} = 1$ and $\bar{s}_j = 0$ otherwise. The marking is $\bar{s}$ if and only if machines 1 and 2 are each processing a part, a part is on conveyor 1 awaiting transfer to a machine, no parts are on conveyor 2, and each robot is in its null position.

Denote by $\alpha(n)$ the random index of the nth marking change at which transition e_8 becomes enabled. Using the fact that robot 2 transfers raw parts from conveyor 1 to the lowest-numbered available machine and that transfer of a part from machine 1 has priority over transfer of a part either to or from machine 2, it can be shown that each $\beta(n)$, and hence each $\alpha(n)$, is a.s. finite.

We claim that there exists a collection of positive constants $\{\delta(s^+): s^+ \in G_\alpha\}$ such that (2.33) holds. To see this, fix $n \geq 1$ and suppose, for example, that $S_{\alpha(n)} = s^+$, where $s_4^+ = s_9^+ = s_{11}^+ = s_{21}^+ = s_{23}^+ = 1$ and $s_j^+ = 0$ otherwise. Then each machine is processing a part, a part is on conveyor 1 awaiting transfer to a machine, no parts are on conveyor 2, and each robot is returning to its null position. Observe that $R = \max(R_1, R_2)$ is an upper bound on the time for both robots to return to their null positions. Also observe that $S_{\beta(n)} = \bar{s}$ if each robot returns to its null position in the interval $[\zeta_{\alpha(n)}, \zeta_{\beta(n)+1})$ and machine 2 does not finish processing a part in this interval. It follows from Lemma 4.10 in Chapter 3 that, given $\mathcal{F}_{\alpha(n)}$, the conditional probability that the transitions fire in this way is bounded below by

$$\delta(s^+) = \int_R^\infty e^{-qx} \, dF(x; e_8),$$

on the set $\{S_{\alpha(n)} = s^+\}$. The constant $\delta(s^+)$ is the probability that an independent sample A_8 from the clock-setting distribution function $F(\cdot; e_8)$ and an independent sample A_9 from the (exponential) clock-setting distribution function $F(\cdot; e_9)$ satisfy $R < A_8 < A_9$. Note that $\delta(s^+)$ is positive since $\text{ess sup} F(\cdot; e_8) = \text{ess sup} F(\cdot; e_9) = \infty$ by assumption. A similar analysis can be performed for each state $s^+ \in G_\alpha$, and the desired result follows.

EXAMPLE 2.35 (Telephone system). For the system of Example 1.27, suppose that $N > 5$ and $M > 2$, and that the call-length random variables $L_1, L_2, L_3, \ldots, L_N$ have a common distribution function H that is GNBU with lower bound x^*. Also suppose that the waiting-time random variables $A_1, A_2, \ldots, A_N$ are each distributed according to an exponential distribution function with intensity q for some $q > 0$. Finally, suppose that

$$x^* < \text{ess sup} H. \tag{2.36}$$

Denote by $\beta(n) + 1$ the random index of the nth marking change at which transition $e_{2,1} = $ "end of call connected on link 1" fires.

Suppose we wish to show that $P_\mu\{S_{\beta(n)} \in \bar{G} \text{ i.o.}\} = 1$, where $\bar{s} \in \bar{G}$ if and only if $\bar{s}_{3,1} = 1$ and $\bar{s}_{3,m} = 0$ for $2 \leq m \leq K$. The marking is an element of $\bar{G}$ if and only if a call is connected on link 1 and all other links are idle.

Denote by $\alpha(n)$ the random index of the nth marking change at which transition $e_{2,1}$ becomes enabled. Using the fact that a placed call always is connected on the lowest available link, it is not hard to show that each $\beta(n)$, and hence each $\alpha(n)$, is a.s. finite.

We claim that there exists a collection of positive constants $\{\,\delta(s^+)\colon s^+ \in G_\alpha\,\}$ such that (2.33) holds. To see this, fix $n \geq 1$ and suppose, for example, that $S_{\alpha(n)} = s^+$, where $s_{2,1,1}^+ = s_{2,2,1}^+ = s_{2,3,2}^+ = s_{2,4,2}^+ = 2$ and $s_{1,j}^+ = 1$ for $5 \leq j \leq N$. That is, at time $\zeta_{\alpha(n)}$ lines 1 and 2 are connected on link 1, lines 3 and 4 are connected on link 2, and no other lines are connected. Clearly, $S_{\beta(n)} \in \bar{G}$ if the call underway on link 2 completes before time $\zeta_{\beta(n)+1}$ and no calls are placed in the interval $[\zeta_{\alpha(n)}, \zeta_{\beta(n)+1})$. It follows that

$$P_\mu\left\{\, S_{\beta(n)} \in \bar{G} \,\big|\, \mathcal{F}_{\alpha(n)}\,\right\}$$
$$\geq P_\mu\Big\{\, C_{\alpha(n),2,2} \leq C_{\alpha(n),2,1},\ C_{\alpha(n),1,j} > C_{\alpha(n),2,1} \text{ for } 5 \leq j \leq N,$$
$$\text{and } C_{\nu(n,j),1,j} > C_{\alpha(n),2,1} \text{ for } j = 3,4 \,\big|\, \mathcal{F}_{\alpha(n)}\,\Big\}$$

on the set $\{\,S_{\alpha(n)} = s^+\,\}$, where $\nu(n,j)$ $(1 \leq j \leq 4)$ is the random index of the first marking change after $\alpha(n)$ at which transition $e_{1,j}$ becomes enabled. A straightforward application of Lemmas 4.10 and 4.19 in Chapter 3 shows that the right side of the above inequality is bounded below by

$$\delta(s^+) = \int_{x^*}^\infty \gamma(x)\, dH(x),$$

on the set $\{\,S_{\alpha(n)} = s^+\,\}$, where

$$\gamma(x) = \left(1 - \sup_{y \geq 0} \frac{\overline{H}(x+y)}{\overline{H}(y)}\right) e^{-(N-2)qx}$$

for $x \geq 0$. The GNBU assumption on H implies that γ is positive on (x^*, ∞), and the positivity condition in (2.36) implies that $\int_{x^*}^\infty dH > 0$. It follows that $\delta(s^+) > 0$. A similar analysis can be performed for each state $s^+ \in G_\alpha$, and the desired result follows.

EXAMPLE 2.37 (Cyclic queues with feedback). For the network of Example 1.4 in Chapter 2, suppose that the service-time distribution at center 1 is GNBU and that the essential supremum of the service-time distribution at center 2 is infinite. Represent this system by an SPN as in Example 2.6 in Chapter 4, and denote by $\beta(n)+1$ the random index of the nth marking change at which transition $e_2 =$ "service completion at center 2" fires. Also denote by $\alpha(n)$ the random index of the nth marking change at which e_2 becomes enabled. It is easy to see that transition e_1 fires infinitely often with probability 1, and an application of the Borel–Cantelli lemma (Proposition 1.3 in the Appendix) shows that, with probability 1, infinitely many service completions at center 1 result in a job moving from center 1 to center 2. It follows that transition e_2 fires infinitely often with probability 1, and hence every $\alpha(n)$ and $\beta(n)$ is a.s. finite.

Suppose we wish to show that $P_\mu\{ S_{\beta(n)} = \bar{s} \text{ i.o.} \} = 1$, where $\bar{s} = (0, N)$. We claim that there exists a collection of positive constants $\{ \delta(s^+): s^+ \in G_\alpha \}$ such that (2.33) holds. To see this, fix $n \geq 1$ and suppose, for example, that $S_{\alpha(n)} = s^+$, where $s^+ = (m, N - m)$ with $m > 0$. Clearly, $S_{\beta(n)} = \bar{s}$ if all m jobs at center 1 complete service and move to center 2 during the interval $[\zeta_{\alpha(n)}, \zeta_{\beta(n)+1})$. It follows that

$$P_\mu\left\{ S_{\beta(n)} = \bar{s} \mid \mathcal{F}_{\alpha(n)} \right\}$$

$$\geq P_\mu\left\{ C_{\alpha(n),1} + \sum_{l=1}^{m-1} C_{\alpha(n,1,l),1} < C_{\alpha(n),2} \right.$$

$$\left. \text{and } S_{\nu(n,1,l),1} = m - l \text{ for } 1 \leq l \leq m \;\middle|\; \mathcal{F}_{\alpha(n)} \right\},$$

on the set $\{ S_{\alpha(n)} = s^+ \}$, where $\alpha(n, 1, l)$ is the random index of the lth marking change after $\alpha(n)$ at which transition e_1 becomes enabled and $\nu(n, 1, l)$ is the random index of the lth marking change after $\alpha(n)$ at which e_1 fires. Recall that p is the probability that a job moves to center 2 upon completion of service at center 1. An argument similar to the proof of Lemma 4.10 in Chapter 3 shows that the right side of the above inequality is bounded below by

$$\delta(s^+) = p^m \inf_{y \geq 0} P\left\{ (A_1 - y) + A_2 + \cdots + A_m < B | A_1 > y \right\},$$

on the set $\{ S_{\alpha(n)} = s^+ \}$, where the random variables $A_1, A_2, \ldots, A_m$ are i.i.d. according to $F(\,\cdot\,; e_1)$ and B is distributed according to $F(\,\cdot\,; e_2)$. It follows from Lemma 2.16 that $\delta(s^+) > 0$. A similar analysis can be performed for each state $s^+ \in G_\alpha$, and the desired result follows.

EXAMPLE 2.38 (Token ring). We can weaken the positivity condition used to establish recurrence for the marking $\bar{s}$ in Example 2.17. (Recall that the marking $\bar{s}$ corresponds to the state in which all ports have a packet awaiting transmission and the ring token is propagating from port N to port 1.) The idea is to use Lemma 4.19 in Chapter 3 rather than Lemma 4.10 in that chapter. Specifically, denote by $R_{j,1} = \sum_{i=j}^{N} R_i$ the time for the token to propagate from port j to port 1, and suppose that each interarrival distribution F_j satisfies

$$\text{ess inf } F_j < R_{j,1}; \tag{2.39}$$

cf. (2.18). Also suppose that each F_j is NBU. As in Example 2.17, let $\beta(n)+1$ be the random index of the nth marking change at which transition $e_{3,1} =$

"observation of ring token by port 1" fires. Unlike Example 2.17, set $\alpha(n) = \beta(n-1)$ for $n \geq 1$. Fix $n \geq 1$ and denote by $\nu(n,j)$ the first time at or after $\zeta_{\alpha(n)}$ at which the ring token begins to propagate from port j to the next port. Observe that $S_{\beta(n)} = \bar{s}$ if each transition $e_{1,j}$ fires in the interval $[\zeta_{\nu(n,j)}, \zeta_{\beta(n)+1}]$, and thus

$$P_\mu \left\{ S_{\beta(n)} = \bar{s} \mid \mathcal{F}_{\alpha(n)} \right\} \geq P_\mu \left\{ C_{\nu(n,j),1,j} \leq R_{j,1} \text{ for } 1 \leq j \leq N \mid \mathcal{F}_{\alpha(n)} \right\}.$$

We now bound the term on the right:

$$P_\mu \left\{ C_{\nu(n,j),1,j} \leq R_{j,1} \text{ for } 1 \leq j \leq N \mid \mathcal{F}_{\alpha(n)} \right\}$$

$$= E_\mu \left[P_\mu \left\{ C_{\nu(n,j),1,j} \leq R_{j,1} \text{ for } 1 \leq j \leq N \mid \mathcal{F}_{\nu(n,N)} \right\} \Big| \mathcal{F}_{\alpha(n)} \right]$$

$$= E_\mu \left[P_\mu \left\{ C_{\nu(n,N),1,N} \leq R_{N,1} \mid \mathcal{F}_{\nu(n,N)} \right\} \right.$$

$$P_\mu \left\{ C_{\nu(n,j),1,j} \leq R_{j,1} \text{ for } 1 \leq j \leq N-1 \mid \mathcal{F}_{\nu(n,N)} \right\} \Big| \mathcal{F}_{\alpha(n)} \right]$$

$$\geq F_N(R_{N,1})$$

$$E_\mu \left[P_\mu \left\{ C_{\nu(n,j),1,j} \leq R_{j,1} \text{ for } 1 \leq j \leq N-1 \mid \mathcal{F}_{\nu(n,N)} \right\} \Big| \mathcal{F}_{\alpha(n)} \right]$$

$$\geq F_N(R_{N,1}) \, P_\mu \left\{ C_{\nu(n,j),1,j} \leq R_{j,1} \text{ for } 1 \leq j \leq N-1 \mid \mathcal{F}_{\alpha(n)} \right\} \text{ a.s.,}$$

where the second equality is a consequence of Lemma 4.19 in Chapter 3 and the first inequality is, in the usual way, a consequence of Lemma 4.10 in Chapter 3 and the NBU assumption on each F_j. Iterating the above calculations, we obtain (2.5) with $\delta = \prod_{j=1}^{N} F_j(R_{j,1})$. The positivity condition in (2.39) ensures that $\delta > 0$.

Notes

Our discussion of ϕ-irreducibility and Harris recurrence follows Meyn and Tweedie (1993a); see also Glynn and Meyn (1996) and Haas (1999a, 1999c). In particular, the proof of Proposition 1.13 can be found in these references. Proposition 1.12 is due to Sean Meyn; see Haas (1999c). A proof of Proposition 1.10 can be found in Asmussen (1987a, Section VI.3). The function v that appears in the drift conditions is sometimes called a "stochastic Lyapunov function" in analogy to the ordinary Lyapunov functions that are used to establish stability for systems governed by nonlinear differential equations. Extensions of stability results to continuous-time Markov processes can be found in papers by Meyn and Tweedie (1993b, 1993c).

Some of the results in the literature require that a 1-step drift criterion hold for a chain $\{ Z_n : n \geq 0 \}$. If an m-step drift criterion ($m > 1$) holds with a distance function v, then a 1-step drift criterion holds with distance function $w(z) = E_z [v(Z_0) + v(Z_1) + \cdots + v(Z_{m-1})]$; see Haas (1999c).

Some early results on stability for general discrete-event systems can be found in the work of König et al. (1967, 1974). These authors consider finite-state irreducible GSMPs in which events are never cancelled and in which each clock-setting distribution function has finite mean and a density that is positive on $(0, \infty)$. They show that such GSMPs converge in total variation—see Definition 1.38 in the Appendix—to a unique stationary distribution, and hence are "Harris ergodic" as defined in Section 5.1.1. Sigman (1990a) establishes a drift criterion for closed networks of queues; this work inspired the drift results in the current chapter. There is a large literature concerned with specialized techniques for stability analysis of specific types of discrete-event systems such as "polling" systems and multiclass networks of queues; see, for example, Altman et al. (1992) and Dai (1995).

Our discussion of the positive density conditions follows Haas (1999a, 1999b, 1999c). In these papers a variant of Theorem 1.22 is given in which the requirement that each clock-setting distribution function be an element of $\mathcal{G}^+$ is weakened to require only that each distribution have a finite rth moment for some $r \geq 1$. The resulting (weaker) drift condition is then

$$E_{(s,c)} \left[g_r(S_m^+, C_m^+) - g_r(S_0^+, C_0^+) \right] \leq -\beta g_{r-1}(s, c)$$

for $(s, c) \in \Sigma^+ - H_b$, where $g_r(s, c) = 1 + \max_{1 \leq i \leq M} c_i^r$.

The SPN representation of the telephone system model originally appeared in Haas and Shedler (1991).

When verifying Assumption PD, it typically is straightforward to verify the positive density and moment conditions on the clock-setting distributions and the positivity requirement on the speeds, since the modeller specifies the clock-setting distributions and speeds. It then remains to determine whether the marking set is finite and whether the SPN is irreducible. (These properties also need to be verified when computing steady-state performance measures analytically or numerically for more tractable SPNs such as nets with exponential clock-setting distributions.) When the marking set G is specified explicitly, determining whether $|G| < \infty$ is trivial. In practice, however, G is often defined implicitly as the set of markings reachable (in the sense of the relation $\rightsquigarrow$ in Section 2.4) from some specified set of initial markings; it can then be nontrivial to determine whether G is finite. Under various restrictions on the form of the new-marking probabilities and speeds, both finiteness and irreducibility can be checked, at least in principle, by constructing "coverability graphs" using an algorithm similar to that given in Section 4.2.1 of Peterson (1981). This approach is applicable, for example, to deterministic SPNs—see Section 2.4—having no inhibitor arcs. In general, however, the problem of determining finiteness and irreducibility can be difficult: the marking set can be so large that the computational costs of the coverability analysis are prohibitive or there may exist no algorithm that is guaranteed to terminate. The problem of determining whether $|G| < \infty$, for example, is "undecidable" over the class of all SPNs. On the other hand, there are many SPN models of practical interest for which finiteness and irreducibility can be verified based on the analyst's understanding of the system under study. In such models the difficulty of determining time-average limits arises not so much from the size or complexity of the marking set or SPN graph, but rather from the fact that the clock-setting distributions are nonexponential.

Iglehart and Shedler (1983) originally proposed the use of the geometric trials lemma together with NBU distributional assumptions to establish recurrence. This approach was extended and applied in a variety of contexts by Haas and Shedler (1985a, 1986, 1987a, 1987b, 1989b, 1992, 1993b). A good introduction to NBU distributions, failure rates, and related concepts can be found in Barlow and Proschan (1975).

The sample path conditions in Theorems 2.21 and 2.29 can be combined. The resulting sample path condition consists of a set of inequalities as in (2.22), with each inequality involving sums of residual clock readings as in (2.30); see Haas and Shedler (1987a, 1987b) for examples.

6

Regenerative Simulation

A regenerative stochastic process has the characteristic property that there exists an infinite sequence of random times at which the process probabilistically restarts. As discussed in Section 6.1, the essence of regeneration is that the evolution of the process between any two successive regeneration points is an independent probabilistic replica of the process in any other such "cycle." Under mild regularity conditions, time-average limits for a regenerative process are well defined and finite, provided that the regenerative cycle length has finite mean. The value of a time-average limit is determined by the expected behavior of the process in a single regenerative cycle—a fact that has important implications for simulation analysis. Under some additional regularity conditions, the time-average limit can also be interpreted as a steady-state or limiting mean. Most of these results extend to the setting of "od-equilibrium" and "od-regenerative" processes. Such processes are similar to regenerative processes in that sample paths can be decomposed into identically distributed cycles, but differ in that adjacent cycles need not be independent.

In Section 6.2 we give conditions on the new-marking probabilities, clock-setting distributions, and other building blocks of an SPN under which there exist regeneration points for the marking process $\{ X(t) : t \geq 0 \}$ or the underlying chain $\{ (S_n, C_n) : n \geq 0 \}$ or both. These conditions further guarantee both the existence and finiteness of a large class of time-average limits. Our key assumption is that there exist a distinguished marking $\bar{s}$ and a distinguished set of transitions $\bar{E}$ such that the marking process probabilistically restarts whenever the marking is $\bar{s}$ and the transitions in $\bar{E}$ fire simultaneously. The random times at which this probabilistic restart occurs correspond to the successive times at which the underlying chain hits a distinguished set of states. The results in Chapter 5 can be used to show that the chain hits the distinguished set infinitely often with probability 1, so that each regeneration point is a.s. finite. Extensions of these results can be used to show that integrals or sums of the output process over a regenerative cycle—as well as the cycle length itself—have finite moments.

By exploiting the special structure of a regenerative process, we can obtain strongly consistent point estimates and asymptotic confidence intervals for time-average limits based on simulation of a finite portion of a single sample path. The resulting "regenerative method" for analysis of simulation output is presented in Section 6.3. We also outline extensions of the basic method that deal with excessive bias in the estimator, simulation up to a specified time, a priori precision requirements, estimation of nonlinear functions of time-average limits, estimation of gradients of time-average limits with respect to model parameters, and dependence between adjacent cycles.

6.1 Regenerative Processes

In this section we formally define the regenerative property and give conditions under which time-average limits for regenerative processes are well defined and finite. We then extend these results to processes with one-dependent cycles.

6.1.1 Definition of a Regenerative Process

We first consider processes that evolve over continuous time. For the sequence of random times $\{T_k: k \geq 0\}$ defined below, set $\tau_k = T_k - T_{k-1}$ for $k \geq 1$.

Definition 1.1. The stochastic process $\{X(t): t \geq 0\}$ with state space S is a *regenerative process* in continuous time if there exists an increasing sequence $0 \leq T_0 < T_1 < T_2 < \cdots$ of a.s. finite random times such that the post-T_k process $\{X(T_k + t): t \geq 0;\ \tau_{k+l}: l \geq 1\}$

(i) is distributed as the post-T_0 process $\{X(T_0 + t): t \geq 0;\ \tau_l: l \geq 1\}$, and

(ii) is independent of the pre-T_k process $\{X(t): 0 \leq t < T_k;\ \tau_1, \ldots, \tau_k\}$

for $k \geq 1$.

The sequence $\{T_k: k \geq 0\}$ of regeneration points is a (possibly delayed) renewal process—see Section A.2.3 in the Appendix—that decomposes sample paths of $\{X(t): t \geq 0\}$ into i.i.d. *cycles*; the kth cycle is $\{X(t): T_{k-1} \leq$

$t < T_k$ }. The random variable τ_k defined above is the length of the kth cycle.

When $T_0 = 0$ the process $\{ X(t) : t \geq 0 \}$ is called a *nondelayed* regenerative process; otherwise, it is called a *delayed* regenerative process. For a delayed regenerative process $\{ X(t) : t \geq 0 \}$, the "0th cycle" $\{ X(t) : 0 \leq t < T_0 \}$ need not have the same distribution as the other cycles. Similarly, the length of this cycle—denoted by τ_0—need not have the same distribution as τ_1, τ_2, and so forth.

Remark 1.2. Checking whether a stochastic process $\{ X(t) : t \geq 0 \}$ satisfies Definition 1.1 amounts to verifying whether

$$P\{ X(T_k + t_1) \in A_1, \ldots, X(T_k + t_n) \in A_n,$$

$$\tau_{k+1} \leq u_1, \ldots, \tau_{k+m} \leq u_m \mid X(t) : 0 \leq t < T_k \}$$

$$= P\{ X(T_0 + t_1) \in A_1, \ldots, X(T_0 + t_n) \in A_n, \tag{1.3}$$

$$\tau_1 \leq u_1, \ldots, \tau_m \leq u_m \} \text{ a.s.}$$

for all $k, m, n \geq 1$, $t_1, \ldots, t_n \geq 0$, $u_1, \ldots, u_m \geq 0$, and $A_1, A_2, \ldots, A_n \subseteq S$. If the state space S is finite or countably infinite, then (1.3) need only be verified for sets $A_1, A_2, \ldots, A_n$ such that each A_i is of the form $A_i = \{ s_i \}$ for some $s_i \in S$. Similarly, if S is a subinterval of $\Re_+$, then we can restrict attention to sets $A_1, A_2, \ldots, A_n$ such that each A_i is of the form $A_i = [0, a_i] \cap S$ for some $a_i > 0$. Analogous simplifications apply when S is a subset of a Cartesian product: if, for example, $S \subseteq S_1 \times S_2$, where S_1 is finite or countably infinite and S_2 is a subinterval of $\Re_+$, then (1.3) need only be verified for sets $A_1, A_2, \ldots, A_n$ such that each A_i is of the form $A_i = \{ s_i \} \times ([0, a_i] \cap S_2)$ with $s_i \in S_1$ and $a_i > 0$.

If, as often happens, each regeneration point T_k is a stopping time[1] with respect to $\{ X(t) : t \geq 0 \}$, then the cycle lengths $\{ \tau_k : k \geq 1 \}$ are determined by the process $\{ X(t) : t \geq 0 \}$, and it suffices to show that

$$P\{ X(T_k + t_1) \in A_1, \ldots, X(T_k + t_n) \in A_n \mid X(t) : 0 \leq t < T_k \}$$

$$= P\{ X(T_0 + t_1) \in A_1, \ldots, X(T_0 + t_n) \in A_n \} \text{ a.s.}$$

for $k, n \geq 1$, $t_1, \ldots, t_n \geq 0$, and $A_1, A_2, \ldots, A_n \in \mathcal{S}$, where—as discussed above—$\mathcal{S}$ is an appropriate class of subsets of S.

Remark 1.4. If $\{ X(t) : t \geq 0 \}$ is a regenerative process in continuous time, then $\{ f(X(t)) : t \geq 0 \}$ is a regenerative process in continuous time for

---

[1]Let $\{ X(t) : t \geq 0 \}$ be a continuous-time stochastic process with sample paths that are right-continuous and have limits from the left. A real-valued random variable T is said to be a *stopping time* with respect to $\{ X(t) : t \geq 0 \}$ if the occurrence or nonoccurrence of the event $\{ T \leq t \}$ is completely determined by $\{ X(u) : 0 \leq u \leq t \}$ for $t \geq 0$.

any function f. In contrast, the Markov property is not preserved under arbitrary mappings.

EXAMPLE 1.5 (Continuous-time Markov chain). Consider an irreducible CTMC $\{ X(t)\colon t \geq 0 \}$ with a finite state space S and initial state $s \in S$. Let T_k be the kth time at which the chain hits state s. As discussed in Section 3.4, each state of the chain—and in particular state s—is hit infinitely often with probability 1, so that each T_k is a.s. finite. Moreover, each T_k is a stopping time with respect to the CTMC. It follows immediately from the strong Markov property for CTMCs that

$$P\{ X(T_k + t_1) = s_1, \ldots, X(T_k + t_n) = s_n \mid X(t)\colon 0 \leq t < T_k \}$$

$$= P\{ X(t_1) = s_1, \ldots, X(t_n) = s_n \} \text{ a.s.}$$

for $k \geq 1$, $n \geq 1$, and $t_1, t_2, \ldots, t_n \geq 0$. Thus the random times $\{ T_k\colon k \geq 0 \}$ form a sequence of regeneration points for the process $\{ X(t)\colon t \geq 0 \}$, and the CTMC is a nondelayed regenerative process.

The successive times $\{ T'_k\colon k \geq 0 \}$ at which the CTMC makes a transition from state s (to some other state) also form a sequence of regeneration points for $\{ X(t)\colon t \geq 0 \}$—the regenerative property again follows from the strong Markov property. Observe that $P\{ X(T'_k) = \cdot \} = W(s, \cdot)$ for $k \geq 0$, where W is the transition matrix of the embedded jump chain (see Section 3.4.1). All the foregoing results also hold for an irreducible positive recurrent CTMC with a countably infinite state space.

We now consider discrete-time processes. For the sequence of random indexes $\{ \theta(k)\colon k \geq 0 \}$ defined below, set $\tau_k = \theta(k) - \theta(k-1)$ for $k \geq 1$.

Definition 1.6. The stochastic process $\{ Z_n\colon n \geq 0 \}$ with state space Γ is a *regenerative process* in discrete time if there exists an increasing sequence $0 \leq \theta(0) < \theta(1) < \theta(2) < \cdots$ of a.s. finite random times such that the post-$\theta(k)$ process $\{ Z_{\theta(k)+n}, \tau_{k+n+1}\colon n \geq 0 \}$

(i) is distributed as the post-$\theta(0)$ process $\{ Z_{\theta(0)+n}, \tau_{n+1}\colon n \geq 0 \}$, and

(ii) is independent of the pre-$\theta(k)$ process $\{ Z_0, \ldots, Z_{\theta(k)-1}; \tau_1, \ldots, \tau_k \}$

for $k \geq 1$.

As for regenerative processes in continuous time, the sequence $\{ \theta(k)\colon k \geq 0 \}$ of regeneration points is a (possibly delayed) discrete-time renewal process that decomposes sample paths of $\{ Z_n\colon n \geq 0 \}$ into i.i.d. cycles; the random variable τ_k is the length of the kth cycle. Observe that each τ_k takes values in the positive integers.

EXAMPLE 1.7 (Discrete-time Markov chain). Consider an irreducible DTMC $\{ Z_n\colon n \geq 0 \}$ with a finite state space Γ. Fix a state $z \in \Gamma$ and let $\theta(k)$ be the random index of the kth state transition at which the chain hits z.

As in Example 1.5, each $\theta(k)$ is a.s. finite, and it follows from the strong Markov property that the random indices $\{\theta(k): k \geq 0\}$ form a sequence of regeneration points for $\{Z_n: n \geq 0\}$. In analogy to Example 1.5, we note that the random indices $\{\theta(k) + 1: k \geq 0\}$ also form a sequence of regeneration points for $\{Z_n: n \geq 0\}$.

EXAMPLE 1.8 (Waiting time in a single-server queue). Consider a queueing system with one single-server service center. Jobs arrive according to a renewal process and are served according to a first-come, first-served queuing discipline. The server is never idle when there are jobs in the system. Successive service times are i.i.d. and independent of the arrival process. Jobs are numbered in arrival order, and we assume that job 0 arrives at time 0. Denote by U_{n+1} the time between the arrival of job n and job $n+1$, by V_n the service time for job n, and by D_n the waiting time in queue for job n. Under our assumptions, $\{U_n: n \geq 1\}$ and $\{V_n: n \geq 0\}$ each form a sequence of i.i.d. random variables, and the U_i's are independent of the V_i's. The waiting times obey the following recursive relationship: $D_0 = 0$ and

$$D_{n+1} = (D_n + V_n - U_{n+1})^+ \tag{1.9}$$

for $n \geq 0$, where $x^+ = \max(x, 0)$. It follows from (1.9) and the assumptions on the sequences $\{U_n: n \geq 1\}$ and $\{V_n: n \geq 0\}$ that $\{D_n: n \geq 0\}$ is a discrete-time Markov chain with state space $\Re_+$. We say that a *busy period* starts whenever a job arrives to an empty service center. Denote by $\theta(n)$ the number of the job that initiates the nth busy period, so that $D_{\theta(n)} = 0$ for $n \geq 0$. Provided that $E[V_1] < E[U_1]$, each $\theta(n)$ is a.s. finite, and it then follows from the strong Markov property that the random indices $\{\theta(k): k \geq 0\}$ form a sequence of regeneration points for $\{D_n: n \geq 0\}$.

We assume henceforth and without further comment that the state space S in Definition 1.1 is always a subset of d-dimensional Euclidean space $\Re^d$ for some $d \geq 1$, and similarly for the state space Γ in Definition 1.6.

6.1.2 Stability of Regenerative Processes

We first give conditions under which time-average limits for a continuous-time or discrete-time regenerative process are well defined and finite. We then give further conditions under which a time-average limit can also be interpreted as a steady-state or limiting mean.

Time-Average Limits

Consider a regenerative stochastic process $\{X(t): t \geq 0\}$ with state space S and regeneration points $\{T_k: k \geq 0\}$. As before, denote by τ_k the length of the kth regenerative cycle. For each real-valued function f defined on S, set

$$Y_k(f) = \int_{T_{k-1}}^{T_k} f(X(u))\, du \tag{1.10}$$

for $k \geq 0$. (Take $T_{-1} = 0$.) Also define the function $|f|$ by setting $|f|(s) = |f(s)|$ for $s \in S$, so that

$$Y_k(|f|) = \int_{T_{k-1}}^{T_k} |f(X(u))|\, du$$

for $k \geq 1$. It follows from the definition of a regenerative process that the sequence $\{ (Y_k(f), \tau_k) : k \geq 1 \}$ consists of i.i.d. random pairs.[2] Set

$$r(f) = \frac{E[Y_1(f)]}{E[\tau_1]} \tag{1.11}$$

and observe that $r(f)$ is well defined and finite if $r(|f|) < \infty$.

Theorem 1.12. *Suppose that $E[\tau_1] < \infty$. Then $r(|f|) < \infty$ and*

$$\lim_{t \to \infty} \frac{1}{t} \int_0^t f(X(u))\, du = r(f) \quad a.s. \tag{1.13}$$

for any real-valued function f such that $Y_0(|f|) < \infty$ a.s. and $E[Y_1(|f|)] < \infty$.

Remark 1.14. Observe that

$$Y_k(|f|) \leq \tau_k \sup_{s \in S} |f(s)|$$

for $k \geq 0$ and any real-valued function f. Thus if f is bounded or the state space S is finite, then for $q \geq 0$ we have $E[Y_1^q(|f|)] < \infty$ whenever $E[\tau_1^q] < \infty$. Moreover, $Y_0(|f|) < \infty$ a.s. because $\tau_0 = T_0 < \infty$ a.s. by definition.

Remark 1.15. Suppose that $E[\tau_1] < \infty$ and f is nonnegative. Then the convergence in (1.13) holds without any further conditions, provided that we allow $r(f)$ to be infinite.

PROOF. Fix a function f such that $Y_0(|f|) < \infty$ a.s. and $E[Y_1(|f|)] < \infty$. Clearly, the contribution of $\{ f(X(t)) : 0 \leq t \leq T_0 \}$ to the time-average limit is a.s. negligible, so assume without loss of generality that $T_0 = 0$. We have

$$\frac{1}{n} \sum_{k=1}^{n} Y_k(f) \to E[Y_1(f)] \quad a.s.,$$

---

[2]For a delayed regenerative process, the random variable $Y_0(f)$ need not have the same distribution as $Y_1(f)$, $Y_2(f)$, and so forth. For a nondelayed regenerative process, $Y_0(f)$ is identically zero.

$$\frac{1}{n}\sum_{k=1}^{n} Y_k(|f|) \to E\left[Y_1(|f|)\right] \text{ a.s.},$$

and

$$\frac{1}{n}\sum_{k=1}^{n} \tau_k \to E\left[\tau_1\right] \text{ a.s.}$$

as $n \to \infty$ by the strong law of large numbers (SLLN) for i.i.d. random variables. The desired result now follows from Theorem 2.9(v) in Chapter 3; in the theorem, take $Z(t) = \int_0^t f(X(u))\,du$ for $t \ge 0$ and $\Delta_k = \tau_k$ for $k \ge 1$, and use the fact that $\sup_{T_{k-1} \le t \le T_k} |Z(t) - Z(T_{k-1})| \le Y_k(|f|)$ for $k \ge 1$. □

For a discrete-time regenerative process $\{Z_n : n \ge 0\}$ with state space Γ, an analog to Theorem 1.12 can be obtained by applying Theorem 1.12 to the continuous-time process $\{X(t) : t \ge 0\}$, where $X(t) = Z_{\lfloor t \rfloor}$ and $\lfloor x \rfloor$ is, as before, the greatest integer less than or equal to x. (This trick often can be used to obtain results for discrete-time processes from corresponding results for continuous-time processes.) We state the resulting theorem for ease of reference. Suppose that the random indices $\{\theta(k) : k \ge 0\}$ form a sequence of regeneration points for the process $\{Z_n : n \ge 0\}$. As before, set $\tau_k = \theta(k) - \theta(k-1)$ for $k \ge 0$. (Take $\theta(-1) = 0$.) For each real-valued function f defined on Γ, set

$$Y_k(f) = \sum_{j=\theta(k-1)}^{\theta(k)-1} f(Z_j) \tag{1.16}$$

for $k \ge 0$, and set

$$r(f) = \frac{E\left[Y_1(f)\right]}{E\left[\tau_1\right]}. \tag{1.17}$$

Theorem 1.18. *Suppose that $E\left[\tau_1\right] < \infty$. Then $r(|f|) < \infty$ and*

$$\lim_{n\to\infty} \frac{1}{n}\sum_{j=0}^{n-1} f(Z_j) = r(f) \text{ a.s.}$$

for any real-valued function f such that $Y_0(|f|) < \infty$ a.s. and $E\left[Y_1(|f|)\right] < \infty$.

Limiting Distributions

When a time-average limit $r(f)$ exists for a regenerative process, it is natural to ask whether the process has a limiting distribution and, if so, whether $r(f)$ can be interpreted as a steady-state or limiting mean. Theorems 1.20 and 1.25 show that under mild regularity conditions the answer to these questions is affirmative, provided that the regenerative cycle length has finite mean.

For a real-valued function f defined on S, denote by $D(f)$ the subset of points of S at which f is discontinuous. Recall from Section A.1.8 that we write $X(t) \Rightarrow X$ as $t \to \infty$ if and only if $\lim_{t \to \infty} P\{X(t) \le x\} = P\{X \le x\}$ for all x at which the function $F(x) = P\{X \le x\}$ is continuous.

Definition 1.19. The real-valued random variable X is said to be *periodic* with period d if d is the largest real number such that

$$\sum_{n=0}^{\infty} P\{X = nd\} = 1.$$

If no such number d exists, then X is said to be *aperiodic*. If X is aperiodic, then the distribution function of X also is said to be aperiodic.

In the following, $Y_k(f)$ is given by (1.10) and $r(f)$ is given by (1.11).

Theorem 1.20. *Suppose that the cycle length τ_1 is aperiodic with $E[\tau_1] < \infty$ and that $\{X(t) : t \ge 0\}$ has right-continuous sample paths. Then there exists a random variable X such that*

(i) $X(t) \Rightarrow X$ *as* $t \to \infty$,

(ii) $f(X(t)) \Rightarrow f(X)$ *as* $t \to \infty$ *for any real-valued function f such that $P\{X \in D(f)\} = 0$,*

(iii) $E[f(X)] = r(f)$ *for any real-valued function f such that $E[Y_1(|f|)] < \infty$ or $E[|f(X)|] < \infty$, and*

(iv) $\lim_{t \to \infty} E[f(X(t))] = E[f(X)]$ *for any real-valued function f such that $\sup_{s \in S} |f(s)| < \infty$ and $P\{X \in D(f)\} = 0$.*

The proof of the assertions in (i) and (iii) uses the key renewal theorem (Proposition 2.16 in the Appendix) and is beyond the scope of the current discussion. The assertion in (ii) follows immediately from the assertion in (i) and the continuous mapping theorem (Proposition 1.42 in the Appendix). The assertion in (iv) follows from the assertion in (ii) and the uniform integrability of $\{f(X(t)) : t \ge 0\}$—see Proposition 1.50 in the Appendix. As discussed in Section A.1.8, $P\{X \in D(f)\} = 0$ for any function f whenever, as with the marking process of an SPN, the state space of $\{X(t) : t \ge 0\}$ is finite or countably infinite.

Remark 1.21. The first assertion of the theorem is that $\{X(t) : t \ge 0\}$ has a limiting distribution. The form of this distribution follows from the ratio formula

$$E[f(X)] = \frac{E[Y_1(f)]}{E[\tau_1]} \tag{1.22}$$

in the third assertion of the theorem. In particular, fix a subset $A \subseteq S$ and take $f = 1_A$ in (1.22). Then the limiting probability that $X(t) \in A$ as

$t \to \infty$ is equal to the expected time that $\{X(t): t \geq 0\}$ spends in the set A during a regenerative cycle divided by the expected length of the cycle.

Remark 1.23. Theorem 1.20 as given above is well suited to the SPN applications that are the focus of our discussion. There exist many variants of this result, however. In the final assertion of the theorem, for example, the requirement that f be bounded on S can be replaced by the weaker requirement that the process $\{f(X(t)): t \geq 0\}$ be uniformly integrable— see Definition 1.49 in the Appendix. As another example, the requirement that the sample paths of $\{X(t): t \geq 0\}$ be right-continuous can be replaced by the requirement that the distribution function of τ_1 be "spread out" as defined in Section A.2.3. Under this latter condition, it can be shown that $\{X(t): t \geq 0\}$ converges to X in total variation. As discussed in Section A.1.8, convergence in total variation is stronger than convergence in distribution.

We next give an analog of Theorem 1.20 for a discrete-time regenerative process $\{Z_n: n \geq 0\}$ with state space Γ.

Definition 1.24. The integer-valued random variable X is said to be *periodic in discrete time* with period d if $d \geq 2$ and d is the largest integer such that

$$\sum_{n=0}^{\infty} P\{X = nd\} = 1.$$

If no such integer d exists, then X is said to be *aperiodic in discrete time*.

In the following, $Y_k(f)$ is given by (1.16) and $r(f)$ is given by (1.17).

Theorem 1.25. *Suppose that the cycle length τ_1 is aperiodic in discrete time with $E[\tau_1] < \infty$. Then there exists a random variable Z such that*

(i) $Z_n \Rightarrow Z$ as $n \to \infty$,

(ii) $f(Z_n) \Rightarrow f(Z)$ as $n \to \infty$ for any real-valued function f such that $P\{Z \in D(f)\} = 0$,

(iii) $E[f(Z)] = r(f)$ for any real-valued function f such that $E[Y_1(|f|)] < \infty$ or $E[|f(Z)|] < \infty$, and

(iv) $\lim_{n \to \infty} E[f(Z_n)] = E[f(Z)]$ for any real-valued function f such that $\sup_{z \in \Gamma} |f(z)| < \infty$ and $P\{Z \in D(f)\} = 0$.

6.1.3 Processes with Dependent Cycles

When considering the behavior of the underlying or embedded chain of an SPN or the properties of a sequence of delays in an SPN having a regenerative marking process—see Chapter 8—we are led to consider processes in which there can be some limited dependence between cycles. Such processes also arise when studying certain functions of a regenerative process—see Example 1.30 below. Fortunately, as discussed in this subsection, most of the key stability results for regenerative processes hold in this broader setting.

Specifically, we consider stochastic processes with sample paths that can be decomposed into cycles that are identically distributed—in fact, stationary—and one-dependent.[3] Such processes are called *od-regenerative processes*. Time-average limits for an od-regenerative process are well defined and finite provided that the cycle length has finite mean. An *od-equilibrium* process is an od-regenerative process in which the cycle lengths (though not necessarily the cycles themselves) are mutually independent. Under mild conditions, a time-average limit for an od-equilibrium process can also be interpreted as a limiting or steady-state mean. Thus od-equilibrium processes enjoy the same long-run stability properties as regenerative processes. We focus on processes in discrete time, since our primary application of the results in this section is to the underlying or embedded chain of an SPN or to a sequence of delays in an SPN.

OD-Regenerative Processes

We start with the following definition. As before, set $\tau_k = \theta(k) - \theta(k-1)$ for $k \geq 1$.

Definition 1.26. The stochastic process $\{Z_n : n \geq 0\}$ with state space Γ is an *od-regenerative process* in discrete time if there exists an increasing sequence $0 \leq \theta(0) < \theta(1) < \theta(2) < \cdots$ of a.s. finite random times such that the post-$\theta(k)$ process $\{Z_{\theta(k)+n}, \tau_{k+n+1} : n \geq 0\}$

(i) is distributed as the post-$\theta(0)$ process $\{Z_{\theta(0)+n}, \tau_{n+1} : n \geq 0\}$ for $k \geq 1$, and

(ii) is independent of the pre-$\theta(k-1)$ process $\{Z_0, Z_1, \ldots, Z_{\theta(k-1)-1}; \tau_1, \tau_2, \ldots, \tau_{k-1}\}$ for $k \geq 2$.

The random indices $\{\theta(k) : k \geq 0\}$ are called *od-regeneration points* for the process $\{Z_n : n \geq 0\}$ and serve to decompose sample paths of $\{Z_n : n \geq 0\}$ into one-dependent stationary (o.d.s.) cycles. The quantity τ_k is the length of the kth such cycle.

Time-average limits exist for an od-regenerative process under the same conditions as for an ordinary regenerative process. Let $\{Z_n : n \geq 0\}$ be an od-regenerative process with state space Γ and od-regeneration points $\{\theta(k) : k \geq 0\}$. For a real-valued function f defined on Γ, define $Y_k(f)$

---

[3]As discussed in Section A.2.2, a sequence of random variables $\{X_n : n \geq 0\}$ is *stationary* if $(X_0, X_1, \ldots, X_k)$ and $(X_n, X_{n+1}, \ldots, X_{n+k})$ are identically distributed for all $k, n \geq 0$. The sequence is *one-dependent* if X_{n+j} is independent of $\{X_0, X_1, \ldots, X_n\}$ for each $n \geq 0$ and $j > 1$.

$(k \geq 1)$ as in (1.16) and $r(f)$ as in (1.17). Observe that the sequence $\{ (Y_k(f), \tau_k) \colon k \geq 1 \}$ consists of o.d.s. random vectors.

Theorem 1.27. *Suppose that* $E[\tau_1] < \infty$. *Then* $r(|f|) < \infty$ *and*

$$\lim_{n \to \infty} \frac{1}{n} \sum_{j=0}^{n-1} f(Z_j) = r(f) \ a.s.$$

for any real-valued function f *such that* $Y_0(|f|) < \infty$ *a.s. and* $E[Y_1(|f|)] < \infty$.

The proof of this result is essentially the same as for Theorem 1.12, except that we use the SLLN for one-dependent and identically distributed (o.i.d.) random variables—see Proposition 2.7 in the Appendix.

In general, od-regeneration points $\{ \theta(k) \colon k \geq 0 \}$ do not form a renewal process in discrete time, so that results as in Theorem 1.25 cannot be extended to this setting.

OD-Equilibrium Processes

As with an od-regenerative process, set $\tau_k = \theta(k) - \theta(k-1)$ for $k \geq 1$.

Definition 1.28. The stochastic process $\{ Z_n \colon n \geq 0 \}$ with state space Γ is an *od-equilibrium process* in discrete time if there exists an increasing sequence $0 \leq \theta(0) < \theta(1) < \theta(2) < \cdots$ of a.s. finite random times such that, for $k \geq 1$, the post-$\theta(k)$ process $\{ Z_{\theta(k)+n}, \tau_{k+n+1} \colon n \geq 0 \}$

(i) is distributed as the post-$\theta(0)$ process $\{ Z_{\theta(0)+n}, \tau_{n+1} \colon n \geq 0 \}$,

(ii) is independent of the pre-$\theta(k-1)$ process $\{ Z_0, Z_1, \ldots, Z_{\theta(k-1)-1}; \tau_1, \tau_2, \ldots, \tau_{k-1} \}$, and

(iii) is independent of τ_k.

The random indices $\{ \theta(k) \colon k \geq 0 \}$ are called *od-equilibrium points* for the process $\{ Z_n \colon n \geq 0 \}$ and serve to decompose sample paths of $\{ Z_n \colon n \geq 0 \}$ into o.d.s. cycles. The definition of an od-equilibrium process is almost identical to Definition 1.26, except for the additional requirement in (iii). This latter condition ensures that the cycle lengths are i.i.d. and hence that the sequence of points $\{ \theta(k) \colon k \geq 0 \}$ is a renewal process in discrete time.

EXAMPLE 1.29 (Discrete-time Markov chain). Consider a recurrent DTMC $\{ X_n \colon n \geq 0 \}$ and let $\{ \theta(k) \colon k \geq 0 \}$ be the successive times that the chain jumps out of a specified state s. Fix an integer $l \geq 1$ and set $\tilde{\theta}(k) = \theta(k) + l$ for $k \geq 0$. Then the random indices $\{ \tilde{\theta}(k) \colon k \geq 0 \}$ typically form a sequence of od-equilibrium points for the process $\{ X_n \colon n \geq 0 \}$. To see this, observe that, as discussed in Example 1.7, the random indices $\{ \theta(k) \colon k \geq 0 \}$ form a sequence of regeneration points for the chain, and thus the cycle lengths

$\{\tilde{\theta}(k) - \tilde{\theta}(k-1): k \geq 1\} = \{\theta(k) - \theta(k-1): k \geq 1\}$ are i.i.d.. Moreover, the $\tilde{\theta}(k)$-cycles are identically distributed. For each k, however, there is, in general, no probabilistic restart at time $\tilde{\theta}(k)$, so that the adjacent cycles demarcated by $\tilde{\theta}(k)$ are typically dependent. Nonadjacent $\tilde{\theta}(k)$-cycles are always separated by at least one point $\theta(l)$ and are therefore independent.

EXAMPLE 1.30 (Pairwise mapping of a regenerative process).   Let $\{Z_n : n \geq 0\}$ be a regenerative process with state space Γ and regeneration points $\{\theta(k): k \geq 0\}$. For a real-valued function f defined on $\Gamma \times \Gamma$, set $W_n = f(Z_n, Z_{n+1})$ for $n \geq 0$. Although the cycles of the process $\{W_n: n \geq 0\}$ defined by the points $\{\theta(k): k \geq 0\}$ clearly are identically distributed, they may not be independent—indeed, $W_{\theta(k)-1}$ and $W_{\theta(k)}$ may both depend explicitly on $Z_{\theta(k)}$. Observe, however, that for $k \geq 2$ the post-$\theta(k)$ process $\{W_{\theta(k)+n}, \tau_{k+n+1}: n \geq 0\}$ is determined by $\{Z_{\theta(k)+n}, \tau_{k+n+1}: n \geq 0\}$ whereas $U_k = \{W_0, W_1, \ldots, W_{\theta(k-1)-1}; \tau_1, \tau_2, \ldots, \tau_k\}$ is determined by $\{Z_0, Z_1, \ldots, Z_{\theta(k-1)}; \tau_1, \tau_2, \ldots, \tau_k\}$. It follows from the regenerative structure of $\{Z_n: n \geq 0\}$ that the post-$\theta(k)$ process is independent of U_k. Thus the random indices $\{\theta(k): k \geq 0\}$ form a sequence of od-equilibrium points for $\{W_n: n \geq 0\}$. Note that if $Z_{\theta(k)} \equiv z$ for some $z \in \Gamma$ and each $k \geq 0$, then $\{W_n: n \geq 0\}$ is, in fact, a regenerative process.

Since od-equilibrium processes are a subclass of od-regenerative processes, Theorem 1.27 applies. Thus—under mild regularity conditions—time-average limits of an od-equilibrium process are well defined and finite provided that the cycle length has finite mean. Moreover, since the points $\{\theta(k): k \geq 0\}$ form a renewal process, the proof of Theorem 1.25 applies essentially without change to establish the following result for an od-equilibrium process $\{Z_n: n \geq 0\}$. In the theorem, we define $Y_k(f)$ $(k \geq 1)$ as in (1.16) and $r(f)$ as in (1.17).

Theorem 1.31. *Suppose that the cycle length τ_1 is aperiodic in discrete time with $E[\tau_1] < \infty$. Then there exists a random variable Z such that*

(i) *$Z_n \Rightarrow Z$ as $n \to \infty$,*

(ii) *$f(Z_n) \Rightarrow f(Z)$ as $n \to \infty$ for any real-valued function f such that $P\{Z \in D(f)\} = 0$,*

(iii) *$E[f(Z)] = r(f)$ for any real-valued function f such that $E[Y_1(|f|)] < \infty$ or $E[|f(Z)|] < \infty$, and*

(iv) *$\lim_{n\to\infty} E[f(Z_n)] = E[f(Z)]$ for any real-valued function f such that f is bounded and $P\{Z \in D(f)\} = 0$.*

Perhaps the most important examples of od-equilibrium processes are Harris recurrent Markov chains. Proposition 1.32 asserts that any Harris recurrent chain is an od-equilibrium process and gives a representation of the invariant measure of the chain in terms of cycles. The proposition also asserts that the cycle length has moments of all orders, provided that the chain satisfies a geometric drift condition.

Proposition 1.32. *Let* $\{Z_n : n \geq 0\}$ *be a Harris recurrent Markov chain with state space* Γ *and initial distribution* μ. *Then there exists at least one sequence* $\{\theta(k) : k \geq 0\}$ *of od-equilibrium points for* $\{Z_n : n \geq 0\}$. *For any such sequence, the measure* π_0 *defined for* $A \subseteq \Gamma$ *by*

$$\pi_0(A) = E_\mu \left[\sum_{n=\theta(0)}^{\theta(1)-1} 1_A(Z_n) \right].$$

is an invariant measure for the chain. If, moreover, the stability conditions (1.14) and (1.15) in Chapter 5 hold for some choice of B, v, *and* β, *and if the initial distribution* μ *satisfies* $\int v(z) \, \mu(dz) < \infty$, *then the cycle length* $\tau_1 = \theta(1) - \theta(0)$ *satisfies* $E_\mu[e^{r\tau_1}] < \infty$ *for sufficiently small* $r > 0$ *(and hence has finite moments of all orders).*

Remark 1.33. Observe that a Harris recurrent chain $\{Z_n : n \geq 0\}$ is positive Harris recurrent if and only if $\pi_0(\Gamma) = E_\mu[\tau_1] < \infty$, in which case the measure π given by

$$\pi(A) = \frac{\pi_0(A)}{\pi_0(\Gamma)} = \frac{E_\mu[Y_1(1_A)]}{E_\mu[\tau_1]} = \frac{E_\mu\left[\sum_{n=\theta(0)}^{\theta(1)-1} 1_A(Z_n)\right]}{E_\mu[\tau_1]} \qquad (1.34)$$

for $A \subseteq \Gamma$ is the unique invariant probability measure of the chain. If, moreover, the chain is aperiodic in the sense of Section 5.1.1, then τ_1 is aperiodic in the sense of Definition 1.19, and it follows from Theorem 1.31 that $Z_n \Rightarrow Z$, where Z is distributed according to π. Thus a Harris ergodic chain converges in distribution to a unique invariant probability measure.

The proof of Proposition 1.32 rests on the rather deep fact that for a ϕ-irreducible chain there exists a set $C \subseteq \Gamma$ such that $\phi(C) > 0$ and

$$P^r(z, \cdot) = b\lambda(\cdot) + (1-b)Q(z, \cdot), \qquad z \in C, \qquad (1.35)$$

for some $r \geq 1$, $b \in (0, 1]$, probability distribution λ, and transition kernel Q—indeed, it can be shown that any set $A \subseteq \Gamma$ with $\phi(A) > 0$ contains such a "C-set." It follows from the Harris recurrence that C is hit infinitely often with probability 1. The decomposition in (1.35) permits construction of a version of the chain together with a sequence $\{\theta(k) : k \geq 0\}$ of random indices that serve as od-equilibrium points. The construction uses a sequence $\{I_n : n \geq 0\}$ of i.i.d. Bernoulli random variables with $P_\mu\{I_n = 1\} = 1 - P_\mu\{I_n = 0\} = b$. The idea is to generate successive states of the chain according to the initial distribution μ and transition kernel P until the first time $M \geq 0$ such that $Z_M \in C$. If $I_M = 1$, then generate Z_{M+r} according to λ; if $I_M = 0$, then generate Z_{M+r} according to $Q(Z_M, \cdot)$. Next, generate the intermediate states $Z_{M+1}, Z_{M+2}, \ldots, Z_{M+r-1}$ according to the appropriate conditional distribution (conditioned on the endpoint values Z_M and Z_{M+r}). Now iterate this procedure starting from state Z_{M+r}. Denote by $\theta(0), \theta(1), \ldots$ the successive times at which the state of the chain is generated according to λ. Using the strong Markov property as in (1.6) in Chapter 3, it is straightforward to show that the cycles formed by the points $\{\theta(k): k \geq 0\}$ are identically distributed and have i.i.d. lengths; each cycle consists of at least r state transitions. By construction, each $Z_{\theta(k)}$ depends at most on $Z_{\theta(k)-1}, Z_{\theta(k)-2}, \ldots, Z_{\theta(k)-r+1}$ (via the conditioning described above). It follows that the post-$\theta(k)$ process $\{Z_{\theta(k)+n}, \tau_{k+n+1}: n \geq 0\}$ is independent of $\{Z_n: 0 \leq n \leq \theta(k) - r\}$ and $\{\tau_l: 0 \leq l \leq k\}$, so that the cycles are one-dependent. Observe that when (1.35) holds with $r = 1$, then the random indices $\{\theta(k): k \geq 0\}$ form a sequence of regeneration points for the chain. Indeed, it can be shown that (1.35) *must* hold with $r = 1$ for a sequence of regeneration points to exist.

The final result in this section can be viewed as a partial converse to Proposition 1.32.

Proposition 1.36. *Suppose that there exists a sequence $\{\theta(k): k \geq 0\}$ of od-regeneration points for a Markov chain and $E_\mu\left[\theta(1) - \theta(0)\right] < \infty$. Then the chain is positive Harris recurrent.*

PROOF. Denote by Γ the state space of the chain and by μ the initial distribution. Suppose that $\pi(A) > 0$ for a fixed set $A \subseteq \Gamma$, where π is defined by (1.34). It follows from Theorem 1.27 that $\lim_{n\to\infty}(1/n)\sum_{j=0}^{n-1} 1_A(Z_j) = \pi(A) > 0$ a.s., and hence

$$P_\mu\{Z_n \in A \text{ i.o.}\} = P_\mu\left\{\sum_{n=0}^{\infty} 1_A(Z_n) = \infty\right\} = 1.$$

Thus the chain $\{Z_n: n \geq 0\}$ is Harris recurrent with recurrence measure π. It then follows from Remark 1.33 that the chain is actually positive Harris recurrent since $E_\mu\left[\theta(1) - \theta(0)\right] < \infty$. $\qquad\square$

6.2   Regeneration and Stochastic Petri Nets

In this section we give conditions on the building blocks of an SPN under which there exists a sequence of regeneration points for the marking process or the underlying chain or both. Theorem 2.2 gives general sufficient conditions for such regenerative structure. Theorems 2.24, 2.31, 2.36, and 2.44 refine these conditions when Assumption PD of Chapter 5 holds or a geometric trials criterion is satisfied. These results also give conditions under which integrals or sums of the output process over a regenerative cycle—as in (1.10) or (1.16)—have finite moments. In particular, these results give conditions under which the cycle length has finite moments.

Throughout this section, we consider an SPN with marking set G, timed marking set S, transition set E, immediate transition set E', marking process $\{X(t): t \geq 0\}$, and underlying chain $\{(S_n, C_n): n \geq 0\}$. Recall that the chain takes values in $\Sigma = \bigcup_{s \in G}(\{s\} \times C(s))$, that ζ_n is the epoch (in continuous time) of the nth marking change, and that $E_n^* = E^*(S_n, C_n)$ is the set of transitions that fire simultaneously and trigger the $(n+1)$st marking change.

6.2.1 General Conditions for Regenerative Structure

For a marking $\bar{s} \in G$ and set of transitions $\bar{E} \subseteq E(\bar{s})$, denote by $\{\theta(k): k \geq 0\}$ the indices of the successive marking changes at which the marking is $\bar{s}$ and the transitions in $\bar{E}$ fire simultaneously: $\theta(-1) = 0$ and

$$\theta(k) = \inf\{n > \theta(k-1): S_{n-1} = \bar{s} \text{ and } E_{n-1}^* = \bar{E}\} \qquad (2.1)$$

for $k \geq 0$. In accordance with our usual notation, we denote by $O(s'; \bar{s}, \bar{E})$ the set of transitions in $E - \bar{E}$ that are enabled both before and after a marking change from $\bar{s}$ to s' triggered by the simultaneous firing of the transitions in $\bar{E}$.

Theorem 2.2. Let $\bar{s} \in G$ and $\bar{E} \subseteq E(\bar{s})$. Suppose that

$$P_\mu\{(S_n, E_n^*) = (\bar{s}, \bar{E}) \text{ i.o.}\} = 1.$$

Also suppose that for each s' such that $p(s'; \bar{s}, \bar{E}) > 0$, either

(a) $O(s'; \bar{s}, \bar{E}) = \varnothing$, or

(b) $O(s'; \bar{s}, \bar{E}) \neq \varnothing$ and the clock for each transition $e_i \in O(s'; \bar{s}, \bar{E})$ is always set according to an exponential distribution with fixed intensity $v(e_i)$.

Then the random times $\{\zeta_{\theta(k)}: k \geq 0\}$ defined via (2.1) form a sequence of regeneration points for $\{X(t): t \geq 0\}$. If, in particular, the condition in (a) holds for all s' such that $p(s'; \bar{s}, \bar{E}) > 0$, then the random indices $\{\theta(k): k \geq 0\}$ form a sequence of regeneration points (in discrete time) for $\{(S_n, C_n): n \geq 0\}$.

Theorem 2.2 asserts that the successive times at which the marking is $\bar{s}$ and transitions in $\bar{E}$ fire simultaneously form a sequence of regeneration points for the marking process. Heuristically, at each time $\zeta_{\theta(k)}$ the new marking $S_{\theta(k)}$ is generated according to the fixed probability mass function $p(\cdot; \bar{s}, \bar{E})$. The clock for each newly enabled transition $e_i \in N(S_{\theta(k)}; \bar{s}, \bar{E})$ is set according to a distribution function $F(\cdot; S_{\theta(k)}, e_i, \bar{s}, \bar{E})$ that depends on the history of the marking process only through the new marking $S_{\theta(k)}$. The clock for each old transition $e_i \in O(S_{\theta(k)}; \bar{s}, \bar{E})$ has been set at some previous time according to an exponential distribution with fixed intensity

$v(e_i)$; the memoryless property of the exponential distribution—see Corollary 4.17 in Chapter 3—implies that the remaining time on the clock is exponentially distributed with intensity $v(e_i)$ regardless of the past history of the marking process. Thus, the joint distribution of the new marking and the clock-reading vector is the same at each time $\zeta_{\theta(k)}$. Because the future evolution of the marking process depends only on the new marking and the clock-reading vector, the regenerative property follows. The formal proof of Theorem 2.2 is given at the end of the subsection.

Definition 2.3. A marking $\bar{s} \in G$ is a *single state* if $E(\bar{s}) = \{\bar{e}\}$ for some $\bar{e} \in E$.

Remark 2.4. Observe that the condition in (a) always holds for a single state. Thus, if an SPN has a recurrent single state, then there exists a sequence of regeneration points for both the marking process and the underlying chain. In practice—as illustrated by the examples in the following subsections—regeneration points for SPNs with nonexponential clock-setting distributions are almost always defined in terms of a single state.

EXAMPLE 2.5 (Flexible manufacturing system). For the SPN of Example 2.9 in Chapter 2, the immediate marking $\bar{s} = (0, 0, 0, 1, 1, 1, 0, 0, 1)$ is a single state with $\bar{e} = e_4 =$ "unloading of finished parts and loading of raw parts." When the marking is $\bar{s}$, all machines are idle and three finished parts are awaiting unloading. Suppose that the clock-setting distribution functions satisfy conditions as in Example 1.28 in Chapter 5 or Example 2.2 in Chapter 5, so that $\bar{s}$ is recurrent. Denote by $\{\theta(k) : k \geq 0\}$ the indices of the successive marking changes at which the marking is $\bar{s}$ and transition e_4 fires. Then, by Theorem 2.2, the random indices $\{\theta(k) : k \geq 0\}$ form a sequence of regeneration points for the underlying chain and the random times $\{\zeta_{\theta(k)} : k \geq 0\}$ form a sequence of regeneration points for the marking process.

Remark 2.6. Suppose that the conditions of Theorem 2.2 hold for two initial distributions μ and μ'. Then by Theorem 2.2 the random times $\{\zeta_{\theta(k)} : k \geq 0\}$ defined via (2.1) form a sequence of regeneration points for the marking process under either initial distribution. If the cycle length τ_1 is aperiodic with $E_\mu[\tau_1] < \infty$, then $E_{\mu'}[\tau_1] < \infty$ and Theorem 1.20 implies that $X(t) \Rightarrow X$ under μ and $X(t) \Rightarrow X'$ under μ'. But, setting $T_k = \zeta_{\theta(k)}$ for $k \geq 0$, we see that

$$P_\mu\{X \in A\} = \frac{E_\mu\left[\int_{T_0}^{T_1} 1_A(X(u))\, du\right]}{E_\mu[\tau_1]}$$

$$= \frac{E_{\mu'}\left[\int_{T_0}^{T_1} 1_A(X(u))\, du\right]}{E_{\mu'}[\tau_1]}$$

$$= P_{\mu'}\{X' \in A\}$$

for any set $A \subseteq S$. Thus the limiting distribution of the marking process does not depend on the initial distribution. Similarly, the value of a time-average limit does not depend on the initial distribution. Analogous remarks apply in the discrete-time setting.

Remark 2.7. If the conditions of Theorem 2.2 hold with $\bar{s} \in E'$—as in Example 2.5—then the regeneration points for the marking process may not be detectable from the sample paths of the marking process alone. This is not an issue in practice, however. Indeed—as indicated in Section 3.1.3—a sample path of the marking process is usually generated by first generating a sample path of the underlying chain, and the regeneration points are detectable from the latter sample path.

Remark 2.8. In general, $\{X(t): t \geq 0\}$ is a delayed regenerative process under the conditions of Theorem 2.2. Suppose, however, that the SPN behaves as if at time 0 the marking is $\bar{s}$ and the transitions in $\bar{E}$ fire simultaneously. That is, suppose that the initial distribution of the underlying chain is equal to ψ, where

$$\psi(H) = p(s'; \bar{s}, \bar{E}) \prod_{e_i \in N(s'; \bar{s}, \bar{E})} F(x_i; s', e_i, \bar{s}, \bar{E}) \prod_{e_i \in O(s'; \bar{s}, \bar{E})} \left(1 - e^{-v(e_i)x_i}\right)$$

(2.9)

for all sets

$$H = \{s'\} \times \{(c_1', \ldots, c_M') \in C(s'): 0 \leq c_i' \leq x_i \text{ for } 1 \leq i \leq M\}. \quad (2.10)$$

Then we can take $\theta(0) = 0$, so that $\{X(t): t \geq 0\}$ is a nondelayed regenerative process.

Remark 2.11. If the marking process of an SPN is regenerative, then—since the regeneration points are a.s. increasing by definition—there exists $\delta > 0$ such that $P_\mu\{\tau_1 > \delta\} > 0$. It follows that the expected cycle length is positive. Moreover, the Borel–Cantelli lemma implies that $P_\mu\{\tau_k > \delta \text{ i.o.}\} = 1$, so that the lifetime of the marking process is a.s. infinite.

Remark 2.12. Let $\bar{s}$, $\bar{E}$, and $\{\theta(k): k \geq 0\}$ be as in Theorem 2.2. Suppose that $P_\mu\{(S_n, E_n^*) = (\bar{s}, \bar{E}) \text{ i.o.}\} = 1$ and the condition in Theorem 2.2(b) holds for all s' such that $p(s'; \bar{s}, \bar{E}) > 0$. Although the random times $\{\zeta_{\theta(k)}: k \geq 0\}$ form a sequence of regeneration points for the marking process, the random indices $\{\theta(k): k \geq 0\}$ do not form a sequence of regeneration points for the underlying chain $\{(S_n, C_n): n \geq 0\}$. To see this, fix $k \geq 0$ and observe that, for $e_i \in O(S_{\theta(k)}; \bar{s}, \bar{E})$, the clock reading $C_{\theta(k),i}$ is completely determined by $\{(S_n, C_n): 0 \leq n \leq \theta(k) - 1\}$. It follows that, in general, the cycles of the underlying chain formed by the $\theta(k)$'s are not mutually independent (or even m-dependent for some fixed $m \geq 1$). Interestingly, it can be shown—by taking expectations in (2.18) and using the strong Markov property—that the cycles are identically distributed.

Remark 2.13. Let $\bar{s} \in G$ and $\bar{E} \subseteq E(\bar{s})$, and define $\{\theta(k): k \geq 0\}$ as in (2.1). Also let $\bar{S}' \subseteq G$ be such that $p(\bar{s}'; \bar{s}, \bar{E}) > 0$ for $\bar{s}' \in \bar{S}'$. Denote by $\tilde{\theta}(k)$ the index of the kth marking change at which the old marking is $\bar{s}$, the set of transitions that trigger the marking change is $\bar{E}$, and the new marking is an element of $\bar{S}'$; thus $\{\tilde{\theta}(k): k \geq 0\}$ is a random subsequence of $\{\theta(k): k \geq 0\}$. Suppose that $P_\mu\{(S_n, E_n^*) = (\bar{s}, \bar{E}) \text{ i.o.}\} = 1$ and that $O(\bar{s}'; \bar{s}, \bar{E}) = \varnothing$ for $\bar{s}' \in \bar{S}'$. Then a straightforward modification of the proof of Theorem 2.2 shows that the random indices $\{\tilde{\theta}(k): k \geq 0\}$ form a sequence of regeneration points for the underlying chain. Moreover, if $W_n = f(S_n, C_n, S_{n+1}, C_{n+1})$ for some function f and all $n \geq 0$, and if $\bar{S}' = \{\bar{s}'\}$ for some $\bar{s}' \in G$, then the latter random indices also form a sequence of regeneration points for the process $\{W_n: n \geq 0\}$; see Remark 1.30. Similar observations hold in continuous time for the marking process. Virtually all the results in this section can be modified in a straightforward manner to encompass regeneration points of the form $\{\tilde{\theta}(k): k \geq 0\}$.

Remark 2.14. Let $\bar{E} \subseteq E$ and let $\bar{G} \subset G$ be a set of markings such that

- $\bar{E} \subseteq E(\bar{s})$ for all $\bar{s} \in \bar{G}$, and

- $p(\,\cdot\,; \bar{s}, \bar{E}) = p(\,\cdot\,; \bar{s}', \bar{E})$ for $\bar{s}, \bar{s}' \in \bar{G}$, and

- $P_\mu\{S_n \in \bar{G} \text{ and } E_n^* = \bar{E} \text{ i.o.}\} = 1$.

Set $\theta(-1) = 0$ and

$$\theta(k) = \inf\left\{n > \theta(k-1): S_{n-1} \in \bar{G} \text{ and } E_{n-1}^* = \bar{E}\right\} \qquad (2.15)$$

for $k \geq 0$. Then the conclusions of Theorem 2.2 hold for the sequence $\{\theta(k): k \geq 0\}$ if either (1) the condition in Theorem 2.2(a) holds for all $\bar{s} \in \bar{G}$, or (2) for each $s' \in G$ is such that $p(s'; \bar{s}, \bar{E}) > 0$ for some $\bar{s} \in \bar{G}$, the clock for each transition $e_i \in E(s')$ is always set according to an exponential distribution with fixed intensity $v(e_i)$.

Remark 2.16. If there exists a sequence $\{\theta(k): k \geq 0\}$ of regeneration points for the underlying chain as in Theorem 2.2, then there exists a sequence of regeneration points for the embedded chain $\{(S_n^+, C_n^+): n \geq 0\}$. This latter sequence is defined as follows. Recall from (1.14) in Chapter 3 that $\gamma(n)$ $(n \geq 0)$ is the index of the nth marking change at which the new marking is timed. Set $\alpha(k) = \inf\{n \geq \theta(k): S_n \in S\}$ for $k \geq 0$. Then define $\theta^+(k)$ for $k \geq 0$ via the relation $\gamma(\theta^+(k)) = \alpha(k)$, so that $(S_{\theta^+(k)}^+, C_{\theta^+(k)}^+) = (S_{\alpha(k)}, C_{\alpha(k)})$ for each k. A straightforward modification of the proof of Theorem 2.2 shows that the random indices $\{\theta^+(k): k \geq 0\}$ form a sequence of regeneration points for the embedded chain.

PROOF OF THEOREM 2.2. Each $\theta(k)$ is a stopping time with respect to the underlying chain, and both $\{X(t): t \geq \zeta_{\theta(k)}\}$ and $\{\tau_n: n > k\}$ are completely determined by the process $\{ (S_n, C_n) : n \geq \theta(k) \}$. To prove the first assertion of the theorem, it therefore suffices to show that

$$P_\mu \{ (S_{\theta(k)}, C_{\theta(k)}) \in H_0, \ldots, (S_{\theta(k)+n}, C_{\theta(k)+n}) \in H_n$$
$$| X(t) : 0 \leq t < \zeta_{\theta(k)} \} \qquad (2.17)$$
$$= P_\psi \{ (S_0, C_0) \in H_0, \ldots, (S_n, C_n) \in H_n \} \text{ a.s.}$$

for $k, n \geq 0$ and subsets $H_0, \ldots, H_n \subseteq \Sigma$ of the form (2.10), where ψ is defined as in (2.9); cf. Remark 1.2. To establish (2.17), fix $k \geq 0$ and consider an arbitrary but fixed set H of the form (2.10). Recall from Section 3.4.2 the definition of the partial history $\mathcal{F}_n$ of the underlying chain up to the nth marking change, and of the modified partial history $\tilde{\mathcal{F}}_n$ given by $\tilde{\mathcal{F}}_n = \mathcal{F}_n - \{ S_n \}$. Observe that $\theta(k)$ is a stopping time with respect to the increasing sequence of modified partial histories $\{ \tilde{\mathcal{F}}_n : n \geq 0 \}$. Using the definition of $\theta(k)$ together with Corollary 4.18 in Chapter 3, we find that

$$P_\mu \{ (S_{\theta(k)}, C_{\theta(k)}) \in H \mid \tilde{\mathcal{F}}_{\theta(k)} \} = \psi(H)$$
$$= P_\psi \{ (S_0, C_0) \in H \} \text{ a.s..} \qquad (2.18)$$

A straightforward inductive argument using the strong Markov property then shows that

$$P_\mu \{ (S_{\theta(k)}, C_{\theta(k)}) \in H_0, \ldots, (S_{\theta(k)+n}, C_{\theta(k)+n}) \in H_n \mid \tilde{\mathcal{F}}_{\theta(k)} \} \qquad (2.19)$$
$$= P_\psi \{ (S_0, C_0) \in H_0, \ldots, (S_n, C_n) \in H_n \} \text{ a.s.}$$

for $n \geq 0$ and subsets $H_0, \ldots, H_n \subseteq \Sigma$ of the form (2.10). Because the process $\{ X(t) : 0 \leq t < \zeta_{\theta(k)} \}$ is completely determined by $\tilde{\mathcal{F}}_{\theta(k)}$, (2.17) follows from (2.19) by a simple application of Proposition 1.30 in the Appendix.

To prove the second assertion, observe that each $\theta(k)$ is a stopping time with respect to the underlying chain. By the strong Markov property for the underlying chain and the specific form of the transition kernel P—see (1.9) in Chapter 3—we have

$$P_\mu \{ (S_{\theta(k)}, C_{\theta(k)}) \in H \mid \mathcal{G}_k \} = P((S_{\theta(k)-1}, C_{\theta(k)-1}), H)$$
$$= P_\psi \{ (S_0, C_0) \in H \} \text{ a.s.}$$

for $H \subseteq \Sigma$, where $\mathcal{G}_k = \{ (S_0, C_0), \ldots, (S_{\theta(k)-1}, C_{\theta(k)-1}) \}$. An inductive argument then shows that

$$P_\mu \{ (S_{\theta(k)}, C_{\theta(k)}) \in H_0, \ldots, (S_{\theta(k)+n}, C_{\theta(k)+n}) \in H_n \mid \mathcal{G}_k \}$$
$$= P_\psi \{ (S_0, C_0) \in H_0, \ldots, (S_n, C_n) \in H_n \} \text{ a.s.}$$

for $n \geq 0$ and $H_0, \ldots, H_n \subseteq \Sigma$. The desired result now follows from a discrete-time analog of Remark 1.2. $\qquad \square$

6.2.2 SPNs with Positive Clock-Setting Densities

Using Theorem 2.2 and the results in Section 5.1, we obtain Theorem 2.24 below, which is applicable when Assumption PD holds—that is, when the marking set is finite, the SPN is irreducible, all speeds are positive, and the clock-setting distributions for the timed transitions have convergent LaPlace–Stieltjes transforms and density components that are positive and continuous on an interval of the form $(0, \bar{x}]$.

To prepare for Theorem 2.24, we first introduce the notion of a "polynomially dominated" function. Recall from Section 3.1.2 that $\Sigma^+ = \{(s, c) \in \Sigma : s \in S\}$ is the state space of the embedded chain $\{(S_n^+, C_n^+) : n \geq 0\}$, and set

$$\tilde{g}_q(s, c) = \begin{cases} 1 + \max_{1 \leq i \leq M} c_i^q & \text{if } (s, c) \in \Sigma^+; \\ 1 & \text{if } (s, c) \in \Sigma - \Sigma^+ \end{cases}$$

for $s \in G$, $c = (c_1, c_2, \ldots, c_M) \in C(s)$, and $q \geq 0$. As in Chapter 5, write $\tilde{f} = O(\tilde{g})$ for real-valued functions $\tilde{f}$ and $\tilde{g}$ defined on Σ if (with $0/0 = 0$)

$$\sup_{(s,c) \in \Sigma} |\tilde{f}(s, c)| / |\tilde{g}(s, c)| < \infty.$$

Definition 2.20. A real-valued function $\tilde{f}$ defined on Σ is *polynomially dominated* if $\tilde{f} = O(\tilde{g}_q)$ for some $q \geq 0$.

Thus a function $\tilde{f}$ is polynomially dominated if $|\tilde{f}|$ is bounded above on Σ^+ by a polynomial function of the maximum clock reading and is bounded above on $\Sigma - \Sigma^+$ by a constant.

EXAMPLE 2.21 (Holding-time function). Suppose that there exists $r > 0$ such that $r(s, e) \geq r$ for all $s \in S$ and $e \in E(s)$—such an r exists, for example, if S is finite and all speeds are positive. Recall the definition of the holding-time function t^* from (1.7) in Chapter 3, and observe that

$$t^*(s, c) = \min_{\{i : \, e_i \in E(s)\}} c_i / r(s, e_i) \leq \max_{\{i : \, e_i \in E(s)\}} c_i / r \leq r^{-1} \left(1 + \max_{1 \leq i \leq M} c_i\right)$$

for $s \in S$ and $c = (c_1, c_2, \ldots, c_M) \in C(s)$. Because, trivially, $t^*(s, c) = 0 < 1/r$ for $(s, c) \in \Sigma - \Sigma^+$, we see that $t^*(s, c) \leq r^{-1} \tilde{g}_1(s, c)$ for $(s, c) \in \Sigma$ and hence t^* is polynomially dominated.

For a sequence of random indices $\{\theta(k) : k \geq 0\}$ defined as in (2.1), set

$$Y_k(f) = \int_{\zeta_{\theta(k-1)}}^{\zeta_{\theta(k)}} f(X(u)) \, du \tag{2.22}$$

for each real-valued function f defined on S and

$$\tilde{Y}_k(\tilde{f}) = \sum_{j=\theta(k-1)}^{\theta(k)-1} \tilde{f}(S_j, C_j) \tag{2.23}$$

for each real-valued function $\tilde{f}$ defined on Σ.

Theorem 2.24. *Let $\bar{s} \in S$ and $\bar{e} \in E(\bar{s})$. Suppose that Assumption PD holds. Also suppose that for each s' such that $p(s'; \bar{s}, \bar{e}) > 0$ either*

(a) $O(s'; \bar{s}, \bar{e}) = \varnothing$ *or*

(b) $O(s'; \bar{s}, \bar{e}) \neq \varnothing$ *and the clock for each transition $e_i \in O(s'; \bar{s}, \bar{e})$ is always set according to an exponential distribution with fixed intensity $v(e_i)$.*

Then

(i) *the random times $\{ \zeta_{\theta(k)} : k \geq 0 \}$ defined via (2.1) with $\bar{E} = \{ \bar{e} \}$ form a sequence of regeneration points for the marking process $\{ X(t) : t \geq 0 \}$,*

(ii) *$E_\mu [Y_1^r(|f|)] < \infty$ for $r \geq 0$ and any real-valued function f defined on S, where $Y_1(f)$ is defined by (2.22), and*

(iii) *$E_\mu [\tilde{Y}_1^r(|\tilde{f}|)] < \infty$ for $r \geq 0$ and any polynomially dominated function $\tilde{f}$ defined on Σ, where $\tilde{Y}_1(\tilde{f})$ is defined by (2.23).*

If, in particular, the condition in (a) holds for all s' such that $p(s'; \bar{s}, \bar{e}) > 0$, then also

(iv) *the random indices $\{ \theta(k) : k \geq 0 \}$ form a sequence of regeneration points for $\{ (S_n, C_n) : n \geq 0 \}$.*

We defer the proof of the theorem to the end of the subsection.

Remark 2.25. Under the conditions of Theorem 2.24 the cycle lengths $\tau_1 = \zeta_{\theta(1)} - \zeta_{\theta(0)}$ and $\tilde{\tau}_1 = \theta(1) - \theta(0)$ for the marking process and underlying chain each have finite moments of all orders. This assertion follows by taking $f \equiv 1$ and $\tilde{f} \equiv 1$ in the theorem.

Remark 2.26. Observe that $E_\mu [\tilde{Y}_1^r(|\tilde{f}|)] < \infty$ for $r \geq 0$ and any polynomially dominated function $\tilde{f}$ even when—as discussed in Remark 2.12— the random indices $\{ \theta(k) : k \geq 0 \}$ do not form a sequence of regeneration points for the underlying chain $\{ (S_n, C_n) : n \geq 0 \}$.

Remark 2.27. Suppose that the conditions of Theorem 2.24 are satisfied. Because each clock-setting distribution has a density component that is continuous and positive on an interval of the form $(0, \bar{x})$, the time τ_1 between successive regeneration points of the marking process is aperiodic. Moreover, the marking process has right-continuous sample paths by definition. Thus Theorem 1.20 applies, so that time-average limits can also be viewed as limiting or steady-state means. When applying Theorem 1.20(iv), observe that $\sup_{s \in S} f(s) < \infty$ for any real-valued function f defined on S, because S is finite by hypothesis.

$$e_{1,j} = \text{stoppage of machine } j$$
$$e_{2,j} = \text{start of repair for machine } j$$
$$e_3 = \text{end of repair}$$

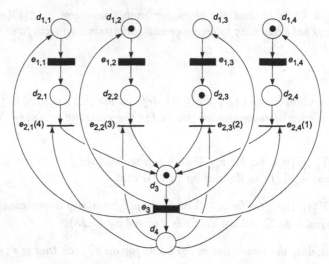

Figure 6.1. SPN representation of machine repair system (four machines).

EXAMPLE 2.28 (Machine repair). Consider a group of N (≥ 1) machines (numbered $1, 2, \ldots, N$) under the care of a single repairperson. Whenever a machine stops and the repairperson is idle, the repairperson immediately starts to repair the machine. Whenever the repairperson completes a repair and at least one machine is stopped, the repairperson immediately starts to repair the lowest-numbered stopped machine; if no machines are stopped, then the repairperson becomes idle. The successive times (lifetimes) between end of repair and the next stoppage of machine j are i.i.d according to a gamma distribution, and the successive times for the repairperson to repair (and restart) machine j are i.i.d. according to a uniform distribution on $[0, u_j]$ for some constant $u_j \in (0, \infty)$.

This system can be specified as an SPN with timed and immediate transitions and a finite marking set; see Figure 6.1 for $N = 4$. Each place contains at most one token. There is a token in place $d_{1,j}$ if and only if machine j is running and a token in place $d_{2,j}$ if and only if machine j is stopped and awaiting repair. There is a token in place d_3 if and only if the repairperson is repairing a machine and a token in place d_4 if and only if the repairperson is idle. All speeds for enabled transitions are equal to 1. Each timed transition $e_{1,j}$ and immediate transition $e_{2,j}$ is deterministic. Priorities are displayed for each transition $e_{2,j}$; these priorities are used to model the service discipline described above. Whenever transition $e_3 =$ "completion of repair" fires, it removes a token from place d_3 and deposits a token in place d_{1,j^*}, where j^* is the unique index such that neither place d_{1,j^*} nor d_{2,j^*} contains any tokens just before the firing of e_3. Thus the repairperson becomes available to repair another machine and the machine that has just completed repair starts running.

Observe that the marking $\bar{s}$ in which $\bar{s}_{1,1} = \cdots = \bar{s}_{1,N} = 0$, $\bar{s}_{2,1} = \bar{s}_4 = 0$, $\bar{s}_{2,2} = \cdots = \bar{s}_{2,N} = 1$, and $\bar{s}_3 = 1$ is a single state with $E(\bar{s}) = \{e_3\}$. (All machines are stopped and a repair of machine 1 is underway whenever the marking is $\bar{s}$.) Moreover, the SPN is irreducible. To see this, let $\hat{s}$ be the marking in which all machines are running. Then, for $s, s' \in S$, an easy argument shows that $s \rightsquigarrow \hat{s}$ and $\hat{s} \rightsquigarrow s'$, so that $s \rightsquigarrow s'$. Each clock-setting distribution function for a timed transition has a convergent LaPlace–Stieltjes transform in a neighborhood of the origin and a density function that is positive and continuous on the interval $(0, \bar{x}]$, where $\bar{x} = \min_{1 \leq j \leq N} u_j$. Thus Assumption PD holds and the conditions of Theorem 2.24 are satisfied.

EXAMPLE 2.29 (Producer–consumer system with nonpreemptive priority). For the system of Example 2.1 in Chapter 2 with buffer capacities B_1 and B_2, suppose that the creation-time random variables A_1 and A_2 are each distributed according to a truncated normal distribution on $[0, \infty)$. Also suppose that the transmission-time random variables L_1 and L_2 are each distributed according to a beta distribution. For the SPN in Figure 2.4, observe that the marking $\bar{s} = (0, B_1 - 1, 1, 0, B_2, 0, 0)$ is a single state with $E(\bar{s}) = \{e_3\}$, where $e_3 =$ "end of transmission to consumer 1." There are B_1 items in buffer 1, B_2 items in buffer 2, and a transmission to consumer 1 is in progress whenever the marking is $\bar{s}$. Setting $\tilde{s} = (B_1, 0, 0, B_2, 0, 0, 1)$, it is straightforward to show that $s \rightsquigarrow \tilde{s}$ and $\tilde{s} \rightsquigarrow s'$ for any $s, s' \in G$, so that the SPN is irreducible. It follows that Assumption PD holds and the conditions of Theorem 2.24 are satisfied.

EXAMPLE 2.30 (Telephone system). For the system of Example 1.27 in Chapter 5, suppose that successive durations of calls placed at line i are i.i.d. according to a uniform distribution on $[0, u]$ for some $u > 0$ and the successive times from the end of a call placed or received at line i to the next call placed at line i are i.i.d. according to an exponential distribution with intensity q for some $q > 0$. Consider the SPN given in Figure 5.5, and let $\bar{G}$ be the set of markings in which there is a call connected on link 1 and all other links are idle. Set $\bar{e} = e_{2,1} =$ "end of call connected on link 1," and observe that the pair $(\bar{G}, \{\bar{e}\})$ satisfies the conditions given in Remark 2.14. Example 1.27 in Chapter 5 shows that Assumption PD holds. It then follows from Corollary 1.26 in Chapter 5 and Remark 2.14 that

- The random times $\{\zeta_{\theta(k)}: k \geq 0\}$ defined via (2.15) form a sequence of regeneration points for the marking process of the SPN.

- $E_\mu[Y_1^r(|f|)] < \infty$ for $r \geq 0$ and any real-valued function f defined on S, where $Y_1(f)$ is defined by (2.22).

The assumption in Theorem 2.24 that $\bar{s}$ is a timed marking can be relaxed. In particular, we have the following result, the proof of which is sketched at the end of the subsection.

Theorem 2.31. *Let $\bar{s} \in S'$ and $\bar{E} = E(\bar{s}) \cap E'$. Suppose that Assumption PD holds. Also suppose that for each s' such that $p(s'; \bar{s}, \bar{E}) > 0$ either*

(a) $O(s'; \bar{s}, \bar{E}) = \varnothing$ *or*

(b) $O(s'; \bar{s}, \bar{E}) \neq \varnothing$ *and the clock for each transition $e_i \in O(s'; \bar{s}, \bar{E})$ is always set according to an exponential distribution with fixed intensity $v(e_i)$.*

Then

(i) *the random times $\{\zeta_{\theta(k)} : k \geq 0\}$ defined via (2.1) form a sequence of regeneration points for the marking process $\{X(t) : t \geq 0\}$,*

(ii) *$E_\mu[Y_1^r(|f|)] < \infty$ for $r \geq 0$ and any real-valued function f defined on S, where $Y_1(f)$ is defined by (2.22), and*

(iii) *$E_\mu[\tilde{Y}_1^r(|\tilde{f}|)] < \infty$ for $r \geq 0$ and any polynomially dominated function $\tilde{f}$ defined on Σ, where $\tilde{Y}_1(\tilde{f})$ is defined by (2.23).*

If, in particular, the condition in (a) holds for all s' with $p(s'; \bar{s}, \bar{E}) > 0$, then also

(iv) *the random indices $\{\theta(k) : k \geq 0\}$ form a sequence of regeneration points for $\{(S_n, C_n) : n \geq 0\}$.*

We conclude this subsection by giving the proof of Theorem 2.24. To this end, we need the following lemma, which follows immediately from Corollary 1.26 in Chapter 5 and Proposition 1.32. In the lemma, we take $\theta(-1) = 0$.

Lemma 2.32. *Suppose that Assumption PD holds. Then there exists at least one sequence $\{\theta^+(k) : k \geq 0\}$ of od-equilibrium points for the embedded chain $\{(S_n^+, C_n^+) : n \geq 0\}$. For any such sequence, the cycle length $\tilde{\tau}_k^+ = \theta^+(k) - \theta^+(k-1)$ has finite moments of all orders for $k \geq 0$.*

PROOF OF THEOREM 2.24. The sequence $\{\theta(k) - 1 : k \geq 0\}$ corresponds to the successive times at which the chain $\{(S_n, C_n) : n \geq 0\}$ hits the set $A = \{(s, c) \in \Sigma : s = \bar{s} \text{ and } E^*(s, c) = \{\bar{e}\}\}$. By Corollary 1.26 in Chapter 5, there exists $\bar{x} > 0$ such that the embedded chain $\{(S_n^+, C_n^+) : n \geq 0\}$ is positive Harris recurrent with recurrence measure $\bar{\phi}$ given by (1.17) in Chapter 5. Clearly, $\bar{\phi}(A) > 0$, so that the embedded chain—and hence the underlying chain—hits the set A infinitely often with probability 1 and each $\theta(k)$ is a.s. finite. The assertions in (i) and (iv) now follow from Theorem 2.2. The assertion in (ii) for a specified function f follows from the assertion in (iii) with $\tilde{f}(s,c) = f(s)t^*(s,c)$. The remainder of the proof is therefore devoted to establishing the assertion in (iii). To this end, fix $r > 0$ and a polynomially dominated function $\tilde{f}$; without loss of generality, suppose that $\tilde{f}$ is nonnegative. Also suppose for ease of exposition that $\theta(0) = 0$.

We first establish the assertion in (iii) when the condition in (a) holds for all s' with $p(s'; \bar{s}, \bar{e}) > 0$, so that $\{\theta(k): k \geq 0\}$ is a sequence of regeneration points for the underlying chain. Write $\tilde{Y}_1(\tilde{f}) = \tilde{Y}_1^+(\tilde{f}) + \tilde{Y}_1'(\tilde{f})$, where

$$\tilde{Y}_1^+(\tilde{f}) = \sum_{n=\theta(0)}^{\theta(1)-1} \tilde{f}(S_n, C_n) 1_S(S_n)$$

and

$$\tilde{Y}_1'(\tilde{f}) = \sum_{n=\theta(0)}^{\theta(1)-1} \tilde{f}(S_n, C_n) 1_{S'}(S_n).$$

Because

$$E_\mu[\tilde{Y}_1^r(\tilde{f})] \leq c_r E_\mu[(\tilde{Y}_1^+(\tilde{f}))^r] + c_r E_\mu[(\tilde{Y}_1'(\tilde{f}))^r]$$

for a finite constant c_r depending only on r—see (1.12) in the Appendix for a discussion of the "c_r-inequality"—it suffices to show that $\tilde{Y}_1^+(\tilde{f})$ and $\tilde{Y}_1'(\tilde{f})$ each have finite moments of all orders.

We first consider $\tilde{Y}_1^+(\tilde{f})$. Recall from Remark 2.16 that the regeneration points $\{\theta(k): k \geq 0\}$ for the underlying chain induce a sequence of regeneration points $\{\theta^+(k): k \geq 0\}$ for the embedded chain, and set $\tilde{\tau}_1^+ = \theta^+(1) - \theta^+(0)$. Using the Cauchy–Schwarz inequality, we have

$$E_\mu[(\tilde{Y}_1^+(\tilde{f}))^r] = E_\mu\left[\left(\sum_{n=\theta^+(0)}^{\theta^+(1)-1} \tilde{f}(S_n^+, C_n^+)\right)^r\right]$$

$$\leq E_\mu\left[(\tilde{\tau}_1^+)^r \max_{\theta^+(0)\leq n\leq\theta^+(1)-1} \tilde{f}^r(S_n^+, C_n^+)\right]$$

$$\leq E_\mu^{1/2}[(\tilde{\tau}_1^+)^{2r}] E_\mu^{1/2}\left[\max_{\theta^+(0)\leq n\leq\theta^+(1)-1} \tilde{f}^{2r}(S_n^+, C_n^+)\right]$$

$$\leq E_\mu^{1/2}[(\tilde{\tau}_1^+)^{2r}] E_\mu^{1/2}[\tilde{Y}_1^+(\tilde{f}^{2r})].$$

$$(2.33)$$

It therefore suffices to show that $\tilde{\tau}_1^+$ has finite moments of all orders and that $E_\mu[\tilde{Y}_1^+(\tilde{f}^{2r})] < \infty$. The finiteness of the moments of $\tilde{\tau}_1^+$ follows from Lemma 2.32, since the regeneration points $\{\theta^+(k): k \geq 0\}$ are also od-equilibrium points. To show that $\tilde{Y}_1^+(\tilde{f}^{2r})$ has finite mean, observe that the function $\tilde{f}^{2r}$ is polynomially dominated, so that $\pi^+(\tilde{f}^{2r}) < \infty$ by Corollary 1.26 in Chapter 5—here π^+ is the invariant probability measure of the embedded chain. The finiteness of $E_\mu[\tilde{Y}_1^+(\tilde{f}^{2r})]$ then follows from the ratio formula

$$\pi^+(\tilde{f}^{2r}) = \frac{E_\mu[\tilde{Y}_1^+(\tilde{f}^{2r})]}{E_\mu[\tilde{\tau}_1^+]};$$

see Remark 1.33.

We now consider $\tilde{Y}_1'(\tilde{f})$. Recall from (1.14) in Chapter 3 that $\{\gamma(n) : n \geq 0\}$ are the indices of the successive marking changes at which the new marking is timed—since $\theta(0) = 0$ by assumption, we have $\theta^+(0) = \gamma(0)$. For $k \geq 0$, denote by U_k the reward (as measured by $\tilde{f}$) that the underlying chain accumulates during the sojourn in the set $\Sigma - \Sigma^+$ that ends at the $\gamma(k)$th marking change:

$$U_k = \sum_{n=\gamma(k-1)+1}^{\gamma(k)-1} \tilde{f}(S_n, C_n).$$

Then $\tilde{Y}_1'(\tilde{f}) = \sum_{k=0}^{\tilde{\tau}_1^+ - 1} U_k$. Because the function $\tilde{f}$ is polynomially dominated by hypothesis, and hence bounded on $\Sigma - \Sigma^+$,

$$\tilde{Y}_1'(\tilde{f}) \leq \psi \sum_{k=0}^{\tilde{\tau}_1^+ - 1} M_k,$$

where $\psi = \sup_{(s,c)\in\Sigma-\Sigma^+} \tilde{f}(s,c) < \infty$ and $M_k = \gamma(k) - \gamma(k-1) - 1$ is the length of the kth sojourn in $\Sigma - \Sigma^+$. Because $\tilde{\tau}_1^+$ has finite moments of all orders, it suffices to show that

$$E_\mu\left[\sum_{k=0}^{\tilde{\tau}_1^+ - 1} M_k^{2r}\right] < \infty, \tag{2.34}$$

for then the finiteness of $E_\mu[(\tilde{Y}_1'(\tilde{f}))^r]$ follows by a computation analogous to (2.33). To establish (2.34), define a vector $H_k = (H_{k,1}, H_{k,2}, \ldots, H_{k,M})$ that, in effect, records for each transition e the most recent distribution used to set the clock for e between the $\gamma(k-1)$st and $\gamma(k)$th marking change. Specifically, set

$$H_{k,i} = \begin{cases} (S_{\xi(k,i)}; S_{\xi(k,i)-1}, E^*_{\xi(k,i)-1}) & \text{if } \xi(k,i) > 0; \\ (\Delta, \Delta, \varnothing) & \text{if } \xi(k,i) = 0 \end{cases}$$

for $1 \leq i \leq M$, where

$$\xi(k,i) = \sup\left\{\gamma(k-1) < j \leq \gamma(k) : e_i \in N(S_j; S_{j-1}, E^*_{j-1})\right\}$$

for $1 \le i \le M$ and $n \ge 1$—set $\xi(n, i) = 0$ if the supremum is taken over an empty set. Denote by $\mathcal{H}$ the state space of the process $\{H_n : n \ge 1\}$, and fix $s', s \in S$ and $h \in \mathcal{H}$. It follows from (3.10) in Chapter 3 that there exist constants $a \in (0, \infty)$ and $\rho \in [0, 1)$ such that

$$P_\mu \left\{ M_k > n \mid S_{\gamma(k-1)} = s \right\} \le a\rho^n$$

for $s \in S$. Thus

$$P_\mu \left\{ M_k > n \mid S_{\gamma(k-1)} = s, S_{\gamma(k)} = s', H_k = h \right\}$$
$$\le \frac{P_\mu \left\{ M_k > n \mid S_{\gamma(k-1)} = s \right\}}{P_\mu \left\{ S_{\gamma(k)} = s', H_k = h \mid S_{\gamma(k-1)} = s \right\}}$$
$$\le \frac{a\rho^n}{u(s', s, h)},$$

where $u(s', s, h) = P_\mu \left\{ S_{\gamma(k)} = s', H_k = h \mid S_{\gamma(k-1)} = s \right\}$. (The function u is well defined because the latter probability does not depend explicitly on k.) Set $\bar{u} = \min_{s', s, h} u(s', s, h)$, where the minimum is taken over all $s', s \in S$ and $h \in \mathcal{H}$ such that $u(s', s, h)$ is positive, and observe that $\bar{u} > 0$. We then have

$$P_\mu \left\{ M_k > n \mid S_{\gamma(k-1)} = s, S_{\gamma(k)} = s', H_k = h \right\} \le b\rho^n,$$

where $b = a/\bar{u} < \infty$. Fix $q \ge 1$ and use a standard moment inequality—see (1.16) in the Appendix—to obtain

$$E_\mu \left[M_k^q \mid S_{\gamma(k-1)} = s, S_{\gamma(k)} = s', H_k = h \right] \le \beta_q$$

for all $s', s \in S$ and $h \in \mathcal{H}$, where $\beta_q = bq \sum_{n=0}^\infty (n+1)^{q-1} \rho^n < \infty$. Next, set

$$\mathcal{G} = \left\{ \tilde{\tau}_1^+, S_{\gamma(0)}, S_{\gamma(1)}, \ldots, S_{\gamma(\tilde{\tau}_1^+)}, H_1, H_2, \ldots, H_{\tilde{\tau}_1^+} \right\}$$

and observe that, given $\mathcal{G}$, the random variables $M_0, M_1, \ldots, M_{\tilde{\tau}_1^+ - 1}$ are conditionally independent. Moreover, the distribution of each M_k depends on $\mathcal{G}$ only through $S_{\gamma(k-1)}$, $S_{\gamma(k)}$, and H_k. It follows that

$$E_\mu \left[\sum_{k=0}^{\tilde{\tau}_1^+ - 1} M_k^q \right] = E_\mu \left[E_\mu \left[\sum_{k=0}^{\tilde{\tau}_1^+ - 1} M_k^q \mid \mathcal{G} \right] \right]$$
$$= E_\mu \left[\sum_{k=0}^{\tilde{\tau}_1^+ - 1} E_\mu \left[M_k^q \mid \mathcal{G} \right] \right]$$
$$= E_\mu \left[\sum_{k=0}^{\tilde{\tau}_1^+ - 1} E_\mu \left[M_k^q \mid S_{\gamma(k-1)}, S_{\gamma(k)}, H_k \right] \right]$$
$$\le \beta_q E_\mu \left[\tilde{\tau}_1^+ \right]$$

for $q \geq 1$, which implies (2.34).

We now establish the assertion in (iii) when the condition in (b) holds for at least one marking s' such that $p(s'; \bar{s}, \bar{e}) > 0$. As discussed in Remark 2.12, the random indices $\{\theta(k) : k \geq 0\}$ do not, in general, form a sequence of regeneration points—or even of od-equilibrium points—for the underlying chain, so that the previous argument does not apply directly. We can, however, argue as follows. By Lemma 2.32 there exists a sequence of od-equilibrium points for the embedded chain; these points decompose sample paths of the embedded chain into o.d.s. cycles. It is not hard to see that these points also decompose sample paths of the underlying chain into o.d.s. cycles, and hence induce a sequence of od-regeneration points $\{\theta'(k) : k \geq 0\}$ for the underlying chain. Observe that

$$\tilde{Y}_1(\tilde{f}) \leq \sum_{k=0}^{N} \tilde{Z}_k(\tilde{f}),$$

where N is the number of points of the sequence $\{\theta'(k) : k \geq 0\}$ that lie in the interval $[0, \theta(1)]$ and $\tilde{Z}_k(\tilde{f}) = \sum_{n=\theta'(k-1)}^{\theta'(k)-1} \tilde{f}(S_n, C_n)$ for $k \geq 0$. [We take $\theta'(-1) = 0$.] An argument almost identical to the first part of the proof shows that $\tilde{Z}_k(\tilde{f})$ has finite moments of all orders. By a computation analogous to (2.33), it then suffices to show that the random variable N has finite moments of all orders. For $k \geq 0$, set $I_k = 1$ if at least one point of the sequence $\{\theta(k) : k \geq 0\}$ lies in the interval $[\theta'(k-1), \theta'(k)]$; otherwise, set $I_k = 0$. Observe that $I_1, I_2, \ldots$ is an o.i.d. sequence, and set $p = P_\mu \{I_1 = 0\}$. Because each $\theta(k)$ is a.s. finite, it follows that $p < 1$— otherwise, $\sum_{k=0}^{\infty} P_\mu \{I_k = 1\} = 0$, so that $P_\mu \{I_k = 1 \text{ i.o.}\} = 0$ by the first Borel–Cantelli lemma (Proposition 1.2 in the Appendix), which leads to a contradiction. For $k \geq 1$, we have

$$P_\mu \{N > k\} \leq P_\mu \{I_1 = 0, I_3 = 0, \ldots, I_{l(k)} = 0\}$$
$$= P_\mu \{I_1 = 0\} P_\mu \{I_3 = 0\} \cdots P_\mu \{I_{l(k)} = 0\},$$

where $l(k) = k - 1$ if k is even and $l(k) = k$ if k is odd. It follows that

$$P_\mu \{N > k\} \leq p^{\lfloor k/2 \rfloor} \leq c\rho^k$$

for $k \geq 2$, where $c = 1/p$ and $\rho = p^{1/2}$. Because the distribution of N has a geometrically decreasing right tail, N has moments of all orders. $\qquad \square$

To prove Theorem 2.31, use the positive Harris recurrence of the embedded chain to show that $P_\mu \{S_n^+ = s^+ \text{ i.o.}\} = 1$ for a timed marking s^+ such that $s^+ \rightsquigarrow \bar{s}$, where at least one path from s^+ to $\bar{s}$ has no intermediate timed markings. Then use a geometric trials argument to show that $\bar{s}$ is recurrent. Now proceed as in the proof of Theorem 2.24.

6.2.3   SPNs Satisfying Geometric Trials Criteria

Theorems 2.36 and 2.44 below complement Theorems 2.24 and 2.31 and are meant to be used in conjunction with the geometric trials technique developed in Chapter 5. For a fixed set of transitions $\bar{E} \subseteq E$, set $\beta(-1) = -1$ and

$$\beta(n) = \inf \{ k > \beta(n-1) \colon E^*(S_k, C_k) = \bar{E} \} \qquad (2.35)$$

for $n \geq 0$. According to this definition, $S_{\beta(n)}$ is the marking just before the nth marking change at which the transitions in $\bar{E}$ fire simultaneously. For a marking $\bar{s} \in G$ with $\bar{E} \subseteq E(\bar{s})$, define $\{ \theta(k) \colon k \geq 0 \}$ as in (2.1) to be the random indices of the successive marking changes at which the marking is $\bar{s}$ and the transitions in $\bar{E}$ fire simultaneously. Thus $\{ \theta(k) \colon k \geq 0 \}$ is a random subsequence of $\{ \beta(n) + 1 \colon n \geq 0 \}$. Here we take $\beta(0) = -1$ whenever $\theta(0) = 0$—see Remark 2.8. Recall from Section 3.4.2 the definition of $\{ \mathcal{F}_n \colon n \geq 0 \}$, the increasing sequence of partial histories of the underlying chain. Also define $Y_1(f)$ by (2.22).

Theorem 2.36. *Let $\bar{s} \in G$ and $\bar{E} \subseteq E(\bar{s})$. Suppose that each random index $\beta(n)$ defined in (2.35) is a.s. finite. Let $\{ \alpha(n) \colon n \geq 1 \}$ be an increasing sequence of random indices such that each $\alpha(n)$ is a stopping time with respect to $\{ \mathcal{F}_k \colon k \geq 0 \}$ and $\beta(n-1) \leq \alpha(n) < \beta(n)$. Suppose that*

$$P_\mu \{ S_{\beta(n)} = \bar{s} \mid \mathcal{F}_{\alpha(n)} \} > \delta \text{ a.s.}$$

for some $\delta > 0$ and all $n \geq 0$. Also suppose that for each s' with $p(s'; \bar{s}, \bar{E}) > 0$ either

(a) $O(s'; \bar{s}, \bar{E}) = \varnothing$ *or*

(b) $O(s'; \bar{s}, \bar{E}) \neq \varnothing$ *and the clock for each transition $e_i \in O(s'; \bar{s}, \bar{E})$ is always set according to an exponential distribution with fixed intensity $v(e_i)$.*

Then the random times $\{ \zeta_{\theta(k)} \colon k \geq 0 \}$ defined via (2.1) form a sequence of regeneration points for the marking process $\{ X(t) \colon t \geq 0 \}$. Moreover, for any bounded real-valued function f defined on S, the cycle sum $Y_1(|f|)$ has finite mean if

$$\liminf_{n \geq 0} E_\mu \left[\zeta_{\beta(n+1)+1} - \zeta_{\beta(n)+1} \right] < \infty$$

and finite rth moment $(r > 1)$ if

$$\liminf_{n \geq 0} E_\mu \left[\left(\zeta_{\beta(n+1)+1} - \zeta_{\beta(n)+1} \right)^{r+\epsilon} \right] < \infty \qquad (2.37)$$

for some $\epsilon > 0$.

PROOF. For ease of exposition, suppose that $\theta(0) = 0$. By Lemma 2.4 in Chapter 5, $P_\mu \{ S_{\beta(n)} = \bar{s} \text{ i.o.} \} = 1$, so that each $\theta(k)$ is a.s. finite. The first assertion of the theorem then follows from Theorem 2.2.

To prove the remaining assertions, define a sequence of random indices $\{\lambda(k) : k \geq 0\}$ by writing $\theta(k) = \beta(\lambda(k)) + 1$ for $k \geq 0$. Thus the kth regeneration point corresponds to the $\lambda(k)$th time that the transitions in $\bar{E}$ fire simultaneously. Set $\eta_k = \lambda(k) - \lambda(k-1)$ for $k \geq 1$ and $D_n = \zeta_{\beta(n+1)+1} - \zeta_{\beta(n)+1}$ for $n \geq 0$. Observe that the random variables $\{\eta_k : k \geq 1\}$ are i.i.d. and, as shown in (3.7) in Chapter 3,

$$P_\mu\{\eta_1 > k\} \leq (1-\delta)^k$$

for $k \geq 1$, so that η_1 has moments of all orders. It suffices to show that, for $r \geq 1$ and $\epsilon \geq 0$,

$$E_\mu\left[\sum_{n=0}^{\eta_1-1} D_n^{r+\epsilon}\right] < \infty \tag{2.38}$$

whenever (2.37) holds. Indeed, taking $r = 1$ and $\epsilon = 0$ in (2.38) shows that the cycle length $\tau_1 = \zeta_{\theta(1)} - \zeta_{\theta(0)}$ has finite mean; since f is bounded by assumption, the second assertion of the theorem follows (cf. Remark 1.14). If (2.37) holds for some $r > 1$ and $\epsilon > 0$, then—using (2.38) and performing a calculation analogous to (2.33) but based on Hölder's inequality—we find that

$$E_\mu\left[\tau_1^r\right] = E_\mu\left[\left(\sum_{n=0}^{\eta_1-1} D_n\right)^r\right]$$

$$\leq E_\mu\left[\eta_1^r \max_{0 \leq n < \eta_1} D_n^r\right]$$

$$\leq E_\mu^{\epsilon/(r+\epsilon)}\left[\eta_1^{r(r+\epsilon)/\epsilon}\right] E_\mu^{r/(r+\epsilon)}\left[\max_{0 \leq n < \eta_1} D_n^{r+\epsilon}\right]$$

$$\leq E_\mu^{\epsilon/(r+\epsilon)}\left[\eta_1^{r(r+\epsilon)/\epsilon}\right] E_\mu^{r/(r+\epsilon)}\left[\sum_{n=0}^{\eta_1-1} D_n^{r+\epsilon}\right]$$

$$< \infty,$$

and the final assertion of the theorem follows from the boundedness of f.

To establish (2.38), observe that the random indices $\{\lambda(k): k \geq 0\}$ form a sequence of regeneration points for the discrete-time process $\{D_n: n \geq 0\}$. Moreover, η_1 is aperiodic in discrete time. Thus, by Theorem 1.25(i), there exists a nonnegative random variable D such that $D_n \Rightarrow D$ as $n \to \infty$. Theorem 1.25(ii) then implies that $D_n^{r+\epsilon} \Rightarrow D^{r+\epsilon}$. Using a version of Fatou's lemma for convergence in distribution (Proposition 1.48 in the Appendix), we find that

$$E_\mu\left[D^{r+\epsilon}\right] \leq \liminf_{n \to \infty} E_\mu\left[D_n^{r+\epsilon}\right] < \infty,$$

where the last inequality is a restatement of (2.37). Thus

$$E_\mu[D^{r+\epsilon}] = \frac{E_\mu[\sum_{n=0}^{\eta_1-1} D_n^{r+\epsilon}]}{E_\mu[\eta_1]}$$

by Theorem 1.25(iii), and (2.38) follows. □

The following result—and easily derived extensions of this result—can be useful when verifying that (2.37) holds.

Lemma 2.39. *Let $e_i \in E - E'$ and let β be an a.s. finite stopping time with respect to the sequence $\{\mathcal{F}_n : n \geq 0\}$ of partial histories of the underlying chain. Suppose that e_i is simple and that the clock-setting distribution function $F(\cdot; e_i)$ is GNBU. Then $E_\mu[C_{\beta,i}^r] < \infty$ for $r \geq 0$.*

PROOF. Fix $r \geq 1$ and write $F_i(\cdot) = F(\cdot; e_i)$. By hypothesis, F_i is GNBU with some lower bound x^*. Set $\gamma_i(x) = \sup_{y \geq 0} \overline{F}_i(x+y)/\overline{F}_i(y)$ for $x \geq 0$. As in Section 3.4.2, define $Z_{n,i}$ to be the amount of time that has elapsed on the clock for transition e_i between the most recent clock-setting time prior to ζ_n and time ζ_n itself. Using Lemma 4.10 in Chapter 5 and (1.13) in the Appendix, we have

$$E_\mu[C_{\beta,i}^r] = E_\mu\Big[E_\mu[C_{\beta,i}^r \mid \mathcal{F}_\beta]\Big]$$

$$= E_\mu\Big[\int_0^\infty rx^{r-1}\frac{\overline{F}_i(x + Z_{\beta,i})}{\overline{F}_i(Z_{\beta,i})}\,dx\Big]$$

$$\leq \int_0^\infty rx^{r-1}\gamma_i(x)\,dx.$$

Fix $x > x^*$, so that $\gamma_i(x) < 1$. As in the proof of Lemma 2.12 in Chapter 5, $\overline{F}_i(kx + y) \leq \gamma_i^k(x)\overline{F}_i(y)$ for $y \geq 0$ and $k \in \{0, 1, 2, \dots\}$. It follows that $\gamma_i(kx) \leq \gamma_i^k(x)$ for each nonnegative integer k. We can now argue as in the proof of Lemma 2.12 in Chapter 5 to show that

$$\int_0^\infty rx^{r-1}\gamma_i(x)\,dx = \sum_{k=0}^\infty \int_{kx}^{(k+1)x} ry^{r-1}\gamma_i(y)\,dy$$

$$\leq rx^r \sum_{k=0}^\infty (k+1)^{r-1}\gamma_i^k(x)$$

$$< \infty,$$

and the desired result follows. □

Sometimes a discrete-time version of the condition in (2.37) is easier to verify than (2.37) itself. In this connection the following result can be useful.

Theorem 2.40. *Suppose that the conditions of Theorem 2.36 hold. Also suppose that the marking set G is finite and all speeds are positive. Then, for any real-valued function f defined on S, the cycle sum $Y_1(|f|)$ has finite mean if each clock-setting distribution has finite mean and*

$$\liminf_{n \geq 0} E_\mu[\beta(n+1) - \beta(n)] < \infty, \qquad (2.41)$$

and $Y_1(|f|)$ has finite rth moment $(r > 1)$ if each clock-setting distribution has finite rth moment and

$$\liminf_{n \geq 0} E_\mu[(\beta(n+1) - \beta(n))^{r+\epsilon}] < \infty \qquad (2.42)$$

for some $\epsilon > 0$.

PROOF. Fix $r \geq 1$ and $\epsilon \geq 0$ such that $\epsilon > 0$ if $r > 1$ and $\epsilon = 0$ if $r = 1$. For ease of exposition, suppose that each transition is simple and that all speeds are equal to 1. As in the proof of Theorem 2.36, it suffices to show that τ_1 has finite rth moment. Observe that

$$\tau_1 \leq \sum_{i=1}^{M} \sum_{k=1}^{N_i} C_{\eta(i,k),i},$$

where N_i is the number of marking changes in the interval $[\zeta_{\theta(0)}, \zeta_{\theta(1)})$ at which the clock for transition e_i is set and $\eta(i, k)$ is the index of the kth such marking change. An application of the c_r-inequality shows that

$$E_\mu[\tau_1^r] \leq M^{r-1} \sum_{i=1}^{M} E_\mu\left[\left(\sum_{k=1}^{N_i} C_{\eta(i,k),i}\right)^r\right],$$

and so it suffices to show that

$$E_\mu\left[\left(\sum_{k=1}^{N_i} C_{\eta(i,k),i}\right)^r\right] < \infty$$

for $1 \leq i \leq M$. Fix i and observe that $N_i \leq \tilde{\tau}_1$ for $1 \leq i \leq M$, where $\tilde{\tau}_1 = \theta(1) - \theta(0)$. An argument almost identical to the proof of Theorem 2.36 shows that $E_\mu[\tilde{\tau}_1^r] < \infty$, so that each N_i has finite rth moment. Set $\mathcal{G}_k = \{(S_0, C_0), (S_1, C_1), \dots, (S_{\eta(i,k)}, C_{\eta(i,k)})\}$ for $k \geq 1$, and observe that, for each k, the random variable $C_{\eta(i,k),i}$ is determined by $\mathcal{G}_k$ and is independent of $\mathcal{G}_{k-1}$. Moreover, $N_i + 1$ is a stopping time with respect to $\{\mathcal{G}_k : k \geq 1\}$. Finally, it follows from Lemma 4.19 in Chapter 3 that $C_{\eta(i,1),i}, C_{\eta(i,2),i}, \dots$ are i.i.d. with common distribution function $F(\cdot; e_i)$.

Using Proposition 1.20 in the Appendix, we find that

$$E_\mu\left[\left(\sum_{k=1}^{N_i} C_{\eta(i,k),i}\right)^r\right] \leq E_\mu\left[\left(\sum_{k=1}^{N_i+1} C_{\eta(i,k),i}\right)^r\right]$$

$$\leq b_r E_\mu\left[(N_i+1)^r\right] E_\mu\left[C_{\eta(i,1),i}^r\right]$$

$$< \infty,$$

for some constant $b_r < \infty$, and the desired result follows. □

Remark 2.43. The requirement in Theorem 2.40 that G be finite can be replaced by the requirement that $\inf_{s,e} r(s,e) > 0$ and that there be a finite number of distinct clock-setting distribution functions for the timed transitions. Then the conclusion of the theorem holds for any bounded function f.

Theorem 2.44 gives conditions under which the underlying chain is a regenerative process in discrete time. The proof is analogous to that of Theorem 2.36.

Theorem 2.44. *Let $\bar{s} \in G$ and $\bar{E} \subseteq E(\bar{s})$. Suppose that each random index $\beta(n)$ defined in (2.35) is a.s. finite. Let $\{\alpha(n): n \geq 1\}$ be an increasing sequence of random indices such that each $\alpha(n)$ is a stopping time with respect to $\{\mathcal{F}_k: k \geq 0\}$ and $\beta(n-1) \leq \alpha(n) < \beta(n)$. Suppose that*

$$P_\mu\{S_{\beta(n)} = \bar{s} \mid \mathcal{F}_{\alpha(n)}\} > \delta \text{ a.s.}$$

for some $\delta > 0$ and all $n \geq 0$. Also suppose that $O(s'; \bar{s}, \bar{E}) = \varnothing$ for all s' with $p(s'; \bar{s}, \bar{E}) > 0$. Then the random indices $\{\theta(k): k \geq 0\}$ defined via (2.1) form a sequence of regeneration points for the underlying chain $\{(S_n, C_n) : n \geq 0\}$. Moreover, for any bounded real-valued function $\tilde{f}$ defined on Σ, the cycle sum $\tilde{Y}_1(|\tilde{f}|)$ has finite mean if

$$\liminf_{n \geq 0} E_\mu[\beta(n+1) - \beta(n)] < \infty$$

and finite rth moment $(r > 1)$ if

$$\liminf_{n \geq 0} E_\mu[(\beta(n+1) - \beta(n))^{r+\epsilon}] < \infty$$

for some $\epsilon > 0$.

The following result can be useful when verifying that (2.41) and (2.42) hold or, equivalently, when verifying that the conditions of Theorem 2.44 hold—see Example 2.51 below.

Lemma 2.45. *Let* $\{X_n : n \geq 1\}$ *be a sequence of i.i.d. nonnegative random variables and let* Y *be a nonnegative random variable independent of* $\{X_n : n \geq 1\}$. *Set* $N = \inf\{n \geq 1 : X_1 + \cdots + X_n > Y\}$. *Then for* $r \geq 1$ *there exist finite constants* a_r *and* b_r *(depending only on* r*) such that*

$$E[N^r] \leq a_r E[Y^r] + b_r.$$

PROOF. Fix $r \geq 1$ and set $N(t) = \inf\{n \geq 1 : X_1 + \cdots + X_n > t\}$ for $t \geq 0$. Pick $\alpha > 0$ such that $P\{X_1 \geq \alpha\} > 0$ and set

$$\bar{X}_n = \begin{cases} 0 & \text{if } X_n < \alpha; \\ \alpha & \text{if } X_n \geq \alpha \end{cases}$$

for $n \geq 1$. Define $\bar{N}(t)$ analogously to $N(t)$, but in terms of $\{\bar{X}_n : n \geq 0\}$. Clearly, $\bar{X}_n \leq X_n$ for each n, so that $N(t) \leq \bar{N}(t)$ for each t. Fixing $t \geq 0$ and viewing each $\bar{X}_n$ as the time (possibly 0) between a pair of successive "events," we see that $\bar{N}(t)$ can be interpreted as the number of events that occur in the interval $[0, t]$, where an event always occurs at time 0. By construction, events occur only at times $0, \alpha, 2\alpha, \ldots$ and the number of events that occur at each such time has a geometric distribution with mean $q = 1/P\{X_1 \geq \alpha\}$. Thus $\bar{N}(t)$ is distributed as $\sum_{i=1}^{l(t)} G_i$, where $l(t) = \lfloor t/\alpha + 1 \rfloor$ and $\{G_i : i \geq 1\}$ is a sequence of i.i.d. random variables having a common geometric distribution with mean q. Using the c_r-inequality, we have

$$E[N^r(t)] \leq E[\bar{N}^r(t)] = E\left[\left(\sum_{i=1}^{l(t)} G_i\right)^r\right] \leq l^{r-1}(t) E\left[\sum_{i=1}^{l(t)} G_i^r\right] = l^r(t) E[G_i^r].$$

Because $l^r(t) \leq 2^{r-1}(t^r/\alpha^r + 1)$, we have

$$E[N^r(t)] \leq a_r t^r + b_r,$$

where $a_r = 2^{r-1} E[G_i^r]/\alpha^r < \infty$ and $b_r = 2^{r-1} E[G_i^r] < \infty$—the finiteness of a_r and b_r follows from the fact that geometric random variables have finite moments of all orders. Thus

$$E[N^r] = E[E[N^r(Y) \mid Y]] \leq E[a_r Y^r + b_r],$$

and the desired result follows.   □

EXAMPLE 2.46 (Producer–consumer system with nonpreemptive priority). For the system of Example 2.1 in Chapter 2, suppose that the creation-time random variables A_1 and A_2 are each distributed as $Y + a$, where a is a positive constant and Y is an exponential random variable with intensity q for some $q > 0$. Also suppose that the distribution of the transmission-time random variable L_1 has an essential supremum that exceeds $\max((B_1 - 1)a, B_2 a)$. Finally, suppose that the transmission-time random variable L_2 has a GNBU distribution. Consider the SPN representation of the producer–consumer system given in Figure 2.4, and observe that $\bar{s} = (0, B_1 - 1, 1, 0, B_2, 0, 0)$ is a single state with $\bar{e} = e_3 =$ "end of transmission to consumer 1." Recall that the marking is $\bar{s}$ if and only if there are B_1 items in buffer 1—one of which is being transmitted to consumer 1—and B_2 items in buffer 2. As in Example 2.25 in Chapter 5, denote by $\beta(n) + 1$ the random index of the nth marking change $(n \geq 1)$ at which transition e_3 fires and by $\alpha(n)$ the index of the nth marking change at which transition e_3 becomes enabled. It was shown in this example that every $\alpha(n)$ and $\beta(n)$ is a.s. finite and that $P_\mu\{ S_{\beta(n)} = \bar{s} \mid \mathcal{F}_{\alpha(n)} \} > \delta$ for some $\delta > 0$. It follows from Theorems 2.36 and 2.44 that the random indices $\{ \theta(k) : k \geq 0 \}$ defined via (2.1) form a sequence of regeneration points for the underlying chain and the random times $\{ \zeta_{\theta(k)} : k \geq 0 \}$ form a sequence of regeneration points for the marking process.

Fix $n \geq 0$ and consider the time between the end of transmission to consumer 1 at time $\zeta_{\beta(n)+1}$ and the next end of transmission to consumer 1. If, at time $\zeta_{\beta(n)+1}$, buffer 1 contains at least one item awaiting transmission, then immediate transition e_2 fires and another transmission to consumer 1 starts instantaneously, at the $(\beta(n) + 2)$nd marking change. If no items are awaiting transmission, then there is a delay until producer 1 finishes creating an item for transmission. At this point, there may be a further delay if a transmission to consumer 2 is in progress. When transition e_6 fires—ending the latter transmission—transition e_2 fires and a transmission to consumer 1 starts instantaneously. It follows that

$$\zeta_{\beta(n+1)+1} - \zeta_{\beta(n)+1} = 1_{A_2 \cup A_3} C_{\beta(n)+1,1} + 1_{A_3} C_{\nu(n),6}$$
$$+ 1_{A_1} C_{\beta(n)+2,3} + 1_{A_2} C_{\nu(n)+1,3} + 1_{A_3} C_{\lambda(n)+1,3},$$
$$(2.47)$$

where $\nu(n)$ is the index of the first marking change after $\beta(n) + 1$ at which transition $e_1 =$ "creation of item by producer 1" fires, $\lambda(n)$ is the first marking change after $\nu(n)$ at which transition $e_6 =$ "end of transmission to consumer 2" fires, and the events A_1, A_2, and A_3 are given by

$$A_1 = \{ S_{\beta(n)+1,2} > 0 \},$$
$$A_2 = \{ S_{\beta(n)+1,2} = 0 \text{ and } S_{\nu(n),6} = 0 \},$$

and

$$A_3 = \{ S_{\beta(n)+1,2} = 0 \text{ and } S_{\nu(n),6} = 1 \}.$$

In (2.47), we take $1_{A_3} C_{\lambda(n)+1,3} = 0$ if $\lambda(n) = \infty$. Using Lemmas 4.10 and 4.19 in Chapter 3, it can be shown that the quantity $1_{A_1} C_{\beta(n)+2,3} + 1_{A_2} C_{\nu(n)+1,3} + 1_{A_3} C_{\lambda(n)+1,3}$ is distributed as an independent random sample from the distribution of L_1. Because, as discussed in Example 2.25 in

Chapter 5, the distribution of A_1 is GNBU, it follows from Lemma 2.39 that $C_{\beta(n)+1,1}$ has finite moments of all orders. Similarly, $C_{\nu(n),6}$ has finite moments of all orders. Thus if, for example, $E\left[L_1^{r+\epsilon}\right] < \infty$ for some $r > 1$ and $\epsilon > 0$, then an application of the c_r-inequality yields (2.37).

EXAMPLE 2.48 (Manufacturing cell with robots). For the system of Example 3.6 in Chapter 2, denote by R_1 the (constant) time for robot 1 to return to its null position after transfer of a part to conveyor 1. Similarly, denote by R_2 the (constant) time for robot 2 to return to its null position after transfer of a part to machine 1. Suppose that the processing-time random variable L_2 has an exponential distribution with intensity q for some $q > 0$. Also suppose that the distribution function of the processing-time random variable L_1 has an infinite essential supremum. For the SPN representation of the manufacturing cell given in Figure 2.21, let $\bar{s}$ be the unique marking such that $\bar{s}_4 = \bar{s}_9 = \bar{s}_{11} = \bar{s}_{22} = \bar{s}_{24} = 1$ and $\bar{s}_j = 0$ otherwise—the marking is $\bar{s}$ if and only if machines 1 and 2 are each processing a part, a part is on conveyor 1 awaiting transfer to a machine, no parts are on conveyor 2, and each robot is in its null position. Recall that $e_8 =$ "end of processing by machine 1," and observe that $\bar{s}$ and $\bar{E} = \{e_8\}$ satisfy the condition in (b) of Theorem 2.36. As in Example 2.34 in Chapter 5, denote by $\beta(n) + 1$ the random index of the nth marking change at which transition e_8 fires and by $\alpha(n)$ the index of the nth marking change at which transition e_8 becomes enabled. It was shown in this example that every $\alpha(n)$ and $\beta(n)$ is a.s. finite, and that $P_\mu\left\{ S_{\beta(n)} = \bar{s} \mid \mathcal{F}_{\alpha(n)} \right\} > \delta$ for some $\delta > 0$. It follows from Theorem 2.36 that the random times $\left\{ \zeta_{\theta(k)} \colon k \geq 0 \right\}$ defined via (2.1) form a sequence of regeneration points for the marking process. Additional conditions under which (2.37) holds can be obtained in a manner similar to Example 2.46.

EXAMPLE 2.49 (Telephone system). For the system of Example 1.27 in Chapter 5 with K links and N lines, suppose that $K > 2$ and $N > 2K + 2$, and that the call-length random variables $L_1, L_2, L_3, \ldots, L_N$ have a common distribution function H that is GNBU with lower bound x^*. Also suppose that the waiting-time random variables $A_1, A_2, \ldots, A_N$ are each distributed according to an exponential distribution function with intensity q for some $q > 0$. Finally, suppose that $x^* < \text{ess sup } H$. For the SPN representation of the telephone system given in Figure 5.5, denote by $\bar{G}$ the set of markings in which a call is connected on link 1 and all other links are idle. Recall that $e_{2,1} =$ "end of call connected on link 1," and—as mentioned in Example 2.30—the pair $(\bar{G}, \{e_{2,1}\})$ satisfies the conditions in Remark 2.14. As in Example 2.35 in Chapter 5, denote by $\beta(n) + 1$ the random index of the nth marking change at which transition $e_{2,1}$ fires and by $\alpha(n)$ the index of the nth marking change at which transition $e_{2,1}$ becomes enabled. It was shown in this example that every $\alpha(n)$ and $\beta(n)$ is a.s. finite, and that $P_\mu\left\{ S_{\beta(n)} \in \bar{G} \mid \mathcal{F}_{\alpha(n)} \right\} > \delta$ for some $\delta > 0$. In light of Remark 2.14, it can be seen that the random times $\{\zeta_{\theta(k)}: k \geq 0\}$ defined via (2.1) form a sequence of regeneration points for the marking process.

Fix $n \geq 0$ and consider the time between the end, at time $\zeta_{\beta(n)+1}$, of the call connected on link 1 and the next end of a call connected on link 1. This time has two components: the time $D_1(n)$ until the next call is placed on link 1 and the time $D_2(n)$ until this call is completed. Suppose for simplicity that a line is equally likely to place a call to any of the other lines, and denote by I_n the number of idle lines at time $\zeta_{\beta(n)+1}$. Arguing as in previous examples, we find that $D_2(n)$ is distributed as D_2, where D_2 is an independent sample from the distribution H; since H is GNBU, D_2 has finite moments of all orders. We can obtain upper bounds on the moments of $D_1(n)$ by considering an artificial scenario in which the calls that are connected at time $\zeta_{\beta(n)+1}$ (on links other than link 1) never end. In this scenario, each of the $N - I_n$ lines that are busy at time $\zeta_{\beta(n)+1}$ remains busy forever and therefore can never place a call that is connected on link 1. It follows that the random variable $D_1(n)$ is stochastically dominated—see Definition 1.7 in the Appendix—by the time $D_1'(n)$ until the next call is placed on link 1 under the artificial scenario. Given I_n, the random variable $D_1'(n)$ is conditionally distributed as the sum of M exponential random variables with intensity $I_n q$, where M has a geometric distribution with parameter $V_n = (I_n - 1)/N$ and corresponds to the number of busy calls before the next successful call. This assertion follows from the memoryless property of the exponential distribution (as in Corollary 4.17 in Chapter 3) and other standard properties of the exponential distribution. An easy argument using LaPlace–Stieltjes transforms—see Proposition 1.17 in the Appendix—shows that $D_1'(n)$ is exponentially distributed with intensity $V_n I_n q$. Observe that $I_n \geq N - 2(K - 1)$ a.s., so that $V_n I_n \geq l$ a.s., where $l = (N-2K+2)(N-2K+1)/N$. It follows that $D_1'(n)$—and hence $D_1(n)$—is stochastically dominated by D_1, where D_1 is exponentially distributed with intensity lq. Because D_1 has moments of all orders, Proposition 1.15 in the Appendix implies that $D_1(n)$ has moments of all orders. Thus

$$E_\mu\left[\left(\zeta_{\beta(n+1)+1} - \zeta_{\beta(n)+1}\right)^v\right] \leq E\left[(D_1 + D_2)^v\right] \leq 2^{v-1}\left(E\left[D_1^v\right] + E\left[D_2^v\right]\right)$$

for $v \geq 1$, where the rightmost expression is finite, and it follows that, for example, (2.37) holds for $r > 1$ and $\epsilon > 0$.

EXAMPLE 2.50 (Token ring). For the system of Example 2.6 in Chapter 2, suppose that the distribution function F_j of each interarrival-time random variable A_j is NBU. Recall that R_j is the time for the ring token to propagate from port j to the next port, and suppose that ess inf $F_j < R_N$ for $1 \leq j \leq N$. Consider the SPN representation of the token ring given in Figure 2.10, and observe that $\bar{s} = (1, 0, 0, 0, \ldots, 1, 0, 0, 0, 1, 0, 0, 1)$ is a single state with $\bar{e} = e_{3,1} = $ "observation of ring token by port 1." The marking is $\bar{s}$ if and only if all ports have a packet awaiting transmission and the ring token is propagating from port N to port 1. As in

$e_1 = $ completion of service at center 1

$e_2 = $ completion of service at center 2

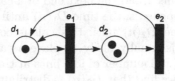

Figure 6.2. SPN representation of cyclic queues with feedback (three jobs).

Example 2.17 in Chapter 5, denote by $\beta(n) + 1$ the random index of the nth marking change at which transition $e_{3,1}$ fires and by $\alpha(n)$ the index of the nth marking change at which transition $e_{3,1}$ becomes enabled. It was shown in this example that every $\alpha(n)$ and $\beta(n)$ is a.s. finite and that $P_\mu\{\, S_{\beta(n)} = \bar{s} \mid \mathcal{F}_{\alpha(n)} \,\} > \delta$ for some $\delta > 0$. It follows from Theorems 2.36 and 2.44 that the random indices $\{\, \theta(k) \colon k \geq 0 \,\}$ defined via (2.1) form a sequence of regeneration points for the underlying chain and the random times $\{\, \zeta_{\theta(k)} \colon k \geq 0 \,\}$ form a sequence of regeneration points for the marking process.

We can use Theorem 2.36 to show that $Y_1(|f|)$ has finite moments. Specifically, observe that

$$E_\mu\left[(\zeta_{\beta(n+1)+1} - \zeta_{\beta(n)+1})^{r+\epsilon} \right] \leq E\left[\left(\sum_{j=1}^{N} (R_j + L_j) \right)^{r+\epsilon} \right]$$

for $r, \epsilon, n \geq 0$ where, as before, the successive times for port j to transmit a packet are i.i.d. as L_j. If, for example, $E[L_j^{r+\epsilon}] < \infty$ for each j and some $r > 1$ and $\epsilon > 0$, then an application of the c_r-inequality yields (2.37). A proof of this result based on Theorem 2.40 is perhaps even easier—as shown in Example 2.17 in Chapter 5, $\beta(n) - \beta(n+1) \leq 4N$ for $n \geq 0$, so that (2.42) holds for $r, \epsilon \geq 0$.

EXAMPLE 2.51 (Cyclic queues with feedback). Consider the closed network of queues of Example 1.4 in Chapter 2, and suppose that successive service times at center i ($i = 1, 2$) are i.i.d. as a positive random variable L_i, where the distribution of L_1 is GNBU and the distribution of L_2 is continuous and has an infinite essential supremum. This system can be represented by an SPN similar to that in Example 2.6 of Chapter 4; see Figure 6.2 for $N = 3$ jobs. For this SPN, $p(s; s, e_1) = 1 - p$ and $p((s_1 - 1, s_2 + 1); s, e_1) = p$ for $s = (s_1, s_2) \in S$.

The marking $\bar{s} = (0, N)$ is a single state with $E(\bar{s}) = \{ e_2 \}$. As in Example 2.37 in Chapter 5, denote by $\beta(n) + 1$ the random index of the nth marking change at which transition e_2 fires and by $\alpha(n)$ the random index of the nth marking change at which e_2 becomes enabled. It was shown in this example that every $\alpha(n)$ and $\beta(n)$ is a.s. finite, and that $P_\mu\{ S_{\beta(n)} = \bar{s} \mid \mathcal{F}_{\alpha(n)} \} > \delta$ for some $\delta > 0$. It follows from Theorems 2.36 and 2.44 that the random indices $\{\theta(k): k \geq 0\}$ defined via (2.1) form a sequence of regeneration points for the underlying chain and the random times $\{ \zeta_{\theta(k)}: k \geq 0 \}$ form a sequence of regeneration points for the marking process.

We now consider moments of cycle sums—in particular, we give conditions on the clock-setting distributions under which (2.37) holds. Observe that the time between two successive service completions at center 2 is the sum of two components: the time from the first service completion at center 2 to the next start of service at center 2, and the time from the next start of service at center 2 to the second service completion at center 2. The second component is simply a center 2 service time. The first component is equal to 0 if two or more jobs are at center 2 just before the first service completion at center 2. Otherwise, the first component is equal to the time from the first service completion at center 2 to the next time at which a job completes service at center 1 and moves to center 2; this time interval equals the residual service time at center 1 (measured at the time of the first service completion at center 2) plus the sum of a random number—possibly 0—of center 1 service times. It follows that

$$E_\mu\left[(\zeta_{\beta(n+1)+1} - \zeta_{\beta(n)+1})^{r+\epsilon} \right] \leq E\left[\left(A_n + C_{\beta(n)+1,1} + \sum_{j=1}^{M(n)} B_{n,j} \right)^{r+\epsilon} \right]$$

for $r, \epsilon, n \geq 0$, where A_n is the first new clock reading generated for transition e_2 after time $\zeta_{\beta(n)}$, $B_{n,j}$ is the jth new clock reading generated for transition e_1 after time $\zeta_{\beta(n)}$, and $M(n)$ is the number—starting at time $\zeta_{\beta(n)}$—of successive service completions at center 1 at which the job completing service joins the tail of the queue at center 1. It follows from Lemma 2.39 that $C_{\beta(n)+1,1}$ has finite moments of all orders. Moreover, since the common distribution of the i.i.d. sequence $\{ B_{n,j}: j \geq 1 \}$ is GNBU—and hence has finite moments of all orders—and since $M(n)$ is independent of the sequence $\{ B_{n,j}: j \geq 1 \}$ and has a geometric distribution with finite moments of all orders, it follows that the sum $B_{n,1} + B_{n,2} + \cdots + B_{n,M(n)}$ has finite moments of all orders; see Remark 1.21 in the Appendix. If, for instance, $E[L_2^{r+\epsilon}] < \infty$ for some $r > 1$ and $\epsilon > 0$, then an application of the c_r-inequality yields (2.37).

Similar arguments establish (2.42). Specifically, consider the number of marking changes between two successive firings of transition e_2—that is, the number of events between two successive completions of service at center 2. If no jobs are at center 2 just after the first service completion, then a random number M of jobs complete service at center 1 before a job moves to center 2 and the next center 2 service begins. During this center 2 service, N_0 additional service completions occur at center 1 for a total of $M + N_0$ events between the two center 2 service completions. If, on the other hand, one or more jobs are at center 2 just after the first service completion, then the total number of events is simply N_0. As noted above, M has a geometric distribution and hence has finite moments of all orders. Moreover, N_0 is stochastically dominated by the random variable $N = \inf\{n \geq 1 : X_1 + \cdots + X_n > Y\}$, where each X_i is an independent sample from the distribution of L_1 and Y is an independent sample from the distribution of L_2. If, for example, $E\left[L_2^{r+\epsilon}\right] < \infty$ for some $r > 1$ and $\epsilon > 0$, then $E\left[N^{r+\epsilon}\right] < \infty$ by Lemma 2.45, and an application of the c_r-inequality yields (2.42).

6.2.4   The Regenerative Variance Constant

Suppose that there exists a sequence $\{\zeta_{\theta(k)} : k \geq 0\}$ of regeneration points for the marking process of an SPN with $E_\mu[\tau_1] < \infty$, where $\tau_1 = \zeta_{\theta(1)} - \zeta_{\theta(0)}$ as usual. Define $Y_k(\cdot)$ by (2.22) and let f be a function defined on S with $Y_0(|f|) < \infty$ a.s. and $E_\mu[Y_1(|f|)] < \infty$. Then, by Theorem 1.12, $\lim_{t \to \infty}(1/t)\int_0^t f(X(u))\, du = r(f)$ with $r(f) = E_\mu[Y_1(f)]/E_\mu[\tau_1]$. As discussed in the next section, the "regenerative variance constant"

$$\sigma^2(f) = \mathrm{Var}_\mu[Y_1(f) - r(f)\tau_1]$$

must be positive and finite for the regenerative method to be applicable. Similarly, the regenerative variance constant

$$\tilde{\sigma}^2(\tilde{f}) = \mathrm{Var}_\mu[\tilde{Y}_1(\tilde{f}) - \tilde{r}(\tilde{f})\tilde{\tau}_1] \tag{2.52}$$

must be positive and finite for the regenerative method to be applicable to the underlying chain of an SPN—here $\tilde{r}(\tilde{f}) = E_\mu[\tilde{Y}_1(\tilde{f})]/E_\mu[\tilde{\tau}_1]$ with $\tilde{Y}_1(\tilde{f})$ given by (2.23) and $\tilde{\tau}_1$ given by $\theta(1) - \theta(0)$.

The variance constant $\sigma^2(f)$ is well defined and finite whenever $E_\mu[\tau_1^2]$ and $E_\mu\left[Y_1^2(|f|)\right]$ are both finite, and similarly for $\tilde{\sigma}^2(\tilde{f})$. To see that the first assertion holds, observe that $E_\mu\left[Y_1^2(f)\right] < \infty$, $E_\mu[Y_1(|f|)] < \infty$, and $r(|f|) < \infty$ whenever $E_\mu\left[Y_1^2(|f|)\right] < \infty$. Because $E_\mu[Y_1(f) - r(f)\tau_1] = 0$, it then follows from the Cauchy–Schwarz inequality that

$$\begin{aligned}
\sigma^2(f) &= E_\mu\left[(Y_1(f) - r(f)\tau_1)^2\right] \\
&\leq E_\mu[Y_1^2(f)] + 2r(|f|)E_\mu[|Y_1(f)\tau_1|] + E_\mu[\tau_1^2] \\
&\leq E_\mu[Y_1^2(f)] + 2r(|f|)E_\mu^{1/2}[Y_1^2(f)]E_\mu^{1/2}[\tau_1^2] + E_\mu[\tau_1^2] \\
&< \infty
\end{aligned}$$

as asserted. By Remark 1.14, $\sigma^2(f) < \infty$ whenever $E_\mu[\tau_1^2] < \infty$ and either f is bounded or the state space S is finite. Analogous observations hold for $\tilde{\sigma}^2(\tilde{f})$. Thus the results in the previous subsections can be used to establish finiteness of the regenerative variance constant. For example, it follows that if the SPN has a single state and Assumption PD holds, then $\sigma^2(f) < \infty$ for any real-valued function f defined on S and $\tilde{\sigma}^2(\tilde{f}) < \infty$ for any polynomially dominated function $\tilde{f}$ defined on Σ.

In practice, the regenerative variance constant is virtually always positive. For example, we have the following result, which shows that under mild conditions, $\sigma^2(f)$ is positive when Assumption PD holds.

Theorem 2.53. *Suppose that the conditions of Theorem 2.24 hold for some marking $\bar{s}$ and transition $\bar{e}$, so that the random indices $\{ \zeta_{\theta(k)} : k \geq 0 \}$ defined via (2.1) with $\bar{E} = \{ \bar{e} \}$ form a sequence of regeneration points for the marking process. Then $\sigma^2(f) > 0$ for any real-valued function f defined on S such that $f(s) \neq f(s')$ for some $s, s' \in S$.*

PROOF. (Sketch) It suffices to show that $Y_1(f)/\tau_1$ is not a.s. equal to a constant. For ease of exposition, suppose that all speeds are equal to 1, that each transition is simple, and that $\theta(0) = 0$. Under the hypotheses of the theorem, there exists $\bar{x} > 0$ such that the clock-setting distribution for each timed transition $e \in E - E'$ can be written in the form

$$F(\,\cdot\,; e) = p_e F_1(\,\cdot\,; e) + (1 - p_e) F_2(\,\cdot\,; e),$$

where $p_e \in (0, 1]$, both F_1 and F_2 are proper distribution functions, and F_1 is absolutely continuous with density function f_1 positive and continuous on $(0, \bar{x})$. We can modify the usual construction of the marking process slightly so that each new clock reading for a timed transition e is generated in two steps. First a Bernoulli random variable X is generated such that $P\{ X = 1 \} = 1 - P\{ X = 0 \} = p_e$. If $X = 1$, then the clock reading is generated as an independent sample from F_1; otherwise, the clock reading is generated as an independent sample from F_2. Let $\{ I_n : n \geq 0 \}$ be indicator random variables such that $I_n = 1$ if, at the nth marking change, the clock reading for each new timed transition e is generated as a sample from $F_1(\,\cdot\,; e)$; otherwise, $I_n = 0$. If there are no new timed transitions at the nth marking change, then $I_n = 1$ by convention.

Fix $k \geq 1$, $\tilde{s}_0, \tilde{s}_1, \ldots \tilde{s}_{k-1} \in G - \{ \bar{s} \}$, and $\tilde{e}_0, \tilde{e}_1, \ldots, \tilde{e}_{k-1} \in E$ such that

$$p(\tilde{s}_0; \bar{s}, \bar{e})p(\tilde{s}_1; \tilde{s}_0, \tilde{e}_0) \cdots p(\tilde{s}_{k-1}; \tilde{s}_{k-2}, \tilde{e}_{k-2})p(\bar{s}; \tilde{s}_{k-1}, \tilde{e}_{k-1}) > 0$$

and $f(\tilde{s}_i) \neq f(\tilde{s}_j)$ for some $0 \leq i, j \leq k$. Such a selection is possible because of the assumed irreducibility of the SPN. Consider the event A given by

$$A = \{ S_k = \bar{s},\ S_j = \tilde{s}_j \text{ for } 0 \leq j < k,\ C_j \leq \bar{x} \text{ for } 0 \leq j \leq k,$$

$$I_j = 1 \text{ for } 0 \leq j \leq k, \text{ and } E^*(S_j, C_j) = \{ \tilde{e}_j \} \text{ for } 0 \leq j < k \}.$$

An inductive argument on k shows that $P_\mu\{ A \} > 0$. It therefore suffices to show that, for an arbitrary fixed constant c, $P_\mu\{ Y_1(f)/\tau_1 = c; A \} = 0$, because then

$$P_\mu\{ Y_1(f)/\tau_1 = c \} = P_\mu\{ Y_1(f)/\tau_1 = c; \Sigma - A \} \leq P_\mu\{ \Sigma - A \} < 1.$$

To this end, set $E_n^* = E^*(S_n, C_n)$ and $t_n^* = t^*(S_n, C_n)$ for $n \geq 0$. Using Lemma 4.10 in Chapter 3 and an inductive argument on k, it can be shown that the joint cumulative distribution function $F(x_0, x_1, \ldots, x_k) = P_\mu \{ t_0^* \leq x_0, t_1^* \leq x_1, \ldots, t_k^* \leq x_k \}$ is absolutely continuous on $(0, \bar{x})^k$, conditional on the event A. Observe that $Y_1(f, \omega)/\tau_1(\omega) = c$ for $\omega \in A$ if and only if $\sum_{j=0}^k c_j t_j^*(\omega) = 0$, where $c_j = f(\tilde{s}_j) - c$. Because $f(\tilde{s}_i) \neq f(\tilde{s}_j)$ for some $0 \leq i, j \leq k$ by assumption, it follows that $(c_0, c_1, \ldots, c_k) \neq (0, 0, \ldots, 0)$, and hence the set

$$ B = \left\{ (t_0, t_1, \ldots, t_k) \in (0, \bar{x})^k : c_0 t_0 + c_1 t_1 + \cdots + c_k t_k = 0 \right\} $$

is a strict linear subspace of $(0, \bar{x})^k$. Thus the event $\{ (t_0^*, t_1^*, \ldots, t_k^*) \in B; A \}$ has P_μ-probability equal to 0, so that $P_\mu \{ Y_1(f)/\tau_1 = c; A \} = 0$. $\square$

In the discrete-time setting, the quantity $\tilde{\sigma}^2(\tilde{f})$ is almost always positive in applications for which $\tilde{f}$ is nonconstant and takes values in a finite set. Typically, the degenerate situations in which $\tilde{\sigma}^2(\tilde{f}) = 0$ can be detected a priori and the associated estimation problem is trivial. For example, consider the machine repair model of Example 2.28. Suppose we wish to estimate $\tilde{r}(\tilde{f}) = \lim_{n \to \infty}(1/n)\sum_{j=0}^{n-1} \tilde{f}(S_j, C_j)$, where $\tilde{f}(s, c) = 1$ if $E^*(s, c) \subset \{ e_{1,1}, e_{1,2}, \ldots, e_{1,N} \}$ and $\tilde{f}(s, c) = 0$ otherwise. Observe that with probability 1 two machines never stop simultaneously and that the number of repairs in a regenerative cycle is always equal to the number of stoppages. Associated with each stoppage and subsequent repair of machine j $(1 \leq j \leq N)$ are three transition firings: transitions $e_{1,j}$, $e_{2,j}$, and e_3 each fire once. It follows that $\tilde{Y}_1(\tilde{f}) = \tilde{\tau}_1/3$ with probability 1. Thus $\tilde{r}(\tilde{f}) = 1/3$, $\tilde{\sigma}^2(\tilde{f}) = \mathrm{Var}_\mu [\tilde{Y}_1(\tilde{f}) - \tilde{r}(\tilde{f})\tilde{\tau}_1] = 0$, and the estimation problem is trivial. As another example, suppose that the marking set G can be partitioned into d disjoint subsets $G_1, G_2, \ldots, G_d$ such that $s' \in G_{i+1}$ whenever $s \in G_i$ and $s \to s'$. (Take $G_{i+1} = G_1$ when $i = d$.) Consider a function $\tilde{f}$ such that $\tilde{f}(s, c) = \tilde{f}(s', c')$ whenever $s, s' \in G_i$ for some $1 \leq i \leq d$. Set $v = \sum_{i=1}^d \tilde{f}(s_i)$, where $s_i \in G_i$ for $1 \leq i \leq d$. Then there exists a positive integer-valued random variable K_1 such that $\tilde{Y}_1(\tilde{f}) = K_1 v$ and $\tilde{\tau}_1 = K_1 d$, so that $\tilde{r}(\tilde{f}) = v/d$ and $\tilde{\sigma}^2(\tilde{f}) = 0$. Again, the estimation problem is trivial.

6.3   The Regenerative Method

The results in Section 6.2 give conditions under which the marking process or underlying chain of an SPN is a regenerative process and integrals or sums over a regenerative cycle have finite moments. In this section we examine the implications of such regenerative structure for the analysis of simulation output.

6.3.1   The Standard Method

We first describe the simplest version of the regenerative method as applied to the marking process and underlying chain of an SPN. Extensions to this "standard regenerative method" are given in the following subsections.

Regenerative Simulation of the Marking Process

Let $\{X(t): t \geq 0\}$ be the marking process of an SPN. Suppose that we have identified a sequence $\{T_k: k \geq 0\}$ of regeneration points and wish to estimate a time-average limit of the form $\lim_{t \to \infty}(1/t) \int_0^t f(X(u))\,du$, where f is a real-valued function defined on S. The regenerative method for analysis of simulation output provides strongly consistent point estimates and asymptotic confidence intervals for time-average limits. For ease of exposition, we assume throughout that $T_0 = 0$, so that $\{X(t): t \geq 0\}$ is a nondelayed regenerative process.

Set $\tau_k = T_k - T_{k-1}$ for $k \geq 1$ as in Section 6.1, so that τ_k is the length of the kth cycle. Also set

$$Y_k(f) = \int_{T_{k-1}}^{T_k} f(X(u))\,du$$

for $k \geq 1$. Typically, each regeneration point coincides with a marking change, so that $T_k = \zeta_{\theta(k)}$ for $k \geq 0$, where $\{\theta(k): k \geq 0\}$ is a sequence of a.s. finite random indices. We can then write

$$Y_k(f) = \sum_{n=\theta(k-1)}^{\theta(k)-1} f(S_n)t^*(S_n, C_n),$$

where $\{(S_n, C_n): n \geq 0\}$ is the underlying chain. Recall from Section 6.1 that the sequence $\{(Y_k(f), \tau_k): k \geq 1\}$ consists of i.i.d. random pairs. Suppose that $E_\mu[\tau_1] < \infty$ and $E_\mu[Y_1(|f|)] < \infty$; Section 6.2 gives conditions on the building blocks of an SPN under which the quantities τ_1 and $Y_1(|f|)$ have finite moments. It then follows from Theorem 1.12 that

$$r(f) = \frac{E_\mu[Y_1(f)]}{E_\mu[\tau_1]} \tag{3.1}$$

is well defined and finite, and

$$\lim_{t \to \infty} \frac{1}{t} \int_0^t f(X(u))\,du = r(f) \quad \text{a.s..}$$

If, in addition, τ_1 is aperiodic, then there exists a random variable X—independent of the initial distribution μ—such that $X(t) \Rightarrow X$ as $t \to \infty$ and $r(f) = E[f(X)]$. That is, the time-average limit $r(f)$ can also be interpreted as a steady-state mean.

To estimate $r(f)$, observe a fixed number n of cycles of $\{X(t): t \geq 0\}$ and measure the quantities $Y_1(f), Y_2(f), \ldots, Y_n(f)$ and $\tau_1, \tau_2, \ldots, \tau_n$. Set

$$\hat{r}(n) = \frac{\bar{Y}(n)}{\bar{\tau}(n)},$$

where

$$\bar{Y}(n) = \frac{1}{n} \sum_{k=1}^{n} Y_k(f)$$

and

$$\bar{\tau}(n) = \frac{1}{n} \sum_{k=1}^{n} \tau_k.$$

Writing

$$\hat{r}(n) = \frac{\bar{Y}(n)/n}{\bar{\tau}(n)/n}$$

and applying the strong law of large numbers (SLLN) for i.i.d. random variables to both numerator and denominator, we see that

$$\lim_{n \to \infty} \hat{r}(n) = r(f) \text{ a.s..}$$

Thus $\hat{r}(n)$ is strongly consistent for $r(f)$.

We now consider the problem of obtaining an asymptotic confidence interval for $r(f)$. Set

$$\begin{aligned}
\sigma^2(f) &= \text{Var}_\mu \left[Y_1(f) - r(f)\tau_1 \right] \\
&= \text{Var}_\mu \left[Y_1(f) \right] - 2r(f)\text{Cov}_\mu \left[Y_1(f), \tau_1 \right] + r^2(f)\text{Var}_\mu \left[\tau_1 \right].
\end{aligned} \tag{3.2}$$

The quantity $\sigma^2(f)$ is the regenerative variance constant discussed in Section 6.2.4. As shown below, the i.i.d. cycle structure of the marking process implies that, for large n, the distribution of the estimator $\hat{r}(n)$ is approximately normal with mean $r(f)$ and variance $\sigma^2(f)/[n\bar{\tau}^2(n)]$. This result cannot be used directly to obtain a confidence interval since $\sigma^2(f)$ is unknown. From n cycles, however, a natural estimator of $\sigma^2(f)$ is given by

$$s^2(n) = s_{11}(n) - 2\hat{r}(n)s_{12}(n) + \hat{r}^2(n)s_{22}(n), \tag{3.3}$$

where

$$s_{11}(n) = \frac{1}{n-1} \sum_{k=1}^{n} \left(Y_k(f) - \bar{Y}(n) \right)^2,$$

$$s_{22}(n) = \frac{1}{n-1} \sum_{k=1}^{n} \left(\tau_k - \bar{\tau}(n) \right)^2,$$

and

$$s_{12}(n) = \frac{1}{n-1} \sum_{k=1}^{n} \left(Y_k(f) - \bar{Y}(n) \right) \left(\tau_k - \bar{\tau}(n) \right).$$

The quantities $s_{11}(n)$, $s_{22}(n)$, and $s_{12}(n)$ are the usual unbiased estimators of $\text{Var}_\mu [Y_1(f)]$, $\text{Cov}_\mu [Y_1(f)\tau_1]$, and $\text{Var}_\mu [\tau_1]$. Estimation of $\sigma^2(f)$ by $s^2(n)$ leads to the desired interval. More precisely, we have the following result.

Theorem 3.4. *Suppose that* $E_\mu[\tau_1^2] < \infty$, $E_\mu[Y_1^2(|f|)] < \infty$, *and* $\sigma^2(f) > 0$. *Then*

$$\lim_{n \to \infty} s^2(n) = \sigma^2(f) \text{ a.s.}$$

and

$$\frac{\sqrt{n}\left(\hat{r}(n) - r(f)\right)}{s(n)/\bar{\tau}(n)} \Rightarrow N(0,1)$$

as $n \to \infty$, *where* $N(0,1)$ *is a standard (mean 0, variance 1) normal random variable.*

PROOF. Write

$$s_{11}(n) = \frac{1}{n-1} \sum_{k=1}^{n} Y_k^2(f) - \frac{n}{n-1} \bar{Y}^2(n)$$

for $n \geq 1$. Applying the SLLN for i.i.d. random variables to each of the two terms on the right, we see that $\lim_{n\to\infty} s_{11}(n) = \text{Var}_\mu [Y_1(f)]$ a.s.. Similar observations apply to $s_{22}(n)$ and $s_{12}(n)$, and the first assertion of the theorem follows. Next, set $Z_k(f) = Y_k(f) - r(f)\tau_k$ for $k \geq 1$ and observe that the sequence $\{ Z_k(f) \colon k \geq 0 \}$ consists of i.i.d. random variables with common mean 0 and common variance $\sigma^2(f)$. As discussed in Section 6.2.4, we have $0 < \sigma^2(f) < \infty$ under the assumptions of the theorem. It then follows from the central limit theorem (CLT) for i.i.d. random variables that $n^{1/2}\bar{Z}(n)/\sigma(f) \Rightarrow N(0,1)$ as $n \to \infty$, where $\bar{Z}(n) = (1/n)\sum_{k=1}^{n} Z_k(f)$. After some simple algebra, we find that

$$\frac{\sqrt{n}\left(\hat{r}(n) - r(f)\right)}{\sigma(f)/\bar{\tau}(n)} \Rightarrow N(0,1)$$

as $n \to \infty$. Because $\lim_{n\to\infty} s^2(n) = \sigma^2(f)$ a.s., it follows that $s(n) \Rightarrow \sigma(f)$ as $n \to \infty$, and the second assertion of the theorem follows from Slutsky's theorem (Proposition 1.43 in the Appendix). $\square$

Fix $p \in (0,1)$ and let z_p be the unique nonnegative real number such that $P\{-z_p \leq N(0,1) \leq z_p\} = p$. Then

$$\lim_{n \to \infty} P_\mu\left\{ \hat{r}(n) - \frac{z_p \, s(n)}{\bar{\tau}(n)\sqrt{n}} \leq r(f) \leq \hat{r}(n) + \frac{z_p \, s(n)}{\bar{\tau}(n)\sqrt{n}} \right\}$$

$$= \lim_{n \to \infty} P_\mu\left\{ -z_p \leq \frac{\sqrt{n}(\hat{r}(n) - r(f))}{s(n)/\bar{\tau}(n)} \leq z_p \right\}$$

$$= p.$$

Thus the random interval with endpoints $\hat{r}(n) \pm z_p \, s(n)/(\bar{\tau}(n)\sqrt{n})$ contains the unknown constant $r(f)$ approximately $100p\%$ of the time when n is large.

Based on the above discussion, we obtain the following estimation procedure.

Algorithm 3.5 (Regenerative method for the marking process)

1. Select a sequence $\{T_k : k \geq 0\}$ of regeneration points for the process $\{X(t) : t \geq 0\}$.

2. Simulate the process $\{X(t) : t \geq 0\}$ and observe a fixed number n of cycles defined by the random times $\{T_k : k \geq 0\}$.

3. Compute the length τ_k of the kth cycle and the quantity $Y_k(f) = \int_{T_{k-1}}^{T_k} f(X(u)) \, du$ for $1 \leq k \leq n$.

4. Form the strongly consistent point estimate $\hat{r}(n) = \bar{Y}(n)/\bar{\tau}(n)$ for $r(f)$.

5. Form the asymptotic $100p\%$ confidence interval

$$\left[\hat{r}(n) - \frac{z_p \, s(n)}{\bar{\tau}(n)\sqrt{n}}, \hat{r}(n) + \frac{z_p \, s(n)}{\bar{\tau}(n)\sqrt{n}} \right] \tag{3.6}$$

for $r(f)$, where $s(n)$ is defined as in (3.3).

Remark 3.7. Observe that $z_p = \Phi^{-1}((1+p)/2)$, where Φ is the distribution function of $N(0,1)$.

Remark 3.8. It is often desirable to compute the confidence interval for $r(f)$ by means of a single pass through the data. If sample path observations have been generated previously and stored on disk, then the use of a single-pass algorithm can substantially reduce the I/O and computational costs of producing the interval estimate. If sample path observations are being generated on the fly, then the use of such an algorithm avoids the need to store the observations. Clearly, the quantities $\bar{Y}(n)$ and $\bar{\tau}(n)$ can easily be computed in one pass. Computation of the variance estimator

$s^2(n)$—that is, computation of $s_{11}(n)$, $s_{22}(n)$, and $s_{12}(n)$—is trickier. For example, computing $s_{11}(n)$ in a single pass by using the representation $(n-1)s_{11}(n) = \sum_{k=1}^{n} Y_k^2(f) - n\bar{Y}^2(n)$ can lead to numerical instability. An alternative approach is to set $w_{11}(1) = w_{22}(1) = w_{12}(1) = 0$ and then recursively set

$$w_{11}(k) = w_{11}(k-1) + \frac{D_1(k)}{k}\frac{D_1(k)}{k-1},$$

$$w_{22}(k) = w_{22}(k-1) + \frac{D_2(k)}{k}\frac{D_2(k)}{k-1},$$

and

$$w_{12}(k) = w_{12}(k-1) + \frac{D_1(k)}{k}\frac{D_2(k)}{k-1}$$

for $k \geq 2$, where

$$D_1(k) = \sum_{j=1}^{k-1} Y_j(f) - (k-1)\,Y_k(f)$$

and

$$D_2(k) = \sum_{j=1}^{k-1} \tau_j - (k-1)\,\tau_k.$$

Then $s_{11}(n) = w_{11}(n)/(n-1)$, $s_{22}(n) = w_{22}(n)/(n-1)$, and $s_{12}(n) = w_{12}(n)/(n-1)$. Finally, compute $s^2(n)$ as in (3.3). The recursions for w_{11} and w_{22} are each numerically stable, because $D_i(k)$ $(i=1,2)$ is computed as the difference between two numbers of similar (and moderate) magnitude and the term that is added to $w_{ii}(n-1)$ to produce $w_{ii}(n)$ is always nonnegative. The main deficiency in this method arises from possible cancellation or roundoff errors in the calculation of $w_{12}(n)$—unlike $w_{11}(n)$ and $w_{22}(n)$, the term that is added to $w_{12}(n-1)$ to produce $w_{12}(n)$ need not always be nonnegative. In practice, however, the method usually produces acceptable results, provided that calculations are performed using double-precision arithmetic.

Regenerative Simulation of the Underlying Chain

The regenerative method for the underlying chain is similar to the regenerative method for the marking process. Suppose that there exists a sequence $\{\theta(k): k \geq 0\}$ of regeneration points for $\{(S_n, C_n): n \geq 0\}$ and that we wish to estimate a time-average limit $\lim_{n\to\infty}(1/n)\sum_{j=0}^{n-1}\tilde{f}(S_j, C_j)$ for some real-valued function $\tilde{f}$ defined on Σ.

Set $\tilde{\tau}_k = \theta(k) - \theta(k-1)$ and

$$\tilde{Y}_k(\tilde{f}) = \sum_{n=\theta(k-1)}^{\theta(k)-1} \tilde{f}(S_n, C_n) \tag{3.9}$$

for $k \geq 1$; the sequence $\left\{ (\tilde{Y}_k(\tilde{f}), \tilde{\tau}_k) : k \geq 1 \right\}$ consists of i.i.d. random pairs. Suppose that $E_\mu[\tilde{\tau}_1] < \infty$ and $E_\mu[\tilde{Y}_1(|\tilde{f}|)] < \infty$, so that

$$\tilde{r}(\tilde{f}) = \frac{E_\mu[\tilde{Y}_1(\tilde{f})]}{E_\mu[\tilde{\tau}_1]}$$

is well defined and finite, and

$$\lim_{n \to \infty} \frac{1}{n} \sum_{j=0}^{n-1} \tilde{f}(S_j, C_j) = \tilde{r}(\tilde{f}) \quad \text{a.s.}. \tag{3.10}$$

If, in addition, $\tilde{\tau}_1$ is aperiodic in discrete time, then $\tilde{r}(\tilde{f})$ can also be interpreted as a steady-state mean.

To estimate $\tilde{r}(\tilde{f})$, observe a fixed number n of cycles of $\{ (S_n, C_n) : n \geq 0 \}$ and measure the quantities $\tilde{Y}_1(\tilde{f}), \tilde{Y}_2(\tilde{f}), \ldots, \tilde{Y}_n(\tilde{f})$ and $\tilde{\tau}_1, \tilde{\tau}_2, \ldots, \tilde{\tau}_n$. Set

$$\hat{r}(n) = \frac{\bar{Y}(n)}{\bar{\tau}(n)},$$

where

$$\bar{Y}(n) = \frac{1}{n} \sum_{k=1}^{n} \tilde{Y}_k(\tilde{f})$$

and

$$\bar{\tau}(n) = \frac{1}{n} \sum_{k=1}^{n} \tilde{\tau}_k. \tag{3.11}$$

As in the regenerative method for the marking process, $\hat{r}(n)$ is strongly consistent for $\tilde{r}(\tilde{f})$.

To obtain an asymptotic confidence interval for $\tilde{r}(\tilde{f})$, set

$$\tilde{\sigma}^2(\tilde{f}) = \text{Var}_\mu[\tilde{Y}_1(\tilde{f}) - \tilde{r}(\tilde{f})\tilde{\tau}_1] \tag{3.12}$$

and

$$s^2(n) = \tilde{s}_{11}(n) - 2\hat{r}(n)\tilde{s}_{12}(n) + \hat{r}^2(n)\tilde{s}_{22}(n), \tag{3.13}$$

where

$$\tilde{s}_{11}(n) = \frac{1}{n-1} \sum_{k=1}^{n} (\tilde{Y}_k(\tilde{f}) - \bar{Y}(n))^2,$$

$$\tilde{s}_{22}(n) = \frac{1}{n-1} \sum_{k=1}^{n} (\tilde{\tau}_k - \bar{\tau}(n))^2,$$

and

$$\tilde{s}_{12}(n) = \frac{1}{n-1} \sum_{k=1}^{n} (\tilde{Y}_k(\tilde{f}) - \bar{Y}(n))(\tilde{\tau}_k - \bar{\tau}(n)).$$

Provided that $E_\mu[\tilde{\tau}_1^2] < \infty$, $E_\mu[\tilde{Y}_1^2(|\tilde{f}|)] < \infty$, and $\tilde{\sigma}^2(\tilde{f}) > 0$, we have

$$\lim_{n\to\infty} s^2(n) = \tilde{\sigma}^2(\tilde{f}) \text{ a.s.}$$

and

$$\frac{\sqrt{n}(\hat{r}(n) - \tilde{r}(\tilde{f}))}{s(n)/\bar{\tau}(n)} \Rightarrow N(0,1).$$

As before, the above CLT leads to a procedure for obtaining an asymptotic confidence interval.

Algorithm 3.14 (Regenerative method for the underlying chain)

1. Select a sequence $\{\theta(k): k \geq 0\}$ of regeneration points for the process $\{(S_n, C_n): n \geq 0\}$.

2. Simulate the process $\{(S_n, C_n): n \geq 0\}$ and observe a fixed number n of cycles defined by the random indices $\{\theta(k): k \geq 0\}$.

3. Compute the length $\tilde{\tau}_k$ of the kth cycle and the quantity $\tilde{Y}_k(\tilde{f}) = \sum_{n=\theta(k-1)}^{\theta(k)-1} \tilde{f}(S_n, C_n)$ for $1 \leq k \leq n$.

4. Form the strongly consistent point estimate $\hat{r}(n) = \bar{Y}(n)/\bar{\tau}(n)$ for $\tilde{r}(\tilde{f})$.

5. Form the asymptotic $100p\%$ confidence interval

$$\left[\hat{r}(n) - \frac{z_p\, s(n)}{\bar{\tau}(n)\sqrt{n}}, \hat{r}(n) + \frac{z_p\, s(n)}{\bar{\tau}(n)\sqrt{n}} \right]$$

for $\tilde{r}(\tilde{f})$, where $s(n)$ is defined as in (3.13) and z_p is the $(1+p)/2$ quantile of the standard normal distribution.

Although a comprehensive treatment of the regenerative method is beyond the scope of the current discussion, we outline some key issues and important extensions in the following subsections. For convenience, we often restrict the discussion to either simulation of the marking process or simulation of the underlying chain; unless otherwise indicated, results obtained in the one setting carry over with obvious modifications to the other.

6.3.2   Bias of the Point Estimator

Suppose that we wish to estimate the quantity $r(f) = E_\mu [Y_1(f)] / E_\mu [\tau_1]$ for a specified function f, based on simulation of the marking process. Although the estimators $\bar{Y}(n)$ and $\bar{\tau}(n)$ are unbiased for $E_\mu [Y_1(f)]$ and $E_\mu [\tau_1]$, it does not follow that the ratio $\hat{r}(n) = \bar{Y}(n)/\bar{\tau}(n)$ is unbiased for $r(f)$. Indeed, the following result can be established.

Proposition 3.15. *Suppose that* $\sup_{s \in S} |f(s)| < \infty$ *and* $E_\mu[\tau_1^4] < \infty$. *Then*

$$E_\mu [\hat{r}(n)] = r(f) - \frac{E_\mu \left[(Y_1(f) - r(f)\tau_1)\tau_1 \right]}{n E_\mu^2 [\tau_1]} + o(n^{-1}). \qquad (3.16)$$

Proposition 3.15 asserts that $\hat{r}(n)$ is biased for $r(f)$, with the bias decreasing at rate n^{-1} as $n \to \infty$. Recall that the *mean-squared error* of the estimator $\hat{r}(n)$ is defined by

$$\text{MSE}_\mu [\hat{r}(n)] = E_\mu \left[(\hat{r}(n) - r(f))^2 \right] = \text{Var}_\mu [\hat{r}(n)] + \text{Bias}_\mu^2 [\hat{r}(n)].$$

Under suitable regularity conditions, it can be shown that the variance of $\hat{r}(n)$ converges to 0 at rate n^{-1}, so that $\text{MSE}_\mu [\hat{r}(n)] = O(n^{-1})$ and the mean-squared error is dominated by the variance as n becomes large. Several alternative estimators for $r(f)$ have been proposed that attempt to reduce the bias when n is small. If we estimate the bias term in (3.16) and subtract this estimate from $\hat{r}(n)$, we obtain the *Tin estimator*:

$$\hat{r}_1(n) = \hat{r}(n) + \frac{1}{n^2} \sum_{k=1}^{n} \frac{(Y_k(f) - \hat{r}(n)\tau_k)\tau_k}{\bar{\tau}^2(n)}.$$

Jackknifing is another well-known technique for reducing the bias of an estimator. In the current setting, the *jackknife estimator* of $r(f)$ is

$$\hat{r}_2(n) = n\hat{r}(n) - \frac{n-1}{n} \sum_{k=1}^{n} \psi_k,$$

where

$$\psi_k = \frac{\sum_{i \neq k} Y_i(f)}{\sum_{i \neq k} \tau_i}.$$

Both $\hat{r}_1(n)$ and $\hat{r}_2(n)$ typically have a bias of $O(n^{-2})$.
    Set

$$s_i^2(n) = \frac{1}{n-1} \sum_{k=1}^{n} \frac{(Y_k(f) - \hat{r}_i(n)\tau_k)^2}{\bar{\tau}^2(n)}$$

for $i = 1, 2$. Also set $\hat{r}_0(n) = \hat{r}(n)$ and $s_0(n) = s(n)$; thus $\hat{r}_0(n)$ and $s_0(n)$ are the standard estimators of $r(f)$ and $\sigma^2(f)$. As usual, let z_p be the

Table 6.1. Simulation Results for Token Ring with Fixed-Sized Packets: Point Estimates and 95% Confidence-Interval Half-Widths for the Long-Run Utilization (True Value = 0.4462)

Estimator	Number of Cycles Simulated					
	2	10	50	100	1000	5000
standard	0.3333	0.4167	0.4209	0.4299	0.4419	0.4457
	±0.2904	±0.1349	±0.0584	±0.0377	±0.0127	±0.0057
Tin	0.3745	0.4362	0.4252	0.4317	0.4421	0.4457
	±0.2585	±0.1236	±0.0571	±0.0374	±0.0127	±0.0057
jackknife	0.4524	0.4430	0.4255	0.4317	0.4421	0.4457
	±0.2833	±0.1200	±0.0570	±0.0374	±0.0127	±0.0057

$(1+p)/2$ quantile of the standard normal distribution. Under the conditions of Theorem 3.4, it can be shown that $\hat{r}_1(n)$ and $\hat{r}_2(n)$ are each strongly consistent for $r(f)$, and

$$\left[\hat{r}_i(n) - \frac{z_p\, s_j(n)}{\sqrt{n}}, \hat{r}_i(n) + \frac{z_p\, s_j(n)}{\sqrt{n}}\right]$$

is an asymptotic $100p\%$ confidence interval for $i, j = 0, 1, 2$.

EXAMPLE 3.17 (Token ring with fixed-sized packets). We illustrate the various estimators discussed so far in the context of a computer network similar to that in Example 2.6 of Chapter 2 with $N = 4$ ports. The key difference from the original example is that, for each port, the time to transmit a packet is a deterministic constant L. Also, for each port, the time for the ring token to propagate to the next port is a deterministic constant R, and the successive times from end of transmission until the arrival of the next packet for transmission are i.i.d. as an exponential random variable with intensity q.

Suppose that this system is modelled by an SPN as in Figure 2.10 of Chapter 2 and consider the successive times $\{T_n : n \geq 0\}$ at which the marking is $\bar{s} = (0, 1, 0, 0, 0, 1, 0, 0, 0, 1, 0, 0, 0, 1, 0, 1)$—so that each port has a packet awaiting transmission—and transition $e_{3,1}$ = "observation of ring token by port 1" fires. Using Lemma 4.19 in Chapter 3, an ad-hoc recurrence argument similar to those in Examples 2.34 and 2.38 in Chapter 5 shows that each T_n is a.s. finite. An application of Theorem 2.2 then shows that the random times $\{T_n : n \geq 0\}$ form a sequence of regeneration points for the marking process. Moreover, it follows from Theorem 2.40 that the regenerative cycle length τ_1 has finite moments of all orders.

We therefore can use the regenerative method to estimate the long-run utilization $r(f) = \lim_{t\to\infty}(1/t)\int_0^t f\big(X(u)\big)\, du$, where $f(s) = 1$ if a transmission is underway when the marking is s and $f(s) = 0$ otherwise. Formally, $f(s) = \max_{1\leq j\leq 4} s_{3,j}$ for $s = (s_{1,1}, s_{1,2}, \ldots, s_{4,4}) \in G$. It is not hard to see that the cycle length τ_1 is aperiodic, so that—by Theorem 1.20—

the quantity $r(f)$ can also be interpreted as the steady-state or limiting probability that a transmission is underway.

Table 6.1 displays point estimates and 95% confidence intervals for $r(f)$ based on the standard, Tin, and jackknife estimators; results are reported for varying simulation run lengths. The parameter values used in the simulation are $L = 0.15$, $R = 0.05$, and $q = 1.0$. For each simulation, the Tin and jackknife point estimators are closer to the true value than the standard estimator, with the jackknife estimator slightly outperforming the Tin estimator. The differences between the point estimators become negligible, however, as the simulation run length becomes large (≥ 50 cycles). The half-widths of the 95% confidence intervals are comparable (at every run length) for the three estimation methods and decrease roughly as $n^{-1/2}$, where n is the number of simulated cycles.

6.3.3 Simulation Until a Fixed Time

One common variation of the basic regenerative method is to simulate the marking process until a fixed (simulated) time t. Point estimates and confidence intervals are computed as in Algorithm 3.5, except that statistics are computed for the random number $n(t)$ of cycles completed by time t. This procedure is justified by Theorem 3.18. In the theorem, $\hat{r}(n)$, $\bar{\tau}(n)$, and $s^2(n)$ are defined as in Theorem 3.4.

Theorem 3.18. *Under the conditions of Theorem 3.4, the estimators $\hat{r}(n(t))$, $\bar{\tau}(n(t))$, and $s^2(n(t))$ are strongly consistent for $r(f)$, $E_\mu[\tau_1]$, and $\sigma^2(f)$, and*

$$\frac{\sqrt{n(t)}\left(\hat{r}(n(t)) - r(f)\right)}{s(n(t))/\bar{\tau}(n(t))} \Rightarrow N(0,1)$$

as $t \to \infty$.

PROOF. Because $E_\mu[\tau_1] < \infty$, it follows that each τ_k is a.s. finite, which implies that $n(t) \to \infty$ a.s. as $t \to \infty$; see Theorem 2.9(ii) in Chapter 3. The first assertion of the theorem now follows immediately from the strong consistency of $\hat{r}(n)$, $\bar{\tau}(n)$, and $s^2(n)$; see Theorem 3.4. Set $Z_k(f) = Y_k(f) - r(f)\tau_k$ for $k \geq 1$. The remainder of the proof proceeds similarly to the proof of Theorem 3.4, except that we apply the random-index CLT for i.i.d. random variables (Proposition 2.5 in the Appendix) to the sequence $\{Z_k : k \geq 0\}$ rather than the ordinary CLT. To apply the random-index CLT, it suffices to show that $n(t)/t$ converges to a positive finite constant a.s. as $t \to \infty$. Because $\lim_{n\to\infty}(1/n)\sum_{k=1}^n \tau_k = E_\mu[\tau_1]$ a.s. by the SLLN for i.i.d. random variables, Theorem 2.9(ii) in Chapter 3 implies that $n(t)/t \to 1/E_\mu[\tau_1] \in (0,\infty)$ a.s. as $t \to \infty$, and the desired result follows.   □

Thus replacement of n by $n(t)$ in Algorithm 3.5 yields a strongly consistent point estimator and an asymptotic confidence interval for $r(f)$. It can be shown that $\text{Bias}_\mu\left[\hat{r}(n(t))\right] = O(t^{-1})$.

An alternative procedure is to continue the simulation until the first regeneration point $T_{n(t)+1}$ after time t. Using renewal theory, it can be shown that $\text{Bias}_\mu\left[\hat{r}(n(t)+1)\right] = O(t^{-2})$, and a bias reduction is obtained. One disadvantage of this bias-reduction procedure is that the length of the simulation is now random, and the additional effort required to simulate the marking process in the interval $[t, T_{n(t)+1}]$ can be nonnegligible.

The following result leads to a low-bias estimator that does not require simulation of the marking process beyond time t. In the following, set $\hat{r}_3(t) = (1/t) \int_0^t f(X(u))\, du$ for $t \geq 0$.

Proposition 3.19. *Suppose that the cycle length τ_1 is aperiodic, $T_0 = 0$, $E_\mu[\tau_1^2] < \infty$, and $E_\mu[Y_1^2(|f|)] < \infty$. Then*

$$E_\mu\left[\hat{r}_3(t)\right] = r(f) - \frac{1}{tE_\mu\left[\tau_1\right]} E_\mu\left[\int_0^{T_1} u\Big(f(X(u)) - r(f)\Big)\, du\right] + o(t^{-1}).$$

Estimating the bias term and using the approximation $1/E_\mu\left[\tau_1\right] \approx n(t)/t$ leads to the estimator

$$\hat{r}_4(t) = \hat{r}_3(t) + \frac{1}{t^2} \sum_{k=1}^{n(t)} \int_{T_{k-1}}^{T_k} (u - T_{k-1})\Big(f(X(u)) - \hat{r}_3(t)\Big)\, du.$$

It can be shown that $\text{Bias}_\mu\left[\hat{r}_4(t)\right] = o(t^{-1})$, provided that τ_1 is aperiodic, $T_0 = 0$, $E_\mu[\tau_1^5] < \infty$, and $E_\mu[Y_1^5(|f|)] < \infty$. In practice, the bias typically is $O(t^{-2})$. Moreover, under the conditions of Theorem 3.4,

$$\frac{\sqrt{n(t)}\big(\hat{r}_4(t) - r(f)\big)}{s(n(t))/\bar{\tau}(n(t))} \Rightarrow N(0,1),$$

so that an asymptotic confidence interval for $r(f)$ can be based on the estimator $\hat{r}_4(t)$; the asymptotic efficiency of the confidence-interval procedures based on $\hat{r}_4(t)$ and $\hat{r}(n(t))$ is the same in that the lengths of the asymptotic confidence intervals are identical.

EXAMPLE 3.20 (Token ring). We compare estimators $\hat{r}_4(t)$ and $\hat{r}(n(t))$ using the SPN and sequence of regeneration points in Example 3.17. Table 6.2 displays point estimates and 95% confidence intervals for the long-run utilization $r(f)$ defined in Example 3.17. For each simulation length, the number of completed cycles is given in parentheses. Observe that the estimator $\hat{r}_4(t)$ is always closer to $r(f)$ than is $\hat{r}(n(t))$, although the difference between the two estimators becomes small as the simulation run length increases.

Table 6.2. Simulation Results for Token Ring: Point Estimates and 95% Confidence-Interval Half-Widths for the Long-Run Utilization (True Value = 0.4462)

estimator	Simulation Length t					
	2.0	5.0	10.0	100.0	1000.0	5000.0
	(3)	(7)	(14)	(139)	(1389)	(6808)
$\hat{r}(n(t))$	0.2727	0.4074	0.4098	0.4321	0.4430	0.4465
	±0.2916	±0.2264	±0.1132	±0.0306	±0.0110	±0.0048
$\hat{r}_4(t)$	0.3450	0.4495	0.4237	0.4345	0.4432	0.4465
	±0.2916	±0.2264	±0.1132	±0.0306	±0.0110	±0.0048

Note: The number of completed cycles is given in parentheses.

6.3.4 Estimation to Within a Specified Precision

The goal of a simulation often is to estimate the quantity $r(f)$ to within a specified precision with a specified probability; equivalently, we wish to obtain a confidence interval for $r(f)$, where both the confidence level and length of the interval are specified a priori. Specifically, suppose that we wish to estimate $r(f)$ to within $\pm\epsilon g(r(f))$ with probability approximately equal to p, where the parameters ϵ and p are specified a priori and the function g (assumed positive and continuous) determines the type of precision criterion. For example, if $g(x) = 1$ for all x, then we have an *absolute* precision requirement; that is, we wish to estimate $r(f)$ to within $\pm\epsilon$ with probability p. If $g(x) = x$, then we have a *relative* precision requirement; that is, we wish to estimate $r(f)$ to within $\pm 100\epsilon\%$. If $g(x) = \max(x, d)$ for some $d > 0$, then we have a hybrid precision requirement in which we estimate $r(f)$ to within a relative precision if $r(f)$ is "large" (greater than d) and to within an absolute precision if $r(f)$ is "small" (less than d). Approaches to this problem include the use of pilot runs and sequential estimation procedures.

Pilot Runs

Denote by n^* the (unknown) number of cycles required to satisfy the aforementioned precision criterion, and suppose that the precision parameter ϵ is small enough so that n^* is relatively large. In particular, suppose that n^* is large enough so that $s^2(n^*) \approx \sigma^2(f)$ and $\bar{\tau}(n^*) \approx E_\mu[\tau_1]$. Set the half-width of the confidence interval in (3.6) equal to $\epsilon g(r(f))$ and use the foregoing approximations to obtain the expression

$$n^* \approx \frac{z_p^2\, \sigma^2(f)}{E_\mu^2[\tau_1]\, \epsilon^2 g^2(r(f))}. \tag{3.21}$$

From (3.21), we derive the following two-stage procedure. Choose a small number n_0 and create a short pilot run by simulating the marking process for n_0 cycles. Estimate $\sigma^2(f)$, $E_\mu[\tau_1]$, and $r(f)$ by $s^2(n_0)$, $\bar{\tau}(n_0)$, and $\hat{r}(n_0)$. Substitute these estimates into (3.21) to obtain a value for n^*. Then simulate the marking process for n^* cycles and use Algorithm 3.5 to obtain the final point and interval estimates.

Sequential Procedures

The idea behind a sequential procedure is to run the simulation until the precision criterion appears to be satisfied. More precisely, set

$$N(\epsilon) = \min \left\{ n \geq 2 : s(n) > 0 \text{ and } \frac{z_p\, s(n)}{\bar{\tau}(n)g\big(\hat{r}(n)\big)\sqrt{n}} < \epsilon \right\}$$

for $\epsilon > 0$ and $p \in (0,1)$. (We have suppressed the dependence of $N(\epsilon)$ on p in our notation.) The sequential procedure consists of simulating precisely $N(\epsilon)$ cycles and then computing point estimates and confidence intervals as in Algorithm 3.5. Observe that the number $N(\epsilon)$ of cycles to be simulated is a random variable. Under the conditions of Theorem 3.4, it can be shown by an argument almost identical to the proof of Theorem 3.18 that

$$\frac{\sqrt{N(\epsilon)}\big(\hat{r}(N(\epsilon)) - r(f)\big)}{s\big(N(\epsilon)\big)/\bar{\tau}\big(N(\epsilon)\big)} \Rightarrow N(0,1)$$

as $\epsilon \to 0$. That is, the random interval

$$I = \left[\hat{r}\big(N(\epsilon)\big) - \frac{z_p\, s\big(N(\epsilon)\big)}{\bar{\tau}\big(N(\epsilon)\big)\,\sqrt{N(\epsilon)}}, \hat{r}\big(N(\epsilon)\big) + \frac{z_p\, s\big(N(\epsilon)\big)}{\bar{\tau}\big(N(\epsilon)\big)\,\sqrt{N(\epsilon)}} \right]$$

is an asymptotic $100p\%$ confidence interval for $r(f)$ as $\epsilon \to 0$. Equivalently, the probability that the estimator $\hat{r}\big(N(\epsilon)\big)$ is within $\pm\epsilon g\big(r(f)\big)$ of the unknown constant $r(f)$ converges to the specified value p as the precision parameter ϵ becomes small. A sequential estimation procedure with this property is said to be *asymptotically consistent*. The procedure is also *asymptotically efficient* in that

$$\lim_{\epsilon \to 0} \epsilon^2 N(\epsilon) = \frac{z_p^2 \sigma^2(f)}{E_\mu^2[\tau_1]\, g^2\big(r(f)\big)} \quad \text{a.s..}$$

Heuristically, as the precision requirement becomes increasingly stringent, the number of cycles simulated approaches the "minimal" required number as in (3.21).

In practice, care has to be taken so that the procedure does not terminate too soon. Premature termination can result in actual coverage probabilities that are lower than the nominal probability p. This "undercoverage" problem is particularly acute for larger values of ϵ. There are a number of ways in which undercoverage can be reduced while retaining the desirable properties of asymptotic efficiency and consistency. These include

Table 6.3. Simulation Results for Token Ring with Fixed-Sized Packets: Empirical Coverage Probabilities When Estimating Long-Run Utilization with Nominal Coverage Probability of 95%, Based on 1000 Simulation Repetitions

Stopping Rule	Precision			
	1%	5%	10%	20%
$N(\epsilon)$	0.9230	0.9100	0.8520	0.8030
	(7833)	(305)	(69)	(16)
$N'(\epsilon)$	0.9470	0.9450	0.9380	0.9150
	(8216)	(355)	(97)	(28)
$N''(\epsilon)$	0.9280	0.9090	0.8700	0.8300
	(7847)	(310)	(72)	(19)

Note: The average number of simulated cycles is given in parentheses.

- Requiring that some minimum number of cycles be simulated before termination is allowed

- Requiring that the stopping condition—that is, the requirement that the current confidence-interval half-width be smaller than $\epsilon g(r(f))$—be satisfied not just once, but k times for some $k > 1$

- Simulating $N'(\epsilon)$ cycles, where

$$N'(\epsilon) = \min\left\{ n \geq 2 : s(n) > 0 \text{ and } a_n + \frac{z_p \, s(n)}{\bar{\tau}(n) g(\hat{r}(n)) \sqrt{n}} < \epsilon \right\}$$

and $\{a_n : n \geq 1\}$ is a sequence of constants such that $a_n \downarrow 0$

- Simulating $N''(\epsilon)$ cycles, where

$$N''(\epsilon) = \min\left\{ n \geq 2 : s(n) > 0 \text{ and } \frac{t_{p,n} \, s(n)}{\bar{\tau}(n) g(\hat{r}(n)) \sqrt{n}} < \epsilon \right\}$$

and $\{t_{p,n} : n \geq 1\}$ is a sequence of constants such that $t_{p,n} \downarrow z_p$

A typical choice for the constant a_n is $a_n = 1/n$ and $t_{p,n}$ is often taken to be the $(1+p)/2$ quantile of the Student's t distribution with n degrees of freedom. The foregoing methods can be used in combination.

EXAMPLE 3.22 (Token ring with fixed-sized packets). We compare the stopping rules $N(\epsilon)$, $N'(\epsilon)$, and $N''(\epsilon)$ using the SPN and sequence of regeneration points in Example 3.17. As before, the goal is to estimate the long-run utilization. Table 6.3 displays—for various levels of precision—the empirical coverage probability corresponding to a nominal coverage probability of 95%, based on 1000 simulation replications. Average run lengths (in cycles) are given in parentheses. For example, when the desired precision is ±5% and we use stopping rule $N(\epsilon)$, the average simulation run length is 305 cycles and the empirical coverage probability is only 0.9100. That is, the estimated long-run utilization lies within ±5% of the true value (0.4462) in only 91% of the 1000 simulation repetitions, rather than the desired 95%. The situation is even worse for a desired precision of ±20%: the true coverage probability in this case is approximately 80%. As can be seen from the table, use of either the stopping rule $N'(\epsilon)$ or $N''(\epsilon)$ increases the coverage probability relative to $N(\epsilon)$. For example, use of the stopping rule $N'(\epsilon)$ with a desired precision of ±5% increases the average simulation run length to 355 cycles and increases the coverage to 94.5%. Overall, use of $N'(\epsilon)$ yielded the best results in our experiments.

6.3.5 Functions of Cycle Means

As discussed in Section 3.2, a wide variety of performance measures can be expressed as nonlinear functions of time-average limits of the underlying chain $\{(S_n, C_n) : n \geq 0\}$. In this section we consider the problem of estimating such performance measures in the presence of regenerative structure.

In light of (3.10), it can be seen that—when there exists a sequence $\{\theta(k) : k \geq 0\}$ of regeneration points for the underlying chain—the performance measures discussed in Section 3.2 can be rewritten in the form

$$r = g(\alpha(\tilde{f}_1), \alpha(\tilde{f}_2), \ldots, \alpha(\tilde{f}_l)), \qquad (3.23)$$

where g is a real-valued function defined on $\Re^l$ ($l \geq 1$), $\tilde{f}_1, \tilde{f}_2, \ldots, \tilde{f}_l$ are real-valued functions defined on Σ, and $\alpha(\tilde{f}_i)$ is the cycle mean given by

$$\alpha(\tilde{f}_i) = E_\mu \left[\sum_{j=\theta(0)}^{\theta(1)-1} \tilde{f}_i(S_j, C_j) \right]$$

for $1 \leq i \leq l$. According to this notation, $\alpha(\tilde{f}_i) = E_\mu[\tilde{Y}_1(\tilde{f}_i)]$, where $\tilde{Y}_k(\tilde{f})$ is defined by (3.9).

EXAMPLE 3.24 (Time-average limits). Suppose that the cycle length $\tilde{\tau}_1 = \theta(1) - \theta(0)$ has finite mean and let $\tilde{f}$ be a function such that $\tilde{Y}_0(|\tilde{f}|) < \infty$ a.s. and $E_\mu[\tilde{Y}_1(|\tilde{f}|)] < \infty$. Then $\lim_{n \to \infty}(1/n) \sum_{j=0}^{n-1} \tilde{f}(S_j, C_j)$ has the representation

$$g(\alpha(\tilde{f}_1), \alpha(\tilde{f}_2)) = \frac{\alpha(\tilde{f}_1)}{\alpha(\tilde{f}_2)}, \qquad (3.25)$$

where $\tilde{f}_1 = \tilde{f}$ and $\tilde{f}_2(s, c) = 1$ for $(s, c) \in \Sigma$. Similarly, under suitable conditions a time-average limit of the form $\lim_{t \to \infty}(1/t) \int_0^t f(X(u)) \, du$ can be represented in the form (3.25), where $\tilde{f}_1(s, c) = f(s)t^*(s, c)$ and $\tilde{f}_2(s, c) = t^*(s, c)$ for $(s, c) \in \Sigma$.

EXAMPLE 3.26 (Long-run variance). Suppose that $\tau_1 = \zeta_{\theta(1)} - \zeta_{\theta(0)}$ has finite mean and let the function f satisfy $Y_0(|f|) < \infty$ a.s. and $E_\mu[Y_1(|f|)] < \infty$. Then the long-run average value of the output process, that is, the value of $\lim_{t\to\infty}(1/t)\int_0^t f(X(u))\,du$, is well defined and equal to the quantity $r(f)$ given by (3.1). As in Example 2.18 in Chapter 3, the long-run variance

$$v(f) = \lim_{t\to\infty} \frac{1}{t}\int_0^t \Big(f(X(u)) - r(f)\Big)^2 du$$

may be of interest. If the cycle length τ_1 is aperiodic and $E_\mu[Y_1(f^2)] < \infty$, then $X(t) \Rightarrow X$ as $t \to \infty$ and $v(f)$ can also be interpreted as a steady-state variance: $v(f) = \mathrm{Var}[f(X)]$. Set $\tilde{f}_1(s,c) = f(s)t^*(s,c)$, $\tilde{f}_2(s,c) = f^2(s)t^*(s,c)$, and $\tilde{f}_3(s,c) = t^*(s,c)$ for $(s,c) \in \Sigma$. Provided that $Y_1(\tilde{f}_2) < \infty$, we have the representation $v(f) = g(\alpha(\tilde{f}_1), \alpha(\tilde{f}_2), \alpha(\tilde{f}_3))$, where

$$g(a_1, a_2, a_3) = \left(\frac{a_2}{a_3}\right) - \left(\frac{a_1}{a_3}\right)^2.$$

An Estimation Procedure

To obtain strongly consistent point estimates and asymptotic confidence intervals for a performance measure r as in (3.23), suppose that the function g is differentiable in a neighborhood of $\alpha = (\alpha(\tilde{f}_1), \alpha(\tilde{f}_2), \ldots, \alpha(\tilde{f}_l))$ with partial derivatives $g_1, g_2, \ldots, g_l$. Simulate n cycles of the underlying chain, and compute the quantities

$$\bar{Y}_i(n) = \frac{1}{n}\sum_{k=1}^n \tilde{Y}_k(\tilde{f}_i)$$

for $1 \le i \le l$; for convenience, write $\bar{Y}(n) = (\bar{Y}_1(n), \bar{Y}_2(n), \ldots, \bar{Y}_l(n))$. Form the point estimate

$$\hat{r}(n) = g(\bar{Y}(n)) = g(\bar{Y}_1(n), \bar{Y}_2(n), \ldots, \bar{Y}_l(n)), \qquad (3.27)$$

and denote by $V(n)$ the $l \times l$ sample covariance matrix whose (i,j)th entry is given by

$$V_{i,j}(n) = \frac{1}{n-1}\sum_{k=1}^n \big(\tilde{Y}_k(f_i) - \bar{Y}_i(n)\big)\big(\tilde{Y}_k(f_j) - \bar{Y}_j(n)\big).$$

Denote by $\nabla g = (g_1, g_2, \ldots, g_l)$ the gradient of g, and set[4]

$$\begin{aligned}
w(n) &= \nabla g(\bar{Y}(n))^t\, V(n)\, \nabla g(\bar{Y}(n)) \\
&= \sum_{i=1}^l \sum_{j=1}^l g_i(\bar{Y}(n)) g_j(\bar{Y}(n)) V_{i,j}(n).
\end{aligned} \qquad (3.28)$$

---

[4]Here and elsewhere, all vectors are assumed to be column vectors and x^t denotes the transpose of x.

Finally, let z_p be the $(1+p)/2$ quantile of the standard normal distribution and form the interval

$$I_n = \left[\hat{r}(n) - \frac{z_p \sqrt{w(n)}}{\sqrt{n}}, \hat{r}(n) + \frac{z_p \sqrt{w(n)}}{\sqrt{n}}\right]$$

for r. The following result shows that $\hat{r}(n)$ is strongly consistent for r and that I_n is an asymptotic $100p\%$ confidence interval for r.

Theorem 3.29. *Let $\tilde{f}_1, \tilde{f}_2, \ldots, \tilde{f}_l$ ($l \geq 1$) be real-valued functions defined on Σ such that $\alpha(|\tilde{f}_i|) < \infty$ for $1 \leq i \leq l$, and let g be a real-valued function defined on $\Re^l$ that is differentiable in a neighborhood of $\alpha = (\alpha(\tilde{f}_1), \alpha(\tilde{f}_2), \ldots, \alpha(\tilde{f}_l))$. Then $\lim_{n\to\infty} \hat{r}(n) = r$ and*

$$\frac{\sqrt{n}(\hat{r}(n) - r)}{\sqrt{w(n)}} \Rightarrow N(0,1), \tag{3.30}$$

as $n \to \infty$, where r, $\hat{r}(n)$, and $w(n)$ are defined as in (3.23), (3.27), and (3.28).

PROOF. Set $\tilde{Y}_k = (\tilde{Y}_k(\tilde{f}_1), \tilde{Y}_k(\tilde{f}_2), \ldots, \tilde{Y}_k(\tilde{f}_l))$ for $k \geq 1$, and observe that, by the regenerative property, the sequence $\{\tilde{Y}_k : k \geq 1\}$ consists of i.i.d. random vectors. Applying the SLLN for i.i.d. random variables to this sequence (componentwise), we find that $\bar{Y}(n) \to \alpha$ a.s. as $n \to \infty$. It follows from the differentiability of g at α that g is continuous at α, and the strong consistency of $\hat{r}(n)$ follows immediately. To establish the convergence in (3.30), observe that, by the CLT for $\Re^l$-valued random vectors (Proposition 2.6 in the Appendix),

$$\sqrt{n}(\bar{Y}(n) - \alpha) \Rightarrow N(0,B)$$

as $n \to \infty$, where $N(0,B)$ denotes an l-dimensional normal random vector having mean $(0, 0, \ldots, 0)$ and covariance matrix $B = \|b_{ij}\|$ with $b_{ij} = \text{Cov}_\mu[\tilde{Y}_1(\tilde{f}_i), \tilde{Y}_1(\tilde{f}_j)]$ for $1 \leq i, j \leq l$. Using the delta method—see Proposition 1.45 in the Appendix—it follows that

$$\sqrt{n}(\hat{r}(n) - r) \Rightarrow \nabla g(\alpha)^t N(0,B).$$

By standard properties of the multivariate normal distribution, the random variable $\nabla g(\alpha)^t N(0,B)$ has a (univariate) normal distribution with mean 0 and variance $\rho = \nabla g(\alpha)^t B \nabla g(\alpha)$. Thus

$$\frac{\sqrt{n}(\hat{r}(n) - r)}{\sqrt{\rho}} \Rightarrow N(0,1).$$

As in the proof of Theorem 3.4, the SLLN for i.i.d. random variables can be used to show that $\lim_{n\to\infty} w(n) = \rho$ a.s., and (3.30) follows from Slutsky's theorem. □

EXAMPLE 3.31 (Ratio estimation). Suppose that $r = g(\alpha(\tilde{f}_1), \alpha(\tilde{f}_2))$, where $g(x, y) = x/y$. Then $\hat{r}(n) = \bar{Y}_1(n)/\bar{Y}_2(n)$. Moreover, using the notation in the proof of Theorem 3.29, $\nabla g(x, y) = (1/y, -x/y^2)$, and $\rho = (b_{11} - 2rb_{12} + r^2 b_{22})/\alpha^2(\tilde{f}_2)$. Taking $\tilde{f}_2(s, c) = 1$ for $(s, c) \in \Sigma$ and $\tilde{f}_1 = \tilde{f}$ for a specified function $\tilde{f}$, we see that $\alpha(\tilde{f}_2) = E_\mu[\tilde{\tau}_1]$ and $\rho = \tilde{\sigma}^2(\tilde{f})/E_\mu^2[\tilde{\tau}_1]$, where $\tilde{\sigma}^2(\tilde{f})$ is given by (3.12). Similarly, $w(n)$ coincides with $s^2(n)/\bar{\tau}^2(n)$, where $s^2(n)$ and $\bar{\tau}(n)$ are given by (3.13) and (3.11). Thus the above estimation procedure reduces to the standard regenerative method for the underlying chain. Similarly, taking $\tilde{f}_1(s, c) = f(s)t^*(s, c)$ for $(s, c) \in \Sigma$ and $\tilde{f}_2 = t^*$, we obtain the standard regenerative method for the marking process.

When—as is typical—the function g is nonlinear, the estimator $\hat{r}(n)$ is biased for r, especially for small values of n. Appropriate modifications of the bias-reduction techniques discussed in Section 6.3.2 can be used to handle this problem. For example, we can apply the jackknife method by setting

$$\hat{r}_J(n) = \frac{1}{n} \sum_{i=1}^n J^{(i)}(n),$$

where

$$J^{(i)}(n) = ng(\bar{Y}_1(n), \ldots, \bar{Y}_l(n)) - (n-1)g(\bar{Y}_1^{(i)}(n), \ldots, \bar{Y}_l^{(i)}(n))$$

and

$$\bar{Y}_j^{(i)}(n) = \frac{1}{n-1} \sum_{k \neq i} \tilde{Y}_k(\tilde{f}_j).$$

Under the conditions of Theorem 3.29, it can be shown that $\hat{r}_J(n)$ is strongly consistent for r and

$$\left[\hat{r}_J(n) - \frac{z_p \sqrt{w(n)}}{\sqrt{n}}, \hat{r}_J(n) + \frac{z_p \sqrt{w(n)}}{\sqrt{n}} \right]$$

is an asymptotic $100p\%$ confidence interval for r, where $w(n)$ is defined by[5] (3.28).

Extensions

The foregoing results carry over essentially without change to performance measures of the form $r = g(\alpha(f_1), \alpha(f_2), \ldots, \alpha(f_l))$, where

$$\alpha(f_i) = E_\mu\left[\int_{T_0}^{T_1} f_i(X(u)) \, du \right]$$

---

[5] As in Section 6.3.2, alternative confidence intervals can be obtained by using variance estimators other than $w(n)$.

for $1 \leq i \leq l$ and $T_k = \zeta_{\theta(k)}$ for $k \geq 0$. Indeed, the methods described above can be applied to any performance measure of the form

$$r = g\big(E_\mu[Y_1], E_\mu[Y_2], \ldots, E_\mu[Y_l]\big), \tag{3.32}$$

where each Y_i is a random variable that is completely determined by the behavior of the marking process or underlying chain over a cycle; Y_i need not be the integral (or sum) of a function over the cycle.

EXAMPLE 3.33 (Discounted reward). Suppose that $T_0 = 0$ and that rewards accrue continuously at rate $q(s)$ whenever the marking is $s \in S$. Also suppose that the performance measure r of interest is the β-discounted reward, defined as

$$r = E_\mu\left[\int_0^\infty e^{-\beta u} q\big(X(u)\big)\, du\right].$$

Using the regenerative property, we can write

$$r = E_\mu\left[\int_0^{T_1} e^{-\beta u} q\big(X(u)\big)\, du\right] + E_\mu\left[e^{-\beta T_1} \int_{T_1}^\infty e^{-\beta(u-T_1)} q\big(X(u)\big)\, du\right]$$

$$= E_\mu\left[\int_0^{T_1} e^{-\beta u} q\big(X(u)\big)\, du\right] + E_\mu\left[e^{-\beta T_1}\right] r,$$

so that

$$r = \frac{E_\mu\left[\int_0^{T_1} e^{-\beta u} q\big(X(u)\big)\, du\right]}{1 - E_\mu\left[e^{-\beta T_1}\right]}.$$

Thus r is of the form (3.32), where $g(x, y) = x/(1 - y)$,

$$Y_1 = \int_0^{T_1} e^{-\beta u} q\big(X(u)\big)\, du,$$

and $Y_2 = \exp(-\beta T_1)$.

EXAMPLE 3.34 (Mean time to failure). Suppose that the system modelled by the SPN is considered to be in a failed state when the current marking is an element of a subset $A_f \subset S$; when the marking is an element of $S - A_f$ the system is considered operational. Suppose that the initial marking is an element of $S - A_f$, and define the *time to failure* as $t_f = \inf\{t > 0: X(t) \in A_f\}$. A common measure of system reliability is the *mean time to failure* $r = E_\mu[t_f]$. Suppose that $T_0 = 0$ and $P_\mu\{t_f \leq T_1\} > 0$. Writing $x \wedge y = \min(x, y)$ and using the regenerative property, we find that

$$r = E_\mu[t_f; t_f \leq T_1] + E_\mu[t_f; t_f > T_1]$$

$$= E_\mu[t_f; t_f \leq T_1] + E_\mu[T_1; t_f > T_1] + E_\mu[t_f - T_1; t_f > T_1]$$

$$= E_\mu[t_f \wedge T_1] + E_\mu[t_f - T_1 \mid t_f > T_1] P_\mu\{t_f > T_1\}$$

$$= E_\mu[t_f \wedge T_1] + r\, P_\mu\{t_f > T_1\}.$$

It follows that
$$r = \frac{E_\mu\left[t_f \wedge T_1\right]}{P_\mu\{t_f \leq T_1\}}.$$
Thus r is of the form (3.32), where $g(x,y) = x/y$, $Y_1 = t_f \wedge T_1$ and

$$Y_2 = \begin{cases} 1 & \text{if } t_f \leq T_1; \\ 0 & \text{if } t_f > T_1. \end{cases}$$

If the system is highly reliable, so that r is very large, then specialized variance-reduction techniques must be used when estimating r.

6.3.6 Gradient Estimation

Engineers and systems designers are often interested in studying the sensitivity of long-run system performance to changes in the values of various system parameters. Such parameters might correspond, for example, to processing rates for various machines in a manufacturing cell or routing probabilities in a communication network. If the parameter values are under the designer's control, then a typical goal of the analysis is to find parameter settings that maximize the performance measure of interest. When the system under study is modelled as an SPN, the clock-setting distributions or new-marking probabilities, or both, are specified as functions of a parameter vector λ. A sensitivity analysis is then carried out by estimating the gradient of a time-average limit with respect to λ. Such gradient estimates are also a key ingredient of many simulation-based optimization procedures.

A classical technique for estimating gradients is to use finite-difference approximations. This approach is expensive, requiring multiple simulation runs at different settings of the parameter values. To address this problem, several gradient-estimation methods have been developed that require only a single run; these include the *likelihood ratio method* and *infinitesimal perturbation analysis* (IPA). This subsection contains an introduction to the likelihood ratio method for gradient estimation in the setting of SPNs having a regenerative underlying chain.

An Example

We first motivate the gradient estimation problem in more detail.

EXAMPLE 3.35 (Machine repair). Consider the system of Example 2.28, but now suppose that the successive lifetimes of each machine are i.i.d. according to a uniform distribution on the interval $[0, 1]$ and the successive times for the repairperson to repair (and restart) a machine are i.i.d. according to an exponential distribution with intensity λ. Whenever the repairperson is repairing a machine, costs accrue at rate λ^2. Each stopped machine accrues costs at rate β. Suppose that the system is modelled by an SPN as in Figure 6.1. Then the long-run average cost can be expressed as

$$r(\lambda) = r(\lambda; g) = \lim_{t \to \infty} \frac{1}{t} \int_0^t g(X(u), \lambda) \, du,$$

where $\{ X(t) \colon t \geq 0 \}$ is the marking process of the SPN and

$$g(s, \lambda) = \lambda^2 s_3 + \beta \sum_{j=1}^{N} s_{2,j}$$

for $s = (s_{1,1}, s_{2,1}, \ldots, s_{1,N}, s_{2,N}, s_3, s_4) \in S$. The quantity $r'(\lambda)$—that is, the derivative of $r(\lambda)$ with respect to λ—measures the sensitivity of the long-run average cost with respect to the repair rate, and estimation of this quantity is of intrinsic interest. Moreover, estimates of $r'(\lambda)$ can be used to compute the value λ^* that minimizes the long-run average cost. A classical method for computing λ^* is the *Robbins–Monro algorithm*, which is based on the recursion

$$\lambda_{n+1} = \max(\lambda_n - \frac{a}{n} D_{n+1}, 0)$$

for $n \geq 0$. Here $a > 0$, λ_0 is an arbitrary initial starting value and $\{ D_n \colon n \geq 1 \}$ is a sequence of derivative estimates that satisfy

$$E_\mu[D_{n+1} \mid D_n, \lambda_n, D_{n-1}, \lambda_{n-1}, \ldots, D_0, \lambda_0] = r'(\lambda_n).$$

It can be shown that $\lim_{n \to \infty} \lambda_n = \lambda^*$ a.s. and, moreover, that $n^{1/2}(\lambda_n - \lambda^*) \Rightarrow \sigma N(0, 1)$ as $n \to \infty$ for some $\sigma > 0$. Thus the estimates converge to λ^* with probability 1 and the convergence rate is $O(n^{-1/2})$ in probability.

Likelihood Ratios

We illustrate the main idea underlying the likelihood-ratio method for gradient estimation by means of a very simple example. Consider a random variable X having exponential density function $f(x; \lambda) = \lambda \exp(-\lambda x)$ for some $\lambda > 0$, and suppose that we wish to estimate the derivative of $r(\lambda) = E[g(X, \lambda)]$ with respect to λ, where g is a specified real-valued function defined on $\Re_+ \times \Re_+$. The primary difficulty in estimating the derivative of r is that the parameter λ determines the value of $r(\lambda)$ not only explicitly as an argument of the function g, but also implicitly as a parameter of the distribution of X. This problem can be avoided as follows.

Fix $\lambda_0 > 0$ and observe that

$$
\begin{aligned}
r(\lambda) &= E\left[g(X, \lambda)\right] \\
&= \int_0^\infty g(x, \lambda) f(x; \lambda)\, dx \\
&= \int_0^\infty g(x, \lambda) \frac{f(x; \lambda)}{f(x; \lambda_0)} f(x; \lambda_0)\, dx \\
&= E\left[g(Y, \lambda) L(Y; \lambda)\right],
\end{aligned}
$$

where Y is an exponential random variable with intensity λ_0 and

$$
L(Y; \lambda) = \frac{f(Y; \lambda)}{f(Y; \lambda_0)} = \left(\frac{\lambda}{\lambda_0}\right) e^{-(\lambda - \lambda_0)Y}.
$$

The key point is that λ determines the value of $E\left[g(Y, \lambda) L(Y; \lambda)\right]$ only as an explicit argument of the functions g and L—the distribution of Y does not depend on λ. The function L is a *likelihood ratio*; heuristically, $L(x; \lambda)$ is the likelihood of observing the realized value $X = x$ assuming that X has the distribution $F(\,\cdot\,; \lambda)$ relative to the likelihood of observing this realized value assuming that X has the distribution $F(\,\cdot\,; \lambda_0)$. Write $\tilde{g}(y, \lambda) = g(y, \lambda) L(y; \lambda)$ for $y, \lambda \geq 0$ and formally differentiate with respect to λ to obtain

$$
r'(\lambda_0) = \frac{d}{d\lambda} E\left[\tilde{g}(Y, \lambda)\right]\Big|_{\lambda = \lambda_0} = E[\tilde{g}'(Y, \lambda_0)] = E\left[h(Y)\right],
$$

where

$$
h(y) = g(y, \lambda_0) L'(y; \lambda_0) + g'(y, \lambda_0) = g(y, \lambda_0) \frac{f'(y, \lambda_0)}{f(y, \lambda_0)} + g'(y, \lambda_0)
$$

for $y \geq 0$ and a prime denotes the derivative of the corresponding function with respect to λ. The foregoing derivation uses the fact that $L(Y; \lambda_0) = 1$ and is valid provided that the expectation and derivative operators can be interchanged; such an interchange is permissible under mild conditions on the function g. Thus to estimate $r'(\lambda_0)$, generate n i.i.d. random variables $Y_1, Y_2, \ldots, Y_n$ with common distribution function $F(\,\cdot\,; \lambda_0)$ and form the estimator $D_n = (1/n) \sum_{i=1}^n h(Y_n)$. It follows easily that D_n is unbiased and strongly consistent for $r'(\lambda_0)$. Moreover, standard arguments for i.i.d. random variables show that

$$
I = \left[D_n - \frac{z_p s(n)}{\sqrt{n}}, D_n + \frac{z_p s(n)}{\sqrt{n}}\right]
$$

is an asymptotic $100p\%$ confidence interval for $r'(\lambda_0)$, where

$$
s^2(n) = \frac{1}{n-1} \sum_{i=1}^n \left(h(Y_i) - D_n\right)^2.
$$

Observe that, in the above derivation, the likelihood-ratio representation $r(\lambda) = E\left[g(Y, \lambda)L(Y; \lambda)\right]$ is well defined because $f(x; \lambda_0) > 0$ for $x \geq 0$. For an arbitrary distribution function $F(\,\cdot\,; \lambda)$ with density function $f(\,\cdot\,; \lambda)$, the likelihood-ratio representation is valid provided that

$$f(x; \lambda) = 0 \quad \text{whenever} \quad f(x; \lambda_0) = 0. \tag{3.36}$$

The Likelihood-Ratio Method for SPNs

We now indicate how the foregoing methodology can be extended to the SPN setting. For simplicity, attention is restricted to SPNs with unit speeds in which all transitions are simple. Moreover, λ is a single real-valued parameter, so that the gradient reduces to a simple derivative—in practice, the techniques that we discuss would be used to estimate each element of a gradient vector and the derivatives described below would be computed as partial derivatives.

In the following, only the clock-setting distribution functions, initial-marking distribution, and new-marking probabilities are allowed to depend on λ. Accordingly, we use the notation $F(\,\cdot\,; e, \lambda)$, $\nu_0(s, \lambda)$, and $p(s'; s, E^*, \lambda)$ for these building blocks—in general, the likelihood-ratio method is inapplicable to problems in which G, E, or $r(s, e)$ depends on λ. We also write $\mu(\,\cdot\,; \lambda)$ to indicate the dependence of the initial distribution μ on λ (through both ν_0 and the clock-setting distribution functions). Similarly, we write $P(\,\cdot\,, \,\cdot\,, \lambda)$ to indicate the dependence on λ of the transition kernel of the underlying chain. Finally, we modify our usual notation and write P_λ and E_λ to denote probabilities and expectations when the parameter value is equal to λ.

Suppose that each clock-setting distribution function is absolutely continuous and that Assumption PD holds for each λ in some open interval Λ. In analogy with the assumption in (3.36), we require for each $e \in E - E'$ that the support set $\{x\colon f(x; e, \lambda) > 0\}$ not depend on λ, where $f(\,\cdot\,; e, \lambda)$ is the density of $F(\,\cdot\,; e, \lambda)$. For example, we do not allow $F(x; e, \lambda)$ to be a uniform distribution on the interval $[0, \lambda]$. Similarly, we require that the support set $\{s\colon \nu_0(s, \lambda) > 0\}$ not depend on λ and—for each $s \in G$ and $E^* \subseteq E(s)$—the support set $\{s'\colon p(s'; s, E^*, \lambda) > 0\}$ not depend on λ. Finally, suppose that for all $\lambda \in \Lambda$ there exists a sequence of regeneration points $\{\theta(n)\colon n \geq 0\}$ for the underlying chain $\{(S_n, C_n)\colon n \geq 0\}$ and that the sequence is defined independently of λ. Such a sequence exists, for example, if the SPN has a single state $\bar{s}$ with $E(\bar{s}) = \{\bar{e}\}$—our assumptions guarantee that $\bar{s}$ is recurrent for $\lambda \in \Lambda$, and the desired regeneration points correspond to the successive marking changes at which the marking is $\bar{s}$ and transition $\bar{e}$ fires. For convenience, assume that $\theta(0) = 0$ set $\theta = \theta(1)$.

Fix real-valued functions g_1 and g_2, each defined on $G \times \Re$, and denote by

$$r(\lambda) = r(\lambda; g_1, g_2) = \lim_{t \to \infty} \frac{\int_0^t g_1(X(u), \lambda)\, du}{\int_0^t g_2(X(u), \lambda)\, du} \qquad (3.37)$$

the performance measure of interest when the parameter value is equal to λ. Of course, if $g_2(x, \lambda) \equiv 1$ for all x and λ, then $r(\lambda)$ reduces to a standard time-average limit: $r(\lambda) = \lim_{t\to\infty}(1/t)\int_0^t g_1(X(u), \lambda)\, du$. The goal is to estimate the derivative $r'(\lambda) = dr(\lambda)/d\lambda$. To this end, write $Z_n = (S_n, C_n)$ for $n \geq 0$ and set

$$\tilde{h}_i(Z_n, \lambda) = g_i(S_n, \lambda) t^*(Z_n) \qquad (3.38)$$

for $i = 1, 2$, where t^* is defined as in (1.7) in Chapter 3. Theorem 2.24 implies that $r(\lambda) = \alpha(\tilde{h}_1; \lambda)/\alpha(\tilde{h}_2; \lambda)$ for $\lambda \in \Lambda$, where

$$\alpha(\tilde{h}; \lambda) = E_\lambda \left[\sum_{n=0}^{\theta-1} \tilde{h}(Z_n, \lambda) \right]. \qquad (3.39)$$

Thus

$$r'(\lambda) = \frac{\alpha'(\tilde{h}_1; \lambda)\alpha(\tilde{h}_2; \lambda) - \alpha(\tilde{h}_1; \lambda)\alpha'(\tilde{h}_2; \lambda)}{\alpha^2(\tilde{h}_2; \lambda)}.$$

The quantities $\alpha(\tilde{h}_1; \lambda)$ and $\alpha(\tilde{h}_2; \lambda)$ are straightforward to estimate, so the crux of the problem is the estimation of $\alpha'(\tilde{h}_1; \lambda)$ and $\alpha'(\tilde{h}_2; \lambda)$.

To estimate $\alpha'(\tilde{h}_1; \lambda)$ at a fixed value $\lambda = \lambda_0 \in \Lambda$, we can use a likelihood-ratio approach analogous to the one described in the simpler setting. For $z = (s, c_1, c_2, \ldots, c_M), z' = (s', c_1', c_2', \ldots, c_M') \in \Sigma$ and $\lambda \in \Lambda$, set

$$v(z, \lambda) = \nu_0(s, \lambda) \prod_{e_i \in E(s) \cap (E - E')} f(c_i; e_i, \lambda)$$

and

$$q(z, z', \lambda) = p(s'; s, E^*, \lambda) \prod_{e_i \in N} f(c_i'; e_i, \lambda) \prod_{e_i \in O} 1_{\{c_i - t^*\}}(c_i'),$$

where $E^* = E^*(z)$, $N = N(s'; s, E^*) \cap (E - E')$, $O = O(s'; s, E^*)$, and $t^* = t^*(z)$. The function $v(\cdot, \lambda)$ can be viewed as the "density" of the initial distribution $\mu(\cdot, \lambda)$ and the function $q(z, \cdot, \lambda)$ can be viewed as the "density" of the transition kernel $P(z, \cdot, \lambda)$. Then, fixing $\lambda_0 \in \Lambda$, the natural analog of the likelihood ratio $L(Y; \lambda)$ in the previous example is

$$L(Z_0, Z_1, \ldots, Z_{\theta-1}; \lambda) = \frac{v(Z_0, \lambda)}{v(Z_0, \lambda_0)} \prod_{j=0}^{\theta-2} \frac{q(Z_j, Z_{j+1}, \lambda)}{q(Z_j, Z_{j+1}, \lambda_0)}.$$

Here L represents the relative likelihood of observing the realized values $Z_0, Z_1, \ldots, Z_{\theta-1}$ of the underlying chain over the first regenerative cycle under the parameter values λ and λ_0. Observe that, in the above expression,

$$\frac{v(Z_0, \lambda)}{v(Z_0, \lambda_0)} = \frac{\nu_0(S_0, \lambda)}{\nu_0(S_0, \lambda_0)} \prod_{e_i \in E(S_0) \cap (E - E')} \frac{f(C_{0,i}; e_i, \lambda)}{f(C_{0,i}; e_i, \lambda_0)}$$

and

$$\frac{q(Z_j, Z_{j+1}, \lambda)}{q(Z_j, Z_{j+1}, \lambda_0)} = \frac{p(S_{j+1}; S_j, E^*, \lambda)}{p(S_{j+1}; S_j, E^*, \lambda_0)} \prod_{e_i \in N} \frac{f(C_{j+1,i}; e_i, \lambda)}{f(C_{j+1,i}; e_i, \lambda_0)}$$

for $0 \le j \le \theta - 2$, where $E^* = E^*(Z_j)$ and $N = N(S_{j+1}; S_j, E^*) \cap (E - E')$. We also have $L(Z_0, Z_1, \ldots, Z_{\theta-1}; \lambda_0) = 1$ and

$$L'(Z_0, Z_1, \ldots, Z_{\theta-1}; \lambda_0) = \frac{v'(Z_0, \lambda_0)}{v(Z_0, \lambda_0)} + \sum_{j=0}^{\theta-2} \frac{q'(Z_j, Z_{j+1}, \lambda_0)}{q(Z_j, Z_{j+1}, \lambda_0)}$$

where, as before, a prime denotes the derivative of the corresponding function with respect to λ. In the above expression, we have

$$\frac{v'(Z_0, \lambda_0)}{v(Z_0, \lambda_0)} = \frac{\nu_0'(S_0, \lambda_0)}{\nu_0(S_0, \lambda_0)} + \sum_{e_i \in E(S_0) \cap (E - E')} \frac{f'(C_{0,i}; e_i, \lambda_0)}{f(C_{0,i}; e_i, \lambda_0)} \quad (3.40)$$

and

$$\frac{q'(Z_j, Z_{j+1}, \lambda_0)}{q(Z_j, Z_{j+1}, \lambda_0)} = \frac{p'(S_{j+1}; S_j, E^*, \lambda_0)}{p(S_{j+1}; S_j, E^*, \lambda_0)} + \sum_{e_i \in N} \frac{f'(C_{j+1,i}; e_i, \lambda_0)}{f(C_{j+1,i}; e_i, \lambda_0)} \quad (3.41)$$

for $0 \le j \le \theta - 2$.

Proceeding formally as in the simpler example, we find that $\alpha'(\tilde{h}_1; \lambda_0) = E_{\lambda_0}[Y^{(1)}]$, where

$$Y^{(1)} = \sum_{j=0}^{\theta-1} \tilde{h}_1'(Z_j, \lambda_0)$$

$$+ \left(\sum_{j=0}^{\theta-1} \tilde{h}_1(Z_j, \lambda_0) \right) \left(\frac{v'(Z_0, \lambda_0)}{v(Z_0, \lambda_0)} + \sum_{j=0}^{\theta-2} \frac{q'(Z_j, Z_{j+1}, \lambda_0)}{q(Z_j, Z_{j+1}, \lambda_0)} \right).$$

Thus an unbiased and strongly consistent estimator of $\alpha'(\tilde{h}_1; \lambda_0)$ can be obtained as an average of n i.i.d. copies of $Y^{(1)}$. We focus, however, on a variant of this estimator that often has somewhat better empirical behavior. Observe that, for $0 \le j \le \theta - 2$ and $\lambda \in \Lambda$,

$$E_{\lambda_0} \left[\frac{q(Z_j, Z_{j+1}, \lambda)}{q(Z_j, Z_{j+1}, \lambda_0)} \,\middle|\, Z_j, Z_{j-1}, \ldots, Z_0 \right] = 1 \text{ a.s.},$$

so that

$$E_{\lambda_0}\left[\frac{q'(Z_j, Z_{j+1}, \lambda_0)}{q(Z_j, Z_{j+1}, \lambda_0)} \,\middle|\, Z_j, Z_{j-1}, \ldots, Z_0\right] = 0 \text{ a.s..}$$

A straightforward conditioning argument then establishes the alternative representation $\alpha'(\tilde{h}_1; \lambda_0) = E_{\lambda_0}\left[W^{(1)}\right]$, where

$$W^{(1)} = \sum_{j=0}^{\theta-1} \tilde{h}_1'(Z_j, \lambda_0) + \sum_{j=0}^{\theta-1} \tilde{h}_1(Z_j, \lambda_0)\left(\frac{v'(Z_0, \lambda_0)}{v(Z_0, \lambda_0)} + \sum_{l=0}^{j-1} \frac{q'(Z_l, Z_{l+1}, \lambda_0)}{q(Z_l, Z_{l+1}, \lambda_0)}\right).$$
(3.42)

Observe that $Y^{(1)}$ contains many terms that are equal to 0 in expectation, whereas $W^{(1)}$ replaces each of those terms by its expected value, namely 0.

A formal derivation analogous to the one given above leads to the representation $\alpha'(\tilde{h}_2; \lambda_0) = E_{\lambda_0}\left[W^{(2)}\right]$, where

$$W^{(2)} = \sum_{j=0}^{\theta-1} \tilde{h}_2'(Z_j, \lambda_0) + \sum_{j=0}^{\theta-1} \tilde{h}_2(Z_j, \lambda_0)\left(\frac{v'(Z_0, \lambda_0)}{v(Z_0, \lambda_0)} + \sum_{l=0}^{j-1} \frac{q'(Z_l, Z_{l+1}, \lambda_0)}{q(Z_l, Z_{l+1}, \lambda_0)}\right).$$
(3.43)

The following result summarizes the foregoing discussion.

Proposition 3.44. *Let $\lambda_0 \in \Re$ and let g_1 and g_2 be real-valued functions defined on $G \times \Re$ and differentiable at the point λ_0. Define $r(\lambda) = r(\lambda; g_1, g_2)$ as in (3.37) and suppose that*

(i) *each clock-setting distribution function is absolutely continuous and Assumption PD holds for each λ in a neighborhood Λ of λ_0,*

(ii) *for each clock-setting density function $f(\cdot; e, \lambda)$, the support set $\{x : f(x; e, \lambda) > 0\}$ does not depend on λ,*

(iii) *the support set $\{s : v_0(s, \lambda) > 0\}$ does not depend on λ,*

(iv) *for each $s \in G$ and $E^* \subseteq E(s)$, the support set $\{s' : p(s'; s, E^*, \lambda) > 0\}$ does not depend on λ, and*

(v) *for $\lambda \in \Lambda$ there exists a sequence of regeneration points $\{\theta(n): n \geq 0\}$ (defined independently of λ) for the underlying chain $\{Z_n : n \geq 0\}$.*

Then

$$r'(\lambda_0) = \frac{E_{\lambda_0}[W^{(1)}]\alpha(\tilde{h}_2; \lambda_0) - \alpha(\tilde{h}_1; \lambda_0)E_{\lambda_0}[W^{(2)}]}{\alpha^2(\tilde{h}_2; \lambda_0)},$$

where $\tilde{h}_1$ and $\tilde{h}_2$ are as in (3.38), $\alpha(\tilde{h}; \lambda)$ is defined as in (3.39), and $W^{(1)}$ and $W^{(2)}$ are given by (3.42) and (3.43).

Remark 3.45. It can be shown that the interchange of derivative and expectation in the formal derivation above is valid if

$$E_\mu[e^{r\tilde{\tau}_1}] < \infty$$
(3.46)

for sufficiently small $r > 0$. The condition in (3.46) is implied by Assumption PD. Indeed, if all transitions are timed, then (3.46) follows from Theorem 1.22 in Chapter 5 and Proposition 1.32. If at least one transition is immediate, then the desired result can be established using an argument similar to the proof of Theorem 2.24(iii).

Remark 3.47. The above result actually applies almost unchanged to estimation of the derivative for any long-run performance measure that can be expressed in the form $r(\lambda) = \alpha(\tilde{h}_1; \lambda)/\alpha(\tilde{h}_2; \lambda)$, where $\tilde{h}_1(z, \lambda)$ and $\tilde{h}_2(z, \lambda)$ are each polynomially dominated and differentiable at λ_0. Examples include performance measures of the form

$$r(\lambda) = \lim_{n \to \infty} \frac{\sum_{j=0}^{n} \tilde{h}_1(Z_n, \lambda)}{\sum_{j=0}^{n} \tilde{h}_2(Z_n, \lambda)}$$

and

$$r(\lambda) = \lim_{t \to \infty} \frac{R(t, \lambda)}{t},$$

where $R(t, \lambda)$ is a parameterized version of a long-run average reward as in Section 3.2.3. Indeed, Proposition 3.44 can easily be extended to performance measures of the form $r(\lambda) = v(\alpha_1, \alpha_2, \ldots, \alpha_k, \lambda)$, where v is differentiable, $\alpha_i = \alpha(\tilde{h}_i; \lambda)$ for $1 \leq i \leq k$, and each function $\tilde{h}_i$ is polynomially dominated and differentiable at λ_0.

Set $Q_{1,j} = \sum_{n=\theta(j-1)}^{\theta(j)-1} \tilde{h}_1(Z_n, \lambda)$ and $Q_{2,j} = \sum_{n=\theta(j-1)}^{\theta(j)-1} \tilde{h}_2(Z_n, \lambda)$, and let $Q_{3,j}$ and $Q_{4,j}$ be the respective values of $W^{(1)}$ and $W^{(2)}$ in the jth regenerative cycle. Then a strongly consistent estimate of $r'(\lambda_0)$ is given by

$$\hat{r}'(n) = \frac{\bar{Q}_3(n)\bar{Q}_2(n) - \bar{Q}_1(n)\bar{Q}_4(n)}{\bar{Q}_2^2(n)},$$

where $\bar{Q}_i(n) = (1/n)\sum_{j=1}^{n} Q_{i,j}$ for $1 \leq i \leq 4$. Moreover, because $r'(\lambda_0)$ is of the form (3.32), the techniques of Section 6.3.5 can be used to construct an asymptotic confidence interval for $r'(\lambda_0)$. Specifically, for $q = (q_1, q_2, q_3, q_4) \in \Re^4$, set

$$u_1(q) = -q_4/q_2^2,$$
$$u_2(q) = (2q_1q_4 - q_2q_3)/q_2^3,$$
$$u_3(q) = 1/q_2^2,$$
$$u_4(q) = -q_1/q_2^2,$$

and let $u = (u_1, u_2, u_3, u_4)$. The $\Re^4$-valued function $u(q)$ is the gradient of $f(q) = (q_3q_2 - q_1q_4)/q_2^2$. Write $\bar{Q}(n) = (\bar{Q}_1(n), \bar{Q}_2(n), \bar{Q}_3(n), \bar{Q}_4(n))$—where each $\bar{Q}_i(n)$ is defined as above—and set $U(n) = u(\bar{Q}(n))$. Also let

$V(n)$ be the 4×4 matrix whose (i,j)th entry is

$$V_{i,j}(n) = \frac{1}{n-1} \sum_{k=1}^{n} (Q_{i,k} - \bar{Q}_i(n))(Q_{j,k} - \bar{Q}_j(n)),$$

and set

$$w(n) = U(n)^t V(n) U(n). \tag{3.48}$$

Then

$$\frac{\sqrt{n}(\hat{r}'(n) - r'(\lambda_0))}{\sqrt{w(n)}} \Rightarrow N(0,1).$$

The foregoing results lead to the following algorithm.

Algorithm 3.49 (Regenerative method for gradient estimation)

1. Select a sequence $0 = \theta(0), \theta(1), \theta(2), \ldots$ of regeneration points for the underlying chain $\{ Z_n : n \geq 0 \}$.

2. Simulate a cycle of the underlying chain using parameter value λ_0 and observe $Z_0, Z_1, \ldots, Z_{\theta-1}$.

3. Compute the quantities

$$Q_{1,1} = \sum_{j=0}^{\theta-1} \tilde{h}_1(Z_j, \lambda_0),$$

$$Q_{2,1} = \sum_{j=0}^{\theta-1} \tilde{h}_2(Z_j, \lambda_0),$$

$$Q_{3,1} = \sum_{j=0}^{\theta-1} \tilde{h}_1'(Z_j, \lambda_0)$$
$$+ \sum_{j=0}^{\theta-1} \tilde{h}_1(Z_j, \lambda_0) \left(\frac{v'(Z_0, \lambda_0)}{v(Z_0, \lambda_0)} + \sum_{l=0}^{j-1} \frac{q'(Z_l, Z_{l+1}, \lambda_0)}{q(Z_l, Z_{l+1}, \lambda_0)} \right),$$

$$Q_{4,1} = \sum_{j=0}^{\theta-1} \tilde{h}_2'(Z_j, \lambda_0)$$
$$+ \sum_{j=0}^{\theta-1} \tilde{h}_2(Z_j, \lambda_0) \left(\frac{v'(Z_0, \lambda_0)}{v(Z_0, \lambda_0)} + \sum_{l=0}^{j-1} \frac{q'(Z_l, Z_{l+1}, \lambda_0)}{q(Z_l, Z_{l+1}, \lambda_0)} \right),$$

where $v'(Z_0, \lambda_0)/v(Z_0, \lambda_0)$ is computed according to (3.40) and each $q'(Z_l, Z_{l+1}, \lambda_0)/q(Z_l, Z_{l+1}, \lambda_0)$ is computed according to (3.41).

4. Repeat steps 2 and 3 a total of n times to obtain $\{ Q_{i,j} : 1 \leq i \leq 4, \ 1 \leq j \leq n \}$.

5. Form the strongly consistent point estimate

$$\hat{r}'(n) = \frac{\bar{Q}_3(n)\bar{Q}_2(n) - \bar{Q}_1(n)\bar{Q}_4(n)}{\bar{Q}_2^2(n)}$$

for $r'(\lambda_0)$, where $\bar{Q}_i(n) = (1/n)\sum_{j=1}^{n} Q_{i,j}$ for $1 \leq i \leq 4$.

6. Form the asymptotic $100p\%$ confidence interval

$$\left[\hat{r}'(n) - \frac{z_p\sqrt{w(n)}}{\sqrt{n}}, \hat{r}'(n) + \frac{z_p\sqrt{w(n)}}{\sqrt{n}}\right]$$

for $r'(\lambda_0)$, where $w(n)$ is defined as in (3.48) and z_p is the $(1+p)/2$ quantile of the standard normal distribution.

EXAMPLE 3.50 (Cyclic queues with feedback). We illustrate the likelihood-ratio method for gradient estimation using the closed network of queues in Example 1.4 of Chapter 2. Suppose that successive service times at center i ($i = 1, 2$) are i.i.d. according to an exponential distribution with intensity q_i and that the intensities depend on a parameter $0 < \lambda < 2$ via the relations $q_1 = q_1(\lambda) = \lambda$ and $q_2 = q_2(\lambda) = \lambda^{3/2}$. Also suppose that the routing probability p (with which a job completing service at center 1 moves to center 2) depends on λ via the relation $p = p(\lambda) = 0.5 + \lambda/4$. Finally, suppose that there are $N = 4$ jobs and that the system is modelled by the SPN in Figure 2.2.

Consider the *relative utilization* $r(\lambda)$ of the two servers, which is defined as the long-run ratio of the amount of time that the server at center 1 is busy to the amount of time that the server at center 2 is busy. Formal definition of $r(\lambda)$ is via (3.37), with

$$g_i(s) = \begin{cases} 1 & \text{if } s_i > 0; \\ 0 & \text{otherwise} \end{cases}$$

for $s = (s_1, s_2) \in G$ and $i = 1, 2$. Observe that the marking $\bar{s} = (4, 0)$ is a single state and the successive marking changes at which the marking is $\bar{s}$ and transition $e_1 = $ "service completion at center 1" fires form a sequence of regeneration points for the underlying chain. It is straightforward to verify that the remaining conditions of Proposition 3.44 hold, so that we can use Algorithm 3.49 to estimate $r'(\lambda_0)$, where λ_0 is any fixed parameter value. For $z = (s, c)$ and $i = 1, 2$, we have $\tilde{h}_i(z, \lambda) \equiv \tilde{h}_i(z) = g_i(s)t^*(s, c)$. It follows that $\tilde{h}_1' \equiv 0$ and $\tilde{h}_2' \equiv 0$, so that

$$Q_{3,k} = \sum_{j=\theta(k-1)}^{\theta(k)-1} \tilde{h}_1(Z_j, \lambda_0)R_{j,k}$$

and

$$Q_{4,k} = \sum_{j=\theta(k-1)}^{\theta(k)-1} \tilde{h}_2(Z_j, \lambda_0) R_{j,k}$$

for $k \geq 1$ in Algorithm 3.49, where

$$R_{j,k} = \frac{v'(Z_{\theta(k-1)}, \lambda_0)}{v(Z_{\theta(k-1)}, \lambda_0)} + \sum_{l=\theta(k-1)}^{\theta(k-1)+j-1} \frac{q'(Z_l, Z_{l+1}, \lambda_0)}{q(Z_l, Z_{l+1}, \lambda_0)}.$$

Because the clock-setting densities are given by

$$f(x; e_i, \lambda) = q_i(\lambda) \exp(-q_i(\lambda)x)$$

for $i = 1, 2$, it follows that

$$\frac{f'(x; e_1, \lambda)}{f(x; e_1, \lambda)} = \frac{1 - q_1(\lambda)x}{q_1(\lambda)} = \frac{1 - \lambda x}{\lambda} \stackrel{\text{def}}{=} \psi_1(x, \lambda)$$

and

$$\frac{f'(x; e_2, \lambda)}{f(x; e_2, \lambda)} = \frac{q_2'(\lambda)}{q_2(\lambda)} (1 - q_2(\lambda)x) = \frac{3}{2\lambda} (1 - \lambda^{3/2}x) \stackrel{\text{def}}{=} \psi_2(x, \lambda).$$

Similarly, since for $s = (s_1, s_2)$ and $s' = (s_1', s_2')$

$$p(s'; s, e^*, \lambda) = \begin{cases} p(\lambda) & \text{if } s_1' = s_1 - 1; \\ 1 - p(\lambda) & \text{if } s = s'; \\ 1 & \text{if } s_1' = s_1 + 1; \\ 0 & \text{otherwise,} \end{cases}$$

it follows that

$$\frac{p'(s'; s, e^*, \lambda)}{p(s'; s, e^*, \lambda)} = \begin{cases} \psi_3(\lambda) & \text{if } s_1' = s_1 - 1; \\ \psi_4(\lambda) & \text{if } s = s'; \\ 0 & \text{otherwise,} \end{cases}$$

where

$$\psi_3(\lambda) = \frac{p'(\lambda)}{p(\lambda)} = \frac{1}{\lambda + 2}$$

and

$$\psi_4(\lambda) = \frac{-p'(\lambda)}{1 - p(\lambda)} = \frac{-1}{2 - \lambda}.$$

Thus, for $z = (s, c) = (s_1, s_2, c_1, c_2)$ and $z' = (s', c') = (s_1', s_2', c_1', c_2')$,

$$\frac{v'(z, \lambda)}{v(z, \lambda)} = \begin{cases} \psi_1(c_1, \lambda) + \psi_2(c_2, \lambda) + \psi_3(\lambda) & \text{if } s = (3, 1); \\ \psi_1(c_1, \lambda) + \psi_4(\lambda) & \text{if } s = (4, 0) \end{cases}$$

and

$$\frac{q'(z,z',\lambda)}{q(z,z',\lambda)} = \begin{cases} \psi_3(\lambda) + 1_{\{s'_1>0\}}\psi_1(c'_1,\lambda) \\ \qquad + 1_{\{s_2=0\}}\psi_2(c'_2,\lambda) & \text{if } s'_1 = s_1 - 1; \\ \psi_4(\lambda) + \psi_1(c'_1,\lambda) & \text{if } s = s'; \\ 1_{\{s_1=0\}}\psi_1(c'_1,\lambda) + 1_{\{s'_2>0\}}\psi_2(c'_2,\lambda) & \text{if } s'_1 = s_1 + 1; \\ 0 & \text{otherwise.} \end{cases}$$

Although the foregoing formulas may seem algebraically somewhat complex, the corresponding simulation procedure is straightforward. The simulation of the kth regenerative cycle ($k \geq 1$) is initialized by setting $Q_{i,k} = 0$ for $i = 1,2,3,4$. Then we set the current marking s equal to $(3,1)$ with probability p or $(4,0)$ with probability $1 - p$; these two scenarios correspond to the movement of the job that completes service at center 1 at time $\zeta_{\theta(k-1)}$ either to center 2 or to the tail of the queue at center 1. In the former case we generate new clock readings c_1 and c_2 that correspond to newly scheduled service completions at centers 1 and 2; in the latter case we generate only a single new clock reading c_1. Then we set $R_{0,k} = \psi_1(c_1,\lambda_0) + \psi_2(c_2,\lambda_0) + \psi_3(\lambda_0)$ if $s = (3,1)$ and $R_{0,k} = \psi_1(c_1,\lambda_0) + \psi_4(\lambda_0)$ if $s = (4,0)$. In general, just after the jth marking change ($j \geq 0$) during the kth regenerative cycle, the holding time t^* in the current marking s is computed, and

- $Q_{1,k}$ is incremented by $g_1(s) \cdot t^*$.

- $Q_{2,k}$ is incremented by $g_2(s) \cdot t^*$.

- $Q_{3,k}$ is incremented by $R_{j,k} \cdot g_1(s) \cdot t^*$.

- $Q_{4,k}$ is incremented by $R_{j,k} \cdot g_2(s) \cdot t^*$.

The next marking s' is then computed, and new events corresponding to the marking change from s to s' are scheduled as necessary by generating new clock readings. If a new clock reading A_i is generated for transition e_i ($i = 1,2$), then the quantity $R_{j,k}$ is incremented by $\psi_i(A_i,\lambda_0)$. Moreover, if the marking change from s to s' corresponds to a completion of service at center 1, then $R_{j,k}$ is incremented either by $\psi_3(\lambda_0)$ or by $\psi_4(\lambda_0)$, the former if the job completing service moves to center 2 and the latter if the job joins the tail of the queue at center 1. At this point $R_{j,k}$ has been updated to $R_{j+1,k}$, and the simulation proceeds to the next marking change.

Table 6.4 displays simulation results for the system of cyclic queues when $\lambda_0 = 1.1$. As can be seen, the point estimator for the derivative of the long-run relative utilization is much more variable than the point estimator for the long-run relative utilization itself: the confidence-interval half-widths for $r'(\lambda)$ are over 4 times as long as those for $r(\lambda)$, and the normalized half-widths (i.e., the half-widths divided by the corresponding point estimates)

Table 6.4. Simulation Results for Cyclic Queues with Feedback: Point Estimates and 95% Confidence-Interval Half-Widths for the Long-Run Relative Utilization $r(\lambda_0)$ and Derivative $r'(\lambda_0)$, Where $\lambda_0 = 1.1$ (True Values are $r(\lambda_0) = 1.3533$ and $r'(\lambda_0) = 0.1786$)

estimand	number of cycles simulated ($\times 10^3$)				
	10	50	100	500	1000
$r(\lambda_0)$	1.3508	1.3510	1.3463	1.3513	1.3518
	±0.0259	±0.0115	±0.0081	±0.0037	±0.0026
$r'(\lambda_0)$	0.1241	0.1654	0.1920	0.1781	0.1783
	±0.1124	±0.0477	±0.0346	±0.0159	±0.0111

are almost 40 times as long. This phenomenon often arises in gradient estimation problems.

6.3.7   A Characterization of the Regenerative Method

To clarify the relationship between the regenerative method and the estimation methods introduced in subsequent chapters, we focus on yet another variant of the basic method. In this variant, the marking process is simulated until a fixed (simulated) time t and the point estimator of $r(f)$ is computed as quantity

$$\bar{r}(t) = \frac{1}{t} \int_0^t f\big(X(u)\big)\, du.$$

Suppose that $E_\mu[\tau_1] < \infty$, $Y_0(|f|) < \infty$ a.s., and $E_\mu[Y_1(|f|)] < \infty$. Then, by Theorem 1.12, the estimator $\bar{r}(t)$ is strongly consistent for $r(f)$. Confidence intervals can be based on the following result, which is of independent interest as a fundamental CLT for regenerative processes. In the theorem, $r(f)$ and $\sigma^2(f)$ are defined as in (3.1) and (3.2).

Theorem 3.51. *Suppose that $\{X(t): t \geq 0\}$ is a regenerative process with state space S and regeneration points $\{T_n : n \geq 0\}$, and let f be a real-valued function defined on S. Under the conditions of Theorem 3.4,*

$$\sqrt{t}\big(\bar{r}(t) - r(f)\big) \Rightarrow \sigma_0 N(0,1),$$

as $t \to \infty$, where $\sigma_0^2 = \sigma^2(f)/E_\mu[\tau_1]$.

PROOF. Assume without loss of generality that $T_0 = 0$. As in Section 6.3.3, denote by $n(t)$ the random number of cycles completed by time t, and write

$$\sqrt{t}\big(\bar{r}(t) - r(f)\big) = \frac{1}{\sqrt{t}} \int_0^{T_{n(t)}} \Big(f\big(X(u)\big) - r(f)\Big)\, du$$

$$+ \frac{1}{\sqrt{t}} \int_{T_{n(t)}}^t \Big(f\big(X(u)\big) - r(f)\Big)\, du. \tag{3.52}$$

As in the proof of Theorem 3.18, $n(t)/t \to 1/E_\mu [\tau_1]$ a.s. as $t \to \infty$, and the estimators $\hat{r}(n(t))$, $\bar{\tau}(n(t))$, and $s^2(n(t))$ are strongly consistent for $r(f)$, $E_\mu [\tau_1]$, and $\sigma^2(f)$. Using Theorem 3.18 and Slutsky's theorem, we have

$$
\frac{1}{\sqrt{t}} \int_0^{T_{n(t)}} \Big(f(X(u)) - r(f) \Big) \, du
$$

$$
= \left(\frac{n(t)}{t} \right)^{1/2} s(n(t)) \frac{\sqrt{n(t)} \Big(\hat{r}(n(t)) - r(f) \Big)}{s(n(t))/\bar{\tau}(n(t))}
$$

$$
\Rightarrow \sigma_0 N(0,1)
$$

as $t \to \infty$. It therefore suffices to show that the second term on the right in (3.52) converges to 0 with probability 1. To this end, observe that

$$
\left| \frac{1}{\sqrt{t}} \int_{T_{n(t)}}^t \Big(f(X(u)) - r(f) \Big) \, du \right|
$$

$$
\leq \frac{Y_{n(t)+1}(|f|)}{\sqrt{t}} + |r(f)| \frac{T_{n(t)+1}}{\sqrt{t}}
$$

$$
= \left(\frac{n(t)}{t} \right)^{1/2} \frac{Y_{n(t)+1}(|f|)}{\sqrt{n(t)}} + |r(f)| \left(\frac{n(t)}{t} \right)^{1/2} \frac{T_{n(t)+1}}{\sqrt{n(t)}}.
$$

Because $\lim_{t\to\infty} n(t) = \infty$ a.s., it suffices to show that

$$
\frac{Y_n(|f|)}{\sqrt{n}} \to 0 \text{ a.s.} \qquad \text{and} \qquad \frac{T_n}{\sqrt{n}} \to 0 \text{ a.s.}
$$

as $t \to \infty$. We establish the second convergence result—the proof of the first result is almost identical. Observe that

$$
\lim_{n\to\infty} \frac{1}{n} \sum_{k=1}^n \tau_k^2 = E_\mu \left[\tau_1^2 \right] \text{ a.s.}
$$

because, by the regenerative property, $\{ \tau_k \colon k \geq 0 \}$ is a sequence of i.i.d. variables and $E_\mu[\tau_1^2] < \infty$ by assumption. Thus $\tau_n^2/n \to 0$ a.s. by Theorem 2.9(i) in Chapter 3, and the desired result follows.   □

Using the above result together with the strong consistency of the estimators $s(n)$ and $\bar{\tau}_n$, it follows that

$$
\frac{\sqrt{t}(\bar{r}(t) - r(f))}{\hat{\sigma}_0(t)} \Rightarrow N(0,1)
$$

as $t \to \infty$, where

$$
\hat{\sigma}_0^2(t) = \frac{s^2(n(t))}{\bar{\tau}(n(t))}.
$$

In the usual way, we obtain the following asymptotic $100p\%$ confidence interval for $r(f)$ based on simulation until time t:

$$\left[\bar{r}(t) - \frac{z_p\,\hat{\sigma}_0(t)}{\sqrt{t}}, \ \hat{r}(n) + \frac{z_p\,\hat{\sigma}_0(t)}{\sqrt{t}}\right]. \tag{3.53}$$

As discussed in subsequent chapters, many output processes (not necessarily regenerative) have the property that the time average $\bar{r}(t)$—suitably normalized—converges in distribution to $\sigma N(0,1)$ for some $\sigma^2 \in (0,\infty)$. The crux of the regenerative method is the consistent estimation of the variance constant σ^2. Indeed, observe that the point estimator $\bar{r}(t)$ does not depend on the cycles—it is simply the time average of the output process over the simulated time interval—but the estimator $\hat{\sigma}_0^2(t)$ depends crucially on the cycle structure.

In general, estimation methods can be characterized as being of the "consistent-estimation" type, in which σ^2 is estimated explicitly, or of the "cancellation" type. The latter type of estimation method rests on limit theorems in which σ^2 has been cancelled out, so that explicit estimation is not required. The regenerative method is therefore a consistent-estimation method; some cancellation methods are introduced in the next chapter.

Remark 3.54. Suppose that the marking process $\{X(t)\colon t \geq 0\}$ obeys a CLT with variance constant σ^2, and let $\{\check{X}(t)\colon t \geq 0\}$ be a strictly stationary version—see Section A.2.2—of the marking process. Such a stationary version exists whenever the expected cycle length is finite. Provided that

$$\int_0^\infty \left|\mathrm{Cov}\left[f(\check{X}(0)), f(\check{X}(u))\right]\right| du < \infty, \tag{3.55}$$

we have the representation

$$\sigma^2 = 2\int_0^\infty \mathrm{Cov}\left[f(\check{X}(0)), f(\check{X}(u))\right] du,$$

and σ^2 can also be viewed as a limiting variance:

$$\lim_{t\to\infty} t\,\mathrm{Var}\left[\frac{1}{t}\int_0^t f(\check{X}(u))\,du\right] = \sigma^2.$$

It can be shown that (3.55) holds whenever the distribution function of τ_1 is spread out—see Definition 2.15 in the Appendix—and the quantities $E_\mu\left[\tau_1^2\right]$, $E_\mu\left[Y_1^2(|f|)\right]$, and $E_\mu\left[Y_1(f^2)\right]$ are finite. If, moreover, the bias of $\bar{r}(t)$ for the nonstationary version of the marking process is $O(t^{-1})$—as is typical—then σ^2 is also a limiting variance for the nonstationary version:

$$\lim_{t\to\infty} t\,\mathrm{Var}_\mu\left[\frac{1}{t}\int_0^t f(X(u))\,du\right] = \sigma^2.$$

Analogous results hold for a discrete-time regenerative process $\{X_n : n \geq 0\}$. In the discrete-time setting,

$$\sigma^2 = \mathrm{Var}\left[\check{X}_0\right] + 2\sum_{n=1}^{\infty} \mathrm{Cov}\left[f(\check{X}_0), f(\check{X}_n)\right]$$

and, rather than being spread out, the distribution function of τ_1 must be aperiodic in discrete time.

Remark 3.56. In practice, there may exist more than one sequence of regeneration points for the marking process, raising the question of which sequence is preferable. It can be shown that the variance constant σ_0^2 in Theorem 3.51 is insensitive to the choice of regeneration points. It follows that the length of the confidence interval for $r(f)$ based on a simulation of length t is asymptotically insensitive to the choice of regeneration points as $t \to \infty$. More precisely, let $L_1(t)$ and $L_2(t)$ be the respective (random) lengths of a $100p\%$ confidence interval for $r(f)$ based on a simulation of length t and two different sequences of regeneration points—here $p \in (0, 1)$ is arbitrary but fixed and the confidence interval is computed as in (3.53). Then $L_1(t)/L_2(t) \to 1$ a.s. as $t \to \infty$. Although the asymptotic length of the confidence interval does not depend on the choice of regeneration points, the variance of the estimator $s(n)$—and hence of the confidence-interval length—is extremely sensitive to the choice of regeneration points. Contrary to some folklore, the variance of $s(n)$ is not necessarily minimized by choosing the sequence in which regeneration points occur most frequently. Determination of the optimal choice of regeneration points is an open problem.

6.3.8 Extension to Dependent Cycles

In this subsection we focus on estimation methods for od-regenerative processes in discrete time. In particular, suppose that there exists a sequence of od-regeneration points $\{\theta(k) : k \geq 0\}$ for the process $\{Z_n : n \geq 0\}$ and we wish to estimate a time-average limit $\lim_{n\to\infty}(1/n)\sum_{j=0}^{n-1} f(Z_j)$ for some real-valued function f. We consider two possible approaches: an extension of the standard regenerative method and a "multiple-runs" method.

The former approach is based on a development similar to that of the standard regenerative method. Set $\tau_k = \theta(k) - \theta(k-1)$ and

$$Y_k(f) = \sum_{n=\theta(k-1)}^{\theta(k)-1} f(Z_n)$$

for $k \geq 1$; the sequence $\{(\tau_k, Y_k(f)) : k \geq 1\}$ consists of o.d.s. random pairs. Suppose that $E_\mu[\tau_1] < \infty$ and $E_\mu[Y_1(|f|)] < \infty$. Then, by

Theorem 1.27,

$$r(f) = \frac{E_\mu[Y_1(f)]}{E_\mu[\tau_1]}$$

is well defined and finite, and

$$\lim_{n\to\infty} \frac{1}{n} \sum_{j=0}^{n-1} f(Z_j) = r(f) \text{ a.s..}$$

If, in addition, $\{Z_n : n \geq 0\}$ is in fact an od-equilibrium process and τ_1 is aperiodic in discrete time, then $r(f)$ can also be interpreted as a steady-state mean.

To estimate $r(f)$, observe a fixed number n of cycles of $\{Z_n : n \geq 0\}$ and measure the quantities $Y_1(f), Y_2(f), \ldots, Y_n(f)$ and $\tau_1, \tau_2, \ldots, \tau_n$. Set

$$\hat{r}(n) = \frac{\bar{Y}(n)}{\bar{\tau}(n)},$$

where

$$\bar{Y}(n) = \frac{1}{n} \sum_{k=1}^{n} Y_k(f)$$

and

$$\bar{\tau}(n) = \frac{1}{n} \sum_{k=1}^{n} \tau_k.$$

It follows from the SLLN for o.i.d. random variables that $\hat{r}(n)$ is strongly consistent for $r(f)$.

To obtain an asymptotic confidence interval for $r(f)$, set

$$\sigma^2(f) = \text{Var}_\mu[Y_1(f) - r(f)\tau_1] + 2\text{Cov}_\mu[Y_1(f) - r(f)\tau_1, Y_2(f) - r(f)\tau_2]$$

and

$$s^2(n) = \frac{1}{n-1} \sum_{k=1}^{n} (Y_k(f) - \hat{r}(n)\tau_k)^2$$

$$+ \frac{2}{n-1} \sum_{k=1}^{n-1} (Y_k(f) - \hat{r}(n)\tau_k)(Y_{k+1}(f) - \hat{r}(n)\tau_{k+1}).$$

(3.57)

Provided that $E_\mu[\tau_1^2] < \infty$, $E_\mu[Y_1^2(|f|)] < \infty$, and $\sigma^2(f) > 0$, we have

$$\lim_{n\to\infty} s^2(n) = \sigma^2(f) \text{ a.s.}$$

and
$$\frac{\sqrt{n}\big(\hat{r}(n) - r(f)\big)}{s(n)/\bar{\tau}(n)} \Rightarrow N(0,1).$$

These results are derived as in the proof of Theorem 3.4, but we use the SLLN for o.i.d. random variables (Proposition 2.7 in the Appendix) and the CLT for o.d.s. random variables (Corollary 2.10 in the Appendix).

Algorithm 3.58 (Extended regenerative method)

1. Select a sequence $\{\theta(k): k \geq 0\}$ of od-regeneration points for the process $\{Z_n: n \geq 0\}$.

2. Simulate the process $\{Z_n: n \geq 0\}$ and observe a fixed number n of cycles defined by the random indices $\{\theta(k): k \geq 0\}$.

3. Compute the length τ_k of the kth cycle and the quantity $Y_k(f) = \sum_{n=\theta(k-1)}^{\theta(k)-1} f(Z_n)$ for $1 \leq k \leq n$.

4. Form the strongly consistent point estimate $\hat{r}(n) = \bar{Y}(n)/\bar{\tau}(n)$ for $r(f)$.

5. Form the asymptotic $100p\%$ confidence interval
$$\left[\hat{r}(n) - \frac{z_p\, s(n)}{\bar{\tau}(n)\,\sqrt{n}}, \hat{r}(n) + \frac{z_p\, s(n)}{\bar{\tau}(n)\,\sqrt{n}}\right]$$

for $r(f)$, where $s(n)$ is given by (3.57) and z_p is the $(1+p)/2$ quantile of the standard normal distribution.

The extended regenerative method is almost identical to the standard regenerative method, except that the variance constant reflects the dependence between adjacent cycles.

Under the conditions of this subsection, an alternative estimation procedure based on multiple runs can be used to obtain a strongly consistent point estimate and asymptotic confidence interval for $r(f)$. For convenience, assume that $\theta(0) = 0$. Simulate the process $\{Z_n: n \geq 0\}$ up to the random time $\theta(1)$ to create $\{Z_{n,1}: 0 \leq n < \theta_1(1)\}$, and set

$$Y_{1,1}(f) = \sum_{n=0}^{\theta_1(1)-1} f(Z_{n,1})$$

and $\tau_{1,1} = \theta_1(1)$. Repeat this step m times to create m independent replicates and produce $\{Z_{n,i}: 0 \leq n < \theta_i(1)\}$ for $1 \leq i \leq m$. Then compute point estimates and confidence intervals for $r(f)$ as in the standard regenerative method. Specifically, take $\bar{Y}_M(m) = (1/m)\sum_{i=1}^{m} Y_{1,i}(f)$ and

$\bar{\tau}_M(m) = (1/m) \sum_{i=1}^m \tau_{1,i}$, where $Y_{1,i}(f) = \sum_{n=0}^{\theta_i(1)} f(Z_{n,i})$ and $\tau_{1,i} = \theta_i(1)$ for $1 \le i \le m$. The estimator

$$\hat{r}_M(m) = \frac{\bar{Y}_M(m)}{\bar{\tau}_M(m)}$$

is strongly consistent for $r(f)$. An asymptotic $100p\%$ confidence interval for $r(f)$ is given by

$$\left[\hat{r}_M(n) - \frac{z_p\, s_M(n)}{\bar{\tau}_M(n)\,\sqrt{n}}, \hat{r}_M(n) + \frac{z_p\, s_M(n)}{\bar{\tau}_M(n)\,\sqrt{n}}\right],$$

where

$$s_M^2(m) = \frac{1}{m-1} \sum_{i=1}^m \left(Y_{1,i}(f) - \hat{r}_M(m)\tau_{1,i}\right)^2.$$

The estimator $s_M^2(m)$ is strongly consistent for $\sigma_M^2(f)$, where

$$\sigma_M^2(f) = \text{Var}_\mu\left[Y_{1,1}(f) - r(f)\tau_{1,1}\right] = \text{Var}_\mu\left[Y_1(f) - r(f)\tau_1\right].$$

One way of viewing the above procedure is as follows. Instead of generating a sample path of the original od-regenerative process $\{Z_n : n \ge 0\}$, we generate a sample path of the process $\{Y_n : n \ge 0\}$, where

$$(Y_0, Y_1, \ldots) = (Z_{0,1}, Z_{1,1}, \ldots, Z_{\theta_1(1)-1,1}, Z_{0,2}, Z_{1,2}, \ldots, Z_{\theta_2(1)-1,2}, \ldots).$$

A sample path of the process $\{Y_n : n \ge 0\}$ is obtained by independently simulating cycles of $\{Z_n : n \ge 0\}$ and then "gluing" the cycles together. The process $\{Y_n : n \ge 0\}$ is regenerative and has the same time-average limits as $\{Z_n : n \ge 0\}$. In the multiple-runs method, we simply apply the standard regenerative method to the process $\{Y_n : n \ge 0\}$.

Suppose that—as often occurs in practice—successive cycle quantities are positively correlated:

$$\text{Cov}_\mu\left[Y_1(f) - r(f)\tau_1, Y_2(f) - r(f)\tau_2\right] > 0.$$

Comparing the definitions of $\sigma^2(f)$ and $\sigma_M^2(f)$, we see that the multiple-runs method produces asymptotically narrower confidence intervals than the extended regenerative method. That is, the multiple-runs method has greater *asymptotic efficiency* than the extended regenerative method and is thus the estimation procedure of choice.

Remark 3.59. As with the standard regenerative method, the variance estimators $s^2(n)$ and $s_M^2(m)$ for the extended regenerative method and the multiple-runs method can be computed by means of a single pass through the data. For example, the variance constant $\sigma^2(f)$ in the extended regenerative method can be rewritten as

$$\sigma^2(f) = \text{Var}_\mu\left[Y_1(f)\right] - 2r(f)\text{Cov}_\mu\left[Y_1(f), \tau_1\right] + r^2(f)\text{Var}_\mu\left[\tau_1\right]$$
$$+ 2\text{Cov}_\mu\left[Y_1(f), Y_2(f)\right] - 2r(f)\text{Cov}_\mu\left[Y_2(f), \tau_1\right]$$
$$- 2r(f)\text{Cov}_\mu\left[Y_1(f), \tau_2\right] + 2r^2(f)\text{Cov}_\mu\left[\tau_1, \tau_2\right],$$

and each term on the right side can be estimated using one-pass formulas. If $\{ Z_n : n \geq 0 \}$ is an od-equilibrium process, then $\mathrm{Cov}_\mu [\tau_1, \tau_2] = 0$ in the above expansion.

Remark 3.60. The foregoing development for od-regenerative processes in discrete time carries over in an obvious way to od-regenerative processes in continuous time—replace cycle sums by cycle integrals, and so forth.

Notes

Regenerative processes were originally defined by Smith (1955, 1958). Some important early papers include Brown and Ross (1972) and Miller (1972, 1974a). The books of Çinlar (1975) and Ross (1983) contain readable introductions to renewal theory and include proofs of Proposition 1.20(i) and (iii). Other classic treatments of renewal theory and regenerative processes include Karlin and Taylor (1975) and Asmussen (1987a). Some of the terminology associated with regenerative processes varies among authors. For example, some authors would say that the distribution of the continuous-time cycle length τ_1 is either "arithmetic with span d" or "not arithmetic" rather than saying that τ_1 is either "periodic with period d" or "aperiodic." Similarly, these authors would say that the distribution of the discrete-time cycle length τ_1 is either "arithmetic with span $d > 1$" or "arithmetic with span 1" rather than saying that τ_1 is either "periodic in discrete time with period d" or "aperiodic in discrete time." Asmussen (1987a) uses the terminology "lattice" and "nonlattice" rather than "arithmetic" and "not arithmetic," although the term "lattice" typically is used to describe the distribution of a random variable that takes values in a set of the form $\{ a, a \pm d, a \pm 2d, \dots \}$ with $a \neq 0$ in general.

Inspection of the proof of Theorem 1.12 shows that the moment condition $E[Y_1(|f|)] < \infty$ can be replaced by the weaker condition $E[U_1(f)] < \infty$, where

$$U_k(f) = \sup_{T_{k-1} \leq t \leq T_k} \left| \int_{T_{k-1}}^t f(X(u)) \, du \right|$$

for $k \geq 1$. When $E[\tau_1] < \infty$ and $Y_0(|f|) < \infty$ a.s., Asmussen (1987a) shows that the condition $E[U_1(f)] < \infty$ is in fact necessary and sufficient for the conclusion of the theorem to hold.

Our definition of an od-regenerative process follows Sigman (1990b). Smith (1955) originally defined an equilibrium process as one in which the cycle lengths are i.i.d. and the cycles themselves are stationary—Thorisson (2000) and others call such a process a "wide-sense regenerative" process. Definition 1.28 imposes the additional requirement that the cycles be one-dependent. Processes with dependent cycles are discussed at length in Thorisson (2000), as well as in Fox and Glynn (1987), Glynn (1994), Glynn and

Iglehart (1989), Glynn and Sigman (1992), Kalashnikov (1994), and Sigman and Wolff (1993). In the context of queueing networks, the notion of od-equilibrium structure is closely related to the idea of "renovating" events. For discussions of renovating events and related topics, see, for example, Asmussen and Foss (1993), Borovkov (1984), Borovkov and Foss (1992), Foss and Kalashnikov (1991), Morozov (1994a, 1994b, 1998), and Morozov and Sigovtsev (2000).

Discussions of the decomposition in (1.35) can be found in Asmussen (1987a, Section VI.3), Glynn (1982b), Glynn and L'Ecuyer (1995), Henderson and Glynn (1999a, 2001), and Meyn and Tweedie (1993a)—the treatment in Glynn and L'Ecuyer (1995) is especially detailed. The idea of using (1.35) to establish wide-sense regenerative structure originated in the work of Athreya and Ney (1978) and Nummelin (1978). As discussed in Glynn (1982b), Glynn and L'Ecuyer (1995), and Henderson and Glynn (2001), a sequence of od-equilibrium points for a Harris chain can be obtained *after* generating the sample path of the chain. The idea is to identify the first time M at which the chain $\{ Z_n : n \geq 0 \}$ hits the distinguished set C and then generate a Bernoulli random variable I with success probability $b \, l(Z_{M+r})/p^r(Z_M, Z_{M+r})$, where $l(\cdot)$ and $p^r(z, \cdot)$ are appropriately defined "densities" of $\lambda(\cdot)$ and $P^r(z, \cdot)$. If $I = 1$, then a regeneration point is declared to occur at time $M + r$. Starting at time $M + r$, this process is then repeated to identify successive regeneration points.

The second assertion of Proposition 1.32 follows from Theorem VI.3.2 in Asmussen (1987a) and—as discussed in Haas (1999b)—the final assertion follows from results in Roberts and Tweedie (1999); see also Glynn and L'Ecuyer (1995, Proposition 4). Nummelin (1984, Theorem 4.3) has shown that (1.35) must hold with $r = 1$ for there to exist a sequence of regeneration points for a Harris recurrent Markov chain; see Henderson and Glynn (2001). In practice, the od-equilibrium points for the underlying chain of an SPN usually cannot be identified explicitly and are used primarily as a theoretical device for establishing stability properties. Recent results due to Henderson and Glynn (1999a) indicate, however, that explicit identification may be possible at least for certain simple models.

Numerous authors have established the regenerative property for individual models with special structure; perhaps the most general results of this type are those given for closed networks of queues by Borovkov (1986), Kaspi and Mandelbaum (1992), and Haas and Shedler (1987a); see also Morozov (1994a, 1994b, 1998). Glynn (1989b) gives sufficient conditions for regeneration in GSMPs when each clock-setting distribution has a hazard rate bounded above and below by finite positive constants. Sufficient conditions for regeneration in GSMPs based on geometric trials recurrence criteria are given by Haas and Shedler (1985a, 1987b, 1992) and Iglehart and Shedler (1983). Refinements and extensions of these results in the SPN setting are given by Haas and Shedler (1986, 1989b, 1993b). The bounding technique used to prove Lemma 2.45 is taken from the proof of Proposition 3.2.2 in Ross (1983).

Using results in Glynn and Haas (2002b), the assertion in Remark 2.25 about cycle-length moments can be sharpened. Specifically, consider an irreducible SPN with finite marking set and positive speeds. Suppose that there exists $0 < \bar{x} < \infty$ such that each clock-setting distribution function $F(\,\cdot\,; s', e', s, e^*)$ and $F_0(\,\cdot\,; e', s)$ with $e' \in E - E'$ has a density component that is positive and continuous on $(0, \bar{x})$. Also suppose that there exists a sequence $\{\,\theta(k) \colon k \geq 0\,\}$ of regeneration points for the underlying chain and hence a sequence $\{\,\zeta_{\theta(k)} \colon k \geq 0\,\}$ of regeneration points for the marking process. If each clock-setting distribution has finite qth moment $(q \geq 1)$, then so do the cycle lengths $\tau_1 = \zeta_{\theta(1)} - \zeta_{\theta(0)}$ and $\tilde{\tau}_1 = \theta(1) - \theta(0)$ for the marking process and underlying chain. This assertion remains true when $\{\,\theta(k) \colon k \geq 0\,\}$ is a sequence of od-equilibrium points or od-regeneration points. The foregoing results lead to conditions under which cycle quantities $Y_1(|f|)$ and $\tilde{Y}_1(|\tilde{f}|)$ as in Theorem 2.24 have finite moments of various orders—see, for example, the notes at the end of Chapter 7.

Crane and Iglehart (1975) were the first to focus on the application of regenerative-process theory to analysis of simulation output. Good expository treatments of the regenerative method can be found in Crane and Lemoine (1977) and Shedler (1993). Aspects of the basic theory are also discussed in Glynn and Iglehart (1987, 1993). Chan et al. (1983) discuss the one-pass formulas mentioned in Remark 3.8 for computation of $s_{11}(n)$ and $s_{22}(n)$. These authors also give other numerically stable one-pass and two-pass procedures for computing variance-type quantities. The extended regenerative method and multiple-runs method given in Section 6.3.8 are discussed and compared by Glynn (1994) and also by Haas and Shedler (1996).

The regenerative method can be used to obtain asymptotic confidence intervals under somewhat weaker assumptions than those given in Theorem 3.4. For example, asymptotic confidence intervals for $r(f)$ can be obtained even when $E_\mu[Y_1^2(f)] = E_\mu[\tau_1^2] = \infty$, provided that

1. $E_\mu[Y_1(f) - r(f)\tau_1] = 0$, and

2. $0 < E_\mu[(Y_1(f) - r(f)\tau_1)^2] < \infty$;

see Glynn and Iglehart (1993). The conditions in Theorem 3.51 can similarly be weakened.

As discussed in Remark 3.56, some quantities in regenerative simulation are sensitive to the particular choice of regeneration points and other quantities are not. See Calvin (1994) for a detailed discussion of this issue.

The results referred to in Remark 3.54 follow from a continuous-time version of Lemma 3 on p. 172 of Billingsley (1968), Theorem 5.5 in Glynn (1989a), and Corollary (2.3) in Glynn and Iglehart (1986a).

The sequential estimation procedures discussed in Section 6.3.4 are due to Lavenberg and Sauer (1977). Glynn and Whitt (1992b) give general conditions on a simulation under which such procedures are valid. An early discussion of pilot runs can be found in Cox (1952), where the procedure is called "double sampling" and the observations collected during the pilot run are incorporated into the final estimates.

Iglehart (1975) originally considered using the Tin and jackknife estimators, among others, to reduce the bias of the standard regenerative point estimator. Extensions of the jackknife estimator to general nonlinear functions of cycle means as in Section 6.3.5 can be based, for example, on results in Glynn and Heidelberger (1992). General discussions of the jackknife method for bias reduction can be found in Miller (1974b) and the book of Efron and Tibshirani (1993). The low-bias estimators $\hat{r}(n(t) + 1)$ and $\hat{r}_4(t)$ in Section 6.3.3 are due to Meketon and Heidelberger (1982) and Glynn (1987, 1994), respectively. Henderson and Glynn (2001) extend the results in Meketon and Heidelberger (1982) to the setting of one-dependent cycles. Heidelberger and Lewis (1981) describe a "regression-adjusted" regenerative point estimator that attempts to correct for bias; the authors' "sectioning" approach permits extended formal and graphical analysis of bias, skewness, and departures from normality in the point estimator. Glynn and Heidelberger (1990, 1992) consider bias issues for simulation problems in which the allowable computation cost is bounded. When comparing the asymptotic efficiency of confidence-interval procedures in Section 6.3.3, we have defined relative efficiency in terms of the relative lengths of the confidence intervals. A more comprehensive definition of efficiency would explicitly consider computation costs; see Glynn and Whitt (1992a).

In our empirical comparison of bias-reduction techniques—see Examples 3.17, 3.20, and 3.22—we cite exact values for the utilization of the token ring with fixed-sized packets. These values can be found in Shedler (1993); the results are given in the context of a "patrolling repairman" model that is essentially the same as the token ring model.

Various authors have considered the problem of estimating steady-state quantities other than means, such as quantiles (Iglehart, 1976; Seila, 1982) and central moments (Glynn and Iglehart, 1986b). Fox and Glynn (1989) and Glynn and Heidelberger (1992) discuss the problem of estimating discounted costs in the regenerative setting, and Iglehart and Stone (1983) discuss techniques for exploiting regenerative structure when estimating extreme values. Variance-reduction techniques for estimating the mean time to failure and related performance measures for highly reliable systems are described, for example, in Goyal et al. (1992) and Heidelberger et al. (1994).

Our discussion of gradient estimation follows Glynn (1989c). The proof of Proposition 3.44 follows from results in Glasserman and Glynn (1992) and Glynn and L'Ecuyer (1995). The book of Glasserman (1991) gives a good introduction to IPA methods for gradient estimation. Expository treatments of the likelihood-ratio method—also known as the "score-function" method—can be found in the books of Rubinstein and Melamed (1998) and Rubinstein and Shapiro (1993). Nakayama and Shahabuddin (1998) study conditions on the building blocks of a GSMP under which the likelihood-ratio method can be used to estimate gradients of finite-time performance measures. See Andradottir (1998) for an overview of how gradient estimates are used when optimizing a stochastic system via simulation.

The basic Robbins–Monro algorithm described in Section 6.3.6 can exhibit poor performance in practice, and there are ongoing efforts to improve both the rate of convergence and the stability of the algorithm. Moreover, the basic algorithm has been extended to handle feasibility constraints and multiple decision parameters. Ólafsson and Shi (1999) describe an alternative optimization approach that exploits regenerative structure. Closely related to these optimization procedures are methods for "ranking and selection" as surveyed in Goldsman and Nelson (1998). Heidelberger and Iglehart (1979) and Iglehart (1977), in particular, describe the application of the regenerative method to comparing the steady-state performance of two or more systems.

Heidelberger (1979) and Iglehart and Lewis (1979) discuss schemes for reducing the variance of the standard point estimator for the regenerative method, and Glynn (1982a) gives methods for improving the coverage of the standard confidence interval. Some other variance-reduction techniques related to regenerative simulation are given in Calvin and Nakayama (2000) and Henderson and Glynn (1999b). Asmussen (1987b) considers the behavior of the regenerative method when simulating queues in heavy traffic.

7
Alternative Simulation Methods

The previous chapter concerns SPNs in which regeneration points exist for the marking process or underlying chain or both. For such nets, regenerative methods often can be used to obtain strongly consistent point estimates and asymptotic confidence intervals for time-average limits of the form $r = \lim_{t \to \infty} \bar{r}(t)$, where $\bar{r}(t) = (1/t) \int_0^t f(X(u)) \, du$ for some function f. This chapter deals with methods for estimation of time-average limits when regenerative methods are not applicable. This situation can occur either because there is no apparent sequence of regeneration points or because regenerations occur so infrequently that the method is impractical—in Section 7.1 we give examples of both types of scenario.

The discussion centers on SPNs for which Assumption PD of Chapter 5 holds. For such nets, the limit r is well defined and finite, and the time average $\bar{r}(t)$ obeys a CLT; that is, $\bar{r}(t)$—suitably normalized—converges in distribution to a normal random variable with mean r and variance σ^2 for some $\sigma^2 \in (0, \infty)$. Indeed, we show (Theorem 2.17) that under Assumption PD a stronger convergence result holds: the output process $\{ f(X(t)) : t \geq 0 \}$ obeys a *functional central limit theorem* (FCLT). That is, the associated cumulative (i.e., time-integrated) process, centered about the deterministic function $g(t) = rt$ and suitably compressed in space and time, converges in distribution to a Brownian motion as the degree of compression increases. The ordinary CLT for $\bar{r}(t)$ can be viewed as a consequence of this FCLT—see Section A.2.5 for a general discussion of FCLTs. To establish the FCLT, we show that, under Assumption PD, the underlying chain has od-regenerative structure—see Lemma 2.5 below—and the desired result then follows from an FCLT for od-regenerative processes.

In Section 7.2 we consider estimation methods that are based on "standardized time series" (STS). The idea is to use the foregoing FCLT to derive a limit theorem for $\bar{r}(t)$ similar to the ordinary CLT, but in which the variance constant σ^2 has been "cancelled out." This approach contrasts with the regenerative method, which consistently estimates σ^2 to obtain a confidence interval for r—see Section 6.3.7. The method of batch means (with the number of batches independent of the simulation run length) is probably the best known STS method. An extension of the basic batch-means method can be used to obtain point estimates and confidence intervals for nonlinear functions (such as ratios) of time-average limits. We emphasize that although the validity of STS methods hinges on the existence of od-regenerative structure, these methods do not require explicit identification of the od-regeneration points; it suffices merely to show that they exist.

Section 7.3 is concerned with methods that consistently estimate σ^2 but do not require an explicit sequence of regeneration points. In general, such methods yield confidence intervals with lengths that are asymptotically shorter and less variable than the lengths of confidence intervals produced by cancellation methods. Besides requiring that Assumption PD hold, we also require that the SPN of interest be "aperiodic," which implies (Corollary 3.5) that the underlying chain is Harris ergodic as defined in Section 5.1.1. We first consider the problem of estimating time-average limits of the underlying chain, and assume that a CLT holds in discrete time with variance constant $\tilde{\sigma}^2$. Our focus is on estimators of $\tilde{\sigma}^2$ that have a "localized quadratic-form" representation. Using results from the literature together with properties of Harris ergodic chains, it is often possible to show that a specified quadratic-form estimator is consistent for $\tilde{\sigma}^2$ when the initial distribution is the invariant distribution π, so that the underlying chain is stationary. If, however, a localized quadratic-form estimator is consistent for $\tilde{\sigma}^2$ when the initial distribution is π, then (Theorem 3.15) the estimator is consistent for $\tilde{\sigma}^2$ under any initial distribution—we establish this assertion using a coupling argument. Some specific quadratic-form estimators for which consistency can be established include those produced by the method of "variable" batch means (in which the number of batches is an increasing function of the simulation run length), as well as by certain "spectral" methods—see Theorems 3.20 and 3.26. The foregoing results can be extended to establish consistent estimation methods for nonlinear functions of time-average limits of the underlying chain, and this extension in turn leads to consistent estimation methods for time-average limits of the marking process.

7.1 Limitations of the Regenerative Method

In this section we give two examples for which the regenerative method is inapplicable. In the first example, there is no apparent sequence of regeneration points.

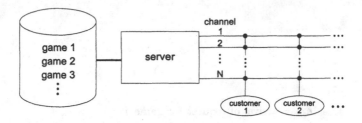

Figure 7.1. Interactive video-on-demand system.

EXAMPLE 1.1 (Interactive video on demand). Consider a system with a video-game server and $N > 1$ channels for the playing of multiperson interactive video games; see Figure 7.1. Games are played one at a time on each channel with two or more customers participating in each game. A customer participates in at most one game at a time. Once a game starts, none of the participants leaves the game and no additional customers join the game. When the game ends, all the participants immediately disconnect from the system. There are $M \geq 1$ different games stored at the server, and more than one instance of a game may be played simultaneously and independently (on different channels). Customers submit requests to play specified games. A customer request is either accepted or rejected by the system after a bounded random delay and according to the following mechanism. The server maintains M buffers; buffer j ($1 \leq j \leq M$) has finite capacity $B > 1$ and contains requests for game j. Associated with each channel i is a positive integer $L(i) \leq B$. Game j may start on channel i only if channel i is *available* for game j, that is, only if no game is currently underway on the channel and at least $L(i)$ requests are in buffer j. Whenever buffer j contains B requests and a new request for game j arrives, the arriving request is immediately rejected. Whenever buffer j is empty and a request for game j arrives, the request is placed in buffer j and a bounded random-length waiting period begins—up to $B - 1$ additional requests for game j may arrive and be placed in buffer j during this waiting period. If, at the end of the waiting period, at least one channel is available for game j, then all the requests in buffer j are accepted. Specifically, the corresponding customers are notified of their acceptance, a channel is selected randomly and uniformly from the set of available channels, and there is an immediate start of game j on the selected channel. If no channels are available at the end of the waiting period, then all the requests in buffer j are rejected. In either case, buffer j instantaneously becomes empty. The successive times between requests for game j are i.i.d. as a random variable A_j with continuous distribution function, the successive lengths of the waiting periods for game j are i.i.d. as a bounded random variable W_j, and the successive playing times of game j are i.i.d. as a random variable T_j.

$e_{j,1}$ = arrival of request for game j

$e_{j,2}$ = end of waiting period for game j

$e_{j,3}$ = rejection of request for game j

$e_{i,4}$ = end of game on channel i

$e_{i,5}$ = disconnection of customer from channel i

$e_{j,i,1}$ = start of game j on channel i

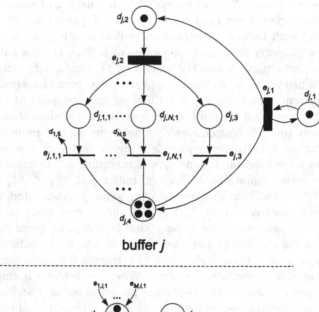

buffer j

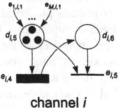

channel i

Figure 7.2. SPN representation of video-on-demand system.

This system can be specified as an SPN with finite marking set, unit speeds, $N+2M$ timed transitions, and $(N+1)M+N$ immediate transitions. The SPN consists of N subnets corresponding to the N channels and M subnets corresponding to the M buffers; Figure 7.2 displays subnets for a generic channel i and a generic buffer j. All the places except $\{ d_{j,4} : 1 \leq j \leq M \}$ and $\{ d_{i,5} : 1 \leq i \leq N \}$ contain either zero or one token. Place $d_{j,1}$ $(1 \leq j \leq M)$ always contains one token, reflecting the fact the arrival process of requests for game j is always active. Place $d_{j,2}$ contains a token if and only if a waiting period for game j is underway. Place $d_{j,3}$ contains a token if and only if a request for game j is being rejected due to lack of available channels. Place $d_{j,4}$ contains k (≥ 0) tokens if and only buffer j contains k requests. Place $d_{i,5}$ contains k tokens if and only if k customers are playing a game together on channel i. Place $d_{i,6}$ contains a token if and only if customers are disconnecting from channel i. Place $d_{j,i,1}$ $(1 \leq j \leq M$ and $1 \leq i \leq N)$ contains a token if and only if game j is about to start on channel i.

Whenever $d_{j,4}$ contains less than B tokens and transition $e_{j,1} =$ "arrival of request for game j" fires, a token is deposited in place $d_{j,4}$. If $d_{j,4}$ contains zero tokens just before the firing of $e_{j,1}$, then a token is also deposited in place $d_{j,2}$ when $e_{j,1}$ fires, and a waiting period for game j starts. Whenever place $d_{j,4}$ contains B tokens and transition $e_{j,1}$ fires, no tokens are removed from or deposited in any of the places in the net, and the arriving request for game j is rejected. For $1 \leq j \leq M$, denote by $\mathcal{I}_j(s)$ the set of channels available to game j when the marking is s: $\mathcal{I}_j(s) = \{ 1 \leq i \leq N : L(i) \leq s_{j,4}$ and $s_{i,5} = 0 \}$. Denote by $|\mathcal{I}_j(s)|$ the number of elements in $\mathcal{I}_j(s)$. When the marking is s and transition $e_{j,2} =$ "end of waiting period for game j" fires, a token is removed from place $d_{j,2}$. If $|\mathcal{I}_j(s)| > 0$, then a token is also deposited in exactly one of the places $\{ d_{j,i,1} : i \in \mathcal{I}_j(s) \}$; a token is deposited in place $d_{j,i,1}$ $(i \in \mathcal{I}_j(s))$ with probability $1/|\mathcal{I}_j(s)|$. If $|\mathcal{I}_j(s)| = 0$, then a token is deposited in place $d_{j,3}$. That is, game j starts on a randomly selected available channel; if no channel is available, then the video server starts rejecting the requests in buffer j. When the marking is s and transition $e_{j,i,1} =$ "start of game j on channel i" fires, a token is removed from place $d_{j,4}$ and a token is deposited in place $d_{i,5}$. If $s_{j,4} = 1$, then a token is also removed from place $d_{j,i,1}$. Thus when place $d_{j,4}$ contains k (> 0) tokens and a token is deposited in place $d_{j,i,1}$, transition $e_{j,i,1}$ fires k times in succession, removing all the tokens in place $d_{j,4}$—as well as the token in place $d_{j,i,1}$—and depositing k tokens in place $d_{i,5}$; this sequence of firings causes transition $e_{i,4} =$ "end of game on channel i" to become enabled, and k customers start to play game j on channel i. Transition $e_{j,3} =$ "rejection of request for game j" behaves in a similar manner: when place $d_{j,4}$ contains k tokens and a token is deposited in place $d_{j,3}$, transition $e_{j,3}$ fires k times in succession, removing the token in place $d_{j,3}$ and all the tokens in place $d_{j,4}$. Thus the k requests in buffer j are all rejected and there is no start of game j. When place $d_{i,5}$

Figure 7.3. An SPN with extremely long cycles.

contains k tokens and transition $e_{i,4}$ = "end of game on channel i" fires, a token is deposited in place $d_{i,6}$ and immediate transition $e_{i,5}$ = "disconnection of customer from channel i" becomes enabled. Transition $e_{i,5}$ then fires k times in succession, removing the token in place $d_{i,6}$ and all the tokens in place $d_{i,5}$. With probability 1 timed transitions never fire simultaneously; this property is a consequence of our assumption that for each game the distribution function of the time between successive requests is continuous.

All speeds are equal to 1. The clock-setting distribution functions for timed transitions are given by $F(x; s', e_{j,1}, s, E^*) = P\{A_j \le x\}$, $F(x; s', e_{j,2}, s, E^*) = P\{W_j \le x\}$, and $F(x; s', e_{i,4}, s, E^*) = P\{T_l \le x\}$ for $1 \le j \le M$, $1 \le i \le N$, and all s', s, and E^*, where $l = l(s)$ is the unique index such that $1 \le l \le M$ and $s_{l,i,1} = 1$.

Suppose that each A_j has a gamma distribution with noninteger shape parameter, each W_j has a uniform distribution on $[0, w_j]$ for some constant $w_j > 0$, and each T_j has a truncated normal distribution. Also suppose that more than one game is stored at the server so that $M > 1$. Then there is no apparent sequence of regeneration points for the marking process. Indeed, $\{e_{1,1}, e_{2,1}, \ldots, e_{M,1}\} \subseteq E(s)$ for all $s \in S$ and hence the marking process $\{X(t): t \ge 0\}$ does not have a single state.

The second example, though artificial, is characteristic of a class of problems for which the regenerative method is applicable in principle but not in practice.

EXAMPLE 1.2 (SPN with extremely long cycles). Consider an SPN consisting of a single simple transition—recall Definition 1.8 in Chapter 3—and N places, where N is an extremely large number; see Figure 7.3. The marking set is $G = \{s^{(1)}, s^{(2)}, \ldots, s^{(N)}\}$, where $s^{(i)}$ ($1 \le i \le N$) is the unique marking in which place d_i contains exactly two tokens and each other place contains exactly one token. The new-marking probabilities are of the form

$$p(s^{(j)}; s^{(i)}, e) = \frac{1}{N} + \epsilon_{ij},$$

where each real number ϵ_{ij} is small—so that, in particular, each $p(s^{(j)}; s^{(i)}, e)$ is positive—and $p(\cdot\,; s^{(i)}, e) \ne p(\cdot\,; s^{(j)}, e)$ for $i \ne j$. The idea is that, starting in marking $s^{(i)}$, the next marking is selected according to a distribution that is "almost" a uniform distribution over the markings in G. Suppose that we wish to estimate a time-average limit of the form

$\lim_{t\to\infty}(1/t)\int_0^t f(X(u))\,du$ with $f(s^{(i)}) = v + \delta_i$ for $1 \leq i \leq N$, where v is a fixed constant and $\delta_1, \delta_2, \ldots, \delta_N$ are small numbers that sum to 0. Also suppose that each new clock reading for transition e is uniformly distributed on $[0, 1]$. Fix an integer $i_0 \in \{1, 2, \ldots, N\}$ and observe that $s^{(i_0)}$ is a single state. It follows from Theorems 2.24 and 1.12 in Chapter 6 that the successive times $T_0, T_1, \ldots$ at which the marking is $s^{(i_0)}$ and transition e fires form a sequence of regeneration points for the marking process, and

$$\lim_{t\to\infty} \frac{1}{t} \int_0^t f(X(u))\,du = v_0 \text{ a.s.,}$$

where $v_0 \approx v$. It is intuitively clear that this simulation problem is well behaved—initialization effects die out quickly and the time average of the output process converges rapidly to v_0. The expected number of marking changes between successive regenerations, however, is $O(N)$, so that the regenerative method is difficult to apply when N is very large.

Remark 1.3. Actually, an efficient nonstandard version of the regenerative method can, at least in theory, be applied to the SPN in Example 1.2. The idea is to set

$$q(s^{(j)}) = \frac{1}{N} + \min_{1\leq i\leq N} \epsilon_{ij}$$

for $1 \leq j \leq N$ and write $p(s'; s, e) = bp_1(s'; e) + (1 - b)p_2(s'; s, e)$ for $s', s \in G$, where $b = \sum_{s\in G} q(s)$, $p_1(s'; e) = q(s')/b$, and $p_2(s'; s, e) = (p(s'; s, e) - q(s'))/(1 - b)$. Whenever the current marking is s and transition e fires, the next state is generated according to $p_1(\cdot; e)$ with probability b and according to $p_2(\cdot; s, e)$ with probability $1 - b$. The regeneration points for the marking process correspond to the successive times at which the new marking is generated according to p_1. This procedure is similar to the construction—described in Section 6.1.3—of od-equilibrium points for a Harris recurrent Markov chain. Unfortunately, for the real-world analogs of the foregoing SPN model, each e_{ij} typically is computed on the fly according to some complicated algorithm. Determination of a "minorizing" distribution p_1 is then highly nontrivial.

Some simulation problems are inherently badly behaved and are not amenable either to the regenerative method or to alternative estimation methods. For example, suppose that the marking process $\{X(t): t \geq 0\}$ corresponds to the number of jobs in a $GI/G/1$ queue, as in Example 1.1 of Chapter 1. Also suppose that the queue experiences "heavy traffic" in that the expected interarrival time $E[A]$ and expected service time $E[B]$ are such that $E[B]/E[A]$ is slightly less than 1. Time-average limits of functions of the marking process are well defined and finite, and the successive times that a job arrives to an empty queue form a sequence of regeneration points for the process $\{X(t): t \geq 0\}$. The expected time between successive regeneration points is extremely long, however, so that the standard regenerative method is not practical. On the other hand, because the process $\{X(t): t \geq 0\}$ takes an extremely long time to settle into a steady-state regime in heavy traffic, alternative simulation methods are also unlikely to perform well for this problem. In contrast, the problem in Example 1.2, while not amenable to regenerative simulation, can be handled by the methods provided in the following sections.

7.2   Standardized Time Series

In this section, we give SLLNs and FCLTs for the embedded chain and marking process of an SPN. These results form the basis of STS estimation methods, which yield strongly consistent point estimates and asymptotic confidence intervals for time-average limits. For one particular STS method, the method of batch means, we show how to obtain point estimates and confidence intervals for functions of time-average limits.

7.2.1   Limit Theorems

SLLNs and FCLTs for the stochastic processes associated with an SPN can be based on corresponding results for od-regenerative processes. We first discuss these latter results.

Limit Theorems for OD-Regenerative Processes

Let $\{Z_n: n \geq 0\}$ be an od-regenerative process with state space Γ and od-regeneration points $\{\theta(k): k \geq 0\}$. For an R^l-valued function f defined on Γ (where $l \geq 1$), set

$$Y_k(f) = \sum_{n=\theta(k-1)}^{\theta(k)-1} f(Z_n)$$

for $k \geq 1$ and

$$r(f) = \frac{E[Y_1(f)]}{E[\tau_1]},$$

where $\tau_1 = \theta(1) - \theta(0)$ as usual. These definitions are essentially the same as in Section 6.3.8, except that now both $Y_k(f)$ and $r(f)$ are random vectors of length l. Throughout, we write $x < \infty$ for $x = (x_1, x_2, \ldots, x_l)$ if each x_i is finite.

The proof of Theorem 1.27 in Chapter 6 applies essentially without change to establish the following l-dimensional extension.

Theorem 2.1. *Suppose that $E\left[\tau_1\right] < \infty$. Then $r(|f|) < \infty$ and*

$$\lim_{n\to\infty} \frac{1}{n} \sum_{j=0}^{n-1} f(Z_j) = r(f) \quad a.s.$$

for any $\Re^l$-valued function f such that $E\left[Y_1(|f|)\right] < \infty$.

We now state an FCLT for od-regenerative processes. For $l \geq 1$, denote by $C^l[0,1]$ the space of continuous $\Re^l$-valued functions on $[0,1]$ and by $\Rightarrow$ weak convergence on $C^l[0,1]$; see Section A.2.5. Weak convergence on $C^l[0,1]$ generalizes to a sequence of $\Re^l$-valued random functions—that is, a sequence of $\Re^l$-valued stochastic processes—the usual notion of convergence in distribution of a sequence of $\Re^l$-valued random variables.

Given an $\Re^l$-valued function defined on Γ, define a sequence of $C^l[0,1]$-valued random functions $U_1(f), U_2(f), \ldots$ by setting

$$U_n(f)(t) = \frac{1}{\sqrt{n}} \int_0^{nt} \left(f(Z_{\lfloor u\rfloor}) - r(f)\right) du \qquad (2.2)$$

for $0 \leq t \leq 1$ and $n \geq 0$; recall that $\lfloor x \rfloor$ is the greatest integer less than or equal to x. Observe that

$$U_n(f)(t) = V_{i,n} \stackrel{\text{def}}{=} \frac{1}{\sqrt{n}} \sum_{j=0}^{i-1} \left(f(Z_j) - r(f)\right)$$

for $t = i/n$ ($i = 0, 1, \ldots, n$). If $i/n < t < (i+1)/n$ for some i, then the value of $U_n(f)(t)$ is obtained by linearly interpolating between $V_{i,n}$ and $V_{i+1,n}$. Setting $S_n(f) = f(Z_0) + f(Z_1) + \cdots + f(Z_n)$ for $n \geq 0$, we can view each function $U_k(f)(\cdot)$ as a "standardized" version of the time series $\{S_n(f): n \geq 0\}$ in the same way that $n^{-1/2}\left(S_n(f) - r(f)\right)$ can be viewed as a standardized version of the sum $S_n(f)$.

Denote by $W^{(l)} = \{\left(W_1^{(l)}(t), W_2^{(l)}(t), \ldots, W_k^{(l)}(t)\right): 0 \leq t \leq 1\}$ a standard l-dimensional Brownian motion on $[0,1]$; see Section A.2.5. For $l \geq 1$, denote by $\|x\|$ the Euclidean norm of $x = (x_1, x_2, \ldots, x_l)$; that is, $\|x\| = (x_1^2 + x_2^2 + \cdots + x_l^2)^{1/2}$.

Proposition 2.3. *Suppose that $E_\mu\left[\tau_1^2\right] < \infty$ and let f be an $\Re^l$-valued function defined on Γ such that $E_\mu\left[\|Y_1(|f|)\|^2\right] < \infty$. Then there exists an $l \times l$ matrix $Q(f)$ such that $U_n(f) \Rightarrow Q(f)W^{(l)}$ as $n \to \infty$ for any initial distribution of the process, where $U_n(f)$ is defined by (2.2).*

Limit Theorems for SPNs

We now apply the foregoing results in the SPN setting. Let $\{X(t): t \geq 0\}$ be the marking process of an SPN with marking set G, timed marking set S, and transition set E. Also let Σ be the state space of the underlying chain

$\{(S_n, C_n): n \geq 0\}$. Recall from Definition 2.20 in Chapter 6 the notion of a polynomially dominated function. Given a sequence $\{\theta(k): k \geq 0\}$ of od-regeneration points for the underlying chain and a real-valued function $\tilde{f}$ defined on Σ, set

$$\tilde{Y}_k(\tilde{f}) = \sum_{n=\theta(k-1)}^{\theta(k)-1} \tilde{f}(S_n, C_n) \tag{2.4}$$

for $k \geq 1$.

Lemma 2.5. *Suppose that Assumption PD holds. Then there exists a sequence $\{\theta(k): k \geq 0\}$ of od-regeneration points for the underlying chain $\{(S_n, C_n): n \geq 0\}$. Moreover, $\tilde{Y}_1(|\tilde{f}|)$ has finite moments of all orders for any polynomially dominated real-valued function $\tilde{f}$ defined on Σ.*

The idea of the proof is as follows. There exists at least one sequence of od-equilibrium points for the embedded chain $\{(S_n^+, C_n^+): n \geq 0\}$—see Lemma 2.32 in Chapter 6. As discussed in the proof of Theorem 2.24(iii) in Chapter 6, these points also decompose sample paths of the underlying chain into o.d.s. cycles and induce a sequence $\{\theta(k): k \geq 0\}$ of od-regeneration points for the underlying chain. The remaining assertion follows by an argument almost identical to the proof of Theorem 2.24(iii) in Chapter 6.

Using Proposition 1.36 in Chapter 6, we obtain the following corollary.

Corollary 2.6. *Suppose that Assumption PD holds. Then the underlying chain $\{(S_n, C_n): n \geq 0\}$ is positive Harris recurrent.*

Remark 2.7. If Assumption PD holds for an SPN having no immediate transitions, then the od-regeneration points in Lemma 2.5 are also od-equilibrium points—that is, the cycle lengths are not only stationary, but also mutually independent. In general, however, the sequence $\{\theta(k): k \geq 0\}$ decomposes sample paths of the underlying chain into cycles with lengths that are stationary and one-dependent. To see that these assertions hold, consider the $(k+1)$st such cycle, which is demarcated by the random indices $\theta(k)$ and $\theta(k+1)$. As discussed above, these random indices correspond to od-equilibrium points $\theta^+(k)$ and $\theta^+(k+1)$, for the embedded chain. Recall that the construction of these od-equilibrium points rests on a decomposition of the r-step transition kernel of the embedded chain for some $r \geq 1$—see (1.35) in Chapter 6—and that the process $\{(S_n^+, C_n^+): n \geq \theta^+(k)\}$ depends on the history of the embedded chain through $\{(S_n^+, C_n^+): \theta^+(k) - r \leq n < \theta^+(k)\}$. It follows that the process $\{(S_n, C_n): n \geq \theta(k)\}$ depends on the history of the underlying chain through $\mathcal{H}_k = \{(S_n, C_n): \beta(k) \leq n < \theta(k)\}$, where $\beta(k)$ is the random index of the underlying chain that corresponds to the random index $\theta^+(k) - r$ of the embedded chain. In general, therefore, information about $\tilde{\tau}_k$ yields information about $\theta(k) - \beta(k)$, which yields information about $\mathcal{H}_k$, which yields information about $\{ (S_n, C_n) : n \geq \theta(k) \}$ and hence about $\tilde{\tau}_{k+1}$. Thus the cycle lengths $\tilde{\tau}_k$ and $\tilde{\tau}_{k+1}$ are, in general, dependent. If, however, there are no immediate transitions, then the embedded and underlying chains coincide, so that $\theta^+(k) = \theta(k)$ and $\beta(k) = \theta(k) - r$ for $k \geq 0$, and hence $\tilde{\tau}_k = r + \beta(k) - \theta(k-1)$. By construction, $\tilde{\tau}_{k+1}$ is independent of $\beta(k) - \theta(k-1)$, and hence of $\tilde{\tau}_k$. Indeed, the od-regeneration points in Lemma 2.5 may also be od-equilibrium points even in the presence of immediate transitions, provided that the random variable $\theta(k) - \beta(k)$ is a.s. equal to a fixed constant. This latter condition holds, for example, when there exists $m \geq 1$ such that the SPN visits exactly m immediate markings between each successive visit to the set of timed markings.

Using Lemma 2.5, Theorem 2.1, and Proposition 2.3, we obtain the following SLLN and FCLT for processes of the form $\{ \tilde{f}(S_n, C_n) : n \geq 0 \}$, where $\tilde{f} = (\tilde{f}_1, \tilde{f}_2, \ldots, \tilde{f}_l)$ is an $\Re^l$-valued function defined on Σ. For such a function, set

$$\tilde{r}(\tilde{f}) = \frac{E_\mu [\tilde{Y}_1(\tilde{f})]}{E_\mu [\tilde{\tau}_1]}, \tag{2.8}$$

where $\tilde{Y}_k(\tilde{f})$ is defined as in (2.4) and $\tilde{\tau}_k = \theta(k) - \theta(k-1)$. The function $\tilde{f}$ is said to be *polynomially dominated* if each $\tilde{f}_j$ is polynomially dominated in the sense of Definition 2.20 in Chapter 6.

Theorem 2.9. *Suppose that Assumption PD holds, so that there exists a sequence $\{ \theta(k) : k \geq 0 \}$ of od-regeneration points for the underlying chain $\{ (S_n, C_n) : n \geq 0 \}$. Then $\tilde{r}(|\tilde{f}|) < \infty$ and*

$$\lim_{n \to \infty} \frac{1}{n} \sum_{j=0}^{n-1} \tilde{f}(S_j, C_j) = \tilde{r}(\tilde{f}) \quad a.s.$$

for any polynomially dominated $\Re^l$-valued function $\tilde{f}$ defined on Σ, where $\tilde{r}(\tilde{f})$ is defined as in (2.8).

Remark 2.10. When, as discussed in Remark 2.7, the od-regeneration points in Lemma 2.5 are also od-equilibrium points, the quantity $\tilde{r}(\tilde{f})$ can be interpreted not only as a time-average limit, but also as a steady-state mean, provided that an additional aperiodicity condition holds. To define this condition, we recall the notation $s \to s'$ and $s \rightsquigarrow s'$ from Definition 4.9 in Chapter 4. A *d-cycle* of an SPN is a finite collection $\{ G_1, G_2, \ldots, G_d \}$ of disjoint subsets of G such that $s' \in G_{i+1}$ whenever $s \in G_i$ and $s \to s'$. (Take $G_{d+1} = G_1$.) The *period* of the SPN is the largest d for which a d-cycle exists; the SPN is called *aperiodic* if $d = 1$ and *periodic* (with period d) if $d > 1$. Theorem 3.4 in the following section asserts that, in the presence of Assumption PD, aperiodicity of the SPN implies aperiodicity of the underlying chain in the sense of Section 5.1.1. It then follows—see

Remark 1.33 in Chapter 6—that there exists a random vector (S, C) such that $(S_n, C_n) \Rightarrow (S, C)$ as $n \to \infty$ for any initial distribution μ. Moreover, $E[\tilde{f}(S, C)] = \tilde{r}(\tilde{f})$ for any polynomially dominated $\Re^l$-valued function $\tilde{f}$ defined on Σ.

Suppose that there exists a sequence $\{\theta(k): k \geq 0\}$ of od-regeneration points for the underlying chain $\{(S_n, C_n): n \geq 0\}$. For an $\Re^l$-valued function $\tilde{f}$ defined on Σ, define a sequence of $C^l[0, 1]$-valued random functions $\tilde{U}_1(\tilde{f}), \tilde{U}_2(\tilde{f}), \ldots$ by setting

$$\tilde{U}_n(\tilde{f})(t) = \frac{1}{\sqrt{n}} \int_0^{nt} \left(\tilde{f}(S_{\lfloor u \rfloor}, C_{\lfloor u \rfloor}) - \tilde{r}(\tilde{f}) \right) du \qquad (2.11)$$

for $0 \leq t \leq 1$ and $n \geq 1$, where $\tilde{r}(\tilde{f})$ is given by (2.8).

Theorem 2.12. *Suppose that Assumption PD holds, so that there exists a sequence $\{\theta(k): k \geq 0\}$ of od-regeneration points for the underlying chain, and let $\tilde{f}$ be a polynomially dominated $\Re^l$-valued function defined on Σ. Then there exists an $l \times l$ matrix $Q(\tilde{f})$ such that $\tilde{U}_n(\tilde{f}) \Rightarrow Q(\tilde{f})W^{(l)}$ as $n \to \infty$ for any initial distribution μ.*

SLLNs and FCLTs for processes of the form $\{f(X(t)): t \geq 0\}$ can be obtained from the corresponding results for the underlying chain. Recall from (1.7) in Chapter 3 the definition of the holding-time function t^*. When there exists a sequence $\{\theta(k): k \geq 0\}$ of od-regeneration points for the underlying chain, set

$$r(f) = \frac{E_\mu[\tilde{Y}_1(ft^*)]}{E_\mu[\tilde{Y}_1(t^*)]} \qquad (2.13)$$

for each $\Re^l$-valued function f defined on S, where $(ft^*)(s, c) = f(s)t^*(s, c)$ for $(s, c) \in \Sigma$ and $\tilde{Y}_k(\tilde{f})$ is defined as in (2.4).

Theorem 2.14. *Suppose that Assumption PD holds, so that there exists a sequence $\{\theta(k): k \geq 0\}$ of od-regeneration points for the underlying chain. Then $r(|f|) < \infty$ and*

$$\lim_{t \to \infty} \frac{1}{t} \int_0^t f(X(u)) \, du = r(f) \quad a.s.$$

for any $\Re^l$-valued function f defined on S, where $r(f)$ is given by (2.13).

PROOF. Because the function t^* is polynomially dominated, so is the function $|ft^*|$. Thus, by Theorem 2.9,

$$\lim_{n \to \infty} \frac{1}{n} \sum_{j=0}^{n-1} |(ft^*)(S_j, C_j)| = \frac{E_\mu[\tilde{Y}_1(|ft^*|)]}{E_\mu[\tilde{\tau}_1]} \quad a.s.,$$

$$\lim_{n\to\infty} \frac{1}{n} \sum_{j=0}^{n-1} (ft^*)(S_j, C_j) = \frac{E_\mu[\tilde{Y}_1(ft^*)]}{E_\mu[\tilde{\tau}_1]} \text{ a.s.,}$$

and

$$\lim_{n\to\infty} \frac{1}{n} \sum_{j=0}^{n-1} t^*(S_j, C_j) = \frac{E_\mu[\tilde{Y}_1(t^*)]}{E_\mu[\tilde{\tau}_1]} \text{ a.s.,}$$

where the three limits are well defined and finite. The desired result now follows from Theorem 2.9(iv) in Chapter 3. □

Remark 2.15. Observe that, under the conditions of Theorem 2.14, the time-average limit $r(f)$ also has the representation

$$r(f) = \frac{\pi(ft^*)}{\pi(t^*)},$$

where π is the invariant probability measure of the underlying chain and the notation $\pi(f)$ is defined as in Section 5.1.1.

For an $\Re^l$-valued function f defined on S, set

$$U_\nu(f)(t) = \frac{1}{\sqrt{\nu}} \int_0^{\nu t} \Big(f(X(u)) - r(f)\Big)\, du \qquad (2.16)$$

for $0 \le t \le 1$ and $\nu > 0$. Just as the discrete-time SLLN in Theorem 2.9 can be used in Theorem 2.14 to obtain a continuous time SLLN, the discrete-time FCLT in Theorem 2.12 can be used to obtain an FCLT in continuous time.

Theorem 2.17. *Suppose that Assumption PD holds and let f be an arbitrary $\Re^l$-valued function defined on S. Then there exists an $l \times l$ matrix $Q(f)$ such that $U_\nu(f) \Rightarrow Q(f)W^{(l)}$ as $\nu \to \infty$ for any initial distribution μ.*

We omit the proof, which is similar to that of Theorem 3.51 in Chapter 6 but uses Proposition 2.25 in the Appendix instead of the random-index CLT.

Remark 2.18. The SLLNs and FCLTs in Theorems 2.9, 2.12, 2.14, and 2.17 can be established under weaker moment conditions than those in Assumption PD. In particular, the SLLN for the marking process requires only finite first moments and the FCLT requires only finite second moments—see the notes at the end of the chapter.

Remark 2.19. When an SPN is not irreducible but the remaining conditions in Assumption PD hold, the quantity $\lim_{t\to\infty}(1/t)\int_0^t f(X(u))\, du$ is, in general, equal to a random variable whose distribution depends on the initial distribution μ. The limit theorems in this subsection and the estimation methods in subsequent subsections still apply, however, provided that the marking process is restricted to an irreducible closed subset of G—that is, a subset $B \subset G$ such that $s \rightsquigarrow s'$ for all $s, s' \in B$ and $s \not\rightsquigarrow s'$ for all $s \in B$ and $s' \in G - B$.

7.2.2 STS Methods

Under Assumption PD, time-average limits of the form

$$r(f) = \lim_{t \to \infty} \frac{1}{t} \int_0^t f(X(u)) \, du \qquad (2.20)$$

and

$$\tilde{r}(\tilde{f}) = \lim_{n \to \infty} \frac{1}{n} \sum_{j=0}^{n-1} \tilde{f}(S_n, C_n)$$

are well defined and finite, where f and $\tilde{f}$ are real-valued functions defined on S and Σ, respectively, and $\tilde{f}$ is polynomially dominated. Moreover, we can obtain point estimates and confidence intervals for such limits using STS methods.

STS Methods in Continuous Time

Fix a real-valued function f and set

$$\bar{Y}_\nu(t) = \frac{1}{\nu} \int_0^{\nu t} f(X(u)) \, du$$

for $0 \le t \le 1$ and $\nu > 0$. Also set $\hat{r}_\nu = \bar{Y}_\nu(1)$. By Theorem 2.14, the point estimator $\hat{r}_\nu$ is strongly consistent for $r(f)$.

To obtain asymptotic confidence intervals for $r(f)$, we proceed as follows. Denote by $C[0, 1]$ the set of continuous real-valued functions defined on $[0, 1]$. For a mapping ξ from $C[0, 1]$ to $\Re$, let $D(\xi)$ be the set of discontinuity points for ξ. That is, $x \in D(\xi)$ if $\lim_{n \to \infty} \xi(x_n) \ne \xi(x)$ for some sequence $x_1, x_2, \ldots \in C[0, 1]$ with $\lim_{n \to \infty} \sup_{0 \le t \le 1} |x_n(t) - x(t)| = 0$. Next, denote by Ξ the set of mappings from $C[0, 1]$ to $\Re$ such that $\xi \in \Xi$ if and only if

(i) $\xi(ax) = a\xi(x)$ for $a \in \Re_+$ and $x \in C[0, 1]$.

(ii) $\xi(x - be) = \xi(x)$ for $b \in \Re$ and $x \in C[0, 1]$, where $e(t) = t$ for $0 \le t \le 1$.

(iii) $P\{\xi(W) > 0\} = 1$.

(iv) $P\{W \in D(\xi)\} = 0$.

Here $W = \{W(t): 0 \leq t \leq 1\}$ is a standard one-dimensional Brownian motion. Fix a mapping $\xi \in \Xi$ and set $\xi_\nu = \xi(\bar{Y}_\nu)$. Theorem 2.17 (with $l = 1$) guarantees the existence of a nonnegative constant $\sigma(f)$ such that

$$U_\nu(f) \Rightarrow \sigma(f)W \qquad (2.21)$$

as $\nu \to \infty$, where $U_\nu(f)$ is given by (2.16). We focus throughout on the nondegenerate case in which $\sigma(f) > 0$; as discussed in Section 6.2.4, $\sigma(f)$ typically is positive provided that $f(s) \neq f(s')$ for some $s, s' \in S$. The convergence in (2.21), the properties in (i), (ii), and (iv), and the continuous mapping theorem (Proposition 1.42 in the Appendix) together imply that

$$\sqrt{\nu}\xi_\nu = \xi\left(\sqrt{\nu}(\bar{Y}_\nu - r(f)e)\right) = \xi(U_\nu(f)) \Rightarrow \sigma(f)\xi(W)$$

and

$$\sqrt{\nu}(\hat{r}_\nu - r(f)) = U_\nu(f)(1) \Rightarrow \sigma(f)W(1) \qquad (2.22)$$

as $\nu \to \infty$, where $\Rightarrow$ denotes ordinary convergence in distribution and the two sequences converge jointly. Using the property in (iii), it follows that

$$\frac{\hat{r}_\nu - r(f)}{\xi_\nu} \Rightarrow \frac{\sigma(f)W(1)}{\sigma(f)\xi(W)} = \frac{W(1)}{\xi(W)} \qquad (2.23)$$

as $\nu \to \infty$. Choosing z_p so that $P\{-z_p \leq W(1)/\xi(W) \leq z_p\} = p$ (where $0 < p < 1$), we obtain the asymptotic $100p\%$ confidence interval

$$[\hat{r}_\nu - \xi_\nu z_p, \hat{r}_\nu + \xi_\nu z_p] \qquad (2.24)$$

for $r(f)$. A key feature of this confidence interval is that there is no need (as in regenerative simulation) to consistently estimate the variance constant $\sigma^2(f)$ that appears in the CLT in (2.22); the variance constant has been "cancelled out" in (2.23).

Of course, to choose appropriate values of z_p we need to determine the distribution of $W(1)/\xi(W)$. Observe in this connection that by definition of Brownian motion, $W(1)$ has a standard (mean 0, variance 1) normal distribution. Moreover, it can be shown that $W(1)$ is independent of $\xi(W)$.

Different choices of the mapping ξ lead to different estimation procedures.

EXAMPLE 2.25 (Batch means with fixed number of batches). Fix $b \geq 2$ and take

$$\xi(x) = \left[\frac{b}{b-1}\sum_{i=1}^{b}\left(x(i/b) - x((i-1)/b) - x(1)/b\right)^2\right]^{1/2}. \qquad (2.26)$$

It can be shown that when ξ is defined by (2.26), the conditions in (i)–(iv) hold and the limiting random variable $W(1)/\xi(W)$ has a Student's t distribution with $b - 1$ degrees of freedom. Thus, setting

$$\bar{X}_\nu(i) = \bar{X}_\nu(i; f) = \frac{1}{\nu/b}\int_{(i-1)\nu/b}^{i\nu/b} f(X(u))\, du \qquad (2.27)$$

for $1 \leq i \leq b$ and $\nu > 0$, we find that the interval in (2.24) is an asymptotic $100p\%$ confidence interval for $r(f)$ when z_p is the $(1+p)/2$ quantile of the Student's t distribution with $b-1$ degrees of freedom and

$$\xi_\nu = \frac{1}{\sqrt{b}} \left[\frac{1}{b-1} \sum_{i=1}^{b} \left(\bar{X}_\nu(i) - \frac{1}{b} \sum_{j=1}^{b} \bar{X}_\nu(j) \right)^2 \right]^{1/2}.$$

According to the foregoing formulas, the batch means confidence interval based on simulation of the marking process over a time interval $[0, \nu]$ is obtained by decomposing the sample path of the process into b disjoint "batches" (intervals) of length ν/b—typically, $10 \leq b \leq 30$. The "batch mean" $\bar{X}_\nu(i)$ is the average of $\{ f(X(t)) : t \geq 0 \}$ over the ith such interval, and the random variable ξ_ν is equal to $b^{-1/2}$ times the sample standard deviation of the batch means. Observe that, in practice, $\bar{X}_\nu(i)$ can be computed from a sample path of the chain $\{ (S_n, C_n) : n \geq 0 \}$ using the formula

$$\bar{X}_\nu(i) = \frac{1}{\nu/b} \sum_{n=N(t_{i-1})}^{N(t_i)} f(S_n)\left[\min(\zeta_{n+1}, t_i) - \max(\zeta_n, t_{i-1})\right],$$

where $t_j = j\nu/b$ for $0 \leq j \leq b$, ζ_n is the time of the nth marking change, and $N(t)$ is the number of marking changes in $(0, t]$.

In forming the confidence interval, the batch means $\{ \bar{X}_\nu(i) : 1 \leq i \leq b \}$ are treated as if they are independent, normally distributed random variables. The intuition underlying this approximation rests on the plausible assumption that observations of the output process $\{ f(X(t)) : t \geq 0 \}$ at widely separated time points are essentially independent. When the batch length is large, most of the observations within a batch are far apart in time from observations in the other batches, so that batch means should be "almost" independent. Because each batch mean is an average of many observations, it is also plausible that a CLT holds, so that each batch mean is approximately normally distributed. The discussion in this section shows that, under Assumption PD, the error in the confidence interval arising from the independence and normality assumptions indeed becomes negligible as the length of each of the b batches—equivalently, the length ν of the simulation—becomes large.

EXAMPLE 2.28 (STS area method). Fix $m \geq 1$ and take

$$\xi(x) = \left[\sum_{i=0}^{m-1} \left(\int_0^1 (\Psi_i \circ x)(t)\, dt \right)^2 \right]^{1/2},$$

where $(\Psi_i \circ x)(t) = x\big((i+t)/m\big) - (1-t)x(i/m) - tx\big((i+1)/m\big)$. It can be verified that ξ satisfies the conditions in (i)–(iv) and that $W(1)/\xi(W)$

is distributed as $\sqrt{12}T_m$, where T_m has a Student's t distribution with m degrees of freedom. Thus the interval in (2.24) is an asymptotic $100p\%$ confidence interval for $r(f)$ when z_p is the $(1+p)/2$ quantile of the Student's t distribution with m degrees of freedom and

$$\xi_\nu^2 = 12 \sum_{i=0}^{m-1} A_i^2,$$

where

$$A_i = \frac{1}{\nu} \int_0^1 \left(Z_i(\lambda u) - uZ_i(\lambda) \right) du$$

with $\lambda = \nu/m$ and

$$Z_i(u) = \int_{i\lambda}^{i\lambda+u} f\big(X(t)\big)\, dt$$

for $0 \leq u \leq \lambda$. To obtain a representation of A_i more amenable to computation, observe that

$$A_i = \frac{1}{\lambda \nu} \int_0^\lambda Z_i(v)\, dv - \frac{Z_i(\lambda)}{2\nu}$$

$$= \frac{1}{\lambda \nu} \left(\lambda Z_i(\lambda) - \int_0^\lambda u\, dZ_i(u) \right) - \frac{Z_i(\lambda)}{2\nu}$$

$$= \frac{1}{\nu} \int_0^\lambda \left(\frac{1}{2} - \frac{u}{\lambda} \right) dZ_i(u)$$

$$= \frac{1}{\nu} \int_0^\lambda \left(\frac{1}{2} - \frac{u}{\lambda} \right) f\big(X(i\lambda + u)\big)\, du,$$

where the second equality follows from an integration by parts. Setting $t_j = j\lambda$ for $j \geq 0$, we can also express each A_i in terms of the underlying chain:

$$A_i = \frac{1}{2\nu\lambda} \sum_{n=N(t_i)}^{N(t_{i+1})} \left[\lambda f(S_n)(u_{i,n} - l_{i,n}) - f(S_n)(u_{i,n}^2 - l_{i,n}^2) \right],$$

where $u_{i,n} = \min(\zeta_n, t_{i+1})$, $l_{i,n} = \max(\zeta_n, t_i)$, and, as before, $N(t)$ is the number of marking changes in $(0, t]$.

EXAMPLE 2.29 (STS maximum method). Fix $m \geq 1$ and take

$$\xi(x) = \left[\sum_{i=0}^{m-1} \big((\Psi_i \circ x)(t_i^*)\big)^2 / \big(t_i^*(1 - t_i^*)\big) \right]^{1/2},$$

where $\Psi_i \circ x$ is as in Example 2.28 and t_i^* is the smallest value in $[0, 1]$ such that $(\Psi_i \circ x)(t_i^*) \geq (\Psi_i \circ x)(t)$ for $0 \leq t \leq 1$. It can be verified that

ξ satisfies the conditions in (i)–(iv) and that $W(1)/\xi(W)$ is distributed as $T_{3m}/\sqrt{3}$, where T_{3m} has a Student's t distribution with $3m$ degrees of freedom. Setting $\lambda = \nu/m$ and

$$B_i(t) = \frac{1}{\nu} \int_{i\lambda}^{(i+t)\lambda} f(X(u))\,du - \frac{t}{\nu} \int_{i\lambda}^{(i+1)\lambda} f(X(u))\,du$$

for $0 \le i < m$ and $0 \le t \le 1$, we see that the interval in (2.24) is an asymptotic $100p\%$ confidence interval for $r(f)$ when z_p is the $(1+p)/2$ quantile of the Student's t distribution with $3m$ degrees of freedom and

$$\xi_\nu^2 = \frac{1}{3} \sum_{i=0}^{m-1} A_i^2,$$

where

$$A_i = \frac{B_i(t_i^*)}{\left(t_i^*(1 - t_i^*)\right)^{1/2}}$$

and t_i^* is the smallest value in $[0, 1]$ that maximizes $B_i(\cdot)$.

STS Methods in Discrete Time

Now consider a fixed polynomially dominated real-valued function $\tilde{f}$, and set

$$\hat{r}_n = (1/n) \sum_{j=0}^{n-1} \tilde{f}(S_j, C_j)$$

for $n \ge 0$. By Theorem 2.9, the point estimator $\hat{r}_n$ is strongly consistent for $\tilde{r}(\tilde{f})$. Asymptotic confidence intervals for $\tilde{r}(\tilde{f})$ are obtained as follows. Observe that, by Theorem 2.12, $\tilde{U}_n(\tilde{f}) \Rightarrow \tilde{\sigma}(\tilde{f})W$ as $n \to \infty$, where $\tilde{U}_n(\tilde{f})$ is defined by (2.11) and $\tilde{\sigma}(\tilde{f})$ is a nonnegative finite constant. Suppose that $\tilde{\sigma}(\tilde{f}) > 0$, as is typical when $\tilde{f}$ is nonconstant and takes values in a finite set—see Section 6.2.4. Then the discrete-time version of any method based on standardized time series can be used to obtain asymptotic confidence intervals for $\tilde{r}(\tilde{f})$. That this assertion holds can be seen by applying the derivation of the STS method in continuous time to the process $X(t) = \tilde{f}(S_{\lfloor t \rfloor}, C_{\lfloor t \rfloor})$.

For example, the method of batch means can be applied by fixing $b \ge 2$ and simulating the underlying chain for $n = bm$ state transitions, where $m \ge 1$. The sample path is then decomposed into b batches of length m. Finally, a $100p\%$ confidence interval for $\tilde{r}(\tilde{f})$ is computed as described previously, except that each $\bar{X}_\nu(i)$ is defined in terms of a sum rather than an integral:

$$\bar{X}_\nu(i) = \frac{1}{m} \sum_{j=0}^{m-1} \tilde{f}(S_{im+j}, C_{im+j}).$$

Similarly, after fixing $m \geq 1$ and simulating the underlying chain for $n = lm$ state transitions, the STS area method yields an asymptotic $100p\%$ confidence interval for $\tilde{r}(\tilde{f})$ of the form $[\hat{r}_n - \xi_n z_p, \hat{r}_n + \xi_n z_p]$. Here

$$\hat{r}_n = \frac{1}{n} \sum_{j=0}^{n-1} \tilde{f}(S_j, C_j),$$

z_p is the $(1+p)/2$ quantile of the Student's t distribution with m degrees of freedom, and

$$\xi_\nu^2 = 12 \sum_{i=0}^{m-1} A_i^2,$$

with

$$A_i = \frac{1}{n} \sum_{j=0}^{l-1} \left(\frac{1}{2} - \frac{j}{l} - \frac{1}{2l} \right) \tilde{f}(S_{il+j}, C_{il+j}).$$

Alternatively, the STS maximum method yields an asymptotic $100p\%$ confidence interval for $\tilde{r}(\tilde{f})$ of the form $[\hat{r}_n - \xi_n z_p, \hat{r}_n + \xi_n z_p]$. Here

$$\hat{r}_n = \frac{1}{n} \sum_{j=0}^{n-1} \tilde{f}(S_j, C_j),$$

z_p is the $(1+p)/2$ quantile of the Student's t distribution with $3m$ degrees of freedom, and

$$\xi_\nu^2 = \frac{1}{3} \sum_{i=0}^{m-1} A_i^2,$$

where each A_i is defined as follows. Set

$$B_i(t) = \frac{1}{n} \sum_{j=0}^{\lfloor lt \rfloor - 1} \tilde{f}(S_{il+j}, C_{il+j}) - \frac{t}{n} \sum_{j=0}^{l-1} \tilde{f}(S_{il+j}, C_{il+j})$$

for $0 \leq t \leq 1$, and denote by k_i^* the smallest value of k in $\{0, 1, \ldots, l\}$ such that $B_i(k_i^*/l) \geq B_i(k/l)$ for k in $\{0, 1, \ldots, l\}$. Then

$$A_i = \frac{B_i(k_i^*/l)}{\left(k_i^*/l \right) \left(1 - (k_i^*/l) \right)^{1/2}}.$$

7.2.3 Functions of Time-Average Limits

Let f_1 and f_2 be real-valued functions defined on S such that $f_1(s) \neq f_1(s')$ for some $s, s' \in S$ and similarly for f_2. Also suppose that f_1 and f_2 are linearly independent in that $a_1 f_1(s) + a_2 f_2(s) = 0$ for all $s \in S$ only if

$a_1 = a_2 = 0$. Under Assumption PD we can obtain point estimates and confidence intervals for the limiting ratio

$$r = r(f_1, f_2) = \lim_{t \to \infty} \frac{\int_0^t f_1(X(u)) \, du}{\int_0^t f_2(X(u)) \, du}. \tag{2.30}$$

Some performance measures of this type are given in Section 3.2.1 and in Example 2.37 below. Because the bias of naive point estimators can be large, especially when the run length is small, we develop a point estimator based on combining the batch-means method given in Section 7.2.2 with the jackknife technique discussed in earlier chapters. Fix $b \geq 2$ and set

$$J_\nu(i) = b \frac{\sum_{j=1}^b \bar{X}_\nu(j; f_1)}{\sum_{j=1}^b \bar{X}_\nu(j; f_2)} - (b-1) \frac{\sum_{j \neq i} \bar{X}_\nu(j; f_1)}{\sum_{j \neq i} \bar{X}_\nu(j; f_2)}$$

for $1 \leq i \leq b$, where $\bar{X}_\nu(i; f)$ is defined as in (2.27). Then set

$$\hat{r}_\nu^{(J)} = \frac{1}{b} \sum_{i=1}^b J_\nu(i). \tag{2.31}$$

The following result shows that $\hat{r}_\nu^{(J)}$ is strongly consistent for r.

Theorem 2.32. *If Assumption PD holds, then* $\hat{r}_\nu^{(J)} \to r$ *a.s. as* $\nu \to \infty$.

PROOF. Set

$$\bar{Y}_\nu(f_j)(t) = \frac{1}{\nu} \int_0^{\nu t} f_j(X(u)) \, du$$

for $0 \leq t \leq 1$, $\nu > 0$, and $j = 1, 2$. Also set $\Lambda_i^b(x) = x(i/b) - x((i-1)/b)$ for $x \in C[0, 1]$ and $1 \leq i \leq b$. Fix i and j. By Theorem 2.14, $\lim_{\nu \to \infty} \bar{Y}_\nu(f_j)(1) = r(f_j)$ a.s., where $r(f_j)$ is given by (2.13). It is known that such an SLLN implies a *functional* SLLN:

$$\lim_{\nu \to \infty} \sup_{0 \leq t \leq 1} |\bar{Y}_\nu(f_j)(t) - r(f_j)t| = 0 \text{ a.s.};$$

that is, $\bar{Y}_\nu(f_j) \to r(f_j)e$ a.s. in $C[0, 1]$ as $\nu \to \infty$, where $e(t) = t$ for $0 \leq t \leq 1$. Since Λ_i^b is a continuous mapping from $C[0, 1]$ to $\Re$, it follows that $\Lambda_i^b(\bar{Y}_\nu(f_j)) \to \Lambda_i^b(r(f_j)e)$ a.s. as $\nu \to \infty$. By definition of Λ_i^b and $\bar{Y}_\nu(f_j)$, this convergence is equivalent to $\lim_{\nu \to \infty} \bar{X}_\nu(i; f_j) = r(f_j)$ a.s.. It then follows easily that $\lim_{\nu \to \infty} J_\nu(i) = r$ a.s. for $1 \leq i \leq b$, and hence $\hat{r}_\nu^{(J)} \to r$ a.s. as $\nu \to \infty$.   □

To obtain confidence intervals, observe that by Theorem 2.17 there exists a 2×2 matrix $Q(f)$ such that

$$U_\nu(f) \Rightarrow Q(f)W^{(2)} \tag{2.33}$$

as $\nu \to \infty$ for any initial distribution μ, where $U_\nu(f)$ is defined by (2.16) and $W^{(2)}$ is a standard two-dimensional Brownian motion. An extension of the argument in Section 6.2.4 shows that $Q(f)$ is nonsingular except in degenerate cases, and we assume that $Q(f)$ is nonsingular throughout. Using the convergence in (2.33) together with an argument similar in spirit of the proof of Theorem 2.32, it can be shown that

$$\frac{\sqrt{b}(\hat{r}_\nu^{(J)} - r)}{s_\nu^{(J)}} \Rightarrow T_{b-1}$$

as $\nu \to \infty$, where T_{b-1} denotes a random variable having a Student's t distribution with $b - 1$ degrees of freedom and

$$s_\nu^{(J)} = \left[\frac{1}{b-1}\sum_{i=1}^{b}\left(J_\nu(i) - \hat{r}_\nu^{(J)}\right)^2\right]^{1/2}.$$

Thus

$$\left[\hat{r}_\nu^{(J)} - \frac{s_\nu^{(J)}z_p}{\sqrt{b}}, \hat{r}_\nu^{(J)} + \frac{s_\nu^{(J)}z_p}{\sqrt{b}}\right] \tag{2.34}$$

is an asymptotic $100p\%$ confidence interval for r, where z_p is the $(1+p)/2$ quantile of the Student's t distribution with $b - 1$ degrees of freedom.

Arguments analogous to those given above lead to point and interval estimators for discrete-time ratios of the form

$$\tilde{r}(\tilde{f}_1, \tilde{f}_2) = \lim_{n\to\infty}\frac{\sum_{j=0}^{n}\tilde{f}_1(S_j, C_j)}{\sum_{j=0}^{n}\tilde{f}_2(S_j, C_j)}, \tag{2.35}$$

provided that Assumption PD holds and $\tilde{f}_i$ is polynomially dominated for $i = 1, 2$. The estimation formulas are almost identical to the continuous-time formulas, except that each $\bar{X}_\nu(i)$ is defined as a sum rather than an integral.

Remark 2.36. Observe that, by virtue of Theorem 2.14, a time-average limit $r(f)$ as in (2.20) can also be viewed as a time-average limit $\tilde{r}(\tilde{f}_1, \tilde{f}_2)$ as in (2.35) with $\tilde{f}_1(s, c) = f(s)t^*(s, c)$ and $\tilde{f}_2(s, c) = t^*(s, c)$ for $(s, c) \in \Sigma$. Thus a jackknifed batch-means estimator of $r(f)$ is available as an alternative to the simple batch-means estimator given in Section 7.2.2. Little is known about the relative performance of these estimators—in the experiments reported in Example 3.35 below, the jackknifed batch-means estimator had a longer expected confidence-interval length and a higher empirical coverage probability than the simple estimator when the simulation run length was small. The performance of the estimators was quite similar for large run lengths.

The foregoing batch-means methodology can be extended to permit estimation not just of ratios but also of arbitrary differentiable real-valued functions of time-average limits. Specifically, suppose that we wish to estimate $r = g(\alpha_1, \alpha_2, \ldots, \alpha_l)$ for some $l \geq 1$ and a real-valued function g defined on $\Re^l$, where

$$\alpha_i = \lim_{t \to \infty} \frac{1}{t} \int_0^t f_i(X(u)) \, du$$

for $1 \leq i \leq l$ and each f_i is a real-valued function defined on S. Also suppose that each f_i is nonconstant and $f_1, f_2, \ldots, f_l$ are linearly independent. Then a strongly consistent point estimator and asymptotic $100p\%$ confidence interval for r are given by (2.31) and (2.34) as before, except that

$$J_\nu(i) = bg\left(\bar{A}_1, \bar{A}_2, \ldots, \bar{A}_l\right) - (b-1)g\left(\bar{A}_1^{(i)}, \bar{A}_2^{(i)}, \ldots, \bar{A}_l^{(i)}\right),$$

where

$$\bar{A}_j = \frac{1}{b} \sum_{l=1}^b \bar{X}_\nu(l; f_j)$$

and

$$\bar{A}_j^{(i)} = \frac{1}{b-1} \sum_{l \neq i} \bar{X}_\nu(l; f_j).$$

An analogous extension is valid in the discrete-time setting.

EXAMPLE 2.37 (Interactive video on demand). For the system of Example 1.1, suppose that more than one game is stored at the server and that the system is modelled by an SPN as in Figure 7.2. Also suppose, as before, that each interrequest-time random variable A_j has a gamma distribution with noninteger shape parameter, each waiting-time random variable W_j has a uniform distribution on $[0, w_j]$, and each playing-time random variable T_j has a truncated normal distribution. As discussed previously, there is no apparent sequence of regeneration points for the marking process of the SPN.

Let $\bar{s} \in S$ be the unique marking such that each of places $d_{1,1}, d_{2,1}, \ldots,$ $d_{M,1}$ contains one token and no other place contains a token; all buffers are empty and no games are underway when the marking is $\bar{s}$. For $s, s' \in G$, it is not hard to see that $s \rightsquigarrow \bar{s}$ and $\bar{s} \rightsquigarrow s'$, so that $s \rightsquigarrow s'$. Thus the SPN is irreducible. Since the marking set G is finite and all speeds are positive, it follows from the form of the clock-setting distribution functions that Assumption PD holds with $u = \min_{1 \leq j \leq M} w_j$. The methods of Section 7.2.2 and the current section therefore can be used to obtain strongly consistent point estimates and asymptotic confidence intervals for time-average limits of the form (2.20), (2.30), or (2.35). A number of pertinent performance characteristics can be expressed as limits of this type. For example, suppose that a customer pays an amount v per unit of game time. Then the long-run average rate at which the system generates revenue can be expressed as a limit of the form

$$r(f) = \lim_{t\to\infty} \frac{1}{t} \int_0^t f(X(u))\, du,$$

where $f(s) = (s_{1,5} + \cdots + s_{N,5})v$. The long-run relative utilization of channel i $(1 \le i \le N)$ can be expressed as a limit of the form

$$r(f_1, f_2) = \lim_{t\to\infty} \frac{\int_0^t f_1(X(u))\, du}{\int_0^t f_2(X(u))\, du},$$

where $f_1(s) = 1_{\{1,2,\ldots,B\}}(s_{i,5})$ and $f_2(s) = \sum_{l=1}^N 1_{\{1,2,\ldots,B\}}(s_{l,5})$. The long-run fraction of requests for game j that get immediately rejected can be expressed as a limit of the form

$$\tilde{r}(\tilde{f}_1, \tilde{f}_2) = \lim_{n\to\infty} \frac{\sum_{j=0}^n \tilde{f}_1(S_j, C_j)}{\sum_{j=0}^n \tilde{f}_2(S_j, C_j)},$$

where

$$\tilde{f}_1(s, c) = \begin{cases} 1 & \text{if } E^*(s,c) = \{e_{j,1}\} \text{ and } s_{j,4} = B; \\ 0 & \text{otherwise} \end{cases}$$

and

$$\tilde{f}_2(s, c) = \begin{cases} 1 & \text{if } E^*(s,c) = \{e_{j,1}\}; \\ 0 & \text{otherwise}. \end{cases}$$

Here the set-valued function E^* is defined as in (1.8) in Chapter 3.

7.2.4 Extensions

It can be shown that the augmented chain $\{(S_n, C_n, S_{n+1}, C_{n+1}): n \ge 0\}$ inherits the stability properties of the underlying chain, and the foregoing STS methods can be extended to yield point estimates and confidence intervals for limits such as

$$\tilde{r}(\tilde{f}_1, \tilde{f}_2) = \lim_{n\to\infty} \frac{\sum_{j=0}^n \tilde{f}_1(S_j, C_j, S_{j+1}, C_{j+1})}{\sum_{j=0}^n \tilde{f}_2(S_j, C_j, S_{j+1}, C_{j+1})}.$$

For example, consider the SPN model of the interactive video-on-demand system in Example 1.1 together with the sequence of times at which there is an end of waiting period for game j $(1 \le j \le M)$. Suppose that we wish to estimate the long-run fraction of these times at which there is an immediate start of game j on channel i $(1 \le i \le N)$. This long-run fraction can be expressed as a limit of the above form with

$$\tilde{f}_1(s, c, s', c') = \begin{cases} 1 & \text{if } s_{j,2} = 1 \text{ and } s'_{j,i,1} = 1; \\ 0 & \text{otherwise} \end{cases}$$

and

$$\tilde{f}_2(s,c,s',c') = \begin{cases} 1 & \text{if } s_{j,2} = 1 \text{ and } s'_{j,2} = 0; \\ 0 & \text{otherwise.} \end{cases}$$

This long-run performance measure needs to be expressed in terms of the augmented chain rather than the usual underlying chain because a game always starts on a channel chosen randomly and uniformly from among the available channels.

In light of the results in Section 3.2, it is apparent that the estimation methods in this section can be adapted to handle a variety of performance measures of interest. For example, consider the SPN in Example 1.1 and suppose that each participant in a game pays a fixed amount u at the start of the game. The revenue $R(t)$ generated by the system in the interval $[0, t]$ can then be represented as in (2.13) in Chapter 3, where $q(s) \equiv 0$ and

$$v(s,c) = \begin{cases} u & \text{if } E^*(s,c) = \{\, e_{j,i,1} \,\} \text{ for some } 1 \le j \le M \text{ and } 1 \le i \le N; \\ 0 & \text{otherwise.} \end{cases}$$

By Theorem 2.14 in Chapter 3, the long-run average rate at which the system generates revenue can be expressed as a limit of the form (2.35), where $\tilde{f}_1 = v$ and $\tilde{f}_2 = t^*$. Such a limit can then be handled by the methods of Section 7.2.3.

7.3   Consistent Estimation Methods

Consider an SPN with an underlying chain $\{\, (S_n, C_n) : n \ge 0 \,\}$ having state space Σ, together with a real-valued function $\tilde{f}$ defined on Σ, such that

$$\lim_{n \to \infty} \bar{r}(n; \tilde{f}) = \tilde{r}(\tilde{f}) \text{ a.s.} \tag{3.1}$$

for some finite constant $\tilde{r}(\tilde{f})$ and

$$\frac{\sqrt{n}\big(\bar{r}(n; \tilde{f}) - \tilde{r}(\tilde{f})\big)}{\tilde{\sigma}(\tilde{f})} \Rightarrow N(0,1) \tag{3.2}$$

as $n \to \infty$ for some constant $\tilde{\sigma}(\tilde{f}) \in (0, \infty)$, where

$$\bar{r}(n; \tilde{f}) = \frac{1}{n} \sum_{j=0}^{n-1} \tilde{f}(S_j, C_j). \tag{3.3}$$

Suppose that we can find an estimator V_n that is *consistent* for the variance constant $\tilde{\sigma}^2(\tilde{f})$ in (3.2), that is, an estimator V_n that converges to $\tilde{\sigma}^2(\tilde{f})$ in probability as $n \to \infty$ or, equivalently,[1] $V_n \Rightarrow \tilde{\sigma}^2(\tilde{f})$ as $n \to \infty$. Then an application of Slutsky's theorem shows that

$$\frac{\sqrt{n}(\bar{r}(n; \tilde{f}) - \tilde{r}(\tilde{f}))}{V_n^{1/2}} \Rightarrow N(0, 1),$$

so that

$$\left[\bar{r}(n; \tilde{f}) - \frac{z_p V_n^{1/2}}{\sqrt{n}}, \bar{r}(n; \tilde{f}) + \frac{z_p V_n^{1/2}}{\sqrt{n}}\right]$$

is an asymptotic $100p\%$ confidence interval for $\tilde{r}(\tilde{f})$, where z_p is the $(1+p)/2$ quantile of the standard normal distribution. This section is concerned with methods for obtaining point estimates and confidence intervals based on consistent estimation of the variance constant. Recall from Section 6.3.7 that the regenerative method is one such "consistent estimation method." Our emphasis in this section is on alternative methods that do not require regenerative structure. As mentioned previously, the lengths of confidence intervals based on consistent estimation methods are, asymptotically, both smaller in expectation and less variable than the lengths of confidence intervals based on cancellation methods (such as STS methods).

Because we focus throughout on SPNs that satisfy Assumption PD, Theorems 2.9 and 2.12 imply that the SLLN and CLT in (3.1) and (3.2) hold for any polynomially dominated function $\tilde{f}$—here we use the fact that, as discussed in Section A.2.5, an ordinary CLT holds whenever an FCLT holds. For the marking process, continuous-time analogs of (3.1) and (3.2) follow from Theorems 2.14 and 2.17. As mentioned in Remark 2.18, the foregoing SLLNs and CLTs can be established under weaker moment conditions than those in Assumption PD.

This section is concerned with conditions on the building blocks of an SPN under which various "quadratic-form" estimators of the variance constant are consistent. We first show that if Assumption PD holds and the SPN is aperiodic, then the underlying chain is aperiodic and hence Harris ergodic. We then show that if a "localized" quadratic-form variance estimator is consistent when applied to a stationary version of a Harris ergodic underlying chain, then the estimator is consistent when applied to any specified version of the chain. The idea is to couple the specified version with a stationary version, that is, to construct the two versions on a common probability space such that they coincide after some a.s. finite random time. Finally, we establish consistency for some specific variance estimators—including the variable batch-means estimator and certain spectral estimators—under arbitrary initial conditions. Our strategy is to invoke known results that establish the consistency of these estimators for stationary processes and then apply the foregoing coupling argument.

---

[1]See Proposition 1.39 in the Appendix.

7.3.1   Aperiodicity and Harris Ergodicity

Recall (1) the definition of an aperiodic SPN from Remark 2.10 and (2) the definition of an aperiodic chain in Section 5.1.1. The following result relates these two notions of aperiodicity.

Theorem 3.4. *Let $\{(S_n, C_n): n \geq 0\}$ be the underlying chain of an aperiodic SPN. If Assumption PD holds, then $\{(S_n, C_n): n \geq 0\}$ is aperiodic.*

As discussed in Section 5.1.1, a positive Harris recurrent Markov chain that is also aperiodic is called *Harris ergodic*. Corollary 3.5 is an immediate consequence of Theorem 3.4 and Corollary 2.6.

Corollary 3.5. *Let $\{(S_n, C_n): n \geq 0\}$ be the underlying chain of an aperiodic SPN. If Assumption PD holds, then $\{(S_n, C_n): n \geq 0\}$ is Harris ergodic.*

To establish Theorem 3.4, we need two preliminary lemmas. The first, Lemma 3.6, is a well-known number-theoretic result.

Lemma 3.6. *Let L be a countably infinite set of nonnegative integers that is closed under addition, and suppose that the elements in L have greatest common divisor 1. Then the set L contains all integers greater than some n_0.*

Lemma 3.7. *Suppose that Assumption PD holds for an aperiodic SPN with underlying chain $\{(S_n, C_n): n \geq 0\}$, and let $A \subseteq \Sigma$ satisfy $\bar{\phi}(A) > 0$, where $\bar{\phi}$ is defined as in (1.17) of Chapter 5. Then*

$$P_{(s,c)} \{ (S_{i+nd}, C_{i+nd}) \in A \text{ for some } n \geq 1 \} > 0$$

for each $(s, c) \in \Sigma$, $d \in \{1, 2, \dots\}$, and $i \in \{0, 1, \dots, d-1\}$.

PROOF. Fix i, d, and (s, c). For ease of exposition suppose that all speeds are equal to 1, and without loss of generality assume that the set A is of the form $A = \{s'\} \times H$ for some $s' \in S$ and $H \subseteq C(s')$. Consider all possible paths of the form $s' \to s_1 \to \cdots \to s_m = s'$ and denote by L the set of lengths of these paths. (We allow intermediate visits to s' along a path.) The set L is closed under addition and, by the aperiodicity of the SPN, the elements in L have greatest common divisor 1. By Lemma 3.6, the set L contains all integers greater than some n_0. It follows that for each state $s_0 \in S$ and integer $l \geq 0$ there exists a path $s_0 \to s_1 \to \cdots \to s_m = s'$ such that $l + m = i + nd$ for some $n \geq 0$. Fix such a path for each s_0 and l, and denote the length of the path by $m(s_0, l)$. An argument similar to the proof of Lemma 1.29 in Chapter 5 shows that for each $s_0 \in S$ and $l \geq 0$ there exist $B = B(s_0, l) \subseteq \{s_0\} \times C(s_0)$ and $\delta_1 = \delta_1(s_0, l) > 0$ such that $\bar{\phi}(B) > 0$ and

$$\inf_{(\bar{s}, \bar{c}) \in B} P^{m(s_0, l)} ((\bar{s}, \bar{c}), A) \geq \delta_1.$$

Next, set
$$\Sigma_\epsilon = \left\{ (\bar{s}, \bar{c}) \in \Sigma : \bar{c} \in [0, \epsilon)^M \right\}$$
for $\epsilon > 0$. Arguing as in the proof of Lemma 1.30 in Chapter 5, it can be shown that there exists a finite partition $\mathcal{Q}$ of Σ into disjoint subsets and a constant $\epsilon > 0$ such that for each $Q \in \mathcal{Q}$ and $l \geq 0$ there exist a state $s_0 = s_0(Q)$, a real number $\delta_2 = \delta_2(Q, l) > 0$, and an integer $b = b(Q) \geq 1$ such that
$$\inf_{(\bar{s}, \bar{c}) \in Q \cap \Sigma_\epsilon} P^b\big((s, c), B(s_0, l)\big) \geq \delta_2. \tag{3.8}$$
Write $m^*(Q, l) = m\big(s_0(Q), l\big)$ and $\delta_1^*(Q, l) = \delta_1\big(s_0(Q), l\big)$. Also, for $n \geq 0$, denote by Q_n the unique element $Q \in \mathcal{Q}$ such that $(S_n, C_n) \in Q$. Finally, arguing as in the proof of Lemma 1.33 in Chapter 5, it can be shown that for any $(\bar{s}, \bar{c}) \in \Sigma$ and $\epsilon > 0$ there exist $\delta_3 = \delta_3(\epsilon, \bar{s}, \bar{c}) > 0$ and $k = k(\epsilon, \bar{s}, \bar{c}) < \infty$ such that
$$P^k\big((\bar{s}, \bar{c}), \Sigma_\epsilon\big) \geq \delta_3.$$
Choose $\epsilon > 0$ so that (3.8) holds, and set
$$J = k(\epsilon, s, c) + b(Q_{k(\epsilon,s,c)}) + m^*\big(Q_{k(\epsilon,s,c)}, k(\epsilon, s, c) + b(Q_{k(\epsilon,s,c)})\big).$$
A straightforward conditioning argument then shows that
$$P_{(s,c)}\left\{ (S_J, C_J) \in A \right\} \geq \delta,$$
where
$$\delta = \delta(\epsilon, s, c) = \min_{Q \in \mathcal{Q}} \delta_1^*\big(Q, k(\epsilon, s, c) + b(Q)\big)$$
$$\cdot \min_{Q \in \mathcal{Q}} \delta_2\big(Q, k(\epsilon, s, c) + b(Q)\big) \cdot \delta_3(\epsilon, s, c) > 0.$$
The desired result now follows because, by construction, $J - i$ is divisible by d.   □

PROOF OF THEOREM 3.4. Suppose that, contrary to the statement of the theorem, the SPN is aperiodic and Assumption PD holds, but $\{ (S_n, C_n) : n \geq 0 \}$ is periodic with period d. Let $\Sigma_1, \Sigma_2, \ldots, \Sigma_d \subset \Sigma$ be the disjoint subsets in the d-cycle for the underlying chain, and assume without loss of generality that the initial distribution μ satisfies $\mu(\Sigma_1) = 1$. There must exist $s \in G$ such that $A_i(s) = \big(s \times C(s)\big) \cap \Sigma_i \neq \varnothing$ and $A_j(s) = \big(s \times C(s)\big) \cap \Sigma_j \neq \varnothing$ for some $i \neq j$, with both $A_i(s)$ and $A_j(s)$ having positive $\bar{\phi}$-measure. Otherwise, the SPN would be periodic with period d and sets $S_1, S_2, \ldots, S_d$ in the d-cycle given by
$$S_i = \{ s \in G : s \times C(s) \subseteq \Sigma_i \}$$
for $1 \leq i \leq d$. Since $\bar{\phi}\big(A_i(s)\big) > 0$, it follows from Lemma 3.7 that
$$P_\mu\left\{ (S_{j-1+nd}, C_{j-1+nd}) \in A_i(s) \text{ for some } n \geq 0 \right\} > 0.$$
Thus $A_i(s) \cap \Sigma_j \neq \varnothing$, contradicting the assumed disjointness of Σ_i and Σ_j.   □

7.3.2   Consistent Estimation in Discrete Time

Let $\bar{r}(n; \tilde{f})$ be defined as in (3.3). As discussed previously, if Assumption PD holds, then for any polynomially dominated function $\tilde{f}$ there exist constants $\tilde{r}(\tilde{f})$ and $\tilde{\sigma}(\tilde{f})$ such that $\lim_{n \to \infty} \bar{r}(n; \tilde{f}) = \tilde{r}(\tilde{f})$ a.s. and $\sqrt{n}(\bar{r}(n; \tilde{f}) - \tilde{r}(\tilde{f})) \Rightarrow \tilde{\sigma}(\tilde{f})N(0,1)$ as $n \to \infty$. Our goal is to find estimators consistent for $\tilde{\sigma}^2(\tilde{f})$. We assume that $\tilde{\sigma}^2(\tilde{f}) > 0$, and the estimators that we consider are of the form

$$V_n = V_n(\tilde{f}) = \sum_{i=0}^{n} \sum_{j=0}^{n} \tilde{f}(S_i, C_i) \tilde{f}(S_j, C_j) q_{i,j}^{(n)}, \qquad (3.9)$$

where each $q_{i,j}^{(n)}$ is a finite constant and $q_{i,j}^{(n)} = q_{j,i}^{(n)}$ for all i, j. As discussed in Section 7.3.3, this class of *quadratic-form* estimators includes both batch means and spectral estimators.

When Assumption PD holds, it follows from Corollary 2.6 and the discussion in Section 5.1.1 that there exists an invariant probability measure π for the underlying chain $\{ (S_n, C_n) : n \geq 0 \}$. By applying general results on consistent variance estimation for stationary processes, it can sometimes be established that $V_n(\tilde{f}) \Rightarrow \tilde{\sigma}^2(\tilde{f})$ for a specified estimator $V_n(\tilde{f})$ when the initial distribution of the underlying chain is π. The following two propositions are useful in this connection and are obtained by direct application of some well-known results for Harris ergodic chains.

Proposition 3.10. *Let $\{ (S_n, C_n) : n \geq 0 \}$ be the underlying chain of an aperiodic* SPN, *and let $\tilde{f}$ be a polynomially dominated real-valued function defined on Σ. Suppose that Assumption PD holds, so that there exists an invariant distribution π for the chain and $\{ \tilde{f}(S_n, C_n) : n \geq 0 \}$ obeys a* CLT *with variance constant $\tilde{\sigma}^2(\tilde{f})$. Then $\tilde{\sigma}^2(\tilde{f})$ has the representation*

$$\tilde{\sigma}^2(\tilde{f}) = \lim_{n \to \infty} n \mathrm{Var}_\pi \left[\frac{1}{n} \sum_{j=0}^{n-1} \tilde{f}(S_j, C_j) \right]. \qquad (3.11)$$

Proposition 3.12. *Suppose that Assumption PD holds for an aperiodic* SPN. *Then there exist $\rho \in (0, 1)$ and $c \in [0, \infty)$ such that*

$$\left| \mathrm{Cov}_\pi \left[\tilde{f}_1(S_0, C_0), \tilde{f}_2(S_k, C_k) \right] \right| \leq c \rho^k$$

for $k \geq 0$ and any polynomially dominated functions $\tilde{f}_1$ and $\tilde{f}_2$.

When the consistency of $V_n(\tilde{f})$ for $\tilde{\sigma}^2(\tilde{f})$ can be established under initial distribution π, the key problem is then to show that $V_n(\tilde{f}) \Rightarrow \tilde{\sigma}^2(\tilde{f})$ even when the initial distribution μ is not equal to π. Coupling arguments are often employed to extend convergence results from a stationary to a nonstationary setting, and use of this approach leads to Theorem 3.15 below. To state the theorem, we need the following definition.

Definition 3.13. A quadratic-form estimator V_n is said to be *localized* if there exist a constant $a_1 \in (0, \infty)$ and sequences $\{a_2(n): n \geq 0\}$ and $\{m(n): n \geq 0\}$ of nonnegative constants with $a_2(n) \to 0$ and $m(n)/n \to 0$ such that

$$|q_{i,j}^{(n)}| \leq \begin{cases} a_1/n & \text{if } |i - j| \leq m(n); \\ a_2(n)/n & \text{if } |i - j| > m(n). \end{cases} \tag{3.14}$$

A localized estimator has the property that, as more and more observations of the output process are obtained, the influence of any one observation on the value of the estimator becomes negligible.

Theorem 3.15. *Let $\{(S_n, C_n): n \geq 0\}$ be the underlying chain of an aperiodic* SPN, *and let $\tilde{f}$ be a polynomially dominated real-valued function defined on Σ. Suppose that Assumption PD holds, so that there exists an invariant distribution π for the chain and $\{\tilde{f}(S_n, C_n): n \geq 0\}$ obeys a* CLT *with variance constant $\tilde{\sigma}^2(\tilde{f})$. If a localized quadratic-form estimator $V_n(\tilde{f})$ satisfies $V_n(\tilde{f}) \Rightarrow \tilde{\sigma}^2(\tilde{f})$ when the initial distribution is π, then $V_n(\tilde{f}) \Rightarrow \tilde{\sigma}^2(\tilde{f})$ for any initial distribution.*

PROOF. Fix an arbitrary initial distribution μ, and write $Z_n = \tilde{f}(S_n, C_n)$ throughout. Corollary 3.5 implies that the underlying chain is Harris ergodic. By Proposition 1.10 in Chapter 5, the chain admits coupling. Thus there exist an a.s. finite random index T and versions $\{Z_n: n \geq 0\}$ and $\{Z_n': n \geq 0\}$ of the chain, all defined on a common probability space $(\Omega, \mathcal{F}, P)$, such that the two versions have respective probability laws P_μ and P_π, and $Z_n = Z_n'$ for $n \geq T$. Denote by V_n and V_n' the quadratic-form estimator computed for the first and second versions of the chain, respectively. We prove the result under the assumption that $\sup_n m(n) = \infty$, the proof for the case $\sup_n m(n) < \infty$ being similar. Moreover, without loss of generality we can assume that $m(n) \to \infty$ as $n \to \infty$; otherwise, replace $m(n)$ by $M(n) = \max_{1 \leq k \leq n} m(k)$ in the following argument and observe that (i) the estimator V_n is localized with respect to $M(n)$ whenever it is localized with respect to $m(n)$ and (ii) $M(n)/n \to 0$ whenever $m(n)/n \to 0$. Suppose that n is sufficiently large so that $n > T + m(n)$. Then we have

$$V_n = \sum_{i=T}^{n} \sum_{j=T}^{n} Z_i' Z_j' q_{i,j}^{(n)} + R_1(n) + R_2(n) + R_3(n),$$

where

$$R_1(n) = \sum_{i=0}^{T-1} \sum_{j=0}^{T-1} Z_i Z_j q_{i,j}^{(n)},$$

$$R_2(n) = \sum_{i=0}^{T-1} \sum_{j=T}^{n} Z_i Z_j' q_{i,j}^{(n)},$$

and

$$R_3(n) = \sum_{i=T}^{n} \sum_{j=0}^{T-1} Z_i' Z_j q_{i,j}^{(n)}.$$

Clearly, $R_1(n) \to 0$ a.s. as $n \to \infty$ since each $q_{i,j}^{(n)}$ is $O(1/n)$. Assume without loss of generality that $a_1 \geq \sup_n a_2(n)$ in Definition 3.13, and observe that

$$|R_2(n)| \leq \sum_{i=0}^{T-1} |Z_i| \left(\frac{a_1}{n} \sum_{j=T}^{T+m(n)} |Z_j'| + \frac{a_2(n)}{n} \sum_{j=T+m(n)}^{n} |Z_j'| \right).$$

Under our hypotheses, we have

$$\lim_{n\to\infty} \frac{1}{m(n)} \sum_{j=T+1}^{T+m(n)} |Z_j'| = E[|Z_1'|] < \infty \text{ a.s.}$$

by Theorem 2.9, so that

$$\lim_{n\to\infty} \frac{a_1}{n} \sum_{j=T+1}^{T+m(n)} |Z_j'| = \lim_{n\to\infty} \left(\frac{a_1 m(n)}{n} \right) \frac{1}{m(n)} \sum_{j=T+1}^{T+m(n)} |Z_j'|$$

$$= 0 \cdot E[|Z_1'|]$$

$$= 0 \text{ a.s..}$$

Similarly, since $a_2(n) \to 0$, we have

$$\lim_{n\to\infty} \frac{a_2(n)}{n} \sum_{j=T+m(n)}^{n} |Z_j'| \leq \lim_{n\to\infty} \frac{a_2(n)}{n} \sum_{j=0}^{n} |Z_j'| = 0 \cdot E[|Z_1'|] = 0 \text{ a.s.},$$

so that $R_2(n) \to 0$ a.s.. An almost identical argument shows that $R_3(n) \to 0$ a.s.. In a similar manner, we have

$$V_n' = \sum_{i=T}^{n} \sum_{j=T}^{n} Z_i' Z_j' q_{i,j}^{(n)} + R'(n),$$

where $R'(n) \to 0$ a.s.. Thus $V_n' - V_n \to 0$ a.s., and the desired result follows from the converging-together lemma (Proposition 1.44 in the Appendix). $\square$

7.3.3 Applications to Batch-Means and Spectral Methods

In this section we discuss some specific estimators of $\tilde{\sigma}^2(\tilde{f})$ that satisfy the conditions of the previous subsection.

Batch Means

For a stationary process $\{Z_n: n \geq 0\}$ with variance constant

$$\tilde{\sigma}^2 = \lim_{n \to \infty} n\mathrm{Var}\left[\frac{1}{n}\sum_{j=0}^{n-1} Z_j\right], \tag{3.16}$$

the standard (discrete time) batch means estimator of $\tilde{\sigma}^2$ based on b batches of length m is given by

$$V_n^{(B)} = \frac{m}{b-1}\sum_{j=1}^{b}(\bar{X}_n(j) - \bar{X}_n)^2 \tag{3.17}$$

for $n = bm$ (the case that we always consider), where

$$\bar{X}_n(j) = \frac{1}{m}\sum_{i=(j-1)m}^{jm-1} Z_i \tag{3.18}$$

is the jth batch mean $(1 \leq j \leq b)$ and $\bar{X}_n = (1/b)\sum_{j=1}^{b}\bar{X}_n(j)$. The reasoning behind this estimator is the same as in Section 7.2.2: for large n, write

$$\tilde{\sigma}^2 \approx n\mathrm{Var}\left[\frac{1}{n}\sum_{j=0}^{n-1} Z_j\right] = \frac{1}{n}\mathrm{Var}\left[\sum_{j=1}^{b} m\bar{X}_n(j)\right] = \frac{m^2}{n}\mathrm{Var}\left[\sum_{j=1}^{b} \bar{X}_n(j)\right].$$

Assuming that, to a good approximation, the batch means are i.i.d., we have

$$\frac{m^2}{n}\mathrm{Var}\left[\sum_{j=1}^{b} \bar{X}_n(j)\right] \approx \frac{bm^2}{n}\mathrm{Var}[\bar{X}_n(1)] = m\mathrm{Var}[\bar{X}_n(1)].$$

Estimating $\mathrm{Var}[\bar{X}_n(1)]$ by the sample variance of the batch means yields the estimator in (3.17).

If the number of batches remains fixed as the simulation run length n increases—as in Section 7.2.2—then $V_n^{(B)}$ is not consistent for $\tilde{\sigma}^2$ in general, and batch means is a cancellation method. We therefore assume that $b = b(n)$ and $m = m(n)$ with $b(n) \to \infty$ and $m(n) \to \infty$ as $n \to \infty$. Typically, $b(n) = O(n^a)$ and $m(n) = O(n^{1-a})$ for some $a \in (0,1)$; one popular choice is $a = 2/3$, which results in variance estimators that have small mean-squared errors.

By expanding the formula for the mean-squared error of V_n and using inequalities of the Cauchy–Schwarz type to bound the resulting terms, the following general result can be established.

Proposition 3.19. *Let* $\{Z_n : n \geq 0\}$ *be a stationary process with finite variance constant* $\tilde{\sigma}^2$, *and let* $V_n^{(B)}$ *be the batch-means estimator of* $\tilde{\sigma}^2$. *Suppose that* $E[Z_n^{12}] < \infty$ *and, for* $p, q, k \geq 0$, $\left|\text{Cov}\left[Z_0^p, Z_k^q\right]\right| \leq c(p, q)k^{-9/2}$, *where* $c(p, q) \in (0, \infty)$. *Also suppose that* $m(n) \to \infty$ *and* $b(n) \to \infty$ *as* $n \to \infty$. *Then*

$$\lim_{n \to \infty} E\left[\left(V_n^{(B)} - \tilde{\sigma}^2\right)^2\right] = 0,$$

and hence $V_n^{(B)} \Rightarrow \tilde{\sigma}^2$.

Note that the final assertion of the proposition is a consequence of the fact that L^2-convergence implies convergence in probability—see Proposition 1.39(v) in the Appendix.

In the SPN setting and for a specified function $\tilde{f}$, the estimator $V_n^{(B)}$ is defined as in (3.17) and (3.18), with $Z_n = \tilde{f}(S_n, C_n)$ for $n \geq 0$. The following result gives conditions on $\tilde{f}$ and on the building blocks of an SPN under which $V_n^{(B)}$ is consistent for $\tilde{\sigma}^2(\tilde{f})$.

Theorem 3.20. *Let* $\{(S_n, C_n) : n \geq 0\}$ *be the underlying chain of an aperiodic SPN, and let* $V_n^{(B)}$ *be given by (3.17) and (3.18) with* $Z_n = \tilde{f}(S_n, C_n)$, *where* $\tilde{f}$ *is a polynomially dominated real-valued function defined on* Σ. *Suppose that Assumption PD holds, so that* $\{\tilde{f}(S_n, C_n) : n \geq 0\}$ *obeys a CLT with variance constant* $\tilde{\sigma}^2(\tilde{f})$. *Also suppose that the number of batches* $b = b(n)$ *and batch length* $m = m(n)$ *satisfy* $b(n) \to \infty$ *and* $m(n) \to \infty$ *as* $n \to \infty$. *Then* $V_n^{(B)} \Rightarrow \tilde{\sigma}^2(\tilde{f})$ *as* $n \to \infty$.

PROOF. There exists an invariant distribution π for the underlying chain $\{(S_n, C_n) : n \geq 0\}$ and, by Proposition 3.10, $\tilde{\sigma}^2(\tilde{f})$ can be represented as a limiting variance of the form (3.11). Proposition 1.13 in Chapter 5 implies that $E_\pi[|\tilde{f}^q(S_n, C_n)|] < \infty$ for $q \geq 0$, and Proposition 3.12 implies that $|\text{Cov}_\pi[\tilde{f}^p(S_0, C_0), \tilde{f}^q(S_k, C_k)]| = o(k^l)$ for $l \in \{1, 2, \dots\}$ and $p, q \geq 0$. We therefore can apply Proposition 3.19 to show that $V_n^{(B)} \Rightarrow \tilde{\sigma}^2(\tilde{f})$ when the initial distribution is π. Observe that $V_n^{(B)}$ can be written in the form (3.9), with

$$q_{i,j}^{(n)} = \begin{cases} (n+1)^{-1} & \text{if } (S_i, C_i) \text{ and } (S_j, C_j) \text{ are in the} \\ & \text{same batch;} \\ -(b(n-m+1))^{-1} & \text{otherwise.} \end{cases}$$

Clearly, $V_n^{(B)}$ is a localized estimator, and an application of Theorem 3.15 yields the desired result. $\qquad\square$

Spectral Methods

Let $\{Z_n : n \geq 0\}$ be a stationary process with common mean $r = E[Z_0]$ and variance constant $\tilde{\sigma}^2$ as in (3.16). Classical spectral estimators of $\tilde{\sigma}^2$

based on observations $Z_0, Z_1, \ldots, Z_{n-1}$ have the form

$$V_n^{(S)} = \sum_{h=-(m-1)}^{m-1} \lambda(h/m)\hat{R}_h, \qquad (3.21)$$

where

$$\hat{R}_h = \frac{1}{n-|h|} \sum_{i=0}^{n-|h|-1} (Z_i - \bar{Z}_n)(Z_{i+|h|} - \bar{Z}_n) \qquad (3.22)$$

and $\bar{Z}_n = (1/n) \sum_{i=0}^{n-1} Z_i$. The function λ is the "lag window," and we assume throughout that the "window length" $m = m(n)$ satisfies $m(n) \to \infty$ and $m^2(n)/n \to 0$. Here $\hat{R}_h$ $(h \geq 0)$ estimates the lag-h covariance

$$\rho(h) = \text{Cov}\,[Z_0, Z_h]\,.$$

Different choices of the lag window lead to different estimators. Well-known windows include the modified Bartlett window $\lambda(h) = 1 - |h|$, the Hanning window $\lambda(h) = 0.5 + 0.5 \cos(\pi h)$, and the Parzen window $1 - h^2$. In general, we consider the class Λ of windows such that $\lambda \in \Lambda$ if and only if

1. λ is continuous on $[-1, 1]$.

2. $\lambda(x) = \lambda(-x)$.

3. $\lambda(0) = 1$.

4. $\lambda(x) = 0$ for $x \notin [-1, 1]$.

5. $\sup_{-1 \leq x \leq 1} |\lambda(x)| < \infty$.

6. $\lim_{x \to 0} (1 - \lambda(x))/|x|^q = \alpha$ for some $q, \alpha \in (0, \infty)$.

It can be shown that all the foregoing windows belong to Λ.

The following result from the time-series literature gives conditions under which $V_n^{(S)} \Rightarrow \tilde{\sigma}^2$. To state this result, we use the fact that any stationary time series $\{Z_n \colon n \geq 0\}$ can be extended to a two-sided stationary time series $\{Z_n \colon -\infty < n < \infty\}$.

Proposition 3.23. *Let $V_n^{(S)}$ be given by (3.21) and (3.22) with $\lambda \in \Lambda$, and suppose that the window length $m = m(n)$ satisfies $m(n) \to \infty$ and $m^2(n)/n \to 0$. Also suppose that*

$$\sum_{n=0}^{\infty} n^p |\rho(n)| < \infty \qquad (3.24)$$

for $p \geq 0$ and

$$\sum_{i,j,k=-\infty}^{\infty} |\kappa(i,j,k)| < \infty, \qquad (3.25)$$

where

$$\kappa(i,j,k) = E_\pi\left[(Z_n - r)(Z_{n+i} - r)(Z_{n+j} - r)(Z_{n+k} - r)\right]$$
$$- \rho(i)\rho(j-k) - \rho(j)\rho(i-k) - \rho(k)\rho(i-j).$$

Then

$$\lim_{n\to\infty} E\left[(V_n^{(S)} - \tilde\sigma^2)^2\right] = 0,$$

and hence $V_n^{(S)} \Rightarrow \tilde\sigma^2$.

In the SPN setting and for a specified function $\tilde f$, the estimator $V_n^{(S)}$ is defined by (3.21) and (3.22) with $Z_n = \tilde f(S_n, C_n)$ for $n \geq 0$. Theorem 3.26 gives conditions under which $V_n^{(S)}$ is consistent for $\tilde\sigma^2(\tilde f)$.

Theorem 3.26. *Let* $\{(S_n, C_n): n \geq 0\}$ *be the underlying chain of an aperiodic* SPN. *Also let* $V_n^{(S)}$ *be defined by (3.21) and (3.22) with* $\lambda \in \Lambda$ *and* $Z_n = \tilde f(S_n, C_n)$, *where* $\tilde f$ *is a polynomially dominated real-valued function defined on* Σ. *Suppose that Assumption PD holds, so that* $\{\tilde f(S_n, C_n): n \geq 0\}$ *obeys a* CLT *with variance constant* $\tilde\sigma^2(\tilde f)$. *Also suppose that the spectral window length* $m = m(n)$ *satisfies* $m(n) \to \infty$ *and* $m^2(n)/n \to 0$. *Then* $V_n^{(S)} \Rightarrow \tilde\sigma^2(\tilde f)$ *as* $n \to \infty$.

PROOF. (Sketch) Write $W_n = (S_n, C_n)$ for $n \geq 0$ and let π be the invariant probability measure of the underlying chain. A conditioning argument in combination with Proposition 3.12 shows that there exist $\beta \in (0,1)$ and $C \in (0, \infty)$ such that

$$\left| E_\pi\left[\hat f_1(W_n)\hat f_2(W_{n+i})\hat f_3(W_{n+j})\hat f_4(W_{n+k})\right]\right| \leq C\beta^i\beta^{j-i}\beta^{k-j}$$

for all integers $0 \leq i < j < k$ and polynomially dominated real-valued functions $\tilde f_1, \tilde f_2, \tilde f_3, \tilde f_4$, where $\hat f_i(w) = \tilde f_i(w) - \pi(\tilde f_i)$ for $1 \leq i \leq 4$. Similar inequalities can be established for other orderings of i, j, and k (e.g., $k < j \leq 0 < i$), as well as for products of two or three terms. (Indeed, Proposition 3.12 applies to products of two terms.) Using these inequalities, we can then show that (3.24) holds for $p \geq 0$ and that (3.25) holds, where $\rho(n) = \mathrm{Cov}_\pi\left[\tilde f(W_0), \tilde f(W_n)\right]$ and

$$\kappa(i,j,k) = E_\pi\left[\hat f(W_n)\hat f(W_{n+i})\hat f(W_{n+j})\hat f(W_{n+k})\right]$$
$$- \rho(i)\rho(j-k) - \rho(j)\rho(i-k) - \rho(k)\rho(i-j)$$

with $\hat f(w) = \tilde f(w) - \pi(\tilde f)$. Propositions 3.10 and 3.23 then imply that $V_n^{(S)} \Rightarrow \tilde\sigma^2(\tilde f)$ when the initial distribution is π. We can write the estimator $V_n^{(S)}$ in the form (3.9), with

$$q_{i,j}^{(n)} = \frac{\lambda(|i-j|/m)}{n+1} - \frac{1}{2(n+1)^2}\sum_{h=-(m-1)}^{m-1} c_{i,j}^{(n)}(h)\lambda(h/m)$$

and

$$c_{i,j}^{(n)}(h) = I(i \geq |h|) + I(i < n - |h|)$$

$$+ I(j \geq |h|) + I(j < n - |h|) - 2\left(1 - \frac{|h|}{n+1}\right).$$

[Here $I(A)$ is the indicator of the condition in A.] It can be seen by inspection that $V_n^{(S)}$ is a localized estimator so that, by Theorem 3.15, $V_n^{(S)} \Rightarrow \tilde{\sigma}^2(\tilde{f})$ for any initial distribution.   □

7.3.4 Functions of Time-Average Limits

Our development is similar to that in Section 6.3.5. Fix $l \geq 1$ and let $\tilde{f} = (\tilde{f}_1, \tilde{f}_2, \ldots, \tilde{f}_l)$ be an $\Re^l$-valued function defined on Σ that is *polynomially dominated* in the sense that each $\tilde{f}_i$ is polynomially dominated for $1 \leq i \leq l$. If Assumption PD holds, then there exists an l-vector $\tilde{r}(\tilde{f}) = (\tilde{r}(\tilde{f}_1), \tilde{r}(\tilde{f}_2), \ldots, \tilde{r}(\tilde{f}_l))$ such that $\bar{r}(n; \tilde{f}) \to \tilde{r}(\tilde{f})$ a.s., where

$$\bar{r}(n; \tilde{f}) = \frac{1}{n} \sum_{j=0}^{n-1} \tilde{f}(S_n, C_n)$$

as before—of course, $\bar{r}(n; \tilde{f})$ is now an l-vector. We consider estimation methods for quantities of the form

$$r = g(\tilde{r}(\tilde{f})) = g(\tilde{r}(\tilde{f}_1), \tilde{r}(\tilde{f}_2), \ldots, \tilde{r}(\tilde{f}_l)), \qquad (3.27)$$

where $g: \Re^l \mapsto \Re$ is differentiable in a neighborhood of $\tilde{r}(\tilde{f})$.

Since differentiability implies continuity, it follows from the a.s. convergence of $\bar{r}(n; \tilde{f})$ to $\tilde{r}(\tilde{f})$ that the estimator $r_n = g(\bar{r}(n; \tilde{f}))$ is strongly consistent for r. To obtain an asymptotic confidence interval for r, we start with the following result, which is a consequence of Theorem 2.9, Proposition 3.10, and the Cramér–Wold theorem (Proposition 1.46 in the Appendix).

Theorem 3.28. *Let $\{(S_n, C_n): n \geq 0\}$ be the underlying chain of an aperiodic* SPN, *and let $\tilde{f} = (\tilde{f}_1, \tilde{f}_2, \ldots, \tilde{f}_l)$ be a polynomially dominated $\Re^l$-valued function defined on Σ. Suppose that Assumption PD holds, so that there exists an invariant measure π for the chain and $\bar{r}(n; \tilde{f}) \to \tilde{r}(\tilde{f})$ a.s. for some finite l-vector $\tilde{r}(\tilde{f})$. Then*

$$\sqrt{n}(\bar{r}(n; \tilde{f}) - \tilde{r}(\tilde{f})) \Rightarrow N(0, W)$$

as $n \to \infty$, where $N(0, W)$ is a multivariate normal random vector with covariance matrix $W = \|w_{s,t}\|$ given by

$$w_{s,t} = \lim_{n \to \infty} n\mathrm{Cov}_\pi[\bar{r}(n; \tilde{f}_s), \bar{r}(n; \tilde{f}_t)]. \qquad (3.29)$$

When the conclusion of Theorem 3.28 holds, an application of the delta method—see Proposition 1.45 in the Appendix—shows that

$$\sqrt{n}(r_n - r) \Rightarrow \sigma N(0, 1)$$

as $n \to \infty$, where

$$\sigma^2 = \nabla g(\tilde{r}(\tilde{f}))^t W \nabla g(\tilde{r}(\tilde{f})).$$

Our goal, then, is to consistently estimate σ^2.

Given a quadratic-form variance estimator as in Section 7.3.2 with coefficients $q_{i,j}^{(n)}$, we can define an $l \times l$ matrix $W_n = \|V_n(s,t)\|$, where

$$V_n(s,t) = \sum_{i=0}^{n} \sum_{j=0}^{n} \tilde{f}_s(S_i, C_i) \tilde{f}_t(S_j, C_j) q_{i,j}^{(n)}$$

for $s, t \in \{1, 2, \ldots, l\}$. For the batch-means estimator of the previous subsection, the calculations that establish the convergence $V_n(s,s) \Rightarrow w_{s,s}$ when the initial distribution is π can be modified in a straightforward way to show that $V_n(s,t) \Rightarrow w_{s,t}$ for $s \neq t$. Thus $W_n \Rightarrow W$ when the initial distribution is π and the conditions of Theorem 3.20 hold. Similarly, for the spectral estimators of the previous subsection, $W_n \Rightarrow W$ when the initial distribution is π and the conditions of Theorem 3.26 hold.

The coupling argument used to establish Theorem 3.15 can similarly be extended to obtain Theorem 3.30 below. In the theorem, the matrix W_n is said to be a *localized* estimator of W if and only if each $q_{i,j}^{(n)}$ satisfies (3.14).

Theorem 3.30. *Let* $\{(S_n, C_n): n \geq 0\}$ *be the underlying chain of an aperiodic* SPN, *and let* $\tilde{f}$ *be a polynomially dominated* $\Re^l$-*valued function defined on* Σ. *Suppose that Assumption PD holds, so that there exists an invariant distribution* π *for the chain and* $\{\tilde{f}(S_n, C_n): n \geq 0\}$ *obeys a* CLT *with covariance matrix* W. *If a localized estimator* W_n *satisfies* $W_n \Rightarrow W$ *when the initial distribution is* π, *then* $W_n \Rightarrow W$ *for any initial distribution.*

The foregoing results can be combined to yield confidence intervals for $r = g(\tilde{r}(\tilde{f}))$. Suppose, for example, that W_n is the batch-means estimator of W and the conditions of Theorem 3.20 hold, or that W_n is a spectral estimator of W and the conditions of Theorem 3.26 hold. Set

$$\sigma_n^2 = \nabla g(\bar{r}(n; \tilde{f}))^t W_n \nabla g(\bar{r}(n; \tilde{f}))$$

for $n \geq 1$. Since $\bar{r}(n; \tilde{f}) \to \tilde{r}(\tilde{f})$ a.s. by Theorem 3.28 and $W_n \Rightarrow W$ by Theorem 3.30, it follows from the differentiability (and hence continuity) of g at $\tilde{r}(\tilde{f})$ together with the continuous mapping theorem—see Proposition 1.42 in the Appendix—that $\sigma_n^2 \Rightarrow \sigma^2$. Thus

$$\frac{\sqrt{n}(r_n - r)}{\sigma_n} \Rightarrow N(0, 1),$$

and

$$\left[r_n - \frac{z_p \, \sigma_n}{\sqrt{n}}, r_n + \frac{z_p \, \sigma_n}{\sqrt{n}} \right]$$

is an asymptotic $100p\%$ confidence interval for r, where z_p is the $(1+p)/2$ quantile of the standard normal distribution.

7.3.5 Consistent Estimation in Continuous Time

As before, we assume that the SPN of interest is aperiodic and that Assumption PD holds, so that, by Theorem 2.14, $\bar{r}(t; f) \to r(f)$ a.s. for some finite constant $r(f)$ and any real-valued function f defined on G; here

$$\bar{r}(t; f) = \frac{1}{t} \int_0^t f(X(u)) \, du$$

as before. We now fix f and consider the problem of obtaining an asymptotic confidence interval for $r(f)$. As before, let t^* be the holding-time function and set $(ft^*)(s, c) = f(s)t^*(s, c)$. Using Theorems 2.9 and 2.14, we find that

$$r(f) = \lim_{t \to \infty} \bar{r}(t; f) = \frac{E_\mu[\tilde{Y}_1(ft^*)]}{E_\mu[\tilde{Y}_1(t^*)]} = \frac{E_\mu[\tilde{Y}_1(ft^*)]/E_\mu[\tilde{\tau}_1]}{E_\mu[\tilde{Y}_1(t^*)]/E_\mu[\tilde{\tau}_1]} = \frac{\tilde{r}(ft^*)}{\tilde{r}(t^*)},$$

where $\tilde{r}(\tilde{f}) = \lim_{n \to \infty}(1/n) \sum_{j=0}^{n-1} \tilde{f}(S_n, C_n)$ as before. Thus $r(f)$ can be expressed in the form (3.27) with $\tilde{f} = (ft^*, t^*)$ and $g(x, y) = x/y$. We therefore can apply the methods of the previous subsection.

Let $\|q_{i,j}^{(n)}\|$ be a set of coefficients such that, for any polynomially dominated functions $\tilde{f}_s$ and $\tilde{f}_t$ defined on Σ, the quadratic-form estimator

$$V_n(\tilde{f}_s, \tilde{f}_t) = \sum_{i=0}^n \sum_{j=0}^n \tilde{f}_s(S_i, C_i) \tilde{f}_t(S_j, C_j) q_{i,j}^{(n)}$$

is consistent for $w_{s,t}$, where $w_{s,t}$ is given by (3.29). (The coefficients for the variable batch-means and spectral methods satisfy this condition, for example.) Then

$$\left[\hat{r}_n - \frac{z_p \, \sigma_n}{\sqrt{n}}, \hat{r}_n + \frac{z_p \, \sigma_n}{\sqrt{n}} \right] \tag{3.31}$$

is an asymptotic $100p\%$ confidence interval for $r(f)$, where

$$\hat{r}_n = \frac{\bar{r}(n; ft^*)}{\bar{r}(n; t^*)},$$

$$\sigma_n^2 = \frac{1}{\bar{r}^2(n; t^*)} \left(V_n(1, 1) - 2\hat{r}_n V_n(1, 2) + \hat{r}_n^2 V_n(2, 2) \right), \tag{3.32}$$

and z_p is the $(1+p)/2$ quantile of the standard normal distribution. Here $\bar{r}(n; \tilde{f})$ is defined as in (3.3),

$$V_n(1,1) = \sum_{i=0}^{n}\sum_{j=0}^{n}(ft^*)(S_i, C_i)\,(ft^*)(S_j, C_j)\,q_{i,j}^{(n)},$$

$$V_n(1,2) = \sum_{i=0}^{n}\sum_{j=0}^{n}(ft^*)(S_i, C_i)\,t^*(S_j, C_j)\,q_{i,j}^{(n)},$$

and

$$V_n(2,2) = \sum_{i=0}^{n}\sum_{j=0}^{n}t^*(S_i, C_i)\,t^*(S_j, C_j)\,q_{i,j}^{(n)}.$$

EXAMPLE 3.33 (Variable batch means). For real-valued functions $\tilde{f}$ and $\tilde{g}$ defined on Σ, extend the notation in (3.17) and (3.18) by setting

$$V_n^{(B)}(\tilde{f}, \tilde{g}) = \frac{m}{b-1}\sum_{j=1}^{b}\big(\bar{X}_n(j; \tilde{f}) - \bar{X}_n(\tilde{f})\big)\big(\bar{X}_n(j; \tilde{g}) - \bar{X}_n(\tilde{g})\big),$$

where

$$\bar{X}_n(j; \tilde{h}) = \frac{1}{m}\sum_{i=(j-1)m}^{jm-1}\tilde{h}(S_i, C_i)$$

and

$$\bar{X}_n(\tilde{h}) = \frac{1}{b}\sum_{j=1}^{b}\bar{X}_n(j; \tilde{h})$$

for $\tilde{h} = \tilde{f}, \tilde{g}$. Then, for the method of variable batch means in continuous time, the variance constant σ_n^2 that appears in the confidence-interval formula (3.31) is given by (3.32), with $V_n(1,1) = V_n^{(B)}(ft^*, ft^*)$, $V_n(2,2) = V_n^{(B)}(t^*, t^*)$, and $V_n(1,2) = V_n^{(B)}(ft^*, t^*)$.

EXAMPLE 3.34 (Spectral methods). For real-valued functions $\tilde{f}$ and $\tilde{g}$ defined on Σ, extend the notation in (3.21) and (3.22) by setting

$$V_n^{(S)}(\tilde{f}, \tilde{g}) = \lambda(0)\hat{R}_0(\tilde{f}, \tilde{g}) + \sum_{h=1}^{m-1}\lambda(h/m)\hat{R}_h(\tilde{f}, \tilde{g}) + \sum_{h=1}^{m-1}\lambda(h/m)\hat{R}_h(\tilde{g}, \tilde{f}),$$

where

$$\hat{R}_h(\tilde{f}, \tilde{g}) = \frac{1}{n-h}\sum_{i=0}^{n-h-1}\big(\tilde{f}(S_i, C_i) - \bar{Z}_n(\tilde{f})\big)\big(\tilde{g}(S_{i+h}, C_{i+h}) - \bar{Z}_n(\tilde{g})\big),$$

Table 7.1. Simulation Results for Cyclic Queues with Feedback: Point Estimates and 95% Confidence-Interval Half-Widths for the Long-Run Average Number of Jobs at Center 1 (True Value = 2.5823)

Estimator	Simulation Length				
	50	100	1000	10000	100000
regenerative (jack.)	2.6027	2.5908	2.5775	2.5833	2.5823
	±0.6529	±0.5152	±0.1819	±0.0579	±0.0183
batch means (cont.)	2.6393	2.6059	2.5795	2.5834	2.5823
	±0.5030	±0.4506	±0.1896	±0.0606	±0.0191
batch means (discr.)	2.6507	2.6097	2.5798	2.5835	2.5823
	±0.6022	±0.5044	±0.1919	±0.0611	±0.0191
STS area	2.6393	2.6059	2.5795	2.5834	2.5823
	±0.2362	±0.2645	±0.1780	±0.0598	±0.0193
spectral (Bartlett)	2.6446	2.6058	2.5795	2.5834	2.5823
	±0.5399	±0.4267	±0.1618	±0.0549	±0.0179
spectral (Parzen)	2.6446	2.6058	2.5795	2.5834	2.5823
	±0.5791	±0.4588	±0.1725	±0.0571	±0.0183
spectral (Hanning)	2.6446	2.6058	2.5795	2.5834	2.5823
	±0.5448	±0.4326	±0.1657	±0.0563	±0.0182
var. batch means	2.6437	2.6051	2.5792	2.5834	2.5823
	±0.5544	±0.4292	±0.1629	±0.0550	±0.0179

Note: Each reported number is an average over 500 simulation repetitions.

$$\bar{Z}_n(\tilde{f}) = \frac{1}{n} \sum_{i=0}^{n-1} \tilde{f}(S_i, C_i),$$

and

$$\bar{Z}_n(\tilde{g}) = \frac{1}{n} \sum_{i=0}^{n-1} \tilde{g}(S_i, C_i).$$

Then, for a spectral method in continuous time with lag window λ, the variance constant σ_n^2 that appears in the confidence-interval formula (3.31) is given by (3.32), with $V_n(1,1) = V_n^{(S)}(ft^*, ft^*)$, $V_n(2,2) = V_n^{(S)}(t^*, t^*)$, and $V_n(1,2) = V_n^{(S)}(ft^*, t^*)$.

EXAMPLE 3.35 (Cyclic queues with feedback). We illustrate the various estimation methods discussed in this chapter using the closed network of queues in Example 1.4 of Chapter 2. Similarly to Example 3.50 of Chapter 6, successive service times at center i ($i = 1, 2$) are i.i.d. according to an exponential distribution with intensity q_i, where $q_1 = 1.1$ and $q_2 = (1.1)^{3/2}$. The routing probability (with which a job completing service at center 1 moves to center 2) is $p = 0.775$. There are $N = 4$ jobs, and we model the system using the SPN in Figure 2.2. For this model, there are roughly 1.7 transitions per time unit.

Table 7.2. Simulation Results for Cyclic Queues with Feedback: Empirical Coverage Probabilities when Estimating the Long-Run Average Number of Jobs at Center 1 with Nominal Coverage Probability of 95%, Based on 500 Simulation Repetitions

Estimator	Simulation Length				
	50	100	1000	10000	100000
regenerative	0.798	0.852	0.926	0.946	0.948
batch means (cont.)	0.770	0.860	0.938	0.946	0.949
batch means (discr.)	0.816	0.886	0.942	0.946	0.948
STS area	0.436	0.636	0.930	0.960	0.951
spectral (Bartlett)	0.782	0.842	0.902	0.938	0.947
spectral (Parzen)	0.800	0.854	0.918	0.942	0.946
spectral (Hanning)	0.788	0.844	0.908	0.940	0.948
var. batch means	0.786	0.838	0.904	0.932	0.945

We use the following methods to estimate $r(f)$, the long-run average number of jobs at center 1:

1. *The regenerative method with jackknifing.* We use the variant of the standard regenerative discussed in Section 6.3.2. The regeneration points are the successive times at which there is a completion of service at center 1 with all jobs at center 1. Estimates are based on the number of cycles completed in $[0, t]$, where t is the simulation length. A cycle is approximately 2.5 time units long on average.

2. *The method of batch means in continuous time.* Estimates are based on $b = 20$ batches.

3. *The method of batch means in discrete time.* We use the jackknife technique as in Remark 2.36. Estimates are based on approximately[2] $b = 20$ batches.

4. *The STS area method.* We use a value of $m = 20$.

5. *Spectral methods.* We use the approach described in Example 3.34 with modified-Bartlett, Parzen, and Hanning lag windows. The window length is $m(n) = n^{1/3}$, rounded to the nearest integer.

---

[2]When the length of the simulation is t, denote by $n(t)$ the total number of discrete-time observations: $n(t) = \sup \{ n \geq 0 : t^*(S_0, C_0) + \cdots + t^*(S_{n-1}, C_{n-1}) \leq t \}$. In general, the nominal batch length $n(t)/b$ is not an integer, and so we choose the batch length to be either $m' = \lfloor n(t)/b \rfloor$ or $m' = \lceil n(t)/b \rceil$. The actual number of batches is therefore $b' = \lfloor n(t)/m' \rfloor$, and we choose m' so as to minimize $n(t) - b'm'$, the number of discarded observations.

6. *The method of variable batch means.* We use the approach described in Example 3.33 with the number of batches approximately equal to $b(n) = n^{2/3}$.

Point estimates and 95% confidence-interval half-widths for $r(f)$ are displayed in Table 7.1. Each displayed number represents an average over 500 i.i.d. simulation repetitions. Thus, for example, when the simulation length is $t = 100$ time units, the expected value of the jackknifed regenerative estimator is approximately 2.5908, the bias of the point estimator is approximately $2.5908 - 2.5823 = 0.0085$, and the expected half-width of the 95% confidence interval is 0.5152. Table 7.2 displays empirical coverage probabilities, also based on 500 i.i.d. simulation repetitions. These probabilities correspond to the nominal coverage probability of 95%.

In this example, the various estimation methods yield fairly similar point estimates and confidence intervals. Observe that the empirical probabilities tend to be less than the nominal value, with the undercoverage being particularly severe at small run lengths. This is perhaps to be expected, since the coverage of an asymptotic confidence interval is only guaranteed to be close to the nominal value when the run length is large. When the simulation length is large (10^5 time units), there is essentially no bias in the point estimators, and the coverage of each algorithm is close to the nominal value of 0.95. Moreover, the confidence-interval lengths for the consistent estimation methods are slightly shorter than those for the STS methods.

The performance of each of the nonregenerative methods is in general sensitive to the settings of the various algorithm parameters. For example, the STS area method does not perform well at short run lengths, with coverages of 0.436 and 0.636 at run lengths of 50 and 100 time units, respectively. When the parameter m is decreased from $m = 20$ to $m = 5$, however, the empirical coverage increases dramatically, with respective values of 0.808 and 0.878 and half-widths of 0.6389 and 0.5716—values that are comparable to the other methods. An important open problem is to develop theoretically sound and practically effective methods for setting parameter values.

Notes

The video-on-demand system in Example 1.1 is closely related to the models of noninteractive video-on-demand studied by Aggarwal et al. (1995) and Pyssysalo and Ojala (1995). Example 1.2 was suggested by Peter Glynn. The method discussed in Remark 1.3 for dealing with the problem in Example 1.2 was pointed out by Shane Henderson and is closely related to ideas in Andradottir et al. (1994), Glynn (1989b), and Henderson and Glynn (1999b). Minh (1987) provides a specialized technique for regenerative simulation of GI/G/k queues in heavy traffic.

The discussion of STS methods for SPNs follows Haas (1999a, 1999c). Some basic references on the method of batch means include Conway (1963) and Brillinger (1973). Schmeiser (1982) gives recommendations on the number of batches. Schruben (1983) introduced the general class of STS methods and specialized the approach to yield the STS maximum and STS area methods. Other STS methods include the method of spaced batch means (with the number of batches independent of the simulation run length) as given by Fox et al. (1991) and the STS weighted-area method (Goldsman et al., 1990; Goldsman and Schruben, 1990).

In general, it can be difficult to determine for a specific SPN and function f whether the output process $\{ f(X(t)) : t \geq 0 \}$ obeys an SLLN, so that the time-average limit is well defined. It is even harder to determine whether STS methods can be used to estimate the time-average limit, assuming that it exists. Birkhoff's ergodic theorem gives general conditions under which a stochastic process obeys an SLLN; see, for example, Breiman (1968, Chapter 6) or Durrett (1991, Section 6.2). A key condition of the ergodic theorem is that the process be stationary. Brillinger (1973) establishes the validity of the batch-means method under the stationarity assumption, and Schruben (1983) establishes an analogous result for the STS maximum and STS area methods. In the simulation setting, however, the initial marking and clock readings of an SPN usually cannot be selected so as to ensure stationarity. Glynn and Iglehart (1990) were the first to avoid stationarity conditions by showing that STS methods are applicable under the sole assumption that the output process obeys an FCLT. They also showed that the method of batch means (with the number of batches independent of the simulation run length) can be viewed as an STS method. In addition, they showed that consistent estimation methods lead to shorter and less variable asymptotic confidence-interval lengths than cancellation methods.

The idea behind the proof of the FCLT in Proposition 2.3 is that $\{ Y_k(f) : k \geq 1 \}$ is a sequence of o.d.s. random vectors. It therefore follows from a multivariate extension of Proposition 2.24 in the Appendix that the process $\{ Y_k(f) : k \geq 1 \}$ satisfies an FCLT. A standard random-time-change argument using Proposition 2.25 in the Appendix—see the proof of Theorem 1 in Glynn and Whitt (1987)—then yields the desired result. In an analogous manner, Theorem 2.12 and a random-time-change argument yield Theorem 2.17; see Haas (1999a, 1999c) for details.[3]

The convergence results in Remark 2.10 can be strengthened considerably. See, for example, Meyn and Tweedie (1993a, Chapters 13–16) for the discrete-time case and Meyn and Tweedie (1993b, 1993c) for the continuous-time case.

---

[3]Haas (1999a, 1999c) actually uses a slightly different form of the FCLT for the underlying chain than Theorem 2.12, but the argument is essentially identical.

Our discussion of the general theory of STS methods follows Glynn and Iglehart (1990). The original paper contains some minor errors in the authors' demonstration that the STS area method and STS maximum method follow as special cases of the general theory—certain Student's t statistics are formed without dividing the χ^2 statistic in the denominator by the degrees of freedom. Our presentation of the computationally efficient representation of the quantity A_i in the STS area method follows the discussion in Goldsman et al. (1990). See Glynn and Whitt (1988, Theorem 4) for a proof of the assertion that an SLLN implies a functional SLLN.

The extended batch-means methodology in Section 7.2.3 is due to Muñoz and Glynn (1997). Muñoz and Glynn (2001) extend the methods of Section 7.2.2 to the multivariate setting and provide techniques for constructing a simultaneous confidence region for two or more time-average limits.

For discussions of the method of variable batch means, see, for example, Carlstein (1986), Chien (1989), Chien et al. (1997), Damerdji (1994, 1995), and Song and Schmeiser (1995). It is shown in Damerdji and Goldsman (1995) that many STS methods have variants that are consistent estimation methods. Many authors have studied spectral estimation methods, both within the general setting of time-series analysis (Anderson, 1971; Brockwell and Davis, 1987; Grenander and Rosenblatt, 1984) and within the specific setting of simulation (Bratley et al., 1987; Damerdji, 1991; Heidelberger and Welch, 1981). Song and Schmeiser (1993) give quadratic-form representations for a variety of estimators, including batch-means and spectral estimators. Note that our formulas differ from those in Song and Schmeiser (1993) by a factor of n; the latter paper is concerned with $\text{Var}_\pi\left[\bar{r}(n)\right]$ rather than $n\text{Var}_\pi\left[\bar{r}(n)\right]$.

The method of batch means studied in this chapter has the property that the batches are disjoint. Variants of the basic method allow some degree of overlap between batches. The terminology "overlapping batch means" typically refers to the method in which the batches have maximal overlap. That is, batch 1 consists of observations $Z_0, Z_1, \ldots, Z_{m-1}$, batch 2 consists of observations $Z_1, Z_2, \ldots, Z_m$, and so forth. Song and Schmeiser (1993) give a quadratic-form representation for the overlapping-batch-means estimator. The method of overlapping batch means was first studied by Meketon and Schmeiser (1984). These authors noted that the method of overlapping batch means and the spectral method with the Bartlett window are asymptotically equivalent as the run length becomes large; see also Song and Schmeiser (1993) and Welch (1987).

Damerdji (1991, 1994, 1995) and Damerdji and Goldsman (1995) establish the validity of several consistent estimation methods when the output process of the simulation obeys a strong invariance principle (also called a strong approximation). Roughly speaking, a strong invariance principle is a strengthening of an FCLT in which convergence to a limiting Brownian motion holds with probability 1. The appeal of this approach is that, when a strong invariance principle holds, the estimator of the variance constant not only converges in distribution—see Damerdji (1995)—but often with probability 1; see Damerdji (1991, 1994) and Damerdji and Goldsman (1995). This latter "strong" consistency is needed, for example, to establish the validity of sequential stopping rules for simulations; see Glynn and Whitt (1992b). It is highly nontrivial, however, to establish strong invariance principles for specific estimation methods. For example, it appears difficult—using currently known sufficient conditions for the strong invariance principle—to establish the validity of estimation methods such as the popular version of variable batch means in which the number of batches grows as the 2/3 power of the run length.

In this chapter, SLLNs, FCLTs and CLTs for both the marking process and underlying chain of an SPN are established under Assumption PD— see Theorems 2.9, 2.12, 2.14, and 2.17, as well as the results in (3.1) and (3.2). All of these results can be established under weaker conditions on the moments of the clock-setting distributions—that is, the clock-setting distribution functions need not be elements of $\mathcal{G}^+$. The idea is to adapt certain results established by Glynn and Haas (2002b) for GSMPs. For simplicity, suppose that all transitions are timed. Let t^* be the usual holding-time function and for $u \geq 0$ denote by $\mathcal{H}_u$ the set of real-valued functions $\tilde{f}$ defined on Σ such that $|\tilde{f}(s,c)| \leq a + b\left(t^*(s,c)\right)^u$ for some $a, b \geq 0$. Also suppose that there exists a sequence $\{\theta(k): k \geq 0\}$ of od-regeneration points for the underlying chain, and define $\tilde{Y}(\tilde{f})$ as in (2.4). It follows from Glynn and Haas (2002b) that if each clock-setting distribution function has finite rth moment, where $r = q\max(u,1)$ for some $q \in \{1, 2, \ldots\}$ and $u \geq 0$, then $E_\mu[\tilde{Y}^q(|\tilde{f}|)] < \infty$ for any function $\tilde{f} \in \mathcal{H}_u$. In particular, if each clock-setting distribution function has finite rth moment—where $r = 2\max(u,1)$ for some $u \geq 0$—and satisfies the remaining conditions of Assumption PD, then the output process $\{\tilde{f}(S_n, C_n): n \geq 0\}$ obeys an SLLN, CLT, and FCLT for any $\tilde{f} \in \mathcal{H}_u$. It follows that the continuous-time process $\{f(X(t)): t \geq 0\}$ obeys an SLLN, CLT, and FCLT for any real-valued function f, provided that each clock-setting distribution function has finite second moment and the remaining conditions of Assumption PD hold.

See Billingsley (1986, Theorem A21) for a proof of Lemma 3.6. Propositions 3.10 and 3.12 follow from Theorems 17.5.3 and Theorem 16.1.5 in Meyn and Tweedie (1993a), respectively. The proof of Proposition 3.19 follows from the proof of Corollary 1 in Chien et al. (1997). In this corollary, the requirement that $\left|\text{Cov}\left[Z_0^p, Z_k^q\right]\right| \leq c(p,q)k^{-9/2}$ for $p, q, k \geq 0$ is replaced by the requirement that $\{Z_n: n \geq 0\}$ is ϕ-mixing (see Section A.2.2) with $\phi_k = O(k^{-9})$. Examination of the proof of the corollary shows that the sole purpose of the latter assumption is to bound covariances of the foregoing type, using the inequality

$$\left|\text{Cov}\left[Z_0^p, Z_k^q\right]\right| \leq 2\sqrt{\phi_k E[Z_0^{2p}] E[Z_k^{2q}]},$$

which follows from Lemma 1 in Section 20 of Billingsley (1968). The discussion of spectral estimators follows Anderson (1971), who shows that each of the lag windows mentioned in the text belongs to the class Λ, and whose Theorems 9.4.3 and 9.4.4 jointly imply Proposition 3.23. For some estimation methods, consistency of the variance estimator can be established under weaker conditions than are given here—see Glynn and Haas (2002a).

As discussed in Example 3.35, most of the simulation methods discussed in this chapter have parameters for which values must be chosen: number of batches, batch lengths, lag-window lengths, and so forth. Currently, there is scant theoretical guidance on how to set these parameters. Even the choice of simulation run length is nontrivial. It seems reasonable to use a sequential stopping rule to determine the run length—indeed, several sequential estimation methods have been proposed in the literature. As mentioned above, however, the estimator of the variance constant $\sigma^2(f)$ or $\tilde{\sigma}^2(\tilde{f})$ must be strongly consistent for such methods to be valid. In the absence of regenerative structure, strong consistency of variance estimators has only been established under the assumption that the output process obeys a strong invariance principle as in Damerdji (1991, 1995, 1994). Because strong invariance principles are hard to establish for specific models, the behavior of most sequential estimation methods is not well understood at present. Law and Kelton (2000, Section 9.5.3) discuss the empirical performance of various fixed-length and sequential estimation methods and provide references to a number of experimental studies.

8
Delays

Our discussion up to this point has centered around performance measures that can be expressed as time-average limits—or functions of time-average limits—that involve the marking process or underlying chain of an SPN. Such measures include long-run system reliability, availability, and cost, as well as throughput and discounted cost. Assessment of computer, communication, manufacturing, and transportation systems, however, often involves analysis of long-run delay characteristics. Examples of such characteristics include the long-run average time to produce an item in a flexible manufacturing system, the long-run fraction of queries in a database system that require more than a specified amount of time to compute, and the long-run average revenue generated by telephone traffic under a graduated rate structure. When the system of interest is modelled as an SPN, each of these latter characteristics can be expressed as a time-average limit of the form $\lim_{n \to \infty} (1/n) \sum_{j=0}^{n-1} f(D_j)$, where f is a real-valued function and $D_0, D_1, \ldots$ is a sequence of delays determined by the marking changes of the net. Other delay characteristics—such as the long-run variability in the time required to transmit a message packet from one node to another in a communication network—can be expressed as functions of time-average limits. As with time-average limits defined in terms of the marking process, time-average limits defined in terms of delays often cannot be computed analytically or numerically, but must be estimated using simulation.

A delay in an SPN is computed as the length of a corresponding "delay interval"—that is, a random time interval—whose start (left endpoint) and termination (right endpoint) each coincide with a marking-change epoch.

Sometimes the *limiting average delay* $\lim_{n\to\infty}(1/n)\sum_{j=0}^{n-1} D_j$ can be estimated indirectly, that is, without measuring lengths of individual delay intervals. For general time-average limits of a sequence of delays, however, individual lengths must be measured and then combined to form point and interval estimates. Specification and subsequent measurement of individual delays is a decidedly nontrivial step of the simulation: in general, there can be more than one ongoing delay at a time point and delays need not terminate in the order in which they start.

In Section 8.1 we introduce a recursively generated sequence of real-valued random vectors, determined by the sample paths of the underlying chain, to provide the link between the starts and terminations of individual delay intervals. Heuristically, the nth such "start vector" records the starts of all ongoing delays and newly started delays at the nth marking change. The values of the starts and the order of the starts in the start vector together summarize the history of the net and comprise sufficient information to measure individual delays. At each marking change, the current time may be inserted, old starts may be deleted, and the components of the start vector may be permuted according to a mechanism that depends explicitly on the current marking, new marking, and set of transitions that trigger the marking change. Deleted starts are subtracted from the current time to compute delays. This method for specifying and measuring delays avoids the need to "tag" individual entities in the system (such as customers) by using either distinguishable tokens or additional places and transitions.

Section 8.2 focuses on estimation of delay characteristics when the SPN under study has a recurrent single state, so that there exist sequences of regeneration points for both the underlying chain and the marking process. When the characteristic of interest is the limiting average delay, point and interval estimates can be obtained by directly applying results in Chapters 6 and 7. One means of doing this is to first express—as in Little's law—the limiting average delay in terms of the long-run average length of the start vector and the long-run average number of starts per unit time. For general time-average limits, the situation is usually more complex. When there are no ongoing delays at any regeneration point for the marking process, the sequence of delays is a regenerative process in discrete time. Under suitable moment conditions, the standard regenerative method for analysis of simulation output can then be used to obtain strongly consistent point estimates and asymptotic confidence intervals for time-average limits of the sequence of delays. When there are ongoing delays at each regeneration point, however, the standard regenerative method is not applicable—see Example 2.5. To handle this situation, we construct a sequence of random indices that decomposes sample paths of the sequence of delays into one-dependent stationary (o.d.s.) cycles. The idea is to identify a sequence of regeneration points for the marking process such that all delays that start during a cycle terminate by the end of the next cycle. Under suitable moment conditions, an extension (as in Section 6.3.8) of the standard regenerative method to one-dependent cycles can then be used to estimate general time-average limits—recall that this extended regenerative method is based on a single simulation run. Also as in Section 6.3.8, an estimation method based on multiple runs can be applied—we compare the statistical efficiency of these two methods in Section 8.2.3.

We next consider SPNs to which the foregoing methods cannot be applied, either because there is no apparent sequence of regeneration points for the marking process or underlying chain, or because regenerations occur too infrequently. If Assumption PD holds for such an SPN, then there exists a sequence of od-regeneration points that decomposes sample paths of the underlying chain into o.d.s. cycles; see Chapter 7. In Section 8.3, we show that the output process $\{ f(D_j) : j \geq 0 \}$ inherits this od-regenerative structure under mild regularity conditions on the start-vector mechanism. Moreover, the sum of the output process over a cycle has finite moments of all orders provided that f is polynomially dominated. Unlike in Section 8.2, the cycles of the output process usually cannot be determined explicitly, and neither the regenerative method nor its extensions can be applied. The mere existence of these cycles, however, implies that the output process obeys an FCLT. It then follows as in Chapter 7 that STS methods such as the method of batch means can be used to obtain strongly consistent point estimates and asymptotic confidence intervals for time-average limits. In addition, an extension of the method of batch means can be used to obtain point estimates and confidence intervals for functions of time-average limits.

8.1 Specification and Measurement of Delays

A sequence of delays in an SPN is specified in terms of *starts* $\{ A_j : j \geq 0 \}$ and *terminations* $\{ B_j : j \geq 0 \}$. These nonnegative random variables are defined on the same probability space as the underlying chain $\{ (S_n, C_n) : n \geq 0 \}$. We restrict attention to sequences $\{ A_j : j \geq 0 \}$ and $\{ B_j : j \geq 0 \}$ such that $A_j = \zeta_{\alpha(j)}$ and $B_j = \zeta_{\beta(j)}$ for $j \geq 0$, where $\alpha(j)$ and $\beta(j)$ are a.s. finite random indices. That is, we restrict attention to delays that start and terminate only at marking changes. We also focus on sequences for which the $\alpha(j)$'s are nondecreasing, so that delays are enumerated in start order. The $\beta(j)$'s need not be nondecreasing, however, reflecting the fact that there can be more than one ongoing delay at a time point and delays need not terminate in the order in which they start.

The key challenge when specifying and measuring delays is to link the starts and terminations of individual delay intervals. After briefly discussing methods based on tagging, we introduce the method of start vectors, which is our preferred approach. As always, we restrict attention to SPNs for which

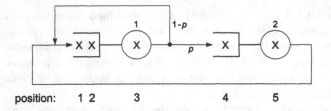

position: 1 2   3   4   5

Figure 8.1. Positions of jobs in cyclic queues with feedback.

the marking process has an infinite lifetime—see Section 3.3—so that

$$P_\mu\Big\{\sup_{n\geq 0}\zeta_n = \infty\Big\} = 1. \tag{1.1}$$

8.1.1 Tagging

Methods based on tagging for measuring individual delays in an SPN with indistinguishable tokens may require a large number of additional places and transitions.

EXAMPLE 1.2 (Cyclic queues with feedback). For the closed network of queues of Example 1.4 in Chapter 2, consider the delay intervals from whenever a job completes service at center 2 (and moves to center 1) to when the job next completes service at center 2, and suppose that we wish to measure the combined sequence of delays for all N jobs. Using additional places and transitions, we can tag each of the jobs and keep track of the jobs as they traverse the network.

We number the jobs from 1 to N and, to specify the position of each job in the network, we conceptually order the jobs in a "job stack." The jobs at center 1 are closer to the top of the stack than the jobs at center 2. At each center, jobs appear in the job stack in the order in which they join the tail of the queue, the latest to join being closest to the top of the job stack. The job at the top of the job stack is said to be in position 1, the next job in position 2, and so forth; see Figure 8.1 for $N = 5$ jobs.

Observe that when a job completes service at center 1 and moves to center 2, the jobs retain their positions. When a job completes service at center 1 and joins the tail of the queue at center 1, it goes into position 1; the position of each other job at center 1 increases by 1. When a job completes service at center 2, it goes into position 1; the position of each other job in the network increases by 1.

We use $2N$ places and N immediate transitions to maintain the position of each job in the network; see Figure 8.2 for $N = 3$ jobs. The transitions have the following interpretation: e_{N+1} = "completion of service at center 1," e_{N+2} = "completion of service at center 2," and e_j = "decrease of position for job j" for $1 \leq j \leq N$. Place d_{2N+i} ($i = 1, 2$) contains n

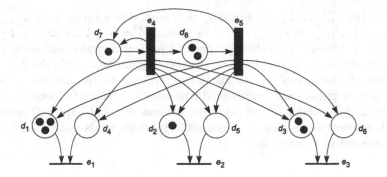

Figure 8.2. SPN for measuring delays in cyclic queues with feedback by tagging (three jobs).

tokens if and only if n jobs are waiting or in service at center i, and place d_j ($1 \leq j \leq N$) contains n tokens if and only if job j is in position n. Place d_{N+j} contains one token if and only if there has just been a completion of service for job j and the job is about to join the tail of the queue at center 1; otherwise, place d_{N+j} contains no tokens.

The idea is as follows. Suppose that the marking is $s = (s_1, s_2, \ldots, s_{2N+2})$ with $s_j = N$ for some j ($1 \leq j \leq N$) and transition e_{N+2} fires—that is, job j is in service at center 2 and there is a completion of service at center 2. Then one token is removed from place d_{2N+2}, and one token is deposited in each of the places $d_{2N+1}, d_1, d_2, \ldots, d_{j-1}, d_{j+1}, \ldots, d_N$, and d_{N+j}. The newly enabled immediate transition e_j then fires repeatedly, removing one token from place d_j each time until exactly one token remains in this place. At the last of these firings, the token in place d_{N+j} is removed so that transition e_j becomes disabled. In this manner the position of job j (which is represented by the number of tokens in place d_j) is set to 1 and the position of each other job is incremented by 1. Now suppose that the marking is $s = (s_1, s_2, \ldots, s_{2N+2})$ with $s_j = s_{2N+1}$ for some $1 \leq j \leq N$ and transition e_{N+1} fires—that is, job j is in service at center 1 and there is a completion of service at center 1. With probability $1 - p$, one token is deposited in place d_{N+j} and in each place d_l such that $1 \leq l \leq N$ and $s_l < s_{2N+1}$ (i.e., such that job l is at center 1); transition e_j then fires repeatedly until exactly one token remains in place d_j. With probability p, one token is removed from place d_{2N+1} and one token is deposited in place d_{2N+2}. In this manner the position of job j is set to 1 and the position of each other job at center 1 is incremented by 1 if job j joins the tail of the queue at center 1, and the positions of the jobs remain unchanged if job j moves to center 2. The foregoing construction heavily uses the fact that new-marking probabilities can explicitly depend on the current marking.

A delay for job j terminates (and the next delay for job j starts) whenever the marking process makes a state transition from $s = (s_1, s_2, \ldots, s_{2N+2})$

to $s' = (s'_1, s'_2, \ldots, s'_{2N+2})$, where $s_j = N$ and $s' = s$ except that $s'_{2N+1} = s_{2N+1} + 1$, $s'_{2N+2} = s_{2N+2} - 1$, $s'_j = 1$, and $s'_l = s_l + 1$ for all $1 \leq l \leq N$ with $l \neq j$. Observe that this approach to measuring individual delays has the undesirable property that—unlike the original SPN in Figure 2.2—the number of places and transitions is proportional to the number of jobs, and the SPN graph must be modified whenever the number of jobs changes. Use of distinguishable tokens (as in the colored SPNs of Chapter 9) provides another means for tagging, but, again, the resulting net is more complicated than the net of Figure 2.2.

8.1.2   Start Vectors

We now give a method for specifying and measuring delays that avoids the need for tagging. The idea is to use a sequence of real-valued random vectors, called *start vectors*, to construct the sequences $\{A_j : j \geq 0\}$ and $\{B_j : j \geq 0\}$. The sequence $\{V_n : n \geq 0\}$ of start vectors is determined by the sample paths of the chain $\{(S_n, C_n) : n \geq 0\}$ and provides the link between the starts and terminations of the individual delay intervals. The nth start vector V_n records the starts of delay intervals for all ongoing delays and newly started delays at time ζ_n, that is, all starts $A_j = \zeta_{\alpha(j)}$ such that $\alpha(j) \leq n < \beta(j)$. Usually (but not necessarily) the positions of the starts in the start vector correspond to the locations in the system of entities, such as jobs or customers, for whom a delay is underway. We assume that the current marking determines the length of the start vector and denote this length by $\psi(s)$ when the current marking is s. Some components of V_n may be equal to -1. As discussed below, lengths are never computed for delay intervals with negative starts, so that negative components of a start vector can be used to ensure that specified deletions do not result in the computation of a delay. The negative components typically serve as placeholders and correspond to entities in the system at time 0 for whom no delay is underway. The initial start vector is a specified vector, denoted $v_0(S_0)$, that is determined by the initial marking S_0 and has components that are equal to 0 or -1. Take $v_0(S_0)$ to be the empty vector $\varnothing$ when $\psi(S_0) = 0$.

Whenever the transitions in the set E^* fire simultaneously and trigger a marking change from s to s', a new start vector is obtained from the current start vector by

1. Inserting the current time at zero or more positions specified by an index vector $i_\alpha(s'; s, E^*)$

2. Deleting components at zero or more positions specified by an index vector $i_\beta(s'; s, E^*)$

3. Permuting the components according to an index vector $i_\pi(s'; s, E^*)$

Components are deleted one at a time in the order in which the indices appear in the vector $i_\beta(s'; s, E^*)$. For each nonnegative component that is deleted, the length of a delay interval is computed by subtracting the deleted component from the current time. These deleted components are the left endpoints of delay intervals for the delays that terminate at the current time. Deleted components equal to -1 are not used to compute lengths of delay intervals and are simply discarded. Observe that a component can be inserted and then immediately deleted—this scenario corresponds to a delay $D_j = 0$, such as when a job arrives at an empty queue and immediately goes into service, thereby avoiding a wait in line.[1]

For a real-valued nonempty vector $v = (v_1, v_2, \ldots, v_k)$ and a nonempty index vector $i = (i_1, i_2, \ldots, i_l)$, denote by $\mathrm{Del}(v, i)$ the vector of length $k - l$ obtained from v by deleting the components at positions $i_1, i_2, \ldots, i_l$. Similarly, denote by $\mathrm{Ins}(v, i, \zeta)$ the vector of length $k + l$ obtained from v by inserting the value $\zeta \in \Re$ to the right of the components at positions $i_1, i_2, \ldots, i_l$. For example, if $v = (v_1, v_2, v_3, v_4, v_5)$ and $i = (0, 2, 2, 3, 5)$, then $\mathrm{Ins}(v, i, \zeta) = (\zeta, v_1, v_2, \zeta, \zeta, v_3, \zeta, v_4, v_5, \zeta)$. If $v = \varnothing$, set $\mathrm{Ins}(v, i, \zeta) = (\zeta, \zeta, \ldots, \zeta)$, where the vector on the right side is of length l. Finally, for a vector v of length k and a vector $i = (i_1, i_2, \ldots, i_k)$ of distinct indices with $1 \le i_1, i_2, \ldots, i_k \le k$, set $\mathrm{Per}(v, i) = (v_{i_1}, v_{i_2}, \ldots, v_{i_k})$ so that $\mathrm{Per}(v, i)$ is the vector of length k obtained from v by permuting the components according to the index vector i. By convention, $\mathrm{Del}(v, \varnothing) = \mathrm{Ins}(v, \varnothing, \zeta) = \mathrm{Per}(v, \varnothing) = v$.

The sequence $\{ V_n : n \ge 0 \}$ is generated recursively. Set $V_0 = v_0(S_0)$, and then set

$$V_n' = \mathrm{Ins}\big(V_{n-1}, i_\alpha(S_n; S_{n-1}, E^*_{n-1}), \zeta_n\big),$$
$$V_n'' = \mathrm{Del}\big(V_n', i_\beta(S_n; S_{n-1}, E^*_{n-1})\big),$$

and

$$V_n = \mathrm{Per}\big(V_n'', i_\pi(S_n; S_{n-1}, E^*_{n-1})\big)$$

for $n \ge 1$. As usual, $E^*_k = E^*(S_k, C_k)$ for $k \ge 0$, so that E^*_{n-1} is the set of transitions that trigger the nth marking change.

Construct the sequence $\{ D_j : j \ge 0 \}$ from the sequence $\{ V_n : n \ge 0 \}$ as follows. Denote by A_j' $(j \ge 0)$ the jth nonnegative component deleted from a start vector in the sequence $\{ V_n : n \ge 0 \}$ and by B_j' the time at which A_j' is deleted. Then $[A_0', B_0'], [A_1', B_1'], [A_2', B_2'], \ldots$ is the sequence of delay intervals, enumerated in order of increasing terminations. If there are no immediate transitions, obtain the sequence $\{ (A_j, B_j) : j \ge 0 \}$ by rearranging the sequence $\{ (A_j', B_j') : j \ge 0 \}$ in order of increasing starts

---

[1]Thus, strictly speaking, V_n records starts for ongoing delays and newly started delays *of positive duration.*

and set $D_j = B_j - A_j$ for $j \geq 0$. If at least one transition is immediate, two delays can start at the same point in continuous time but at different marking changes. In this case we determine[2] for $j \geq 0$ the random index $\alpha'(j)$ for which $A'_j = \zeta_{\alpha'(j)}$ and obtain the sequence $\{ (\alpha(j), A_j, B_j) : j \geq 0 \}$ by rearranging the sequence $\{ (\alpha'(j), A'_j, B'_j) : j \geq 0 \}$ in order of increasing value of the random indices. Then we compute D_j as before.

Denote by $n_\alpha(s'; s, E^*)$ and $n_\beta(s'; s, E^*)$ the lengths of the vectors $i_\alpha(s'; s, E^*)$ and $i_\beta(s'; s, E^*)$, respectively, for each s', s, and E^*. The number of delays that start at time ζ_n is equal to $n_\alpha(S_n; S_{n-1}, E^*_{n-1})$ for $n \geq 1$. Denote by $V_{n,i}$ the ith component of the vector V_n for $1 \leq i \leq \psi(S_n)$, and set

$$K = \inf \{ n \geq 0 : V_{n,i} \neq -1 \text{ for } 0 \leq i \leq \psi(S_n) \}. \qquad (1.3)$$

The number of delays that terminate at time ζ_n is less than or equal to $n_\beta(S_n; S_{n-1}, E^*_{n-1})$ for $1 \leq n \leq K$ and equal to $n_\beta(S_n; S_{n-1}, E^*_{n-1})$ for $n > K$. Similarly, the total number of newly started delays (of positive duration) and ongoing delays at the nth marking change is less than or equal to $\psi(S_n)$ for $0 \leq n < K$ and equal to $\psi(S_n)$ for $n \geq K$.

8.1.3 Examples of Delay Specifications

The following examples illustrate the use of start vectors for specification of delays. As usual, we write $i_\alpha(s'; s, e)$ for $i_\alpha(s'; s, \{e\})$, and so forth.

EXAMPLE 1.4 (Cyclic queues with feedback).   Consider the delay intervals from whenever a job completes service at center 2 to when the job next completes service at center 2, and suppose that we wish to estimate time-average limits of the sequence of delays for all N jobs. The method of start vectors can be used to specify and measure individual delays in the SPN of Figure 2.2—this SPN is much less complicated than the SPN of Figure 8.2.

The start vector V_n records for each of the N jobs in the network the most recent time during the interval $[0, \zeta_n]$ at which there was a completion of service at center 2 and the job moved to center 1. If a job has never moved from center 2 to center 1 during the interval $[0, \zeta_n]$, then the corresponding component of V_n is equal to -1. The components of the start vector are ordered from left to right according to increasing positions—as defined in Example 1.2—of the corresponding jobs in the network.

---

[2]To obtain the sequence $\{ \alpha'(j) : j \geq 0 \}$, use an auxiliary sequence $\{ W_n : n \geq 0 \}$ of random vectors. The components of each W_n are "starts" but expressed as indices of marking changes rather than as points in continuous time. More specifically, set $W_0 = v_0(S_0)$. Then set $W'_n = \text{Ins}(W_{n-1}, i_\alpha(S_n; S_{n-1}, E^*_{n-1}), n)$, $W''_n = \text{Del}(W'_n, i_\beta(S_n; S_{n-1}, E^*_{n-1}))$, and $W_n = \text{Per}(W''_n, i_\pi(S_n; S_{n-1}, E^*_{n-1}))$ for $n \geq 1$. The random index $\alpha'(j)$ is then the jth nonnegative component deleted from a vector in the sequence $\{ W_n : n \geq 0 \}$.

Formally, set $\psi(s) = N$ for $s \in G$. Also set

$$i_\alpha(s'; s, E^*) = \begin{cases} (0) & \text{if } E^* = \{e_2\}; \\ \varnothing & \text{otherwise} \end{cases}$$

and

$$i_\beta(s'; s, E^*) = \begin{cases} (N+1) & \text{if } E^* = \{e_2\}; \\ \varnothing & \text{otherwise.} \end{cases}$$

Thus, whenever there is a completion of service at center 2 and a job moves to the tail of the queue at center 1, the new start vector is obtained from the current start vector by inserting the current time to the left of the first component, deleting the rightmost component,[3] and then subtracting the latter component from the current time to compute a delay if the component is nonnegative. Next, for $s = (s_1, s_2), s' = (s_1', s_2') \in G$ and $E^* \subseteq E(s)$, set $i_\pi(s'; s, E^*) = (s_1, 1, 2, \ldots, s_1 - 1, s_1 + 1, s_1 + 2, \ldots, N)$ if $E^* = \{e_1\}$ and $s_1' = s_1 > 1$. Otherwise, set $i_\pi(s'; s, E^*) = \varnothing$. Thus, whenever there are $s_1 (> 1)$ jobs at center 1 and a job completes service at center 1 and joins the tail of the queue at center 1, the new start vector is obtained from the current start vector by cyclically permuting the first s_1 components. Otherwise, the components are unchanged—in particular, no permutation is needed when $E^* = \{e_2\}$.

Suppose that at time 0 there is a completion of service at center 2 with all jobs at center 2, so that the initial marking is $s_0 = (1, N - 1)$ and a delay starts at time 0. We then set $v_0(s_0) = (0, -1, -1, \ldots, -1)$, where the vector on the right side is of length N. Because $N - 1$ components of $v_0(s_0)$ are equal to -1, there are $N - 1$ marking changes at which there is a completion of service at center 2 and no delay is computed. At the time ζ of each such marking change, the job completing service at center 2 has not previously completed service at center 2 during the interval $[0, \zeta]$ and ζ is not an element of the sequence $\{B_j : j \geq 0\}$ of terminations.

Table 8.1 displays a possible sequence of markings, transitions, and start vectors (in both continuous and discrete time) in a system with $N = 3$ jobs. At time $\zeta_0 = 0$ one job is at center 1 and two jobs are at center 2. Because a delay starts at time 0 by assumption, the leftmost component of both V_0 and W_0 is equal to 0. At time $\zeta_1 = 1.3$ there is a completion of service at center 2 and the marking changes from $s = (1, 2)$ to $s' = (2, 1)$. Because $i_\alpha((2, 1); (1, 2), e_2) = (0)$ and $i_\beta((2, 1); (1, 2), e_2) = (N + 1)$, the vector V_1 is obtained from V_0 by inserting the current time (1.3) to the left of the first component and then deleting the rightmost component (-1). Similarly, the vector W_1 is obtained from W_0 by inserting the index of the

---

[3] Although the length of each start vector V_n is always equal to N for this model, the length of the intermediate vector V_n' is equal to $N + 1$ whenever there is a service completion at center 2 at time ζ_n. Thus the "$N+1$" term in the definition of $i_\beta(s'; s, E^*)$.

Table 8.1. Sequences of Markings, Transitions, and Start Vectors

n	S_n	E_{n-1}^*	ζ_n		V_n	W_n
0	$(1,2)$	–	0	$(A_0 = A_1')$	$(0,-1,-1)$	$(0,-1,-1)$
1	$(2,1)$	$\{e_2\}$	1.3	$(A_1 = A_0')$	$(1.3,0,-1)$	$(1,0,-1)$
2	$(2,1)$	$\{e_1\}$	1.8		$(0,1.3,-1)$	$(0,1,-1)$
3	$(1,2)$	$\{e_1\}$	2.1		$(0,1.3,-1)$	$(0,1,-1)$
4	$(2,1)$	$\{e_2\}$	2.5		$(2.5,0,1.3)$	$(4,0,1)$
5	$(3,0)$	$\{e_2\}$	3.2	$(B_1 = B_0')$	$(3.2,2.5,0)$	$(5,4,0)$
6	$(2,1)$	$\{e_1\}$	3.7		$(3.2,2.5,0)$	$(5,4,0)$
7	$(3,0)$	$\{e_2\}$	3.8	$(B_0 = B_1')$	$(3.8,3.2,2.5)$	$(7,5,4)$

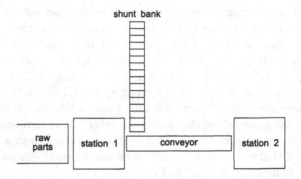

Figure 8.3. Manufacturing flow-line with shunt bank.

current marking change (1) to the left of the first component and then deleting the rightmost component (-1). Since the component deleted from V_0 is equal to -1, it is not used to compute the length of a delay interval. Thus, at time ζ_1 a delay starts but no delay terminates. At time $\zeta_2 = 1.8$ a job completes service at center 1 and joins the tail of the queue at center 1. The start vector V_2 is obtained from V_1 by cyclically permuting the first two components in accordance with the vector $i_\pi\,((2,1);(2,1),e_1)$, and similarly for the vector W_2. At time $\zeta_5 = 3.2$ there is a completion of service at center 2, and a delay terminates. Since ζ_5 is the first time at which a delay terminates and the rightmost components in the vectors V_4 and W_4 are equal to 1.3 and 1, respectively, we have $\beta'(0) = 5$, $\alpha'(0) = 1$, $B_0' = 3.2$, and $A_0' = 1.3$. Similarly, a delay terminates at time $\zeta_7 = 3.8$, and we have $\beta'(1) = 7$, $\alpha'(1) = 0$, $B_1' = 3.8$, and $A_1' = 0$. After rearrangement in order of increasing starts, we obtain $\alpha(0) = 0$, $\beta(0) = 7$, $\alpha(1) = 1$, $\beta(1) = 5$, $[A_0, B_0] = [0, 3.8]$, $[A_1, B_1] = [1.3, 3.2]$, $D_0 = 3.8 - 0 = 3.8$, and $D_1 = 3.2 - 1.3 = 1.9$.

EXAMPLE 1.5 (Manufacturing flow-line with shunt bank).   Consider a manufacturing flow-line with two work stations numbered 1 and 2, a conveyor, and a shunt bank; see Figure 8.3. Parts passing through the flow-line are first processed at station 1 and then moved one at a time by the conveyor to station 2 for further processing. The shunt bank provides temporary storage for parts that have been processed at station 1 but cannot yet be transferred onto the conveyor—parts are transferred to and from the shunt bank in a "last-in, first-out" manner. At most one part can be at a station or on the conveyor at any time, and the shunt bank can store no more than B (≥ 1) parts at a time. Raw parts are always available for processing at station 1. The details of flow-line operation are as follows.

- Whenever a part completes processing at station 1 and the conveyor is unoccupied—that is, no part is on the conveyor and no part is being transferred from the shunt bank to the conveyor—the part at station 1 is instantaneously transferred onto the conveyor and processing of the next raw part begins at station 1. If the conveyor is occupied and fewer than B parts are in the shunt bank, then the part at station 1 is transferred to the shunt bank—upon completion of the transfer, processing of the next raw part begins at station 1. If B parts are in the shunt bank, then the part remains at station 1 and the station becomes blocked. Station 1 remains blocked until the conveyor becomes unoccupied, at which time the part at station 1 is instantaneously transferred to the conveyor and processing of the next raw part begins at station 1.

- Whenever a part is transferred onto the conveyor (from station 1 or from the shunt bank), the conveyor immediately begins to move the part to station 2.

- Whenever there is an end of transfer of a part to the shunt bank and the conveyor is unoccupied, transfer of the part from the shunt bank to the conveyor starts immediately.

- Whenever either (i) a part arrives at station 2 on the conveyor and station 2 is idle or (ii) a part completes processing at station 2 and another part is on the conveyor at the station, the part on the conveyor is instantaneously transferred to station 2 and processing of the part begins.

- Whenever a part is transferred from the conveyor to station 2, station 1 is not blocked, and the shunt bank contains at least one part, transfer of a part from the shunt bank onto the conveyor begins.

The time for the conveyor to move a part from station 1 to station 2 is a deterministic constant. The time for transfer of a part from station 1 to the shunt bank is also a deterministic constant, as is the time for transfer of a part from the shunt bank to the conveyor. The successive times to process a part at station i are i.i.d. as a positive random variable L_i.

e_1 = end of processing of part at station 1

e_2 = transfer of part from station 1 to conveyor

e_3 = end of movement on conveyor of part from station 1 to station 2

e_4 = transfer of part from conveyor to station 2

e_5 = end of processing of part at station 2

e_6 = end of transfer of part from station 1 to the shunt bank

e_7 = start of transfer of part from the shunt bank to the conveyor

e_8 = end of transfer of part from the shunt bank to the conveyor

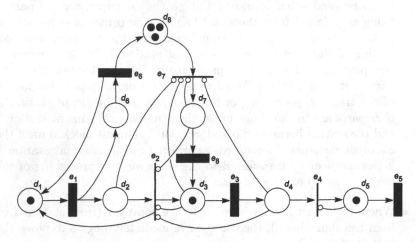

Figure 8.4. SPN representation of manufacturing flow-line with shunt bank.

A part is said to be in stage 1 of the manufacturing process if the part is at station 1; in stage 2 if the part is being transferred from station 1 to the shunt bank; in stage 3 if the part is at the shunt bank; in stage 4 if the part is being transferred from the shunt bank to the conveyor; in stage 5 if the part is on the conveyor; and in stage 6 if the part is being processed at station 2.

This system can be specified as a SPN with a finite marking set; see Figure 8.4. Places $d_1, d_2, \ldots, d_7$ can each contain zero or one token. Place d_8 can contain up to B tokens—the number of tokens in place d_8 corresponds to the number of parts in the shunt bank. Whenever the marking is $s = (s_1, s_2, \ldots, s_8)$ and transition e_1 = "end of processing of part at station 1" fires, a token is removed from place d_1. Moreover, if either $s_3 + s_4 + s_7 = 0$ (the conveyor is unoccupied) or $s_3 + s_4 + s_7 > 0$ and $s_8 = B$ (the conveyor is occupied and B parts are in the shunt bank), then a token is deposited in place d_2; otherwise, a token is deposited in place d_6, so that transfer of a part from station 1 to the shunt bank starts. All other transitions are deterministic. All speeds for enabled transitions are equal to 1.

Consider the delay intervals from whenever there is a start of processing at station 1 for a part to when there is an end of processing at station 2 for the part, and suppose that we wish to estimate time-average limits of the sequence of delays for all parts. The method of start vectors can be used to specify and measure individual delays in the SPN of Figure 8.4. The start vector V_n records, for each part in a stage of the manufacturing process at time ζ_n, the time at which there was a start of processing at station 1 for the part. The components of the start vector are ordered from left to right according to increasing stages of the corresponding parts. Starts corresponding to parts at the shunt bank are ordered from left to right according to increasing arrival times at the shunt bank.

Formally, set $\psi(s) = s_1 + s_2 + \cdots + s_8$ for all $s = (s_1, s_2, \ldots, s_8) \in G$. Also set

$$i_\alpha(s'; s, E^*) = \begin{cases} (0) & \text{if } s_1 = 0 \text{ and } s_1' = 1; \\ \varnothing & \text{otherwise} \end{cases}$$

and

$$i_\beta(s'; s, E^*) \begin{cases} (\psi(s)) & \text{if } E^* = \{e_5\}; \\ \varnothing & \text{otherwise} \end{cases}$$

for $s, s' = (s_1', s_2', \ldots, s_8') \in G$ and $E^* \subseteq E(s)$. Thus, whenever there is a start of processing at station 1, the new start vector is obtained from the current start vector by inserting the current time to the left of the first component; whenever there is an end of processing at station 2, the new start vector is obtained by deleting the rightmost component. Next, set $i_\pi(s'; s, E^*) = (2, 3, \ldots, s_8 + 1, 1, s_8 + 2, s_8 + 3, \ldots, \psi(s))$ if $s_8 > 0$ and either $E^* = \{e_2\}$ or $E^* = \{e_6\}$, that is, if the shunt bank is not empty and there is an end of transfer of a part from station 1 to either the shunt bank or the conveyor. Thus the new start vector is obtained by cyclically permuting the components so that the start for the transferred part appears to the right of the starts for the parts at the shunt bank. Otherwise, set $i_\pi(s'; s, E^*) = \varnothing$, so that the components of the start vector are unchanged.

Suppose that at time 0 there is an end of processing at station 1 and there are no parts at the shunt bank, on the conveyor, or at station 2. Then the initial marking is $s_0 = (0, 1, 0, 0, 0, 0, 0, 0)$, no delays start at time 0, and we set $v_0(s_0) = (-1)$.

EXAMPLE 1.6 (Manufacturing cell with robots).   For the manufacturing cell of Example 3.6 in Chapter 2, a part is said to be in stage 1 of the manufacturing process if the part is being transferred from the loading area to conveyor 1; in stage 2 if the part is on conveyor 1; in stage 3 if the part is being transferred from conveyor 1 to a machine; in stage 4 if the part is at a machine; in stage 5 if the part is being transferred from a machine to conveyor 2; in stage 6 if the part is on conveyor 2; and in stage 7 if the part is being transferred from conveyor 2 to the unloading area.

Consider the delay intervals from whenever (the arm of robot 1 arrives at the loading area and) robot 1 starts to transfer a raw part from the loading area to conveyor 1 to when robot 1 completes transfer of the part to the unloading area. Suppose that we wish to estimate time-average limits of the sequence of delays for all parts. The method of start vectors can be used to specify and measure individual delays in the SPN of Figure 2.21. The start vector V_n records, for each part in a stage of the manufacturing process at time ζ_n, the time at which robot 1 started to transfer the raw part from the loading area to conveyor 1. The components of the start vector are ordered from left to right according to increasing stages of the corresponding parts. If there is a part at each machine—that is, if there are two parts in stage 4—the start corresponding to the part at machine 1 appears to the left of the start corresponding to the part at machine 2.

Formally, set $\psi(s) = s_2 + s_3 + s_4 + s_6 + s_8 + s_9 + s_{10} + s_{11} + s_{12} + s_{14} + s_{16} + s_{17} + s_{18} + s_{20}$ for $s = (s_1, \ldots, s_{24}) \in G$. Also set

$$i_\alpha(s'; s, E^*) = \begin{cases} (0) & \text{if } E^* = \{e_1\}; \\ \varnothing & \text{otherwise} \end{cases}$$

and

$$i_\beta(s'; s, E^*) = \begin{cases} (\psi(s)) & \text{if } E^* = \{e_{16}\}; \\ \varnothing & \text{otherwise} \end{cases}$$

for $s, s' \in G$ and $E^* \subseteq E(s)$. Thus, whenever robot 1 starts to transfer a raw part from the loading area to conveyor 1, the new start vector is obtained from the current start vector by inserting the current time to the left of the first component; whenever robot 1 completes transfer of the part from a machine to the unloading area, the new start vector is obtained by deleting the rightmost component. Next, set $n(s) = s_2 + s_3 + s_4 + s_8$ and $m(s) = s_2 + s_3 + s_4 + s_6 + s_8 + s_9 + s_{10}$, and then set

$$i_\pi(s'; s, E^*) = \big(1, \ldots, n(s) - 1, n(s) + 1, n(s), n(s) + 2, \ldots, \psi(s)\big)$$

if $E^* = \{e_7\}$ and $s_9 + s_{10} = 1$, and

$$i_\pi(s'; s, E^*) = \big(1, \ldots, m(s) - 1, m(s) + 1, m(s), m(s) + 2, \ldots, \psi(s)\big)$$

if $E^* = \{e_{10}\}$ and $s_{11} + s_{12} = 1$. Otherwise, set $i_\pi(s'; s, E^*) = \varnothing$. Thus, whenever there is an end of transfer of a raw part from conveyor 1 to machine 2 with a part at machine 1, the new start vector is obtained from the current start vector by interchanging the components associated with the two parts. A similar interchange occurs whenever there is a start of transfer of a part from machine 1 to conveyor 2 with a part at machine 2. Otherwise the components of the current and new start vectors coincide.

Suppose that at time 0 there are parts only at the loading area and the arm of robot 1 has just left its null position to transfer a raw part from the loading area to conveyor 1. Then the initial marking is $s_0 = (1, 0, 0, \ldots, 0, 1)$, no delays start at time 0, and we set $v_0(s_0) = \varnothing$.

EXAMPLE 1.7 (Token ring). For the system of Example 2.6 in Chapter 2, consider the delay intervals from whenever a packet arrives at a port for transmission until the end of transmission of the packet. Suppose that we wish to estimate time-average limits of the sequence of delays for all ports. The method of start vectors can be used to specify and measure individual delays in the SPN of Figure 2.10. The start vector V_n records, for each packet awaiting or under transmission at time ζ_n, the time at which the packet arrived. The components of the start vector are ordered from left to right according to increasing indices of the arrival ports for the corresponding packets.

Formally, denote by $m(k, s)$ the total number of ongoing delays corresponding to packets at ports 1 through k when the marking is s: $m(0, s) = 0$ and $m(k, s) = s_{1,1} + s_{1,2} + \cdots + s_{1,k}$ for $1 \leq k \leq N$ and $s = (s_{1,1}, \ldots, s_{4,N}) \in G$. Set $\psi(s) = s_{1,1} + s_{1,2} + \cdots + s_{1,N}$ for $s \in G$ and set

$$i_\alpha(s'; s, E^*) = \begin{cases} (m(j-1, s)) & \text{if } E^* = \{e_{1,j}\} \text{ for some } 1 \leq j \leq N; \\ \varnothing & \text{otherwise} \end{cases}$$

and

$$i_\beta(s'; s, E^*) = \begin{cases} (m(j, s)) & \text{if } E^* = \{e_{2,j}\} \text{ for some } 1 \leq j \leq N; \\ \varnothing & \text{otherwise} \end{cases}$$

for $s, s' \in G$ and $E^* \subseteq E(s)$. Thus, whenever m packets are either awaiting or under transmission at ports 1 through $j-1$ (where $1 \leq j \leq N$) and there is an arrival of a packet at port j, the new start vector is obtained from the current start vector by inserting the current time to the right of the mth component; whenever m packets are either awaiting or under transmission at ports 1 through j and there is an end of transmission by port j, the new start vector is obtained by deleting the mth component. Finally, set $i_\pi(s'; s, E^*) = \varnothing$ for all s', s, and E^*—the components of the start vector need never be permuted because the order of the starts in the start vector is determined by the indices of the ports at which the corresponding packets arrived.

Suppose that at time 0 no packets are awaiting or under transmission, and the ring token has just arrived at port 1. Then the initial marking is $s_0 = (0, 1, 0, 1, 0, 1, 0, 0 \ldots, 0, 1, 0, 0)$, no delays start at time 0, and we set $v_0(s_0) = \varnothing$.

EXAMPLE 1.8 (Airport shuttle). Consider an airport shuttle that provides transportation service to N stations numbered $1, 2, \ldots, N$. The shuttle has seats for K (≥ 1) passengers and moves from station to station in a strictly

$e_{1,i}$ = arrival of shuttle at station i

$e_{2,j,i}$ = disembarkment at station i of passengers from station j

$e_{3,i}$ = boarding of passengers at station i

$e_{4,i}$ = arrival of passenger for boarding at station i

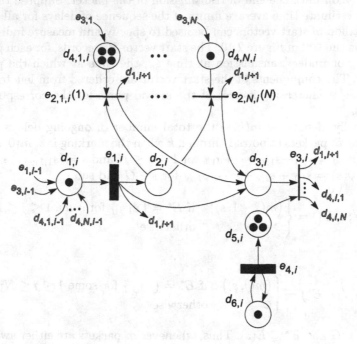

station i

Figure 8.5. SPN representation of airport shuttle.

defined order: $1, 2, \ldots, N, 1, 2, \ldots$. Passengers who wish to board the shuttle at station i arrive at station i according to a renewal process and disembark at station j ($\neq i$) with probability $p_{i,j}$. Passengers who arrive for boarding at station i queue (and subsequently board the shuttle) in the order in which they arrive at the station. Each station i has a finite capacity B_i—passengers arriving at the station when there are already B_i passengers in queue are turned away. At each station, passengers disembark before any waiting passengers board the shuttle. Passengers board and disembark instantaneously. The successive times for the shuttle to travel from station i to station $i + 1$ are i.i.d. as a positive random variable L_i, and the times between successive arrivals of passengers at station i for boarding are i.i.d. as a positive random variable A_i. (When $i = N$ take station $i+1$ as station 1 and when $i = 1$ take station $i - 1$ as station N.)

This transportation system can be specified as an SPN with finite marking set—Figure 8.5 displays the subnet corresponding to a generic station i. Places $d_{1,i}$, $d_{2,i}$, and $d_{3,i}$ ($1 \leq i \leq N$) each contain at most one token. Place $d_{1,i}$ contains a token if and only if the shuttle is travelling from station $i-1$ to station i, place $d_{2,i}$ contains a token if and only if passengers are disembarking at station i, and place $d_{3,i}$ contains one token if and only if passengers are boarding at station i. Place $d_{6,i}$ always contains exactly one token, reflecting the fact that the arrival process of passengers who board at station i is always active. Place $d_{4,j,i}$ ($1 \leq i,j \leq N$ with $i \neq j$) contains k tokens if and only if there are k passengers on the shuttle who boarded at station j and wish to disembark at station i. Place $d_{5,i}$ contains k tokens if and only if k passengers are queued at station i (awaiting the shuttle). All speeds for enabled transitions are equal to 1.

The complete description of the transition-firing mechanism is rather tedious, so we give a brief overview and leave the details to the reader. Consider throughout the subnet corresponding to a fixed station i. Suppose that place $d_{1,i}$ contains a token, the set of places $\{ d_{4,k,l} \colon 1 \leq k, l \leq N$ with $k \neq l \}$ contains a total of m tokens ($m \leq K$), the set of places $\{ d_{4,1,i}, \ldots, d_{4,N,i} \}$ contains a total of n tokens ($1 \leq n \leq m$), and place $d_{5,i}$ contains k tokens ($k \geq 1$). Thus the shuttle is travelling to station i carrying m passengers, n of whom wish to disembark at station i, and k passengers are at station i waiting to board the shuttle. When transition $e_{1,i} =$ "arrival of shuttle at station i" fires, it removes a token from place $d_{1,i}$ and deposits a token in place $d_{2,i}$. The transitions in the set $\{ e_{2,1,i}, \ldots, e_{2,N,i} \}$ then fire a total of n times, removing all tokens from the places in the set $\{ d_{4,1,i}, \ldots, d_{4,N,i} \}$, so that n passengers disembark. At the last of these firings, a token also is removed from place $d_{2,i}$ and a token is deposited in place $d_{3,i}$. Transition $e_{3,i} =$ "boarding of passengers at station i" then fires $l = \min(k, K - m + n)$ times, removing l tokens from place $d_{5,i}$. Whenever transition $e_{3,i}$ fires and removes a token from place $d_{5,i}$, it also deposits a token in exactly one of places $d_{4,i,1}, \ldots, d_{4,i,i-1}, d_{4,i,i+1}, \ldots, d_{4,i,N}$; the token is deposited in place $d_{4,i,j}$ with probability $p_{i,j}$. Moreover, transition $e_{3,i}$ removes a token from place $d_{3,i}$ and deposits a token in place $d_{1,i+1}$ when it fires for the lth time, so that the shuttle begins to travel to station $i+1$.

If no passengers wish to disembark at station i and/or there are no passengers at station i waiting to board the shuttle, then the corresponding stages in the foregoing sequence are skipped. For example, if the set of places $\{ d_{4,1,i}, \ldots, d_{4,N,i} \}$ contains zero tokens and place $d_{5,i}$ contains zero tokens, then transition $e_{1,i}$ removes a token from place $d_{1,i}$ and deposits a token in place $d_{1,i+1}$ when it fires, so that no passengers disembark or board at station i. The behavior of transition $e_{4,i} =$ "arrival of passenger for boarding at station i" is relatively simple. Whenever place $d_{5,i}$ contains less than B_i tokens and transition $e_{4,i}$ fires, a token is deposited in place $d_{5,i}$; whenever place $d_{5,i}$ contains exactly B_i tokens and transition $e_{4,i}$ fires, no tokens are removed or deposited.

To facilitate specification of the new-marking probabilities, we assign priorities to the transitions in the set $\{ e_{2,1,i}, \ldots, e_{2,N,i} \}$; specifically, we set $\mathcal{P}(e_{2,j,i}) = j$ for $1 \leq j \leq N$. Whenever passengers disembark at station i, the enabled transition $e_{2,j,i}$ with the highest priority fires multiple times in succession, then the enabled transition $e_{2,j',i}$ with the second-highest priority fires multiple times in succession, and so forth. Thus we need only explicitly specify singleton new-marking probabilities of the form $p(s'; s, e_{2,j,i})$ in order to formally describe the behavior of the net when passengers disembark.

Consider the delay intervals from whenever a passenger arrives at a station for boarding to when the passenger disembarks, and suppose that we wish to estimate time-average limits of the sequence of delays for all passengers. The method of start vectors can be used to specify and measure individual delays in the SPN of Figure 8.5. The start vector V_n records, for each passenger in the system at time ζ_n, the time at which the passenger arrived at a station for boarding. The components of the start vector are ordered so that

1. Starts corresponding to passengers who originally arrived at station i for boarding appear to the left of starts corresponding to passengers who originally arrived at station j whenever $i < j$.

2. starts corresponding to passengers waiting in queue at station i ($1 \leq i \leq N$) appear to the left of starts corresponding to passengers who originally arrived at station i for boarding and are currently on the shuttle.

3. Starts corresponding to passengers waiting in queue at station i ($1 \leq i \leq N$) appear from left to right in decreasing order of arrival time.

4. Starts corresponding to passengers on the shuttle who originally arrived at station i and wish to disembark at station j appear to the left of the starts corresponding to passengers who originally arrived at station i and wish to disembark at station k whenever $j < k$.

5. Starts corresponding to passengers on the shuttle who originally arrived at station i and wish to disembark at station j appear from left to right in decreasing order of arrival time at station i.

Formal specification of the start-vector mechanism is as follows. When the marking is $s \in G$, denote by $m_j(s)$ the number of passengers currently in the system who originally arrived at station j:

$$m_j(s) = s_{5,j} + \sum_{i \neq j} s_{4,j,i}.$$

Moreover, if the marking s is such that at least one passenger is waiting in queue at station j, let $l_j(s)$ be the position in the start vector corresponding to the next passenger at station j who will board the shuttle:

$$l_j(s) = m_1(s) + \cdots + m_{j-1}(s) + s_{5,j}.$$

Finally, let $n_{j,i}(s)$ be the position of the rightmost of the starts corresponding to those passengers who originally arrived at station j, are currently on the shuttle, and wish to disembark at some station k with $k < i$:

$$n_{j,i}(s) = l_j(s) + s_{4,j,1} + s_{4,j,2} + \cdots + s_{4,j,i-1}.$$

Observe that if no such passengers exist, then $n_{j,i}(s)$ is the position corresponding to the next passenger at station j who will board the shuttle. Using the foregoing notation, set $\psi(s) = \sum_{i=1}^{N} m_i(s)$ for $s \in G$. Also set

$$i_\alpha(s'; s, E^*) = \big(m_1(s) + \cdots + m_{i-1}(s)\big)$$

if $E^* = \{e_{4,i}\}$ for some $1 \le i \le N$; otherwise, set $i_\alpha(s'; s, E^*) = \varnothing$. Next, set

$$i_\beta(s'; s, E^*) = \big(m_1(s) + \cdots + m_{j-1}(s) + s_{5,j} + s_{4,j,1} + s_{4,j,2} + \cdots + s_{4,j,i}\big)$$

if $E^* = \{e_{2,j,i}\}$ for some $1 \le i, j \le N$ with $i \ne j$; otherwise, set $i_\beta(s'; s, E^*) = \varnothing$. Finally, set

$$i_\pi(s'; s, E^*) = \big(1, 2, \ldots, l_j(s) - 1, l_j(s) + 1, l_j(s) + 2, \ldots,$$
$$n_{j,i}(s), l_j(s), n_{j,i}(s) + 1, n_{j,i}(s) + 2, \ldots, \psi(s)\big)$$

if $s'_{4,j,i} = s_{4,j,i} + 1$ for some $1 \le i, j \le N$ with $i \ne j$; otherwise, set $i_\pi(s'; s, E^*) = \varnothing$. Thus, whenever a passenger boards the shuttle at station j and wishes to disembark at station i, the start corresponding to this passenger is moved to the right of the starts corresponding to those passengers who originally arrived at station j, are currently on the shuttle, and wish to disembark at some station k with $k < i$.

Suppose that at time 0 no passengers are in the system and the shuttle is travelling to station 1. Then the initial marking is $s_0 = (s_{1,1}, \ldots, s_{6,N})$, where $s_{1,1} = s_{6,1} = s_{6,2} = \cdots = s_{6,N} = 1$ and all other components of s_0 are equal to 0. No delays start at time 0, and we set $v_0(s_0) = \varnothing$.

The foregoing SPN and start-vector mechanism can also be used to study delays experienced by passengers who board the shuttle at station j and disembark at station i, where i and j are fixed—see Remark 3.18 below.

Remark 1.9. The "loop" airport shuttle system of Example 1.8 is closely related to a "bidirectional" shuttle system. In particular, suppose that the number of stations N can be written in the form $N = 2L$ and that the travel time from station L to station $L + 1$ is identically 0, as is the travel time from station $2L$ to station 1. Also suppose that the interarrival-time random variables A_L and A_{2L} are each a.s. infinite, so that no passengers arrive at stations L or $2L$. Finally, suppose that $p_{i,1} = p_{i,L+1} = 0$ for all i, so that passengers never disembark at stations 1 or $L + 1$. This system coincides with a bidirectional shuttle system having L stations—the shuttle travels in either a "northbound" direction (from station 1 to station L) or a "southbound" direction (from station L to station 1). That is, the shuttle moves from station to station in the order $1, 2, \ldots, L, L-1, \ldots, 1, 2, \ldots$. The idea is to identify station j $(1 \leq j \leq L)$ in the loop shuttle system with the northbound platform of station j in the bidirectional shuttle system, and station $2L - j + 1$ in the loop shuttle system with the southbound platform of station j in the bidirectional shuttle system.

8.2 Regenerative Methods for Delays

In this section we provide methods for estimating general time-average limits of the form $\lim_{n \to \infty} (1/n) \sum_{j=0}^{n-1} f(D_j)$, where the sequence of delays $\{ D_j : j \geq 0 \}$ is determined from the marking changes of an SPN by means of start vectors. We also provide specialized estimation methods in this setting for the limiting average delay $\lim_{n \to \infty} (1/n) \sum_{j=0}^{n-1} D_j$.

Our key assumption is that there exists a sequence of regeneration points for the marking process $\{ X(t) : t \geq 0 \}$ and for the underlying chain $\{ (S_n, C_n) : n \geq 0 \}$. In particular, we suppose throughout that there exists a recurrent single state $\bar{s}$, so that $E(\bar{s}) = \{ \bar{e} \}$ for some $\bar{e} \in E$ and $P_\mu \{ S_n = \bar{s} \text{ i.o.} \} = 1$. The regeneration points then correspond to the successive times at which the marking is $\bar{s}$ and transition $\bar{e}$ fires. That is, if we set $\theta(0) = 0$ and

$$\theta(k) = \inf \left\{ n > \theta(k-1) \colon S_{n-1} = \bar{s} \text{ and } E_{n-1}^* = \{ \bar{e} \} \right\} \qquad (2.1)$$

for $k \geq 1$, then the random indices $\{ \theta(k) : k \geq 0 \}$ form a sequence of regeneration points for $\{ (S_n, C_n) : n \geq 0 \}$ and the random times $\{ \zeta_{\theta(k)} : k \geq 0 \}$ form a sequence of regeneration points for $\{ X(t) : t \geq 0 \}$. Implicit in this definition is the assumption—made for convenience—that the net behaves as if at time 0 the marking is $\bar{s}$ and transition $\bar{e}$ fires. The initial start vector V_0 is defined accordingly: conditional on S_0, compute V_0 by taking a vector of length $\psi(\bar{s})$ with each component equal to -1 and then inserting the current time (0) at positions specified by the index vector $i_\alpha(S_0; \bar{s}, \bar{e})$, deleting components at positions specified by the index vector $i_\beta(S_0; \bar{s}, \bar{e})$, and permuting the components according to the index vector $i_\pi(S_0; \bar{s}, \bar{e})$. We also suppose that the starts $\{ A_j : j \geq 0 \}$, the terminations $\{ B_j : j \geq 0 \}$, and the random index K that is defined by (1.3) satisfy

$$P_\mu \{ K < \infty \} = 1, \qquad (2.2)$$

$$P_\mu \{ A_j < \infty \} = P_\mu \{ B_j < \infty \} = 1 \qquad (2.3)$$

for $j \geq 0$, and

$$P_\mu \{ \lim_{j \to \infty} A_j = \infty \} = 1. \tag{2.4}$$

Observe that, by the a.s. finiteness of the clock readings, $\zeta_K < \infty$ a.s. whenever $K < \infty$ a.s..

There are two basic scenarios to consider, as illustrated by the two types of delays in the following example.

EXAMPLE 2.5 (Cyclic queues with feedback). For the network of queues in Example 1.2, consider the delay intervals from whenever a job completes service at center 2 to when the job next completes service at center 1 and moves to center 2, and suppose that we wish to estimate time-average limits of the sequence of delays for all N jobs. Under suitable assumptions on the distributions of the service-time random variables L_1 and L_2—see Example 2.12 below—the successive random times at which there is a completion of service at center 2 with all other jobs at center 2 form a sequence of regeneration points for the marking process. Observe that there are no ongoing delays at any regeneration point.

In contrast, consider the delay intervals from whenever a job completes service at center 2 to when the job next completes service at center 2 and the sequence of delays for all N jobs. There are at least $N - 1$ ongoing delays at any time point and hence at any regeneration point.

When there are no ongoing delays at any regeneration point for the marking process—see Figure 8.7 below—it is intuitively clear that the regeneration points decompose the delays into i.i.d. blocks. The sequence of delays therefore is a regenerative process in discrete time, and we can estimate time-average limits using methods as in Chapter 6. This scenario holds, for example, whenever

(i) $\psi(\bar{s}) = 0$ or

(ii) all delays are of positive length and $n_\beta(s; \bar{s}, \bar{e}) = \psi(\bar{s})$ for all s such that $p(s; \bar{s}, \bar{e}) > 0$.

The situation is not so simple, however, when there are ongoing delays at each regeneration point, as in Figure 8.6. In the following, we treat these two scenarios in a uniform manner and provide general estimation methods that are applicable under either scenario—we then show that each of these methods reduces in effect to the standard regenerative method when there are no ongoing delays at any regeneration point.

8.2.1 Construction of Random Indices

To obtain point estimates and confidence intervals for time-average limits, we first construct a sequence $\{ \breve{\gamma}(k) : k \geq 0 \}$ of random indices that decomposes sample paths of $\{ D_j : j \geq 0 \}$ into o.d.s. cycles. Extensions of the

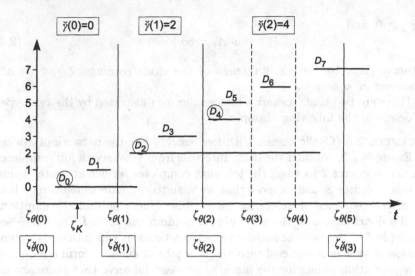

Figure 8.6. Definition of one-dependent cycles.

standard regenerative method as in Section 6.3.8 can then be used to obtain strongly consistent point estimates and asymptotic confidence intervals.

Definition of the Indices

In the following discussion, we assume that there exist fixed index vectors j_α and j_β—of respective lengths $|j_\alpha|$ and $|j_\beta|$—such that

$$i_\alpha(s'; \bar{s}, \bar{e}) = j_\alpha \quad \text{and} \quad i_\beta(s'; \bar{s}, \bar{e}) = j_\beta \qquad (2.6)$$

for all $s' \in G$ with $p(s'; \bar{s}, \bar{e}) > 0$. The condition in (2.6) asserts that, whenever the marking is $\bar{s}$ and transition $\bar{e}$ fires, the number and positions of the starts inserted into and deleted from the current start vector do not depend explicitly on the new marking s'. This condition implies that the start vector contains exactly $\psi(\bar{s}) + |j_\alpha| - |j_\beta|$ components at each time $\zeta_{\theta(k)}$. Moreover, the number of these components that correspond to ongoing delays—and hence the number that correspond to newly started delays—is the same for each time $\zeta_{\theta(k)} > \zeta_K$.

We start with the sequence $\{\, \zeta_{\theta(k)} \colon k \geq 0 \,\}$ of regeneration points for the marking process and recursively construct a subsequence $\{\, \zeta_{\check{\theta}(k)} \colon k \geq 0 \,\}$. The random times $\{\, \zeta_{\check{\theta}(k)} \colon k \geq 0 \,\}$ also form a sequence of regeneration points, but with longer cycles. All delays that start during one of these longer cycles terminate by the end of the next such cycle. To construct the sequence $\{\, \zeta_{\check{\theta}(k)} \colon k \geq 0 \,\}$, take $\check{\theta}(0) = \theta(0) = 0$. Then, given $\check{\theta}(k)$, wait until the first marking change $\check{\nu}(k)$ at which all ongoing delays at the $\check{\theta}(k)$th marking change have terminated, and take as $\check{\theta}(k+1)$ the smallest $\theta(l)$ such that $\theta(l) \geq \check{\nu}(k)$. Equivalently, take as $\check{\theta}(k+1)$ the first $\theta(l)$ after the

$\breve{\theta}(k)$th marking change such that all ongoing delays at the $\theta(l)$th marking change started no sooner than the $\breve{\theta}(k)$th marking change. If there are no ongoing delays at the $\breve{\theta}(k)$th marking change, take as $\breve{\theta}(k+1)$ the smallest $\theta(l)$ such that $\theta(l) > \breve{\theta}(k)$. For $k = 0$, take as $\breve{\theta}(1)$ the smallest $\theta(l)$ such that $\theta(l) \geq K$. To complete the construction, set $\breve{\gamma}(0) = 0$ and

$$\breve{\gamma}(k) = \inf\{ j > \breve{\gamma}(k-1) \colon \alpha(j-1) < \breve{\theta}(m) \leq \alpha(j) \text{ for some } m \geq 0\} \quad (2.7)$$

for $k \geq 1$. These ideas are illustrated in Figure 8.6. In the figure, vertical dashed lines indicate times that are elements of $\{\zeta_{\theta(k)} \colon k \geq 0\}-\{\zeta_{\breve{\theta}(k)} \colon k \geq 0\}$; vertical solid lines indicate times that are elements of $\{\zeta_{\theta(k)} \colon k \geq 0\}\cap\{\zeta_{\breve{\theta}(k)} \colon k \geq 0\}$. The delays $D_{\breve{\gamma}(k)}$ are circled.

Properties of the Construction

The pertinent properties of the foregoing construction are summarized in Theorem 2.8 below—estimation methods for delays rest on these properties. For $k \geq 1$, denote by δ_k the number of delays that start during the interval $[\zeta_{\theta(k-1)}, \zeta_{\theta(k)})$ and set $\tau_k = \zeta_{\theta(k)} - \zeta_{\theta(k-1)}$. Define a real-valued function f to be *polynomially dominated to degree b* (where $b \geq 0$) if $|f(x)| = O(x^b + 1)$.

Theorem 2.8. *Let $\{ D_j \colon j \geq 0 \}$ be a sequence of delays determined from the underlying chain of a marking process using the method of start vectors. Suppose that there exists a recurrent single state $\bar{s}$ and that the conditions in (2.2)–(2.4) and (2.6) hold. Then*

(i) *the random indices $\{\breve{\gamma}(k) \colon k \geq 0\}$ defined by (2.7) form a sequence of od-equilibrium points for $\{ D_j \colon j \geq 0 \}$,*

(ii) *the random indices $\{\breve{\gamma}(k) \colon k \geq 0\}$ also form a sequence of regeneration points for $\{ D_j \colon j \geq 0 \}$, provided that there are no ongoing delays at the $\theta(k)$th marking change for $k \geq 0$, and*

(iii) *the cycle sum $\breve{Y}_1(|f|) = \sum_{j=\breve{\gamma}(0)}^{\breve{\gamma}(1)-1} |f(D_j)|$ has finite rth moment for any real-valued function f that is polynomially dominated to degree b (where $r, b \geq 1$), provided that $E_\mu[\delta_1^{rp}] < \infty$ and $E_\mu[\tau_1^{rbq}] < \infty$ for nonnegative real numbers p and q with $p^{-1} + q^{-1} = 1$.*

We defer the proof until the end of the subsection. It follows from the theorem that $E_\mu[\breve{Y}_1^r(|f|)] < \infty$ whenever f is polynomially dominated to degree b and both $E_\mu[\delta_1^{r(b+1)}]$ and $E_\mu[\tau_1^{r(b+1)}]$ are finite—take $p = b + 1$ and $q = (b+1)/b$.

Remark 2.9. The final assertion of the theorem holds when $b = 0$, $p = 1$, and $q = \infty$, provided that we take $rbq = 0$. Thus, if $E_\mu[\delta_1^r] < \infty$ for some $r \geq 1$, then $E_\mu[\breve{Y}_1^r(|f|)] < \infty$ for any bounded function f. For example, the cycle length $\breve{\delta}_1 = \breve{\gamma}(1) - \breve{\gamma}(0)$ satisfies $E_\mu[\breve{\delta}_1^r] < \infty$ whenever $E_\mu[\delta_1^r] < \infty$ (take $f \equiv 1$).

Remark 2.10. The crux of the final assertion is that the cycles of $\{D_j : j \geq 0\}$ have well-behaved moments whenever the cycles of the underlying chain and marking process have well-behaved moments. Observe that $E_\mu[\delta_1^r] < \infty$ whenever

(i) $E_\mu[(\theta(1) - \theta(0))^r] < \infty$, and

(ii) $\sup_{s', s, E^*} n_\alpha(s'; s, E^*) < \infty$, so that the number of delays that start at a marking change is bounded.

The techniques in Section 6.2 can be used to show that quantities such as $\theta(1) - \theta(0)$ and τ_1 have finite moments.

Remark 2.11. When the condition in (2.6) is violated, there is an additional dependency between the delays in adjacent $\theta(k)$-cycles, and the conclusion of Theorem 2.8 may not hold. Specifically, the number and positions of the starts deleted from the current start vector at the $\breve{\theta}(k)$th marking change may depend explicitly on the new marking $S_{\breve{\theta}(k)}$. Of course, delays that start at or after time $\zeta_{\breve{\theta}(k)}$ also depend on $S_{\breve{\theta}(k)}$. The condition in (2.6) can be dropped, however, if the theorem is slightly modified. The idea is to change the definition of the random indices $\{\breve{\theta}(k): k \geq 0\}$. Specifically, given $\breve{\theta}(k)$, wait until the first marking change $\breve{\nu}(k)$ at which all of the ongoing delays at the $\breve{\theta}(k)$th marking change have terminated, and take as $\breve{\theta}(k+1)$ the smallest $\theta(l)$ such that $\theta(l)$ *is strictly greater than* $\breve{\nu}(k)$. It can then be shown that the conclusion of the theorem holds for the resulting random indices $\{\breve{\gamma}(k): k \geq 0\}$ even when (2.6) does not hold. Of course, the corresponding regenerative cycles for the process $\{D_j : j \geq 0\}$ typically are longer than the original cycles.

Examples

EXAMPLE 2.12 (Cyclic queues with feedback). Suppose that the service-time distribution at center 1 is GNBU and that the essential supremum of the service-time distribution at center 2 is infinite. The marking $\bar{s} = (0, N)$ is a single state with $E(\bar{s}) = \{e_2\}$. As shown in Example 2.37 in Chapter 5, $\bar{s}$ is recurrent, so that each $\theta(k)$ defined by (2.1) is a.s. finite. The regeneration points $\{\zeta_{\theta(k)}: k \geq 0\}$ are the successive random times at which there is a service completion at center 2 with all jobs at center 2.

Consider the delay intervals from whenever a job completes service at center 2 to when the job next completes service at center 2, and suppose that we wish to estimate time-average limits of the sequence of delays for all N jobs. Because the marking $\bar{s}$ is recurrent, (2.2) and (2.3) hold. Moreover, since the marking set G is finite and there are no immediate transitions, it follows from Theorem 3.13 in Chapter 3 that (1.1) holds, and hence that (2.4) holds. The condition in (2.6) holds trivially since the new marking is $s' = (1, N-1)$ whenever the marking is $\bar{s}$ and transition $\bar{e}$ fires. The random indices $\{\tilde{\gamma}(k)\colon k \geq 0\}$ defined by (2.7) therefore form a sequence of od-equilibrium points for the process $\{D_j\colon j \geq 0\}$.

To understand the foregoing result intuitively, observe that at each time $\zeta_{\check{\theta}(k)}$ a delay terminates and a new delay starts for the job that just completed service at center 2. The length of the new delay interval is $D_{\tilde{\gamma}(k)}$. The sequence $\{D_j\colon j \geq \tilde{\gamma}(k)\}$ is determined by $\{(S_n, C_n)\colon n \geq \check{\theta}(k)\}$ according to a mechanism that does not depend on either k or the precise values of the components of $V_{\check{\theta}(k)}$. Hence, the sequence $\{D_j\colon j \geq \tilde{\gamma}(k)\}$ is distributed as $\{D_j\colon j \geq 0\}$. There are $N-1$ ongoing delays at time $\zeta_{\check{\theta}(k)}$, corresponding to the $N-1$ jobs waiting in the queue at center 2 just before time $\zeta_{\check{\theta}(k)}$. Clearly, the delays $\{D_j\colon j \geq \tilde{\gamma}(k)\}$ may depend on these $N-1$ delays, which are a subset of $\{D_j\colon \tilde{\gamma}(k-1) \leq j < \tilde{\gamma}(k)\}$. By construction, however, the terminations that correspond to $\{D_j\colon 0 \leq j < \tilde{\gamma}(k-1)\}$ occur before $\zeta_{\check{\theta}(k)}$. It follows that these latter delays are determined by $\{(S_n, C_n)\colon 0 \leq n < \check{\theta}(k)\}$ and thus are independent of $\{D_j\colon j \geq \tilde{\gamma}(k)\}$.

To show that cycle sums of the form

$$\check{Y}_1(|f|) = \sum_{j=\tilde{\gamma}(0)}^{\tilde{\gamma}(1)-1} |f(D_j)|$$

have finite moments, we can use Theorem 2.8(iii). To apply this result it must be shown that τ_1 and δ_1 have finite moments. Finiteness of the moments of τ_1 and δ_1 can be established using Theorems 2.36, 2.40, and 2.44 in Chapter 6—see Example 2.51 in Chapter 6.

EXAMPLE 2.13 (Manufacturing flow-line with shunt bank). Marking $\bar{s} = (1, 0, 0, 0, 0, 0, 0, 0)$ is a single state with $E(\bar{s}) = \{e_1\}$. Under suitable conditions on the distributions of the processing-time random variables L_1 and L_2, the marking $\bar{s}$ is recurrent and each $\theta(k)$ defined by (2.1) is a.s. finite. The regeneration points $\{\zeta_{\theta(k)}\colon k \geq 0\}$ are the successive random times at which there is an end of processing at station 1 with no other parts in the system.

Consider the delay intervals from whenever there is a start of processing at station 1 for a part to when there is an end of processing at station 2 for the part, and suppose that we wish to estimate time-average limits of the sequence of delays for all parts. The recurrence of $\bar{s}$ implies that (2.2) and (2.3) hold, and it follows from (1.1) that (2.4) holds. The condition in (2.6) holds trivially since the new marking is $s' = (0, 1, 0, 0, 0, 0, 0, 0)$ whenever the marking is $\bar{s}$ and transition $\bar{e} = e_1$ fires. The random indices $\{\tilde{\gamma}(k)\colon k \geq 0\}$ defined by (2.7) therefore form a sequence of od-equilibrium points for the process $\{D_j\colon j \geq 0\}$.

As in the previous example, Theorem 2.8(iii) can be used to show that cycle sums of the form $\check{Y}_1(|f|) = \sum_{j=\tilde{\gamma}(0)}^{\tilde{\gamma}(1)-1} |f(D_j)|$ have finite moments under suitable assumptions on the moments of L_1 and L_2.

Extension to Alternative Regeneration Points

The random indices $\{\theta(k): k \geq 0\}$ used to construct the sequence of od-equilibrium points $\{\check{\gamma}(k): k \geq 0\}$ correspond to the successive times at which the current marking is a single state $\bar{s}$ and a distinguished transition $\bar{e}$ fires. Other choices of $\{\theta(k): k \geq 0\}$ are possible. For example, suppose that there exists a marking $\bar{s}$ with $E(\bar{s}) = \{\bar{e}_0, \bar{e}_1, \ldots, \bar{e}_l\}$ (where $l \geq 1$) such that

$$F(x; s', \bar{e}_i, s, E^*) \equiv F(x; \bar{e}_i) = 1 - \exp(-\lambda_i x)$$

for $1 \leq i \leq l$. Let $\{\theta(k): k \geq 0\}$ be the sequence of random indices that correspond to the successive times at which the marking is $\bar{s}$ and transition $\bar{e}_0$ fires. Suppose that at each time $\zeta_{\theta(k)}$ transitions $\bar{e}_1, \bar{e}_2, \ldots, \bar{e}_l$ are old transitions. Also suppose that (2.2), (2.3), and (2.4) hold and that (2.6) holds with $\bar{e} = \bar{e}_0$. Finally, assume for convenience that each interval $[\zeta_{\theta(k)}, \zeta_{\theta(k+1)}]$ contains at least one start. Then the sequence of random indices $\{\check{\gamma}(k): k \geq 0\}$ constructed from the sequence $\{\theta(k): k \geq 0\}$ decomposes sample paths of the sequence $\{D_j: j \geq 0\}$ into o.d.s. cycles.

To see that this assertion holds, set

$$\mathcal{G}_k = \big(S_{\check{\theta}(k-1)}, t^*_{\check{\theta}(k-1)}, E^*_{\check{\theta}(k-1)},$$

$$S_{\check{\theta}(k-1)+1}, t^*_{\check{\theta}(k-1)+1}, E^*_{\check{\theta}(k-1)+1}, \ldots, S_{\check{\theta}(k)-1}, t^*_{\check{\theta}(k)-1}, E^*_{\check{\theta}(k)-1}\big)$$

for $k \geq 1$, where, as usual, $t^*_n = t^*(S_n, C_n)$ and $E^*_n = E^*(S_n, C_n)$ for $n \geq 0$. It follows from (2.19) in Chapter 6 that the sequence $\{\mathcal{G}_k: k \geq 1\}$ consists of i.i.d. random vectors. Since $\mathcal{G}_k$ and $\mathcal{G}_{k+1}$ completely determine both the number of delays that start during the interval $[\zeta_{\check{\theta}(k-1)}, \zeta_{\check{\theta}(k)})$ and the length of the corresponding delay intervals, the desired result follows. As discussed in Remark 2.12 in Chapter 6, the random indices $\{\theta(k): k \geq 0\}$ do not form a sequence of regeneration points for the chain $\{(S_n, C_n): n \geq 0\}$, but the random times $\{\zeta_{\theta(k)}: k \geq 0\}$ do form a sequence of regeneration points for the marking process $\{X(t): t \geq 0\}$.

EXAMPLE 2.14 (Manufacturing cell with robots). Suppose that the successive times for machine 1 to process a part are i.i.d. according to a distribution function that has support on $(0, \infty)$. Also suppose that the successive times for machine 2 to process a part are i.i.d. according to an exponential distribution. Consider the delay intervals from whenever robot 1 starts to transfer a raw part from the loading area to conveyor 1 to when robot 1 completes transfer of the part to the unloading area, and suppose that we wish to estimate time-average limits for the sequence of delays for all parts. A simple inductive argument shows that (2.2) and (2.3) hold, and it follows from (1.1) that (2.4) holds. Denote by $\bar{s} = (\bar{s}_1, \bar{s}_2, \ldots, \bar{s}_{24})$ the unique marking in which $\bar{s}_4 = \bar{s}_9 = \bar{s}_{11} = \bar{s}_{22} = \bar{s}_{24} = 1$ and $\bar{s}_j = 0$ otherwise, and let $\{\zeta_{\theta(k)}: k \geq 0\}$ be the successive random times at which the marking is $\bar{s}$ and transition $\bar{e} = e_8$ fires. Thus $\zeta_{\theta(k)}$ is the kth successive time at which machine 1 completes processing of a part, machine 2 is processing a part, a raw part is on conveyor 1 awaiting transfer to a machine, no parts are on conveyor 2, and the arm of each robot is in its null position. Observe that $E(\bar{s}) = \{e_8, e_9\}$ and the clock for transition e_9 is always set according to a fixed exponential distribution. The condition in (2.6) holds trivially since the new marking is always equal to a unique fixed marking $\bar{s}'$ whenever the marking is $\bar{s}$ and transition $\bar{e}$ fires. It follows that the sequence of random indices $\{\tilde{\gamma}(k) \colon k \geq 0\}$ constructed from the sequence $\{\theta(k) \colon k \geq 0\}$ decomposes sample paths of the sequence $\{D_j \colon j \geq 0\}$ into o.d.s. cycles.

Using techniques similar to those in the proof of Theorem 2.24 in Chapter 6, we can extend the assertion in Theorem 2.8(iii) to the current setting. Cycle sums of the form $\check{Y}_1(|f|) = \sum_{j=\tilde{\gamma}(0)}^{\tilde{\gamma}(1)-1} |f(D_j)|$ then can be shown to have finite moments under appropriate moment conditions on the processing-time distribution for machine 1; the required arguments are similar to those used in Example 2.51 of Chapter 6.

Proof of Theorem 2.8

To prove Theorem 2.8, we require the following lemma, which concerns a sequence $X, X_1, X_2, \ldots$ of i.i.d. random variables taking values in a set S, along with a set $A \subset S$ with $P\{X \in A\} > 0$. Define a sequence of a.s. finite random indices $\{I(n) \colon n \geq 0\}$ by setting $I(0) = 0$ and $I(n) = \inf\{i > I(n-1) \colon X_i \in A\}$ for $n \geq 1$.

Lemma 2.15. *The random subsequence* $\{X_{I(n)} \colon n \geq 1\}$ *consists of i.i.d. random variables with common distribution given by*

$$P\{X_{I(n)} \in B\} = P\{X \in B \mid X \in A\}$$

for $B \subseteq S$.

PROOF. Fix integers $k \geq 1$ and $1 \leq i_1 < i_2 < \cdots < i_k$, along with subsets $B_1, B_2, \ldots, B_k \subseteq S$, and set $J = \{1, 2, \ldots, i_k\} - \{i_1, i_2, \ldots, i_k\}$. Then

$$P\{X_{I(n)} \in B_n \text{ for } 1 \leq n \leq k \mid I(1) = i_1, \ldots, I(k) = i_k\}$$
$$= P\{X_{i_n} \in B_n \text{ for } 1 \leq n \leq k$$
$$\mid X_{i_n} \in A \text{ for } 1 \leq n \leq k; X_j \in S - A \text{ for } j \in J\}$$
$$= \frac{P\{X_{i_n} \in A \cap B_n \text{ for } 1 \leq n \leq k; X_j \in S - A \text{ for } j \in J\}}{P\{X_{i_n} \in A \text{ for } 1 \leq n \leq k; X_j \in S - A \text{ for } j \in J\}}$$
$$= \prod_{n=1}^{k} \frac{P\{X \in A \cap B_n\}}{P\{X \in A\}}.$$

Multiplying the above equality by $P_\mu\{I(1) = i_1, \ldots, I(k) = i_k\}$ and summing over all possible values of $i_1, \ldots, i_k$ yields the desired result, since k and $B_1, B_2, \ldots, B_k$ are arbitrary. □

We also need the following consequence of Proposition 1.20 in the Appendix. Consider a sequence $\{\mathcal{F}_n: n \geq 1\}$, where each $\mathcal{F}_n$ is a collection of random variables and $\mathcal{F}_n \subset \mathcal{F}_{n+1}$ for $n \geq 1$. Also consider a positive integer-valued random variable N that is a *stopping time* with respect to $\{\mathcal{F}_n: n \geq 1\}$—that is, for each $k \geq 1$ the occurrence or nonoccurrence of the event $\{N \leq k\}$ is determined by the values of the random variables in $\mathcal{F}_k$. Finally, let $S_N = \sum_{n=1}^{N} X_n$, where $\{X_n: n \geq 1\}$ is a sequence of i.i.d. random variables such that X_n is determined by $\mathcal{F}_n$ for $n \geq 1$ and independent of $\mathcal{F}_{n-1}$ for $n \geq 2$. Then for $r \geq 0$ there exists a constant b_r (depending only on r) such that

$$E[|S_N|^r] \leq b_r E[|X_1|^r] E[N^r]. \qquad (2.16)$$

As before, set $\check{\delta}_k = \check{\gamma}(k) - \check{\gamma}(k-1)$, so that $\check{\delta}_k$ is the number of delays that start during the interval $[\zeta_{\check{\theta}(k-1)}, \zeta_{\check{\theta}(k)})$. Proving the first assertion of Theorem 2.8 amounts to showing that the post-$\check{\gamma}(k)$ process $\{(D_{\check{\gamma}(k)+n}, \check{\delta}_{k+n+1}): n \geq 0\}$

(i) is distributed as $\{(D_{\check{\gamma}(0)+n}, \check{\delta}_{n+1}): n \geq 0\}$ for $k \geq 1$,

(ii) is independent of $\{D_0, D_1, \ldots, D_{\check{\gamma}(k-1)-1}; \check{\delta}_1, \check{\delta}_2, \ldots, \check{\delta}_{k-1}\}$ for $k \geq 2$, and

(iii) is independent of $\check{\delta}_k$ for $k \geq 0$.

To this end, suppose that there is at least one ongoing delay at each regeneration point $\zeta_{\theta(k)}$. It follows that there is at least one ongoing delay at each point $\zeta_{\check{\theta}(k)}$, so that $\check{\gamma}(k)$ is simply the index of the first delay that starts after time $\zeta_{\check{\theta}(k)}$: $\check{\gamma}(k) = \inf\{j \geq 0 : \alpha(j) \geq \check{\theta}(k)\}$. Observe that each $\check{\theta}(k)$ is an a.s. finite stopping time with respect to the underlying chain $\{(S_n, C_n): n \geq 0\}$. Moreover, for each k the random variables in the sequence $\{\check{\theta}(l+1) - \check{\theta}(l): l \geq k\}$ are determined by $\{(S_n, C_n): n \geq \check{\theta}(k)\}$ according to a mechanism that does not depend explicitly on either k or the precise values of the components of $V_{\check{\theta}(k)}$—indeed, it can easily be seen that

- The number of ongoing delays and the location of the corresponding starts within the vector $V_{\check{\theta}(k)}$ are determined according to the functions $i_\alpha(S_{\check{\theta}(k)}; \bar{s}, \bar{e})$, $i_\beta(S_{\check{\theta}(k)}; \bar{s}, \bar{e})$, and $i_\pi(S_{\check{\theta}(k)}; \bar{s}, \bar{e})$ and hence are determined by $S_{\check{\theta}(k)}$.

- The number of marking changes until these ongoing delays terminate is determined by the sequence of start-vector insertions, deletions, and permutations that occurs after the $\check{\theta}(k)$th marking change, and this sequence is in turn determined by the evolution of the underlying chain after the $\check{\theta}(k)$th marking change.

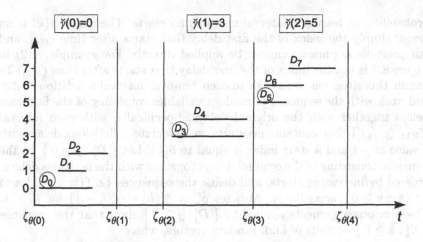

Figure 8.7. Regenerative cycles for delays.

An application of the strong Markov property then shows that the random indices $\{\breve{\theta}(k)\colon k \geq 0\}$ form a sequence of regeneration points for the chain $\{(S_n, C_n)\colon n \geq 0\}$. It follows from the definition of the sequence $\{V_n\colon n \geq 0\}$ that for $k \geq 1$ the post-$\breve{\gamma}(k)$ process $\{(D_{\breve{\gamma}(k)+n}, \breve{\delta}_{k+n+1})\colon n \geq 0\}$ is determined by $\{(S_n, C_n)\colon n \geq \breve{\theta}(k)\}$ according to a mechanism that does not depend explicitly on k or on the precise values of the entries in $V_{\breve{\theta}(k)}$. Hence, by the regenerative property of the chain, the post-$\breve{\gamma}(k)$ process is distributed as the post-$\breve{\gamma}(0)$ process. Thus, the assertion in (i) holds. To see that the assertions in (ii) and (iii) hold, fix $k \geq 2$ and observe that by construction the terminations $B_0, B_1, \ldots, B_{\breve{\gamma}(k-1)-1}$ are all less than or equal to $\zeta_{\breve{\theta}(k)}$. The condition in (2.6) ensures that the number and positions of the starts deleted from the current start vector at time $\zeta_{\breve{\theta}(k)}$ do not depend explicitly on the new marking $S_{\breve{\theta}(k)}$. Thus the collection of random variables $\{D_0, D_1, \ldots, D_{\breve{\gamma}(k-1)-1}; \breve{\delta}_1, \breve{\delta}_2, \ldots, \breve{\delta}_k\}$ is determined completely by the process $\{(S_n, C_n)\colon 0 \leq n < \breve{\theta}(k)\}$. By the regenerative property of the chain, the post-$\breve{\gamma}(k)$ process is independent of $\{D_0, D_1, \ldots, D_{\breve{\gamma}(k-1)-1}; \breve{\delta}_1, \breve{\delta}_2, \ldots, \breve{\delta}_k\}$, as desired. A similar argument shows that the post-$\breve{\gamma}(1)$ process is independent of $\breve{\gamma}(1)$, and the first assertion of the theorem follows.

To prove the second assertion of the theorem, suppose that there are no ongoing delays at any regeneration point $\zeta_{\theta(k)}$, so that $\breve{\theta}(k) = \theta(k)$ for $k \geq 0$. When each interval $[\zeta_{\theta(k)}, \zeta_{\theta(k+1)}]$ contains at least one start, each $\breve{\gamma}(k)$ is, as before, the index of the first delay that starts after time $\zeta_{\theta(k)}$. An argument almost identical to the one given above then shows that the random indices $\{\breve{\gamma}(k)\colon k \geq 0\}$ decompose the sample paths of the sequence $\{D_j\colon j \geq 0\}$ into i.i.d. cycles and hence form a sequence of regeneration points. The situation is more complicated when, with positive probability, at least one interval contains no starts. Then each $\check{\gamma}(k)$ is no longer simply the index of the first delay that starts after time $\zeta_{\theta(k)}$, and our previous arguments cannot be applied directly. For example, $\check{\gamma}(2)$ in Figure 8.7 is not the index of the first delay that starts after time $\zeta_{\theta(2)}$. To handle this situation, we assign to each "empty" interval a fictitious delay and work with the sequence of random variables consisting of the fictitious delays together with the original delays. Specifically, with each interval $[\zeta_{\theta(k)}, \zeta_{\theta(k+1)}]$ that contains no starts we associate a fictitious delay with a value of -1 and a start index α equal to $\theta(k)$. Let $\{ D'_j : j \geq 0 \}$ be the sequence consisting of the original delays together with the fictitious delays, ordered by increasing starts, and define the sequences $\{ \alpha'(j) : j \geq 0 \}$ and $\{ \check{\gamma}'(k) : k \geq 0 \}$ accordingly. Also set $\check{\delta}'_k = \check{\gamma}'(k) - \check{\gamma}'(k-1)$ for $k \geq 1$. The previous arguments applied to $\{ D'_j : j \geq 0 \}$ show that the sequence $\{ X'_k : k \geq 0 \}$ consists of i.i.d. random vectors, where

$$X'_k = (\check{\delta}'_k, D'_{\check{\gamma}'(k)}, D'_{\check{\gamma}'(k)+1}, \ldots, D'_{\check{\gamma}'(k+1)-1}).$$

Observe that the sequence $\{ X_k : k \geq 0 \}$, where

$$X_k = (\check{\delta}_k, D_{\check{\gamma}(k)}, D_{\check{\gamma}(k)+1}, \ldots, D_{\check{\gamma}(k+1)-1}),$$

can be viewed as a random subsequence of $\{ X'_k : k \geq 0 \}$. By Lemma 2.15, the sequence $\{ X_k : k \geq 0 \}$ inherits the i.i.d. property of $\{ X'_k : k \geq 0 \}$, and the desired result follows.

For $k \geq 1$, set $\check{\tau}_k = \zeta_{\check{\theta}(k)} - \zeta_{\check{\theta}(k-1)}$, and recall that $\tau_k = \zeta_{\theta(k)} - \zeta_{\theta(k-1)}$, that $\check{\delta}_k = \check{\gamma}(k) - \check{\gamma}(k-1)$ is the number of delays that start during the interval $[\zeta_{\check{\theta}(k-1)}, \zeta_{\check{\theta}(k)})$, and that δ_k is the number of delays that start during the interval $[\zeta_{\theta(k-1)}, \zeta_{\theta(k)})$. To prove the final assertion of the theorem, it suffices to show that, for $r \geq 1$,

$$E_\mu \left[\check{\delta}_1^r \right] < \infty \qquad \text{whenever} \qquad E_\mu \left[\delta_1^r \right] < \infty \qquad (2.17)$$

and

$$E_\mu \left[\check{\tau}_1^r \right] < \infty \qquad \text{whenever} \qquad E_\mu \left[\tau_1^r \right] < \infty. \qquad (2.18)$$

To see that the desired result follows from (2.17) and (2.18), fix constants r, b, p, and q as in the statement of the theorem, along with a function f that is polynomially dominated to degree b. Observe that

$$\check{Y}_1(|f|) \leq a \check{\delta}_k \left((\check{\tau}_1 + \check{\tau}_2)^b + 1 \right)$$

for some finite nonnegative constant a. Since

$$\left((\check{\tau}_1 + \check{\tau}_2)^b + 1 \right)^r \leq 2^{(b+1)r-2} (\check{\tau}_1^{rb} + \check{\tau}_2^{rb}) + 2^{r-1}$$

by two applications of the c_r-inequality, it follows from Hölder's inequality
that

$$E_\mu \left[\check{Y}_1^r(|f|) \right]$$
$$\leq a' \left(E_\mu \left[\check{\delta}_1^r \check{\tau}_1^{rb} \right] + E_\mu \left[\check{\delta}_1^r \check{\tau}_2^{rb} \right] + E_\mu \left[\check{\delta}_1^r \right] \right)$$
$$\leq 2a' E_\mu^{1/p} \left[\check{\delta}_1^{rp} \right] E_\mu^{1/q} [\check{\tau}_1^{rbq}] + a' E_\mu \left[\check{\delta}_1^r \right],$$

where $a' = 2^{(b+1)r-1}a^r$. Since $E_\mu [\check{\delta}_1^{rp}]$ and $E_\mu[\tau_1^{rbq}]$ are finite by hypothesis,
the desired result follows.

We complete the argument by establishing (2.17); the proof of (2.18) is
similar. First suppose that there are no ongoing delays at any regeneration
point $\zeta_{\theta(k)}$ and that each interval $[\zeta_{\theta(k-1)}, \zeta_{\theta(k)})$ contains at least one start.
Then $\check{\delta}_k = \delta_k$ for $k \geq 1$ and (2.17) follows trivially. Next suppose that there
are no ongoing delays at any regeneration point $\zeta_{\theta(k)}$ and that one or more
intervals of the form $[\zeta_{\theta(k-1)}, \zeta_{\theta(k)})$ contain no starts. For $k \geq 1$, let J_k be
the indicator variable for the event that there is at least one start in the
interval $[\zeta_{\theta(k-1)}, \zeta_{\theta(k)})$:

$$J_k = \begin{cases} 1 & \text{if } \sum_{n=0}^{\infty} 1_{[\theta(k-1),\theta(k))}(\alpha(n)) > 0; \\ 0 & \text{otherwise.} \end{cases}$$

Observe that $P_\mu \{ J_1 = 1 \} > 0$ by (2.4) so that, using Lemma 2.15,

$$E_\mu[\check{\delta}_1^r] = \frac{E_\mu[\delta_1^r J_1]}{P_\mu \{ J_1 = 1 \}} \leq \frac{E_\mu[\delta_1^r]}{P_\mu \{ J_1 = 1 \}} < \infty,$$

as desired. Finally, suppose that there are ongoing delays at each regeneration point $\zeta_{\theta(k)}$, and write

$$\check{\delta}_1 = \sum_{k=1}^{N} \delta_k, \tag{2.19}$$

where N is the number of points of the sequence $\{ \theta(k) \colon k \geq 0 \}$ that lie
in the interval $[\check{\theta}(0), \check{\theta}(1))$. Since $E_\mu [\delta_1^r] < \infty$ by hypothesis, the desired
result follows from (2.16), provided that $E_\mu [N^r] < \infty$—take $X_n = \delta_n$ and

$$\mathcal{F}_n = \{ (S_j, C_j) \colon 0 \leq j \leq \theta(n) - 1 \}.$$

We therefore finish the proof by showing that N has finite moments of all
orders, using an argument similar to the final part of the proof of Theorem 2.24 in Chapter 6. For $k \geq 1$, let $\nu(k)$ be the index of the first marking
change after $\theta(k)$ such that all ongoing delays at the $\theta(k)$th marking change
have terminated, and set $\Lambda_k = \nu(k) - \theta(k)$; if there are no ongoing delays at
time $\zeta_{\theta(k)}$, then set $\Lambda_k = 0$. Also set $\Lambda_0 = K$, where K is defined by (1.3).

Observe that $\{\Lambda_k: k \geq 0\}$ is a sequence of identically distributed random variables. We claim that there exists $m \geq 1$ such that

$$P_\mu\{\Lambda_k < \theta(k+m) - \theta(k)\} > 0$$

for $k \geq 0$; otherwise, since (1.1) implies that $\lim_{k\to\infty} \theta(k) = \infty$ a.s., it follows from Bonferroni's inequality—Proposition 1.1(vi) in the Appendix—that

$$P_\mu\{K = \infty\} = P_\mu\{\Lambda_0 = \infty\} = P_\mu\{\Lambda_0 \geq \theta(m) - \theta(0) \text{ for } m \geq 1\} = 1,$$

contradicting (2.2). For $l \geq 0$, set $J_l = 1$ if $\Lambda_{lm} \geq \theta((l+1)m) - \theta(lm)$; otherwise, set $J_l = 0$. It follows from (2.1) and (2.6) that each J_l is determined by $\{(S_n, C_n): \theta(lm) \leq n < \theta((l+1)m)\}$. Since $\{\theta(k): k \geq 0\}$—and hence $\{\theta(lm): l \geq 0\}$—is a sequence of regeneration points for the underlying chain, it follows that $\{J_l: l \geq 0\}$ is a sequence of i.i.d. random variables with $p = P_\mu\{J_1 = 1\} < 1$. Thus

$$P_\mu\{N > lm\} \leq P_\mu\{J_0 = 1, J_1 = 1, \ldots, J_{l-1} = 1\} = p^l$$

for $l \geq 1$, so that the distribution of N/m has geometrically decreasing tail probabilities and hence N has finite moments of all orders.

8.2.2   The Extended Regenerative Method for Delays

Suppose that we have constructed a sequence of od-equilibrium points $\{\check\gamma(k): k \geq 0\}$ as above and wish to estimate time-average limits of the form $\lim_{n\to\infty}(1/n)\sum_{j=0}^{n-1} f(D_j)$, where f is a real-valued function. Under the moment conditions given below, it follows from Theorem 1.27 in Chapter 6 that such time-average limits exist a.s.. Moreover, the extended regenerative method developed in Section 6.3.8 can be applied in the current setting to obtain strongly consistent point estimates and asymptotic confidence intervals for time-average limits.

As before, set $\check\delta_k = \check\gamma(k) - \check\gamma(k-1)$ for $k \geq 1$, so that $\check\delta_k$ is the length of the kth cycle. Let f be a real-valued function and set

$$\check Y_k(f) = \sum_{j=\check\gamma(k-1)}^{\check\gamma(k)-1} f(D_j)$$

for $k \geq 0$. By the od-equilibrium property, the sequence $\{\check\delta_k: k \geq 1\}$ consists of i.i.d. random variables and the sequence $\{(\check Y_k(f), \check\delta_k): k \geq 1\}$ consists of o.d.s. random vectors. Suppose that $E_\mu[\check\delta_1] < \infty$ and $E_\mu[\check Y_1(|f|)] < \infty$. It then follows from Theorem 1.27 in Chapter 6 that

$$\lim_{n\to\infty} \frac{1}{n} \sum_{j=0}^{n-1} f(D_j) = \frac{E_\mu[\check Y_1(f)]}{E_\mu[\check\delta_1]} \stackrel{\text{def}}{=} r(f) \text{ a.s..}$$

To obtain estimates for the quantity $r(f)$, observe a fixed number n of cycles of $\{D_j\colon j \geq 0\}$ and measure the quantities $\check{Y}_1(f), \check{Y}_2(f), \ldots, \check{Y}_n(f)$ and $\check{\delta}_1, \check{\delta}_2, \ldots, \check{\delta}_n$. Set $\hat{r}(n) = \bar{Y}(n)/\bar{\delta}(n)$, where $\bar{Y}(n) = (1/n)\sum_{k=1}^n \check{Y}_k(f)$ and $\bar{\delta}(n) = (1/n)\sum_{k=1}^n \check{\delta}_k$. Next, take

$$\check{s}^2(n) = \frac{1}{n-1}\sum_{k=1}^n (\check{Y}_k(f) - \hat{r}(n)\check{\delta}_k)^2$$

$$+ \frac{2}{n-1}\sum_{k=1}^{n-1}(\check{Y}_k(f) - \hat{r}(n)\check{\delta}_k)(\check{Y}_{k+1}(f) - \hat{r}(n)\check{\delta}_{k+1})$$

as an estimator of

$$\check{\sigma}^2(f) = \operatorname{Var}_\mu\left[\check{Y}_1(f) - r(f)\check{\delta}_1\right]$$
$$+ 2\operatorname{Cov}_\mu\left[\check{Y}_1(f) - r(f)\check{\delta}_1, \check{Y}_2(f) - r(f)\check{\delta}_2\right].$$

As discussed in Section 6.3.8, we have $\hat{r}(n) \to r(f)$ a.s., $\check{s}^2(n) \to \check{\sigma}^2(f)$ a.s., and

$$\frac{\sqrt{n}(\hat{r}(n) - r(f))}{\check{s}(n)/\bar{\delta}(n)} \Rightarrow N(0,1)$$

as $n \to \infty$, where $N(0,1)$ is a standard normal random variable and $\Rightarrow$ denotes convergence in distribution. These results lead directly to the following estimation procedure.

Algorithm 2.20 (Extended regenerative method for delays)

1. Select a single state $\bar{s}$ and define the corresponding sequence $\{\check{\theta}(k)\colon k \geq 0\}$ of random indices for the underlying chain $\{(S_n, C_n)\colon n \geq 0\}$.

2. Define the corresponding sequence $\{\check{\gamma}(k)\colon k \geq 0\}$ of random indices for the sequence $\{D_j\colon j \geq 0\}$ of delays via (2.7).

3. Simulate the marking process $\{X(t)\colon t \geq 0\}$ and observe a fixed number n of cycles defined by the random indices $\{\check{\gamma}(k)\colon k \geq 0\}$.

4. Compute the length $\check{\delta}_k$ of the kth cycle and the quantity $\check{Y}_k(f) = \sum_{j=\check{\gamma}(k-1)}^{\check{\gamma}(k)-1} f(D_j)$ for $1 \leq k \leq n$.

5. Form the strongly consistent point estimate $\hat{r}(n) = \bar{Y}(n)/\bar{\delta}(n)$ for $r(f)$.

6. Form the asymptotic $100p\%$ confidence interval

$$\left[\hat{r}(n) - \frac{z_p\,\check{s}(n)}{\bar{\delta}(n)\sqrt{n}}, \hat{r}(n) + \frac{z_p\,\check{s}(n)}{\bar{\delta}(n)\sqrt{n}}\right]$$

for $r(f)$, where z_p is the $(1+p)/2$ quantile of the standard normal distribution.

Remark 2.21. If $\check{\delta}_1$ is aperiodic with finite mean, then $D_j \Rightarrow D$ as $j \to \infty$ and $r(f) = E[f(D)]$—see Theorem 1.31 in Chapter 6. Thus, under these conditions the quantity $r(f)$ can be interpreted not only as a time-average limit but also as a steady-state mean.

Remark 2.22. Algorithm 2.20 can be simplified when there are no ongoing delays at any regeneration point $\zeta_{\theta(k)}$. Then $\check{\theta}(k) = \theta(k)$ for $k \geq 0$, and the random indices $\{\check{\gamma}(k) \colon k \geq 0\}$ form a sequence of regeneration points for $\{D_j \colon j \geq 0\}$. It follows that

$$\text{Cov}_\mu[\check{Y}_1(f) - r(f)\check{\delta}_1, \check{Y}_2(f) - r(f)\check{\delta}_2] = 0,$$

and the quantity $\check{s}(n)$ in the confidence interval (3.6) can be replaced by

$$s(n) \overset{\text{def}}{=} \left(\frac{1}{n-1} \sum_{k=1}^{n} (\check{Y}_k(f) - \hat{r}(n)\check{\delta}_k)^2 \right)^{1/2}.$$

The resulting estimation procedure coincides with the standard regenerative method.

Remark 2.23. Strongly consistent point estimates and asymptotic confidence intervals for $r(f)$ can also be based on simulation of the process $\{X(t) \colon t \geq 0\}$ for a fixed (simulated) time u. Compute statistics for the random number $n(u)$ of cycles completed by time u. Then $\hat{r}(n(u)) \to r(f)$ a.s. and

$$\frac{\sqrt{n(u)}\left(\hat{r}(n(u)) - r(f)\right)}{\check{s}(n(u))/\bar{\delta}(n(u))} \Rightarrow N(0,1) \tag{2.24}$$

as $u \to \infty$. The proof of this assertion is similar to that of Theorem 3.18 in Chapter 6 but uses Corollary 2.10 in the Appendix.

8.2.3 *The Multiple-Runs Method*

As an alternative to the extended regenerative method, the multiple-runs method introduced in Section 6.3.8 can be used in the current setting to obtain strongly consistent point estimates and asymptotic confidence intervals for time-average limits of a sequence of delays.

Suppose that the condition in (2.6) holds with j_α and j_β defined so that there is always at least one ongoing delay at each regeneration point $\zeta_{\theta(n)}$. Define sequences $\{\check{\theta}(k) \colon k \geq 0\}$, $\{\check{\gamma}(k) \colon k \geq 0\}$, and $\{(\check{Y}_k(f), \check{\delta}_k) \colon k \geq 1\}$ as before, and suppose that for a fixed real-valued function f we have $E_\mu[\check{\delta}_1] < \infty$ and $E_\mu[\check{Y}_1(|f|)] < \infty$, so that

$$\lim_{n \to \infty} \frac{1}{n} \sum_{j=0}^{n-1} f(D_j) = \frac{E_\mu\left[\check{Y}_1(f)\right]}{E_\mu[\check{\delta}_1]} \overset{\text{def}}{=} r(f) \text{ a.s..}$$

Denote by T the time required to observe the first cycle of the sequence $\{D_j : j \geq 0\}$:

$$T = \max\{B_j : 0 \leq j < \tilde{\gamma}(1)\}.$$

Observe that there is at least one ongoing delay at each time $\zeta_{\check{\theta}(k)}$, so that $\zeta_{\check{\theta}(1)} < T \leq \zeta_{\check{\theta}(2)}$. To obtain estimates for $r(f)$, simulate the process $\{X(t): t \geq 0\}$ up to the random time T to create $\{X_1(t): 0 \leq t \leq T_1\}$ and $\{D_{j,1}: 0 \leq j < \tilde{\gamma}_1(1)\}$. Repeat this step m times to create m independent replicates and produce $\{X_i(t): 0 \leq t \leq T_i\}$ and $\{D_{j,i}: 0 \leq j < \tilde{\gamma}_i(1)\}$ for $1 \leq i \leq m$. Then compute point estimates and confidence intervals for $r(f)$ as in the standard regenerative method, treating the latter sequences as regenerative cycles. The precise algorithm is as follows.

Algorithm 2.25 (Multiple-runs method for delays)

1. Using a fixed number m of independent simulation runs, generate the "cycles" $\{D_{j,i}: 0 \leq j < \tilde{\gamma}_i(1)\}$ for $1 \leq i \leq m$.

2. Compute the length $\check{\delta}_{1,i} = \tilde{\gamma}_i(1)$ of the ith cycle and the quantity

$$\check{Y}_{1,i}(f) = \sum_{j=0}^{\tilde{\gamma}_i(1)-1} f(D_{j,i})$$

   for $1 \leq i \leq m$.

3. Form the strongly consistent point estimate $\hat{r}_M(m) = \bar{Y}_M(m)/\bar{\delta}_M(m)$ for $r(f)$, where

$$\bar{Y}_M(m) = \frac{1}{m}\sum_{i=1}^{m}\check{Y}_{1,i}(f)$$

   and

$$\bar{\delta}_M(m) = \frac{1}{m}\sum_{i=1}^{m}\check{\delta}_{1,i}.$$

4. Compute the quantity

$$\check{s}_M^2(m) = \frac{1}{m-1}\sum_{i=1}^{m}\left(\check{Y}_{1,i}(f) - \hat{r}_M(m)\check{\delta}_{1,i}\right)^2$$

   as an estimator that is strongly consistent for

$$\check{\sigma}_M^2(f) = \mathrm{Var}_\mu\left[\check{Y}_1(f) - r(f)\check{\delta}_1\right].$$

5. Form the asymptotic $100p\%$ confidence interval

$$\left[\hat{r}(n) - \frac{z_p\,\check{s}_M(m)}{\bar{\delta}_M(m)\,\sqrt{m}},\, \hat{r}(n) + \frac{z_p\,\check{s}_M(m)}{\bar{\delta}_M(m)\,\sqrt{m}}\right]$$

   for $r(f)$, where z_p is the $(1+p)/2$ quantile of the standard normal distribution.

Remark 2.26. Analogously to the extended regenerative method, point and interval estimates can be based on the random number of runs completed within a total budget of u units of simulated time. Compute statistics for the random number $m(u) = \inf\{m \geq 0 : \sum_{i=1}^{m} T_i \leq u\}$ of completed runs as in the standard regenerative method. Then, by Theorem 3.18 in Chapter 6, we have $\hat{r}_{\mathrm{M}}(m(u)) \to r(f)$ a.s. and

$$\frac{\sqrt{m(u)}\left(\hat{r}_{\mathrm{M}}(m(u)) - r(f)\right)}{\check{s}_{\mathrm{M}}(m(u))/\bar{\delta}_{\mathrm{M}}(m(u))} \Rightarrow N(0,1) \tag{2.27}$$

as $u \to \infty$.

Remark 2.28. As discussed previously, the variance estimators $\check{s}^2(n)$ and $\check{s}_{\mathrm{M}}^2(m)$ for the extended regenerative method and the multiple-runs method can be computed by means of a single pass through the data—see Remark 3.59 in Chapter 6.

For a fixed value of $p \in (0,1)$, we take the *asymptotic relative efficiency* (ARE) of the extended regenerative and multiple-runs methods to be the limiting ratio of the lengths of the $100p\%$ confidence intervals for $r(f)$ as the simulated time becomes large. Denote by $I(u; p)$ the length of the $100p\%$ confidence interval for $r(f)$ produced by the extended regenerative method based on a budget of u units of simulated time, and let $I_{\mathrm{M}}(u; p)$ be the corresponding length for the multiple-runs method. The central limit theorems in (2.24) and (2.27) imply that

$$I(u; p) = \frac{2 z_p \, \check{s}(n(u))}{\bar{\delta}(n(u)) \, \sqrt{n(u)}}$$

and

$$I_{\mathrm{M}}(u; p) = \frac{2 z_p \, \check{s}_{\mathrm{M}}(m(u))}{\bar{\delta}_{\mathrm{M}}(m(u)) \, \sqrt{m(u)}}.$$

Using Theorem 2.9 in Chapter 3 along with the SLLNs for i.i.d. and o.i.d. random variables, it is straightforward to show that

$$\lim_{u \to \infty} \frac{m(u)}{u} = \frac{1}{E_\mu[T]} \text{ a.s.},$$

$$\lim_{u \to \infty} \frac{n(u)}{u} = \frac{1}{E_\mu[\zeta_{\check{\delta}(1)}]} \text{ a.s.},$$

and

$$\lim_{m \to \infty} \bar{\delta}_{\mathrm{M}}(m) = \lim_{n \to \infty} \bar{\delta}(n) = E_\mu[\check{\delta}_1] \text{ a.s..}$$

These results imply that the ARE is given by

$$\lim_{u \to \infty} \frac{I(u; p)}{I_{\mathrm{M}}(u; p)} = \frac{\check{\sigma}(f)}{\check{\sigma}_{\mathrm{M}}(f)} \left(\frac{E_\mu[\zeta_{\check{\delta}(1)}]}{E_\mu[T]}\right)^{1/2} \text{ a.s.}$$

for all $p \in (0,1)$.

Observe that

$$E_\mu[\zeta_{\breve{\theta}(1)}] \le E_\mu[T] \le 2E_\mu[\zeta_{\breve{\theta}(1)}]$$

and, by the Cauchy–Schwarz inequality,

$$\text{Cov}_\mu\left[\breve{Y}_1(f) - r(f)\breve{\delta}_1, \breve{Y}_2(f) - r(f)\breve{\delta}_2\right] \le \text{Var}_\mu\left[\breve{Y}_1(f) - r(f)\breve{\delta}_1\right].$$

We thus obtain the elementary bounds

$$0 \le \frac{\breve{\sigma}(f)}{\breve{\sigma}_M(f)}\left(\frac{E_\mu[\zeta_{\breve{\theta}(1)}]}{E_\mu[T]}\right)^{1/2} \le \sqrt{3}.$$

If $\text{Cov}_\mu\left[\breve{Y}_1(f) - r(f)\breve{\delta}_1, \breve{Y}_2(f) - r(f)\breve{\delta}_2\right] \ge 0$, then $\breve{\sigma}(f)/\breve{\sigma}_M(f) \ge 1$ and we obtain the sharper bounds

$$\frac{1}{\sqrt{2}} \le \frac{\breve{\sigma}(f)}{\breve{\sigma}_M(f)}\left(\frac{E_\mu[\zeta_{\breve{\theta}(1)}]}{E_\mu[T]}\right)^{1/2} \le \sqrt{3}. \tag{2.29}$$

Recall from Section 6.3.8 that the multiple-runs method is more efficient than the extended regenerative method when these methods are used to estimate time-average limits of the marking process or underlying chain. In contrast, the foregoing bounds suggest that, in the context of estimating time-average limits for delays, neither the extended regenerative method nor the multiple-runs method is more efficient in all situations. The reason for this discrepancy is as follows. As in Section 6.3.8, the point estimator of $r(f)$ in the multiple-runs method typically has lower variance than the estimator in the extended regenerative method. The variance is lower because the cycles in the former method are independent and there are no covariance effects. On the other hand, the multiple-runs method is more expensive to execute, for the following reason. Generation of the kth cycle of the delay process requires simulation of the marking process over an interval of the form $[\zeta_{\breve{\theta}(k)}, T_k]$, where $T_k > \zeta_{\breve{\theta}(k+1)}$. In the multiple-runs method, the marking process must, in effect, be simulated over the interval $[\zeta_{\breve{\theta}(k+1)}, T_k]$ twice—once to generate the kth replicate and once to generate the $(k+1)$st replicate—whereas the extended regenerative method requires simulation over this interval only once. Indeed, the following example shows that the ARE can be arbitrarily close to $\sqrt{2}$ or to $1/\sqrt{2}$. In the latter case, we have $\text{Cov}_\mu\left[\breve{Y}_1(f) - r(f)\breve{\delta}_1, \breve{Y}_2(f) - r(f)\breve{\delta}_2\right] > 0$, and so the lower bound in (2.29) is tight.

EXAMPLE 2.30 (Comparison of the extended regenerative and multiple-runs methods for delays). Consider an SPN with two places and two deterministic timed transitions as in Figure 8.8. Suppose that the two transitions

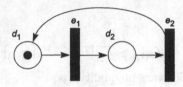

Figure 8.8. SPN for comparison of estimation methods.

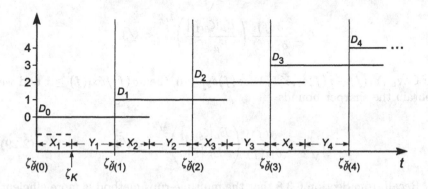

Figure 8.9. Comparison of estimation methods.

never fire simultaneously. Define a sequence of delays via start vectors by setting $\psi((1,0)) = 2$, $\psi((0,1)) = 1$,

$$i_{\alpha(s';s,E^*)} = \begin{cases} (0) & \text{if } E^* = \{e_2\}; \\ \varnothing & \text{otherwise,} \end{cases}$$

$$i_\beta(s';s,E^*) = \begin{cases} (2) & \text{if } E^* = \{e_1\}; \\ \varnothing & \text{otherwise,} \end{cases}$$

and $i_\pi(s';s,E^*) \equiv \varnothing$. To initialize the sequence of start vectors, set $V_0 = (0,-1)$. The marking $s = (0,1)$ is a single state, so that the successive times $\{\zeta_{\theta(k)} : k \geq 0\}$ at which transition e_2 fires form a sequence of regeneration points for the marking process. Define sequences $\{\zeta_{\check{\theta}(k)} : k \geq 0\}$ and $\{\check{\gamma}(k) : k \geq 0\}$ as before—see Figure 8.9. In the figure, X_k denotes the time from the start of the kth cycle to the next firing of transition e_1 and Y_k denotes the time from the firing of e_1 until the end of the kth cycle. Observe that the random variables $X_1, X_2, \ldots$ coincide with the successive new clock readings for transition e_1 and the random variables $Y_1, Y_2, \ldots$ coincide with the successive new clock readings for transition e_2. Take $f(x) = x$ for $x \geq 0$. Because $\check{\delta}_k \equiv 1$ for $k \geq 1$ and $\{(X_k, Y_k) : k \geq 0\}$ is a sequence of i.i.d. pairs with each X_k and Y_k independent, an easy calculation shows that

$$\frac{\breve{\sigma}^2(f)}{\breve{\sigma}_{\mathrm{M}}^2(f)} = \frac{4\mathrm{Var}\,[X_1] + \mathrm{Var}\,[Y_1]}{2\mathrm{Var}\,[X_1] + \mathrm{Var}\,[Y_1]}$$

and

$$\frac{E_\mu[\zeta_{\breve{\theta}(1)}]}{E_\mu\,[T]} = \frac{E\,[X_1] + E\,[Y_1]}{2E\,[X_1] + E\,[Y_1]}.$$

Suppose that, for some $n > 0$,

$$X_1 = \begin{cases} 0 & \text{with probability } (n-1)/n; \\ n & \text{with probability } 1/n, \end{cases}$$

so that $E[X_1] = 1$ and $\mathrm{Var}\,[X_1] = n - 1$, and suppose that $Y_1 = n$ with probability 1. Then the ARE of the extended regenerative and multiple-runs methods is given by

$$\mathrm{ARE} = \sqrt{2}\left(\frac{1+n}{2+n}\right)^{1/2},$$

which converges to $\sqrt{2}$ as n becomes large. On the other hand, if we switch the definitions of X_1 and Y_1, then

$$\mathrm{ARE} = \left(\frac{n+1}{2n+1}\right)^{1/2},$$

which converges to $1/\sqrt{2}$ as n becomes large.

EXAMPLE 2.31 (Cyclic queues with feedback). We compare the extended regenerative and multiple-runs methods for delays using the network of queues in Example 1.2. Successive service times at center i ($i = 1, 2$) are i.i.d. according to an exponential distribution with intensity q_i, where $q_1 = 1.5$ and $q_2 = 1$. The routing probability (with which a job completing service at center 1 moves to center 2) is $p = 0.6$. There are $N = 4$ jobs, and—as in Example 1.4—we model the system using the SPN in Figure 2.2. We consider the delay intervals from whenever a job completes service at center 2 to when the job next completes service at center 2, and estimate the limiting average delay. (In the following section we provide some specialized techniques for estimating this particular performance measure.) Both the extended regenerative and multiple-runs methods are based on the sequence of od-equilibrium points for delays defined in Example 2.12.

Table 8.2 displays estimates of $\breve{\sigma}^2(f)$, $\breve{\sigma}_{\mathrm{M}}^2$, $\mathrm{Cov}_\mu[\breve{Y}_1(f) - r(f)\breve{\delta}_1, \breve{Y}_2(f) - r(f)\breve{\delta}_2]$, $E_\mu[\zeta_{\breve{\theta}(1)}]$, $E_\mu\,[T]$, and the ARE of the two methods, based on 10^6 cycles. As can be seen, the covariance between adjacent cycles is positive, so that the variance constant for the extended regenerative method exceeds the variance constant for the multiple-runs method. The additional expense per cycle for the multiple-runs method (as measured by $E_\mu\,[T] - E_\mu[\zeta_{\breve{\theta}(1)}]$)

Table 8.2. Simulation Results for Cyclic Queues with Feedback: Estimated Quantities for Comparison of Extended Regenerative and Multiple-Runs Methods, Based on 10^6 Cycles

$\check{\sigma}^2(f)$	$\check{\sigma}_M^2(f)$	Cov	$E_\mu[\zeta_{\check{\theta}(1)}]$	$E_\mu[T]$	ARE
145.9	136.5	4.7	11.3	14.3	0.92

Note: "Cov" $= \mathrm{Cov}_\mu[\check{Y}_1(f) - r(f)\check{\delta}_1, \check{Y}_2(f) - r(f)\check{\delta}_2]$.

Table 8.3. Simulation Results for Cyclic Queues with Feedback: Point Estimates and 95% Confidence-Interval Half-Widths for the Limiting Average Delay Through Both Centers (True Value $= 5.2924$)

Number of cycles simulated ($\times 10^3$)				
0.1	1	10	100	1000
5.4969	5.3510	5.2872	5.2905	5.2904
± 0.1990	± 0.0883	± 0.0281	± 0.0088	± 0.0028

is sufficiently large so that the ARE is less than 1 and the extended regenerative method is more efficient overall. Table 8.3 displays typical simulation results based on the extended regenerative method.

In practice, a small pilot run can be used to estimate the ARE and select the more efficient of the two estimation methods.

8.2.4 Limiting Average Delays

Under the moment conditions given below, the limiting average delay

$$r = \lim_{n \to \infty} \frac{1}{n} \sum_{j=0}^{n-1} D_j$$

exists a.s., and both point estimates and confidence intervals for the limiting average delay can be obtained without measuring the lengths of individual delay intervals. Recall that $\psi(s)$ is the length of the start vector when the marking is s and that the number of newly started delays is given by $n_\alpha(s'; s, E^*)$ whenever the transitions in the set E^* fire simultaneously and trigger a marking change from s to s'. Set

$$Z_k = \int_{\zeta_{\theta(k-1)}}^{\zeta_{\theta(k)}} \psi(X(t))\, dt = \sum_{n=\theta(k-1)}^{\theta(k)-1} \psi(S_n) t^*(S_n, C_n),$$

$\tau_k = \zeta_{\theta(k)} - \zeta_{\theta(k-1)}$, and $\delta_k = \sum_{n=\theta(k-1)}^{\theta(k)-1} n_\alpha(S_n; S_{n-1}, E_{n-1}^*)$ for $k \geq 1$. Because Z_k, τ_k, and δ_k are determined by $\{(S_n, C_n): \theta(k-1) \leq n < \theta(k)\}$

for $k \geq 1$, it follows that the sequence $\{(Z_k, \tau_k, \delta_k) : k \geq 1\}$ consists of i.i.d. random vectors. The proof of the following result is given at the end of the subsection.

Theorem 2.32. *Suppose that $E_\mu[Z_1] < \infty$ and $E_\mu[\delta_1] < \infty$ and that (2.2)–(2.4) hold. Then*

$$\lim_{n \to \infty} \frac{1}{n} \sum_{j=0}^{n-1} D_j = \frac{E_\mu[Z_1]}{E_\mu[\delta_1]} \quad a.s.. \tag{2.33}$$

It follows from (2.33) that a version of the standard regenerative method can be used to obtain strongly consistent point estimates and asymptotic confidence intervals for the limiting average delay.

Algorithm 2.34 (Regenerative method for the limiting average delay)

1. Select a single state $\bar{s}$ and define a corresponding sequence $\{\theta(k) : k \geq 0\}$ of random indices as in (2.1).

2. Simulate the marking process $\{X(t) : t \geq 0\}$ and observe a fixed number n of cycles defined by the random times $\{\zeta_{\theta(k)} : k \geq 0\}$.

3. Compute the number of starts δ_k in the kth cycle and the quantity $Z_k = \int_{\zeta_{\theta(k-1)}}^{\zeta_{\theta(k)}} \psi(X(t)) \, dt$ for $1 \leq k \leq n$.

4. Form the strongly consistent point estimate $\hat{r}(n) = \bar{Z}(n)/\bar{\delta}(n)$ for r, where $\bar{Z}(n) = (1/n) \sum_{k=1}^{n} Z_k$ and $\bar{\delta}(n) = (1/n) \sum_{k=1}^{n} \delta_k$.

5. Set

$$s^2(n) = \frac{1}{n-1} \sum_{k=1}^{n} (Z_k - \hat{r}(n)\delta_k)^2$$

and form the asymptotic $100p\%$ confidence interval

$$\left[\hat{r}(n) - \frac{z_p \, s(n)}{\bar{\delta}(n) \sqrt{n}}, \hat{r}(n) + \frac{z_p \, s(n)}{\bar{\delta}(n) \sqrt{n}} \right]$$

for $r(f)$, where z_p is the $(1+p)/2$ quantile of the standard normal distribution.

Remark 2.35. As usual, we can write $s^2(n)$ in the form

$$s^2(n) = s_{11}(n) - 2\hat{r}(n)s_{12}(n) + \hat{r}^2(n)s_{22}(n),$$

where

$$s_{11}(n) = \frac{1}{n-1} \sum_{k=1}^{n} (Z_k - \bar{Z}(n))^2,$$

$$\check{N}_0 = l_{-2,1} + l_{-1,1} \qquad \check{Z}_1 = l_{-2,1} + l_{-1,1} + l_{0,1} + l_{1,1} + l_{2,1}$$
$$\check{N}_1 = l_{0,2} + l_{1,2} \qquad \check{Z}_2 = l_{0,2} + l_{1,2} + l_{3,1} + l_{4,1} + l_{5,1}$$
$$\check{N}_2 = l_{3,2} + l_{4,2} \qquad \check{Y}_1 = l_{0,1} + l_{0,2} + l_{1,1} + l_{1,2} + l_{2,1}$$
$$\check{Y}_2 = l_{3,1} + l_{3,2} + l_{4,1} + l_{4,2} + l_{5,1}$$

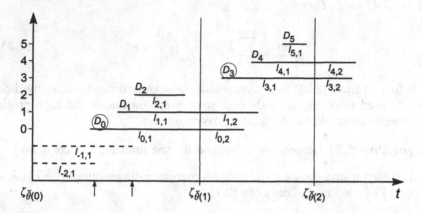

Figure 8.10. Definition of $\check{N}_0$, $\check{N}_1$, $\check{N}_2$, $\check{Z}_1$, $\check{Z}_2$, $\check{Y}_1$, and $\check{Y}_2$.

$$s_{22}(n) = \frac{1}{n-1} \sum_{k=1}^{n} \left(\delta_k - \bar{\delta}(n) \right)^2,$$

and

$$s_{12}(n) = \frac{1}{n-1} \sum_{k=1}^{n} \left(Z_k - \bar{Z}(n) \right) \left(\delta_k - \bar{\delta}(n) \right).$$

We can then use one-pass methods as in Remark 3.8 in Chapter 6 to compute $s^2(n)$ during the course of the simulation.

Remark 2.36. Observe that the Algorithm 2.34 is applicable even when there are ongoing delays at each regeneration point $\zeta_{\theta(k)}$, so that the sequence $\{ D_j : j \geq 0 \}$ does not inherit regenerative structure.

Remark 2.37. The limiting average delay can also be estimated using the extended regenerative method or the multiple-runs method. It can be shown that the point estimators in all three methods have the same asymptotic variability. Algorithm 2.34, however, does not incur the cost of measuring each individual delay and therefore is asymptotically more efficient than the other two algorithms in the sense of providing shorter confidence intervals for equivalent simulation cost.

PROOF OF THEOREM 2.32. We give the proof when (2.6) holds; a modification as in Remark 2.11 can be used to handle the case in which (2.6) does not hold. Define sequences of random indices $\{\check{\theta}(k): k \geq 0\}$ and $\{\check{\gamma}(k): k \geq 0\}$ as in Section 8.2.1. As shown in the proof of Theorem 2.8, the random indices $\{\check{\theta}(k): k \geq 0\}$ form a sequence of regeneration points for the chain $\{(S_n, C_n): n \geq 0\}$ and the random times $\{\zeta_{\check{\theta}(k)}: k \geq 0\}$ form a sequence of regeneration points for the process $\{X(t): t \geq 0\}$. Set

$$\check{Y}_k = \sum_{j=\check{\gamma}(k-1)}^{\check{\gamma}(k)-1} D_j,$$

$$\check{\delta}_k = \check{\gamma}(k) - \check{\gamma}(k-1),$$

$$\check{N}_k = \sum_{j=\check{\gamma}(k-1)}^{\check{\gamma}(k)-1} (B_j - \zeta_{\check{\theta}(k)})^+,$$

and

$$\check{Z}_k = \int_{\zeta_{\check{\theta}(k-1)}}^{\zeta_{\check{\theta}(k)}} \psi(X(t))\, dt$$

for $k \geq 1$. Denote by K_n ($n \geq 0$) the number of components of the start vector V_n that are equal to -1, and set $\check{N}_0 = \sum_{n=0}^{\infty} K_n(\zeta_{n+1} - \zeta_n)$. Observe that

$$\check{Z}_k = \check{N}_{k-1} + \sum_{j=\check{\gamma}(k-1)}^{\check{\gamma}(k)-1} \min(D_j, \zeta_{\check{\theta}(k)} - A_j)$$

for $k \geq 1$. Figure 8.10 illustrates these definitions when there are exactly two ongoing delays at each regeneration point and exactly two components of the initial start vector V_0 are equal to -1. In the figure an arrow pointing to the horizontal axis indicates a marking-change epoch at which a deleted component is equal to -1. Labels of the form $l_{i,j}$ denote lengths of horizontal line segments.

As shown in the proof of Theorem 2.8, $E_\mu[\delta_1] < \infty$ implies $E_\mu[\check{\delta}_1] < \infty$. An almost identical argument shows that $E_\mu[Z_1] < \infty$ implies $E_\mu[\check{Z}_1] < \infty$. Because $0 \leq \check{N}_0 \leq \check{Z}_1$, it then follows that $E_\mu[\check{N}_0] < \infty$. Next observe that $\{\check{N}_n: n \geq 0\}$ is a sequence of i.i.d. random variables, so that $E_\mu[\check{N}_k] < \infty$ for $k \geq 1$. Finally, observe that $\check{Y}_1 = \check{Z}_1 + \check{N}_1 - \check{N}_0$, so that $E_\mu[\check{Y}_1] = E_\mu[\check{Z}_1] < \infty$. As shown in the proof of Theorem 2.8, the sequence $\{\check{\gamma}(k): k \geq 0\}$ of random indices decomposes sample paths of the sequence $\{D_j: j \geq 0\}$ into o.d.s. cycles and

$$\lim_{n \to \infty} \frac{1}{n} \sum_{j=0}^{n-1} D_j = \frac{E_\mu[\check{Y}_1]}{E_\mu[\check{\delta}_1]} \quad \text{a.s..} \tag{2.38}$$

As $E_\mu[\check{Y}_1] = E_\mu[\check{Z}_1]$, however, we have

$$\frac{E_\mu[\check{Y}_1]}{E_\mu[\check{\delta}_1]} = \frac{E_\mu[\check{Z}_1]}{E_\mu[\check{\delta}_1]} = \frac{E_\mu[Z_1]}{E_\mu[\delta_1]}. \tag{2.39}$$

To obtain the second equality, we express the numerator and the denominator on the left side as random sums—cf. (2.19)—and apply Wald's identity [Proposition 1.19(i) in the Appendix]. The desired result now follows from (2.38) and (2.39).   □

As before, let $\tau_k = \zeta_{\theta(k)} - \zeta_{\theta(k-1)}$ for $k \geq 1$. Under the additional assumption that $E_\mu[\tau_1] < \infty$, an alternative proof of Theorem 2.32 can be based on the following version of Little's law.

Proposition 2.40. *Let $\{[A_j, B_j]: j \geq 0\}$ be a sequence of (possibly empty) random intervals such that the A_j's are nondecreasing, and set $D_j = B_j - A_j$ for $j \geq 0$. Denote by $N_A(t)$ the number of A_n's that lie in the interval $[0, t]$, and similarly define $N_B(t)$. Suppose that there exist finite positive constants l and λ such that*

$$\lim_{t \to \infty} \frac{N_A(t)}{t} = \lambda \text{ a.s.}$$

and

$$\lim_{t \to \infty} \frac{1}{t} \int_0^t \left(N_A(u) - N_B(u) \right) du = l \text{ a.s..}$$

Also suppose that $\lim_{j \to \infty} D_j/j = 0$ a.s.. Then there exists a finite positive constant w such that

$$\lim_{n \to \infty} \frac{1}{n} \sum_{j=0}^{n-1} D_j = w \text{ a.s.}$$

and $l = \lambda w$.

To prove Theorem 2.32, denote by $N_A(t)$ and $N_B(t)$ the number of delays that start and terminate, respectively, during the interval $[0, t]$. Observe that $N_A(t) - N_B(t) = \psi(X(t))$ for $t \geq \zeta_K$, where the random index K is defined by (1.3). It follows from the SLLN for i.i.d. random variables together with Theorem 2.9(v) in Chapter 3 that

$$\lambda \overset{\text{def}}{=} \lim_{t \to \infty} \frac{N_A(t)}{t} = \frac{E_\mu[\delta_1]}{E_\mu[\tau_1]} \in (0, \infty) \text{ a.s..} \tag{2.41}$$

Similarly, by Theorem 1.12 in Chapter 6,

$$l \overset{\text{def}}{=} \lim_{t \to \infty} \frac{1}{t} \int_0^t \left(N_A(u) - N_B(u) \right) du$$

$$= \lim_{t \to \infty} \frac{1}{t} \int_0^t \psi(X(u))\, du$$

$$= \frac{E_\mu [Z_1]}{E_\mu [\tau_1]}$$

$$< \infty \text{ a.s..}$$

Provided that $\lim_{j \to \infty} D_j/j = 0$ a.s., the desired result then follows from Proposition 2.40.

To see that $\lim_{j \to \infty} D_j/j = 0$ a.s., define a sequence $\{\check{\theta}(k): k \geq 0\}$ of a.s. finite regeneration points for $\{(S_n, C_n): n \geq 0\}$ as in Section 8.2.1. Set $H_k = \zeta_{\check{\theta}(k+2)} - \zeta_{\check{\theta}(k)}$ for $k \geq 0$. The random variables in the sequence $\{H_k: k \geq 0\}$ are o.i.d. and $E_\mu[H_0] = 2E_\mu[\zeta_{\check{\theta}(1)}] < \infty$; the finiteness follows from (2.18), which is valid in the current setting. Next, set $K(t) = \sup\{k \geq 0: \zeta_{\check{\theta}(k)} \leq t\}$ for $t \geq 0$, and observe that

$$\frac{D_j}{j} \leq \frac{H_{K(A_j)}}{j} = \frac{H_{K(A_j)}}{K(A_j)} \frac{K(A_j)}{A_j} \frac{A_j}{j}$$

for $j \geq 0$. Observe that $\lim_{j \to \infty} A_j = \infty$ a.s. by (2.4) and $\lim_{t \to \infty} K(t) = \infty$ because each $\zeta_{\check{\theta}(k)}$ is a.s. finite—see Theorem 2.9(ii) in Chapter 3. Moreover, (2.41) and the identity $N_A(A_j) = j + 1$ for $j \geq 0$ jointly imply that

$$\lim_{j \to \infty} \frac{A_j}{j} = \lambda^{-1} < \infty \text{ a.s..}$$

It therefore suffices to show that

$$\lim_{k \to \infty} \frac{H_k}{k} = 0 \quad \text{and} \quad \lim_{t \to \infty} \frac{K(t)}{t} < \infty$$

with probability 1. Almost-sure convergence of H_k/k to 0 follows from the SLLN for o.i.d. random variables together with Theorem 2.9(i) in Chapter 3, and the latter theorem also implies that $\lim_{t \to \infty} K(t)/t = 1/E_\mu[\zeta_{\check{\theta}(1)}] < \infty$ a.s..

8.3   STS Methods for Delays

This section deals with STS methods for estimating time-average limits of a sequence of delays. Such methods are useful when there is no apparent (or usable) sequence of regeneration points for the underlying chain or marking process, so that the methods of Section 8.2 are not applicable. When trying to apply STS methods, we face the usual problem: it is highly nontrivial to determine for a specific SPN, start-vector mechanism, and function f whether the output process $\{f(D_j): j \geq 0\}$ obeys an SLLN and whether STS methods are valid.

As in Chapter 7, we focus on SPNs that satisfy Assumption PD, so that there exists a sequence of od-regeneration points that decompose the sample paths of the underlying chain into o.d.s. cycles. In Section 8.3.1 we show that under Assumption PD and some mild regularity conditions on the start-vector mechanism, the output process $\{f(D_j): j \geq 0\}$ inherits the od-regenerative property. Moreover, for suitable functions f the sum of the process over a cycle has finite moments of all orders. Unlike in Section 8.2, the cycles of the output process usually cannot be determined explicitly, and neither the extended regenerative method for delays nor the multiple-runs method can be applied. The mere existence of these cycles, however, implies that the output process obeys an FCLT. As in Chapter 7, it then follows—see Section 8.3.2—that STS methods such as the method of batch means can be used to obtain strongly consistent point estimates and asymptotic confidence intervals for time-average limits and functions of time-average limits.

8.3.1 Stable Sequences of Delays

Consider a sequence of delays $\{D_j: j \geq 0\}$ determined from the underlying chain $\{(S_n, C_n): n \geq 0\}$ of an SPN using the method of start vectors. Lemma 2.5 in Chapter 7 gives conditions under which the underlying chain is an od-regenerative process and cycle sums have finite moments of all orders. As shown below, these stability properties are inherited by the sequence $\{D_j: j \geq 0\}$ provided that the start-vector mechanism is "regular" in the sense of Definition 3.1 below. The idea is to exploit the regularity of the start-vector mechanism and construct a sequence $\{\check{\gamma}(k): k \geq 0\}$ of random indices that decomposes sample paths of $\{D_j: j \geq 0\}$ into o.d.s. cycles.

Regular Start-Vector Mechanisms

To prepare for Definition 3.1, set

$$\mathcal{E}(s) = \begin{cases} \{\{e\}: e \in E(s) \text{ and } r(s,e) > 0\} & \text{if } s \in S; \\ \{E(s) \cap S'\} & \text{if } s \in S' \end{cases}$$

for $s \in G$. Thus, assuming that each timed transition has a continuous clock-setting distribution function, we have $E^* \in \mathcal{E}(s)$ if and only if the transitions in E^* can potentially trigger a marking change when the current marking is s. Denote by $\mathcal{X}$ the set of all infinite-length sequences

$(s^{(0)}, E^{(0)}, s^{(1)}, E^{(1)}, \ldots)$ such that $E^{(k)} \in \mathcal{E}(s^{(k)})$ and $p(s^{(k+1)}; s^{(k)}, E^{(k)})$ > 0 for $k \geq 0$. For an element $x = (s^{(0)}, E^{(0)}, s^{(1)}, E^{(1)}, \ldots) \in \mathcal{X}$, recursively define a sequence of vectors $v_0, v_1, \ldots$ by setting $v_0(x)$ equal to the vector of length $\psi(s^{(0)})$ whose components are all equal to -1 and then setting

$$v_n'(x) = \mathrm{Ins}\big(v_{n-1}(x), i_\alpha(s^{(n)}; s^{(n-1)}, E^{(n-1)}), 0\big),$$
$$v_n''(x) = \mathrm{Del}\big(v_n'(x), i_\beta(s^{(n)}; s^{(n-1)}, E^{(n-1)})\big),$$

and

$$v_n(x) = \mathrm{Per}\big(v_n''(x), i_\pi(s^{(n)}; s^{(n-1)}, E^{(n-1)})\big)$$

for $n \geq 1$. Denote by $\iota(x)$ the smallest integer n such that $v_n(x) = (0, 0, \ldots, 0)$; if such an integer n does not exist, then set $\iota(x) = \infty$.

Definition 3.1. A start-vector mechanism for a specified SPN is said to be *regular* if

 (i) there exists $s \in S$ and $\{e^*\} \in \mathcal{E}(s)$ such that $n_\alpha(s'; s, e^*) > 0$ for all s' with $p(s'; s, e^*) > 0$, and

 (ii) there exists $x = (s^{(0)}, E^{(0)}, s^{(1)}, E^{(1)}, \ldots) \in \mathcal{X}$ such that $\iota(x) < \infty$.

As shown below, the technical condition in (i) ensures that, with probability 1, there are infinitely many starts and, moreover, each o.d.s. cycle contains at least one start. This condition typically holds in practice. In many SPN models, for example, a delay starts whenever some specified timed transition fires—see also Remark 3.8 below. The condition in (ii) is needed to ensure that each delay terminates with probability 1. Roughly speaking, this condition asserts that there exists a finite sequence of marking changes such that (1) timed transitions do not fire simultaneously at any marking change, and (2) all the components of the start vector at the beginning of the sequence are deleted by the end of the sequence.

Construction of Random Indices

We now construct the sequence $\{\tilde{\gamma}(k) \colon k \geq 0\}$ of random indices for the process $\{D_j \colon j \geq 0\}$. For ease of exposition, we describe the construction when there are no immediate transitions—the modifications required for the general case are straightforward. Suppose that Assumption PD holds. Then, as discussed in Section 6.1.3, the transition kernel of the underlying chain satisfies

$$P^r\big((s, c), \cdot\big) = b\lambda(\cdot) + (1 - b)Q\big((s, c), \cdot\big), \qquad (s, c) \in \mathcal{C}, \qquad (3.2)$$

for some $\mathcal{C} \subseteq \Sigma$, $r \geq 1$, $b \in (0, 1]$, λ, and Q. Indeed, any subset $A \subseteq \Sigma$ such that $\bar{\phi}(A) > 0$ contains a subset $\mathcal{C}$ for which a decomposition of the

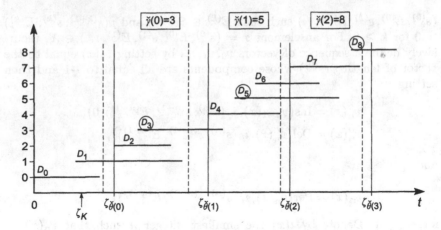

Figure 8.11. Definition of one-dependent cycles (nonregenerative case).

form (3.2) exists—here $\bar{\phi}$ is the recurrence measure defined by (1.17) in Chapter 5. Also as discussed in Section 6.1.3, the decomposition in (3.2) can be used to construct a version of the chain along with[4] a sequence $\{\theta(k): k \geq 0\}$ of od-regeneration points. Recall that the construction uses a sequence $\{I_n: n \geq 0\}$ of i.i.d. Bernoulli random variables and that the post-$\theta(k)$ process is not just independent of the chain prior to $\theta(k-1)$, but is actually independent of the chain prior to $\theta(k) - r$.

We proceed in a manner similar to what we do in Section 8.2.1 and recursively construct a subsequence $\{\breve{\theta}(k): k \geq 0\}$ of $\{\theta(k): k \geq 0\}$ such that the random indices in the subsequence also form a sequence of od-regeneration points for the chain, but with longer cycles. The subsequence retains the "enhanced independence" property referred to above: the post-$\breve{\theta}(k)$ process is independent of the chain prior to $\breve{\theta}(k) - r$, where r is as in (3.2). As before, the point of the construction is that all delays that start during one of these longer cycles terminate by the end of the next such cycle. Indeed, we require that all such delays terminate at least r marking changes before the end of the next cycle.

To initialize the construction of $\{\breve{\theta}(k): k \geq 0\}$, set $\breve{\nu}(-1) = \max(K, M)$, where K is defined by (1.3) and M is the random index of the first marking change at which all newly started delays at time 0 have terminated; when there are no newly started delays at time 0, take $\breve{\nu}(-1) = K$. Then take as $\breve{\theta}(0)$ the smallest $\theta(l)$ such that $\theta(l) \geq \breve{\nu}(-1) + r$. In general, given $\breve{\theta}(k)$, let $\breve{\nu}(k)$ be the index of the first marking change after $\breve{\theta}(k)$ such that

---

[4]As discussed in the notes at the end of Chapter 6, a sequence of od-regeneration points can be obtained after the sample path of the chain has been generated. The simultaneous construction of the od-regeneration points and the sample path, however, is more convenient for our purposes.

all ongoing and newly started delays at the $\check{\theta}(k)$th marking change have terminated. Then take as $\check{\theta}(k+1)$ the smallest $\theta(l)$ such that $\theta(l) \geq \check{\nu}(k)+r$; when there are no ongoing or newly started delays at the $\check{\theta}(k)$th marking change, take as $\check{\theta}(k+1)$ the smallest $\theta(l)$ such that $\theta(l) > \check{\theta}(k)$.

To complete the construction of $\{\check{\gamma}(k): k \geq 0\}$, set $\check{\gamma}(-1) = -1$ and

$$\check{\gamma}(k) = \inf\{ j > \check{\gamma}(k-1): \alpha(j-1) < \check{\theta}(m) < \alpha(j) \text{ for some } m \geq 0\} \quad (3.3)$$

for $k \geq 0$, where we take $\alpha(-1) = 0$. These ideas are illustrated in Figure 8.11. In the figure the delays $D_{\check{\gamma}(k)}$ are circled, and a vertical dashed line to the left of a solid line at time $\zeta_{\check{\theta}(k)}$ indicates the time point $\zeta_{\check{\theta}(k)-r}$. Two complete $\check{\gamma}(k)$-cycles $\{D_3, D_4\}$ and $\{D_5, D_6, D_7\}$ are displayed, as well as the beginning of a third cycle $\{D_8, \ldots\}$. Observe that the delays in the third cycle are determined by the evolution of the marking process from time $\zeta_{\check{\theta}(2)}$ onward and are hence independent of the delays $\{D_j: j \leq 4\}$, which are determined by the evolution of the marking process before time $\zeta_{\check{\theta}(2)-r}$.

Properties of the Construction

We now state our key result. In the following, a real-valued function f is said to be *polynomially dominated* if f is polynomially dominated to degree b—in the sense of Section 8.2.1—for some $b \geq 0$.

Theorem 3.4. *Let* $\{D_j: j \geq 0\}$ *be a sequence of delays determined from the underlying chain of a marking process by means of a regular start-vector mechanism, and suppose that Assumption PD holds. Then*

   (i) the sequence of delays satisfies the conditions in (2.2)–(2.4),

   (ii) the random indices $\{\check{\gamma}(k): k \geq 0\}$ *defined by* (3.3) *form a sequence of od-regeneration points for* $\{D_j: j \geq 0\}$, *and*

   (iii) the cycle sum $\check{Y}_1(|f|) = \sum_{j=\check{\gamma}(0)}^{\check{\gamma}(1)-1} |f(D_j)|$ *has finite moments of all orders for any polynomially dominated function* f.

Remark 3.5. In Section 8.2 the conditions in (2.2)–(2.4) are fundamental hypotheses. In the current section, these conditions are consequences of Assumption PD and the regularity of the start-vector mechanism. Also observe that the random indices $\{\check{\gamma}(k): k \geq 0\}$ in Theorem 2.8 form a sequence of od-equilibrium points for $\{D_j: j \geq 0\}$, but the corresponding random indices in Theorem 3.4 only form a sequence of od-regeneration points.

To prove Theorem 3.4, we need the following result, which extends the inequality in (2.16) to a sequence of o.i.d. random variables.

Lemma 3.6. *Let* $S_N = \sum_{n=1}^{N} X_n$, *where* $\{X_n : n \geq 1\}$ *is a sequence of o.i.d. random variables and* N *is a stopping time with respect to an increasing sequence* $\{\mathcal{F}_n : n \geq 1\}$ *such that* X_n *is determined by* $\mathcal{F}_n$ *for* $n \geq 1$ *and independent of* $\mathcal{F}_{n-2}$ *for* $n \geq 3$. *Then for* $r \geq 0$ *there exists a constant* b_r *(depending only on* r) *such that*

$$E[|S_N|^r] \leq b_r E[|X_1|^r] E[(N+2)^r].$$

PROOF. Fix $r \geq 0$ and observe that $|S_N| \leq \sum_{i=1}^{L} Y_i + \sum_{i=1}^{M} Z_i$, where $L = \lfloor N/2 \rfloor + 1$, $M = \lceil N/2 \rceil + 1$, $Y_i = |X_{2i}|$, and $Z_i = |X_{2i-1}|$. Moreover,

$$E[|S_N|^r] \leq c_r E\left[\left(\sum_{i=1}^{L} Y_i\right)^r\right] + c_r E\left[\left(\sum_{i=1}^{M} Z_i\right)^r\right] \tag{3.7}$$

for some finite constant c_r by (1.12) in the Appendix. Observe that $\{Y_i : i \geq 1\}$ is a sequence of i.i.d. random variables with $E[Y_i^r] = E[|X_1|^r]$ for $i \geq 1$. Also observe that, for $n \geq 1$, the value of Y_n is determined by the values of the random variables in $\mathcal{F}_n' = \mathcal{F}_{2n}$. Moreover, Y_{n+1} is independent of $\mathcal{F}_n'$. Finally, L is a stopping time with respect to $\{\mathcal{F}_n' : n \geq 1\}$, because N is a stopping time with respect to $\{\mathcal{F}_n : n \geq 1\}$ and $L = k$ ($k \geq 1$) if and only if either $N = 2k - 2$ or $N = 2k - 1$. It follows from (2.16) that there exists a constant b_r' (depending only on r) such that

$$E\left[\left(\sum_{i=1}^{L} Y_i\right)^r\right] \leq b_r' E[|X_1|^r] E[L^r].$$

An analogous argument shows that

$$E\left[\left(\sum_{i=1}^{M} Z_i\right)^r\right] \leq b_r' E[|X_1|^r] E[M^r],$$

and the desired result follows from (3.7), since

$$E[L^r + M^r] \leq 2E[(L+M)^r] = 2E[(N+2)^r]. \qquad \square$$

PROOF OF THEOREM 3.4. (Sketch) To establish the first assertion of Theorem 3.4, let s and e^* be as in Definition 3.1(i). Corollary 1.26 in Chapter 5 implies that, under Assumption PD, the underlying chain is Harris recurrent with recurrence measure $\bar{\phi}$ given by (1.17) in Chapter 5. It then follows that there are infinitely many marking changes at which the marking is s and transition e^* fires, so that there are infinitely many starts. Because $n_\alpha(s'; s, e^*)$ is finite for each s', it follows from (1.1) that (2.4) holds. Next, let $x = (s^{(0)}, E^{(0)}, s^{(1)}, E^{(1)}, \ldots)$ be a fixed element of $\mathcal{X}$ such that $\iota(x) < \infty$, and set $Z_n = (S_n, C_n, \ldots, S_{n+l}, C_{n+l})$ for $n \geq 0$, where

$l = \iota(x)$. Set

$$A = \Big\{ (s_0, c_0, \ldots, s_l, c_l) \in \Sigma^l :$$

$$s_k = s^{(k)} \text{ and } E^*(s_k, c_k) = E^{(k)} \text{ for } 0 \leq k \leq l \Big\},$$

and observe that all ongoing and newly started delays at time ζ_n terminate by time ζ_{n+l} whenever $Z_n \in A$. Using the Harris recurrence of the underlying chain together with a geometric trials argument, it can be shown that $P_\mu \{ Z_n \in A \text{ i.o.} \} = 1$. Thus each delay terminates with probability 1. Because each new clock reading is a.s. finite, each ζ_n is a.s. finite, and both (2.2) and (2.3) hold.

The remainder of the proof is similar to that of Theorem 2.8. As before, assume for ease of exposition that there are no immediate transitions. The first step in establishing the second assertion of the theorem is to show that the random indices $\{ \breve{\theta}(k) \colon k \geq 0 \}$ form a sequence of od-regeneration points for the underlying chain, with the enhanced independence property alluded to earlier. To this end, fix $k \geq 1$ and let L_k be the unique a.s. finite integer-valued random variable such that $\breve{\theta}(k) = \theta(L_k)$. Also fix a set $B = B_1 \times B_2$, where $B_1 \subseteq \Sigma$ and $B_2 \subseteq \{0, 1\}$. Finally, let A be any event whose occurrence or nonoccurrence is determined by $\{ (S_n, C_n, I_n) \colon 0 \leq n \leq \breve{\theta}(k) - r \}$, where $\{ I_n \colon n \geq 0 \}$ is the sequence of Bernoulli random variables used in the construction of the sequence $\{ \theta(k) \colon k \geq 0 \}$ and r is as in (3.2). Observe that events of the form $\{ \breve{\theta}(j) - \breve{\theta}(j - 1) \leq x_j \text{ for } 1 \leq j \leq k - 1 \}$ are of this type. We claim that

$$P_\mu \big\{ (S_{\breve{\theta}(k)}, C_{\breve{\theta}(k)}, I_{\breve{\theta}(k)}) \in B \mid L_k = j, A \big\}$$

$$= P_\mu \big\{ (S_{\theta(j)}, C_{\theta(j)}, I_{\theta(j)}) \in B \mid L_k = j, \; A \big\}$$

$$= P_\mu \big\{ (S_{\theta(j)}, C_{\theta(j)}, I_{\theta(j)}) \in B \big\}$$

$$= \lambda'(B),$$

where $\lambda'(B) = \lambda'(B_1 \times B_2) = \lambda(B_1)\eta(B_2)$ with $\eta(B_2) = P \{ I_0 \in B_2 \}$ and λ as in (3.2). Only the second equality requires explanation. To see that this equality holds, observe that the occurrence or nonoccurrence of the event $\{ L_k = j \} \cap A$ is determined by $\mathcal{H}_j = \{ (S_n, C_n, I_n) \colon 0 \leq n \leq \theta(j) - r \}$. The equality then follows because $(S_{\theta(j)}, C_{\theta(j)}, I_{\theta(j)})$ is independent of $\mathcal{H}_j$ by construction. Multiplying the above result by $P_\mu \{ L_k = j \mid A \}$ and summing over j, we obtain

$$P_\mu \big\{ (S_{\breve{\theta}(k)}, C_{\breve{\theta}(k)}, I_{\breve{\theta}(k)}) \in B \mid A \big\} = \lambda'(B),$$

and a simple inductive argument using the strong Markov property then extends the above equality to show that

$$P_\mu \big\{ (S_{\breve{\theta}(k)+n}, C_{\breve{\theta}(k)+n}, I_{\breve{\theta}(k)+n}) \in H_n \text{ for } 0 \leq n \leq m \mid A \big\}$$

$$= P_{\lambda'} \big\{ (S_n, C_n, I_n) \in H_n \text{ for } 0 \leq n \leq m \big\}$$

for any $m \geq 0$ and sets $H_0, H_1, \ldots, H_m \subseteq \Sigma \times \{0,1\}$. Because the random variables in the sequence $\{\check{\theta}(k+1+n) - \check{\theta}(k+n): n \geq 0\}$ are determined by the process $\{(S_n, C_n, I_n): n \geq \zeta_{\check{\theta}(k)}\}$, it follows that the random indices $\{\check{\theta}(k): k \geq 0\}$ form a sequence of od-regeneration points for the underlying chain and that the post-$\check{\theta}(k)$ process is independent of the chain prior to $\check{\theta}(k) - r$.

To complete the proof of the second assertion of the theorem, assume that each interval $[\zeta_{\check{\theta}(k-1)}, \zeta_{\check{\theta}(k)}]$ contains at least one start—because the start-vector mechanism is regular, we can ensure that this condition holds by first choosing $\bar{s} \in S$ and $\bar{e} \in E(s)$ such that $n_\alpha(s'; \bar{s}, \bar{e}) > 0$ for all s' with $p(s'; \bar{s}, \bar{e}) > 0$ and then choosing the set $\mathcal{C}$ in (3.2) such that

$$\mathcal{C} \subseteq \{(s,c) \in \Sigma: s = \bar{s} \text{ and } E^*(s,c) = \{\bar{e}\}\}.$$

An argument almost identical to that in the proof of Theorem 2.8 then shows that the random indices $\{\check{\gamma}(k): k \geq 0\}$ defined by (3.3) form a sequence of od-regeneration points for $\{D_j: j \geq 0\}$.

The proof of the final assertion of the theorem similarly parallels the proof of Theorem 2.8(iii). The main differences are as follows.

- The result in (2.16) is replaced by Lemma 3.6.

- The existence of finite moments for δ_1 and τ_1 is no longer a hypothesis of the theorem, but rather a consequence of Lemma 2.5 in Chapter 7.

- The random variables $\{\Lambda_k: k \geq 0\}$ are defined by first letting $\nu(k)$ be the index of the first state transition after $\theta(k)$ such that all newly started and ongoing delays at the $\theta(k)$th state transition have terminated, and then setting $\Lambda_k = \nu(k) + r - \theta(k)$, where r is as in (3.2). The indicator random variables $\{J_l: l \geq 0\}$ are then i.i.d. as before.

□

Remark 3.8. With some additional work (involving geometric trials arguments), the notion of regularity in Theorem 3.4 can be weakened. Specifically, Definition 3.1(i) can be modified to require only that there exist $s \in S - S'$, $s' \in G$, and $E^* \in \mathcal{E}(s)$ such that $p(s'; s, E^*) > 0$ and $n_\alpha(s'; s, E^*) > 0$. In fact, we can allow s to be immediate. For this latter relaxation, we identify a timed marking s^+ and a transition e^+ such that either (1) $p(s; s^+, e^+) > 0$ or (2) there exist immediate markings $s^{(1)}, s^{(2)}, \ldots, s^{(k)}$ ($k \geq 1$) such that $p(s^{(1)}; s^+, e^+) > 0$ and $s^{(1)} \to s^{(2)} \to \cdots \to s^{(k)} \to s$. We then choose the set $\mathcal{C}$ in (3.2) such that $\mathcal{C} \subseteq \{s^+\} \times C_0(s^+)$, where $C_0(s^+) = \{c^+ \in C(s^+): E^*(s^+, c^+) = \{e^+\}\}$.

A Limit Theorem for Delays

Combining Theorem 3.4 with Theorem 2.1 in Chapter 7 and Proposition 2.3 in Chapter 7, we obtain the following corollary—recall that $C^l[0,1]$ denotes the set of continuous $\Re^l$-valued functions defined on $[0,1]$ and that we write $C[0,1]$ for $C^l[0,1]$.

Corollary 3.9. *Let* $\{D_j : j \geq 0\}$ *be a sequence of delays determined from the underlying chain of a marking process by means of a regular start-vector mechanism, and let* f *be a polynomially dominated* $\Re^l$*-valued function defined on* $\Re_+$ *($l \geq 1$). Suppose that Assumption PD holds. Then*

(i) *there exists a finite constant* $r(f) \in \Re^l$ *such that*

$$\lim_{n \to \infty} \frac{1}{n} \sum_{j=0}^{n-1} f(D_j) = r(f) \quad a.s.,$$

and

(ii) *there exists an* $l \times l$ *matrix* $Q(f)$ *such that* $U_n(f) \Rightarrow Q(f)W^{(l)}$ *as* $n \to \infty$ *for any initial distribution* μ, *where* $\Rightarrow$ *denotes weak convergence on* $C^l[0,1]$, $W^{(l)}$ *is a standard* l*-dimensional Brownian motion, and*

$$U_n(f)(t) = \frac{1}{\sqrt{n}} \int_0^{nt} \left(f(D_{\lfloor u \rfloor}) - r(f) \right) du \tag{3.10}$$

for $0 \leq t \leq 1$ *and* $n \geq 0$.

8.3.2 Estimation Methods for Delays

In this subsection we focus on estimation methods that follow from the foregoing results for sequences of delays.

Standardized Time Series

When the start-vector mechanism is regular and Assumption PD holds, the time-average limit $r(f)$ of the output process $\{f(D_j) : j \geq 0\}$ is well-defined and finite for any polynomially dominated real-valued function f, and we can obtain point estimates and confidence intervals for $r(f)$ using STS methods. To this end, fix f and set

$$\bar{Y}_n(t) = \frac{1}{n} \int_0^{nt} f(D_{\lfloor u \rfloor}) \, du$$

for $0 \leq t \leq 1$ and $n \geq 1$. Also set

$$\hat{r}_n = \bar{Y}_n(1) = \frac{1}{n} \sum_{j=0}^{n-1} f(D_j).$$

By Corollary 3.9(i), the point estimator $\hat{r}_n$ is strongly consistent for $r(f)$. To obtain asymptotic confidence intervals for $r(f)$, we proceed as in Section 7.2.2. Specifically, define the set Ξ of mappings from $C[0,1]$ to $\Re$ as in

Section 7.2.2 and fix $\xi \in \Xi$. Let $\sigma(f)$ be a nonnegative constant such that

$$U_n(f) \Rightarrow \sigma(f)W$$

as $n \to \infty$, where $U_n(f)$ is given by (3.10) and $W = \{W(t): 0 \leq t \leq 1\}$ is a standard one-dimensional Brownian motion. The existence of $\sigma(f)$ follows from Corollary 3.9(ii) with $l = 1$, and we assume throughout that $\sigma(f) > 0$. Finally, set $\xi_n = \xi(\bar{Y}_n)$. Arguing as in Section 7.2.2, we find that

$$\frac{\hat{r}_n - r(f)}{\xi_n} \Rightarrow \frac{\sigma(f)W(1)}{\sigma(f)\xi(W)} = \frac{W(1)}{\xi(W)}, \qquad (3.11)$$

and

$$[\hat{r}_n - \xi_n z_p, \hat{r}_n + \xi_n z_p] \qquad (3.12)$$

is an asymptotic $100p\%$ confidence interval for $r(f)$, where $p \in (0,1)$ and z_p is a positive constant such that $P\{-z_p \leq W(1)/\xi(W) \leq z_p\} = p$.

As discussed in Section 7.2.2, the batch-means confidence interval is obtained as a special case of the interval in (2.24). Fix $b \geq 2$ and suppose that we can write the simulation run length as $n = bm$. Setting

$$\bar{X}_n(i) = \bar{X}_n(i; f) = \frac{1}{m} \sum_{j=(i-1)m}^{im-1} f(D_j) \qquad (3.13)$$

for $1 \leq i \leq b$ and $n > 0$, we find that the interval in (3.12) is an asymptotic $100p\%$ confidence interval for $r(f)$ when z_p is the $(1+p)/2$ quantile of the Student's t distribution with $b - 1$ degrees of freedom and

$$\xi_n = \frac{1}{\sqrt{b}} \left[\frac{1}{b-1} \sum_{i=1}^{b} \left(\bar{X}_n(i) - \frac{1}{b} \sum_{j=1}^{b} \bar{X}_n(j) \right)^2 \right]^{1/2}.$$

That is, the batch-means confidence interval based on simulation of delays $D_0, D_1, \ldots, D_n$ is obtained by decomposing the sequence $f(D_0), f(D_1), \ldots, f(D_n)$ into b disjoint batches of length m. The quantity $\bar{X}_n(i)$ is the ith batch mean, and we treat the batch means as if they were i.i.d. normal random variables when forming a confidence interval—the limit result in (3.11) shows that this approximation becomes increasingly accurate as the simulation run length becomes large.

The other STS methods discussed in Section 7.2.2 are also applicable. If we fix $m \geq 1$ and simulate the underlying chain for $n = lm$ state transitions, then the STS area method yields an asymptotic $100p\%$ confidence interval for $r(f)$ of the form $[\hat{r}_n - \xi_n z_p, \hat{r}_n + \xi_n z_p]$. Here

$$\hat{r}_n = \frac{1}{n} \sum_{j=0}^{n-1} f(D_j),$$

z_p is the $(1+p)/2$ quantile of the Student's t distribution with m degrees of freedom, and

$$\xi_\nu^2 = 12 \sum_{i=0}^{m-1} A_i^2,$$

with

$$A_i = \frac{1}{n} \sum_{j=0}^{l-1} \left(\frac{1}{2} - \frac{j}{l} - \frac{1}{2l} \right) f(D_{il+j}).$$

Alternatively, the STS maximum method yields an asymptotic $100p\%$ confidence interval for $\tilde{r}(\tilde{f})$ of the form $[\hat{r}_n - \xi_n z_p, \hat{r}_n + \xi_n z_p]$. Here

$$\hat{r}_n = \frac{1}{n} \sum_{j=0}^{n-1} f(D_j),$$

z_p is the $(1+p)/2$ quantile of the Student's t distribution with $3m$ degrees of freedom, and

$$\xi_\nu^2 = \frac{1}{3} \sum_{i=0}^{m-1} A_i^2,$$

where each A_i is defined as follows. Set

$$B_i(t) = \frac{1}{n} \sum_{j=0}^{\lfloor lt \rfloor - 1} f(D_{il+j}) - \frac{t}{n} \sum_{j=0}^{l-1} f(D_{il+j})$$

for $0 \le t \le 1$, and denote by k_i^* the smallest value of k in $\{0, 1, \ldots, l\}$ such that $B_i(k_i^*/l) \ge B_i(k/l)$ for k in $\{0, 1, \ldots, l\}$. Then

$$A_i = \frac{B_i(k_i^*/l)}{(k_i^*/l)\left(1 - (k_i^*/l)\right)^{1/2}}.$$

Functions of Time-Average Limits

As in Section 7.2.3, we can use an extension of the batch-means method to estimate functions of time-average limits of a sequence of delays. Specifically, let $\{ D_j : j \ge 0 \}$ be a sequence of delays that is specified by means of a regular start-vector mechanism, and suppose that we wish to obtain strongly consistent point measures and asymptotic confidence intervals for a performance measure of the form

$$r = g(\alpha_1, \alpha_2, \ldots, \alpha_l)$$

$(l \ge 1)$, where g is a real-valued function defined on $\Re^l$,

$$\alpha_i = \lim_{n \to \infty} \frac{1}{n} \sum_{j=0}^{n-1} f_i(D_j)$$

for $1 \leq i \leq l$, and each f_i is a polynomially dominated real-valued function. Also suppose that each f_i is nonconstant and $f_1, f_2, \ldots, f_l$ are linearly independent. Note that, by Corollary 3.9, each α_i is well defined and finite under Assumption PD.

Fix $b \geq 2$ and set

$$J_n(i) = bg\left(\bar{A}_1, \bar{A}_2, \ldots, \bar{A}_l\right) - (b-1)g\left(\bar{A}_1^{(i)}, \bar{A}_2^{(i)}, \ldots, \bar{A}_l^{(i)}\right),$$

for $1 \leq i \leq b$, where

$$\bar{A}_j = \frac{1}{b}\sum_{k=1}^{b} \bar{X}_n(k; f_j)$$

and

$$\bar{A}_j^{(i)} = \frac{1}{b-1}\sum_{k \neq i} \bar{X}_n(k; f_j).$$

and $\bar{X}_n(k; f)$ is defined as in (3.13). Then set $\hat{r}_n^{(J)} = (1/b)\sum_{i=1}^{b} J_n(i)$. The following result shows that $\hat{r}_n^{(J)}$ is strongly consistent for r; the proof is essentially the same as that of Theorem 2.32 in Chapter 7.

Theorem 3.14. *Suppose that Assumption PD holds and that g is differentiable in a neighborhood of $\alpha = (\alpha_1, \alpha_2, \ldots, \alpha_l)$. Then $\hat{r}_n^{(J)} \to r$ a.s. as $n \to \infty$.*

To obtain confidence intervals, observe that by Corollary 3.9(ii) there exists an $l \times l$ matrix $Q(f)$ such that

$$U_n(f) \Rightarrow Q(f)W^{(l)} \tag{3.15}$$

as $n \to \infty$ for any initial distribution μ, where $f = (f_1, f_2, \ldots, f_l)$, $U_n(f)$ is defined by (3.10) and $W^{(l)}$ is a standard l-dimensional Brownian motion. The matrix $Q(f)$ is nonsingular except in degenerate cases, and we assume that $Q(f)$ is nonsingular throughout. As in Section 7.2.3, it follows from the convergence in (3.15) that $\sqrt{b}\left(\hat{r}_n^{(J)} - r\right)/s_n^{(J)} \Rightarrow T_{b-1}$ as $n \to \infty$, where T_{b-1} denotes a random variable having a Student's t distribution with $b-1$ degrees of freedom and $s_n^{(J)} = \left[(1/(b-1))\sum_{i=1}^{b}\left(J_n(i) - \hat{r}_n^{(J)}\right)^2\right]^{1/2}$. Thus

$$\left[\hat{r}_n^{(J)} - \frac{s_n^{(J)}z_p}{\sqrt{b}}, \hat{r}_n^{(J)} + \frac{s_n^{(J)}z_p}{\sqrt{b}}\right]$$

is an asymptotic $100p\%$ confidence interval for r, where z_p is the $(1+p)/2$ quantile of the Student's t distribution with $b-1$ degrees of freedom.

Limiting Average Delays

As in the regenerative setting, the task of measuring individual delays can be avoided when the performance measure of interest is the limiting average delay $\lim_{n \to \infty} (1/n) \sum_{j=0}^{n-1} D_j$. Set $Z_n = (S_n, C_n, S_{n+1}, C_{n+1})$ for $n \geq 0$ and define functions f_1, f_2, and f_3 on $\Sigma \times \Sigma$ by $\tilde{f}_1(s,c,s',c') = \psi(s) t^*(s,c)$, $\tilde{f}_2(s,c,s',c') = n_\alpha(s';s,E^*(s,c))$, and $\tilde{f}_3(s,c,s',c') = t^*(s,c)$. Under Assumption PD, there exists a sequence $\{\theta(k): k \geq 0\}$ of od-regeneration points for the underlying chain, and it is not hard to see that the random indices $\{\theta(2k): k \geq 0\}$ form a sequence of od-regeneration points for the process $\{Z_n: n \geq 0\}$. Moreover, the expected cycle length is finite, and cycle sums of the process $\{\tilde{f}(Z_n): n \geq 0\}$ have finite moments of all orders for $\tilde{f} = \tilde{f}_1, \tilde{f}_2,$ or $\tilde{f}_3$. It then follows from Theorem 1.27 in Chapter 6 and Theorem 2.9 in Chapter 3 that there exist finite nonnegative constants l and λ such that, with probability 1,

$$\lim_{t \to \infty} \frac{1}{t} \int_0^t \psi(X(u)) \, du = \lim_{n \to \infty} \frac{\sum_{j=0}^n \tilde{f}_1(Z_j)}{\sum_{j=0}^n \tilde{f}_3(Z_j)} = l$$

and

$$\lim_{t \to \infty} \frac{N_A(t)}{t} = \lim_{n \to \infty} \frac{\sum_{j=0}^n \tilde{f}_2(Z_j)}{\sum_{j=0}^n \tilde{f}_3(Z_j)} = \lambda,$$

where $N_A(t)$ is the number of delays that start during the interval $[0,t]$. It then follows by a Little's-law argument similar to the alternative proof of Theorem 2.32 that

$$\lim_{n \to \infty} \frac{1}{n} \sum_{j=0}^{n-1} D_j = l/\lambda = \lim_{n \to \infty} \frac{\sum_{j=0}^n \tilde{f}_1(Z_j)}{\sum_{j=0}^n \tilde{f}_2(Z_j)} \quad \text{a.s.;}$$

the only changes in the proof involve the replacement of several cited limit theorems for i.i.d. random variables with corresponding results for o.d.s. random variables. Thus the limiting average delay can be expressed as the limit of a ratio of sums as in Section 7.2.4. As discussed in that section, "jackknifed batch means" methods can be used to obtain point estimates and confidence intervals for such ratios; there is no need to measure individual delays.

8.3.3 Examples

We conclude by illustrating the application of our results to some specific SPN models.

EXAMPLE 3.16 (Cyclic queues with four servers). Consider a closed network of queues with two service centers, two servers at each center, and N (> 4) jobs numbered $1, 2, \ldots, N$. At each center, jobs form a single queue

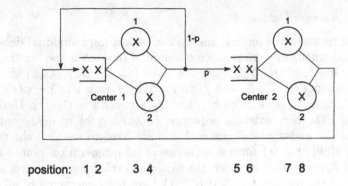

position: 1 2   3 4   5 6   7 8

Figure 8.12. Positions of jobs in cyclic queues with feedback (two servers per center).

and are served by one of two servers, numbered 1 and 2, according to a first-come, first-served queueing discipline; whenever a job joins an empty queue and both servers are idle, there is an immediate start of service by server 1. With fixed probability $p \in (0,1)$, a job that completes service at center 1 moves to center 2 and with probability $1-p$ joins the tail of the queue at center 1. Upon completion of service at center 2, the job moves to center 1. Successive service times by server j at center i are i.i.d. as a positive random variable $L_{i,j}$ with continuous distribution function.

We specify the position of each job in the network by ordering the N jobs in a job stack—see Figure 8.12 for $N = 8$ jobs. The ordering is essentially the same as that in Example 1.2; the main modification is that a job being served by server 1 appears closer to the top of the stack than a job being served by server 2.

This system can be specified as an SPN with a finite marking set; see Figure 8.13. For $i, j \in \{1, 2\}$, transition $e_{i,j} = $ "completion of service by server j at center i" and transition $e_{i,2+j} = $ "start of service by server j at center i." All the transitions are deterministic; we use priorities on the immediate transitions to model the fact that a job is always served by the lowest-numbered available server. All speeds for enabled transitions are equal to 1.

Consider the delay intervals from whenever a job completes service at center 2 (and moves to center 1) to when the job next completes service at center 2. Individual delays can be specified and measured using the method of start vectors in a manner similar to Example 1.4. Set $\psi(s) = N$ for $s \in G$. Also set

$$i_\alpha(s'; s, E^*) = \begin{cases} (0) & \text{if } E^* = \{ e_{2,1} \} \text{ or } \{ e_{2,2} \}; \\ \varnothing & \text{otherwise} \end{cases}$$

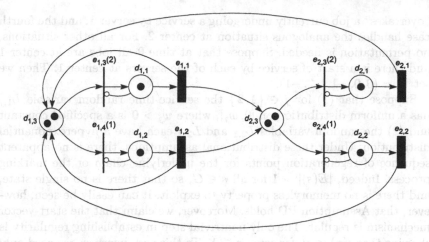

Figure 8.13. SPN representation of cyclic queues with feedback (two servers per center).

and

$$
i_\beta(s'; s, E^*) = \begin{cases}
(s_{1,1} + s_{1,2} + s_{1,3} + s_{2,3} + 2) & \text{if } E^* = \{e_{2,1}\}; \\
(s_{1,1} + s_{1,2} + s_{1,3} + s_{2,3} + s_{2,1} + 2) & \text{if } E^* = \{e_{2,2}\}; \\
\varnothing & \text{otherwise}
\end{cases}
$$

for $s = (s_{1,1}, s_{1,2}, s_{1,3}, s_{2,1}, s_{2,2}, s_{2,3}), s' \in G$, and $E^* \subseteq E(s)$. (The "+2" terms in the above definition are a consequence of the fact that, after an insertion and before a deletion, the start vector temporarily contains $N+1$ components.) Finally, letting $m = s_{1,1} + s_{1,3}$,

(i) set $i_\pi(s'; s, E^*) = (s_{1,3} + 1, 1, 2, \ldots, s_{1,3}, s_{1,3} + 2, s_{1,3} + 3, \ldots, N)$ if $E^* = \{e_{1,1}\}$, $s_{2,3} = s'_{2,3}$, and $s_{1,3} > 0$,

(ii) set $i_\pi(s'; s, E^*) = (m + 1, 1, 2, \ldots, m, m + 2, m + 3, \ldots, N)$ if $E^* = \{e_{1,2}\}$, $s_{2,3} = s'_{2,3}$, and $m > 0$,

(iii) set $i_\pi(s'; s, E^*) = (1, 2, \ldots, s_{1,3} - 1, s_{1,3} + 1, s_{1,3}, s_{1,3} + 2, \ldots, N)$ if $E^* = \{e_{1,4}\}$ and $s_{1,1} = 1$,

(iv) set $i_\pi(s'; s, E^*) = (1, 2 \ldots, N - 2, N, N - 1)$ if $E^* = \{e_{2,4}\}$ and $s_{2,1} = 1$, and

(v) set $i_\pi(s'; s, E^*) = \varnothing$ otherwise.

In the above specification of i_π, the first two cases handle the cyclic permutation that occurs when a job completes service at center 1 and joins the tail of the queue at center 1. The third case handles the situation in which, at center 1, a job starts to undergo a service by server 2 and thereby "overtakes" a job currently undergoing a service by server 1, and the fourth case handles the analogous situation at center 2. For all other situations, no permutation is needed. Suppose that at time 0 all jobs are at center 1 and there is a start of service by each of the servers at center 1. Then we set $V_0 = (0, 0, -1, \ldots, -1)$.

Suppose that (1) for $j \in \{1, 2\}$ the service-time random variable $L_{1,j}$ has a uniform distribution on $[0, w_j]$, where $w_j > 0$ is a specified constant and (2) the random variables $L_{2,1}$ and $L_{2,2}$ each have a hyperexponential distribution. Under these distributional assumptions, there is no apparent sequence of regeneration points for the underlying chain or the marking process. Indeed, $|E(s)| > 1$ for all $s \in G$, so that there is no single state, and there is no memoryless property to exploit. It can easily be seen, however, that Assumption PD holds. Moreover, we claim that the start-vector mechanism is regular. The only nontrivial step in establishing regularity is showing that $\iota(x) < \infty$ for some $x \in \mathcal{X}$. To this end, however, we need only consider a scenario in which all jobs are initially at center 2 and then, at center 2, there is a service completion by server 2 followed by $N - 2$ successive service completions by server 1 and then a final service completion by server 2. Formally, let x be any element $(s^{(0)}, E^{(0)}, s^{(1)}, E^{(1)}, \ldots) \in \mathcal{X}$ such that

$$s^{(0)} = (0, 0, 0, 1, 1, N - 2),$$
$$E^{(0)} = \{e_{2,2}\},$$
$$E^{(1)} = \{e_{1,3}, e_{2,4}\},$$
$$E^{(2)} = \{e_{2,1}\},$$
$$E^{(3)} = \{e_{1,4}, e_{2,3}\},$$
$$E^{(4)} = \{e_{2,1}\},$$
$$E^{(j)} = \{e_{2,3}\} \text{ and } E^{(j+1)} = \{e_{2,1}\} \text{ for } 5 \leq j \leq 3 + 2(N - 4),$$

and
$$E^{(5+2(N-4))} = \{e_{2,2}\}.$$

Then $\iota(x) = 6 + 2(N - 4) < \infty$.

Because the conditions of Theorem 3.4 and Corollary 3.9 are satisfied, any time-average limit of the form

$$r(f) = \lim_{n \to \infty} \frac{1}{n} \sum_{j=0}^{n-1} f(D_j)$$

is guaranteed to exist, provided that the (real-valued) function f is polynomially dominated. Moreover, we can use STS methods as in Section 8.3.2 to obtain strongly consistent point estimates and asymptotic confidence intervals for $r(f)$. For example, if $f(x) = 1_{(u,\infty)}(x)$, then $r(f)$ is the long-run fraction of delays that exceed u time units. If $f(x) = cx$ for some constant c, then $r(f)$ is the long-run average cost due to delays when the cost per time unit of delay is c for each delay.

Using the "jackknifed batch-means" method of Section 8.3.2, we can also estimate functions of time-average limits of the form

$$r = g(\alpha_1, \alpha_2, \ldots, \alpha_l),$$

where $\alpha_i = \lim_{n \to \infty} (1/n) \sum_{j=0}^{n-1} f_i(D_j)$ for $1 \le i \le l$, each f_i is a polynomially dominated real-valued function, and g is a real-valued function differentiable in a neighborhood of $\alpha = (\alpha_1, \alpha_2, \ldots, \alpha_l)$. For example, if $f_1(x) = x$, $f_2(x) = x^2$, and $g(x, y) = y - x^2$, then r is the long-run variance of the sequence of delays. As another example, suppose that $f_1(x) = 1_{(u,\infty)}(x)$ and $f_2(x) = 1_{(v,\infty)}(x)$, where $v > u$, and that $g(x, y) = y/x$. Then r is the long-run fraction of "long" delays that exceed v time units, where a delay is considered "long" if it exceeds u time units.

EXAMPLE 3.17 (Airport shuttle). For the shuttle system of Example 1.8, suppose that each travel-time random variable L_i has a truncated normal distribution and each interarrival-time random variable A_i has a Wald distribution. Clearly, the SPN of Figure 8.5 has no single state, so that neither the marking process nor the underlying chain has an apparent sequence of regeneration points.

Assumption PD holds for this SPN, however. Indeed, the only nontrivial step in establishing Assumption PD is to demonstrate irreducibility. This can be done by showing that $s \rightsquigarrow \bar{s}$ and $\bar{s} \rightsquigarrow s$ for all $s \in G$, where $\bar{s}$ is the unique marking in which no passengers are in the system and the shuttle is travelling to station 1—that is, $\bar{s} = (s_{1,1}, \ldots, s_{6,N})$, where $s_{1,1} = s_{6,1} = s_{6,2} = \cdots = s_{6,N} = 1$ and all other components of $\bar{s}$ are equal to 0.

Not only does Assumption PD hold, but the start-vector mechanism described in Example 1.8 is regular. In Definition 3.1(i), take $s = (s_{1,1}, \ldots, s_{6,N})$ to be any fixed timed marking with $s_{5,1} < B_1$ (so that the number of passengers in queue at station 1 is less than the maximum capacity B_1) and take $e^* = e_{4,1} = $ "arrival of passenger for boarding at station 1." Verification of the condition in Definition 3.1(ii) is similarly straightforward. Let $s = (s_{1,1}, \ldots, s_{6,N})$ be the unique fixed marking in which the shuttle is at station 1, there is only one passenger in the system, and this passenger boarded the shuttle at station 2 and is about to disembark: $s_{4,2,1} = 1$, $s_{4,j,i} = 0$ for $(j, i) \ne (2, 1)$, $s_{5,i} = 0$ for $1 \le i \le N$, and $s_{2,1} = 1$. Next, set $x = (s, \{e^*\}, s')$, where $e^* = e_{2,2,1} = $ "disembarkment at station 1 of passengers from station 2," and s' is the unique marking such that $p(s'; s, e_{2,2,1}) > 0$. Then $\iota(x) = 1$.

Thus the conditions of Theorem 3.4 and Corollary 3.9 are satisfied. We can therefore use STS methods to obtain strongly consistent point estimates and asymptotic confidence intervals for time-average limits defined in terms of the sequence $\{D_j : j \geq 0\}$ given in Example 1.8. We can also estimate functions of such time-average limits by using the jackknifed batch-means techniques of Section 8.3.2.

Remark 3.18. It can be shown that under the conditions of Theorem 3.4, the estimation methods described in this and the previous subsection extend to real-valued functions f that depend both on the delay and on specified states of the marking process during the corresponding delay interval. For example, our methods extend to functions of time-average limits of the form $\tilde{r}(f) = \lim_{n \to \infty}(1/n)\sum_{j=0}^{n-1} f(D_j, S_{\alpha(j)})$ where, as before, $\zeta_{\alpha(j)}$ is the start of jth delay interval. As an application, consider the closed network of queues discussed in this subsection and suppose we wish to estimate the limiting average delay only for those delays that start with an arrival to an empty queue. This limit is of the form $\tilde{r}(f_1)/\tilde{r}(f_2)$, where $f_1(d,s) = d\,1_A(s)$, $f_2(d,s) = 1_A(s)$, and $A = \{(s_{1,1}, \ldots, s_{2,3}) \in G : s_{1,3} = 1 \text{ and } s_{1,1} = s_{1,2} = 0\}$.

As another example, our methods extend to functions of time-average limits of the form $\tilde{r}(f) = \lim_{n \to \infty}(1/n)\sum_{j=0}^{n-1} f(D_j, S_{\beta(j)-1})$. To see the usefulness of this particular extension, recall the airport shuttle of Example 1.8. For fixed i and j, consider the delay intervals as in the example, but only for those passengers that board the shuttle at station j and disembark at station i. Suppose that we wish to estimate the limiting average delay for all such passengers. This limit is of the form $\tilde{r}(f_1)/\tilde{r}(f_2)$, where $f_1(d,s) = d\,1_A(s)$, $f_2(d,s) = 1_A(s)$, and $A = \{(s_{1,1}, \ldots, s_{6,N}) \in G : s_{2,i} = 1, s_{4,j,i} > 0, \text{ and } s_{4,l,i} = 0 \text{ for } j < l \leq N\}$.

Notes

Iglehart and Shedler (1980) and Shedler (1987) use tagging to specify delays in networks of queues—in this work, tagging is implemented by means of a job-stack ordering and an augmented state space. In the setting of queueing systems, delays are sometimes referred to as "passage times" because they correspond to the time for a job to pass from a specified initial position in the network to a specified final position. The tagging approach has been applied in the more general settings of GSMPs (Iglehart and Shedler, 1984; Shedler, 1993) and colored SPNs (Haas and Shedler, 1993b). In the latter setting, tagging is accomplished by means of distinguished tokens. The abovementioned papers systematically develop the use of the standard regenerative method to obtain point estimates and confidence intervals for time-average limits of a sequence of delays—see also Prisgrove and Shedler (1986) for a discussion of regenerative simulation of delays in "symmetric" SPNs.

The extended regenerative method and multiple-runs method for delays are developed in papers by Haas and Shedler (1993a, 1995, 1996); the latter method is based on an idea of Glynn (1994). For a proof of the CLT in (2.24), see Fox and Glynn (1987). The conclusion that neither the extended regenerative method nor the multiple-runs method is always more efficient remains true if—as in Glynn and Whitt (1992a)—we extend the notion of relative asymptotic efficiency to incorporate simulation costs explicitly and to quantify losses due to the discrepancy between $r(f)$ and its estimate. The version of Little's law in Proposition 2.40 is due to Glynn and Whitt (1986), and the assertion in Remark 2.37 of equal asymptotic variability rests on results in Glynn and Whitt (1989).

The discussion in Section 8.3 follows Haas (1999b). As in previous chapters, the moment condition in Assumption PD can be weakened by, for example, adapting results in Glynn and Haas (2002b). For example, it can be shown that the rth moment of a cycle sum of the output process $\{ f(D_j): j \geq 0 \}$ is finite whenever f is polynomially dominated to degree b (≥ 0) and each clock-setting distribution function has finite $r(b + 1)$st moment.

In related work, Glynn (1982b) considers sequences of delays determined by an underlying Harris recurrent Markov chain in which there is at most one ongoing delay at any time point. He establishes the existence of od-regeneration points for such sequences by obtaining a representation of the form $D_j = f(\xi_j)$ for $j \geq 0$, where $\{ \xi_j: j \geq 0 \}$ is a Harris chain and f is a real-valued function.

Under the conditions of Theorem 3.4, the techniques of Muñoz and Glynn (2001) can be used to construct confidence regions for multidimensional limits of the form $\lim_{n \to \infty} (1/n) \sum_{j=0}^{n-1} f(D_j)$, where f is a $\Re^l$-valued function for some $l > 1$.

The topic of delays in SPNs has also been treated by Baccelli et al. (1993), Baccelli and Schmidt (1996), Campos et al. (1989), Molloy (1982), Muppala et al. (1994), Natkin (1985), and Xie et al. (1999), among others. This work has, for the most part, focused on specific types of delays and particular kinds of SPNs—for example, "cycle times" between successive firings of a fixed transition in "stochastic marked graphs" with exponentially distributed firing times. The primary emphasis has been on the development of exact and approximate analytical expressions and bounds for average delays. Baccelli et al. (1993) derive sufficient conditions for time-average convergence of sequences of cycle times in stochastic marked graphs with general firing times.

The manufacturing line with a shunt bank—along with many other stochastic models of manufacturing systems—is discussed in Buzacott and Shanthikumar (1993). The airport shuttle model is adapted from Shedler (1993).

9
Colored Stochastic Petri Nets

Use of the standard set of SPN building blocks to model very large or complex systems can sometimes result in nets that have an enormous number of places and transitions. One popular strategy for obtaining more concise specifications in such cases is to associate "colors" with both tokens and transitions and to work with "colored stochastic Petri nets" (CSPNs). This approach is especially effective when the system under study is composed of many subsystems having a similar structure or behavior.

A CSPN is specified by a finite set of places, a finite number of transitions, and a finite set of colors, along with an "input incidence function" and an "output incidence function." A marking of a CSPN is an assignment of nonnegative integers to the places of the net and represents the number of tokens of each color in each place. Each transition can be simultaneously "enabled in" one or more colors. A transition is enabled in a specified color whenever each "input place" contains a sufficient number of tokens of each color—both the set of input places and the required number of tokens of each color in each input place are specified by the input incidence function.

A transition enabled in a color "fires in" the color by instantaneously removing tokens from input places and depositing tokens in "output places" in a deterministic manner—the set of output places is specified by the output incidence function. A transition may fire in a color only if it is enabled in the color. For each place, the number (possibly zero) of tokens removed from and deposited in the place is specified by the input and output incidence functions; this number depends only on the transition that fires, the firing color, and the identity of the place. The color of a token in a place remains fixed until the token is removed from the place.

A clock is associated with each possible (transition, firing color) pair. Whenever a transition is enabled in a specified color, the corresponding clock reading indicates the remaining time until the transition is scheduled to fire in the color. As with ordinary SPNs, each transition is either immediate or timed. A marking change occurs when one or more clocks run down to 0. When exactly one clock (associated with a transition and color) runs down to 0, the transition fires in the color. When several clocks run down to 0 simultaneously, the corresponding (transition, color) pairs "qualify" to trigger the next marking change. One of these pairs is selected for firing according to a specified probability distribution.

An initial marking is specified at time 0, and initial clock readings are selected according to initial probability distributions. At each subsequent marking change, transitions may become enabled in one or more colors. Whenever a transition becomes enabled in a color, a clock reading is selected according to a fixed probability distribution that depends only on the transition and the color. If a transition is enabled in a color and, at the next marking change, does not fire in the color but remains enabled in the color, then the associated clock continues to run down; if the transition is not enabled in the color after the next marking change, then the associated clock reading is discarded.

Because tokens are removed and deposited in essentially a deterministic manner, CSPNs have somewhat less modelling power than ordinary SPNs—see the discussion at the end of Section 9.1. A wide variety of interesting systems can be modelled within the CSPN framework, however, and the advantages of concisely representing these systems often outweigh the disadvantages due to loss of modelling power.

In Section 9.1 we present the CSPN building blocks, along with examples that illustrate the use of CSPNs for modelling of discrete-event stochastic systems. We also define the marking process of a CSPN; as with ordinary SPNs, the marking process records the marking as it evolves over continuous time and is defined in terms of a general state-space Markov chain that describes the net at successive marking changes. Stability conditions, as well as conditions for the applicability of estimation methods, closely resemble those for ordinary SPNs. We therefore do not describe these results in great detail, but content ourselves with stating some of the key theorems in Section 9.2. More interesting is the study of CSPNs whose behavior is invariant under permutations of the colors. Such "symmetric" CSPNs correspond to systems composed of identical subsystems. In Section 9.3 we describe two ways in which symmetry can be exploited when using regenerative simulation to estimate long-run performance characteristics. The first technique decomposes the sample path of the marking process or underlying chain into independent, nonidentically distributed blocks—the appeal of this approach is that the associated "blocking points" typically occur more frequently than the usual regeneration points. The second technique exploits symmetry to increase statistical efficiency when estimating time-average limits for a sequence of delays in a CSPN.

9.1 The CSPN Model

In this section we present the basic CSPN building blocks, along with illustrative examples of CSPN specifications. We also define the key stochastic processes associated with a CSPN.

9.1.1 Building Blocks

The basic elements of a CSPN graph are

- A finite set $D = \{ d_1, d_2, \ldots, d_L \}$ of *places*

- A finite set $E = \{ e_1, e_2, \ldots, e_M \}$ of *transitions*

- A (possibly empty) set $E' \subset E$ of *immediate transitions*

- A finite set U of *colors* with a fixed enumeration

- *Color domains* $U_D(d) \subseteq U$ for $d \in D$ and $U_E(e) \subseteq U$ for $e \in E$

- An *input incidence function* w^- and an *output incidence function* w^+, each defined on $\bigcup_{e \in E, d \in D}(\{ e \} \times U_E(e) \times \{ d \} \times U_D(d))$ and taking values in the nonnegative integers

For $d \in D$, the color domain $U_D(d) \subseteq U$ is the set of colors that may be assigned to a token in place d. Similarly, for $e \in E$, the color domain $U_E(e) \subseteq U$ is the set of possible firing colors for transition e. Thus there can be a token of color l in place d only if $(d, l) \in \mathcal{D}$, where

$$\mathcal{D} = \bigcup_{d \in D}(\{ d \} \times U_D(d)).$$

Similarly, transition e can fire in color i only if $(e, i) \in \mathcal{E}$, where

$$\mathcal{E} = \bigcup_{e \in E}(\{ e \} \times U_E(e)).$$

Denote by $\mathcal{E}'$ the subset of $\mathcal{E}$ corresponding to the immediate transitions:

$$\mathcal{E}' = \bigcup_{e \in E'}(\{ e \} \times U_E(e)).$$

The input incidence function w^- and the output incidence function w^+ determine when a transition is enabled in a color and the number of tokens removed and deposited when a transition fires in a color. Specifically, transition e is enabled in color i if and only if, for all $(d, l) \in \mathcal{D}$, place d contains at least $w^-(e, i, d, l)$ tokens of color l. Whenever transition e fires in color i, exactly $w^-(e, i, d, l)$ tokens of color l are removed from place d

and exactly $w^+(e, i, d, l)$ tokens of color l are deposited in place d for all $(d, l) \in \mathcal{D}$.

The graphical representation of a CSPN is similar to that of an ordinary SPN: places are drawn as circles, immediate transitions as thin bars, and timed transitions as thick bars. There is a directed arc from place d to transition e if and only if $w^-(e, i, d, l) > 0$ for some $i \in U_E(e)$ and $l \in U_D(d)$, and place d is said to be an *input place* of transition e. Similarly, there is a directed arc from transition e to place d if and only if $w^+(e, i, d, l) > 0$ for some $i \in U_E(e)$ and $l \in U_D(d)$, and place d is said to be an *output place* of transition e. Tokens are drawn as black dots with corresponding colors displayed nearby.

The label on an arc from a transition to a place indicates the number of tokens of each color deposited in the place whenever the transition fires. To indicate the number of tokens of each color that are deposited in a place, we use "formal-sum" notation. For example, an arc from transition e to place d has the label "$1 \cdot i + 3 \cdot j$" if transition e deposits exactly one token of color i and three tokens of color j in place d whenever it fires—that is, for all $l \in U_E(e)$, $w^+(e, l, d, k) = 1$ if $k = i$, $w^+(e, l, d, k) = 3$ if $k = j$, and $w^+(e, l, d, k) = 0$ otherwise. We sometimes abbreviate an expression such as "$1 \cdot i$" simply as "i." The special symbol "I" in a label denotes the firing color of the transition. For example, an arc from transition e to place d has the label "$2 \cdot I$" if, for all $i \in U_E(e)$, transition e deposits two tokens of color i in place d (and no tokens of any other color) whenever it fires in color i. As another example, in a CSPN with $U = \{1, 2, \ldots, N\}$, an arc from transition e to place d has the label "$2 \cdot I + 3 \cdot (I + 1)$" if transition e deposits two tokens of color i and three tokens of color $i + 1$ in place d whenever it fires in color i—for $i = N$, the color $i + 1$ is taken as the color 1. An arc without an explicit label has an implicit label of "$1 \cdot I$" or, equivalently, "I." We use analogous notation for a label on an arc from a place to a transition. For example, an arc from place d to transition e has the label "$1 \cdot 0 + 2 \cdot I$" if, for all $i \in U_E(e)$, transition e is enabled in color i only if place d contains at least one token of color 0 and two tokens of color i. Moreover, transition e removes one token of color 0 and two tokens of color i from place d whenever it fires in color i.

A *marking* of a CSPN is an assignment of token counts, by color, to the places of the net. We represent a marking as a pair (s, u). The first component $s = (s_1, s_2, \ldots, s_L)$ is a vector of nonnegative integers as in an ordinary SPN, where s_j is the number of tokens in place $d_j \in D$. We write $|s| = s_1 + s_2 + \cdots + s_L$. The second component $u = (u_1, u_2, \ldots, u_{|s|}) \in U^{|s|}$ records the color of each of the $|s|$ tokens. In the vector u, the colors of the tokens in place d_i appear to the left of the colors of the tokens in place d_j whenever $i < j$; for each place the colors of the tokens in the place appear from left to right in the order of the fixed enumeration of the set U of colors. Given a marking (s, u), we denote by $u_j(i)$ the number of tokens of color i in place d_j:

$$u_j(i) = \sum_{k=s_1+\cdots+s_{j-1}+1}^{s_1+\cdots+s_j} 1_{\{i\}}(u_k).$$

Observe that $\sum_{i\in U} u_j(i) = s_j$ for $1 \leq j \leq L$. Denote by

$$(s^{(0)}, u^{(0)}) = ((s_1^{(0)}, \ldots, s_L^{(0)}), (u_1^{(0)}, \ldots, u_J^{(0)}))$$

the *initial marking* of the net, where $J = |s^{(0)}|$.

For a marking (s, u), set

$$\mathcal{E}(s, u) = \{(e, i) \in \mathcal{E} : u_j(l) \geq w^-(e, i, d_j, l) \text{ for } (d_j, l) \in \mathcal{D}\}.$$

Transition e is *enabled in color i* when the marking is (s, u) if and only if $(e, i) \in \mathcal{E}(s, u)$; otherwise, transition e is *disabled in color i*.

The marking changes when a transition enabled in a color fires in the color. Whenever the marking is (s, u) and transition e fires in color i, the new marking (s', u') is given by

$$u_j'(l) = u_j(l) - w^-(e, i, d_j, l) + w^+(e, i, d_j, l)$$

for $(d_j, l) \in \mathcal{D}$. Thus, s, u, e, and i uniquely determine (s', u'), and we write $(s', u') = g(s, u, e, i)$. The function g is called the *new-marking function*.

A *clock* is associated with each pair $(e, i) \in \mathcal{E}$. Whenever transition e is enabled in color i, the clock associated with the pair (e, i) records the remaining time until e is scheduled to fire in color i. When the marking is (s, u) and transition e^* fires in color i^*, an a.s. finite clock reading is generated for each pair $(e, i) \in N(s, u, e^*, i^*) = \mathcal{E}(s', u') - (\mathcal{E}(s, u) - \{(e^*, i^*)\})$—here $(s', u') = g(s, u, e^*, i^*)$ is the unique new marking. Denote the *clock-setting distribution function* by $F(\cdot; e, i)$. As with ordinary SPNs, we require that $F(0; e, i) = 1$ for $(e, i) \in \mathcal{E}'$ and $F(0; e, i) = 0$ for $(e, i) \in \mathcal{E} - \mathcal{E}'$, so that immediate transitions always fire instantaneously and timed transitions never fire instantaneously. For $(e, i) \in O(s, u, e^*, i^*) = \mathcal{E}(s', u') \cap (\mathcal{E}(s, u) - \{(e^*, i^*)\})$—where, as above, (s', u') is the unique new marking—transition e is enabled in color i in marking (s, u) and remains enabled in color i in marking (s', u') after e^* fires in color i^*; in this case the old clock reading for (e, i) is kept after the marking change. For $(e, i) \in (\mathcal{E}(s, u) - \{(e^*, i^*)\}) - \mathcal{E}(s', u')$, transition e—which was enabled in color i before transition e^* fired in color i^*—becomes disabled in color i, and the clock reading is discarded. Of course, when transition e^* fires in color i^*, either e^* is enabled in color i^* in the new marking (s', u') and a new clock reading is generated according to $F(\cdot; e^*, i^*)$—cf. the foregoing definition of $N(s, u, e^*, i^*)$—or e^* is not enabled in color i^* and no new clock reading is generated.

(e_1, i) = stoppage of machine i

(e_2, i) = start of repair for machine i

(e_3, i) = end of repair for machine i

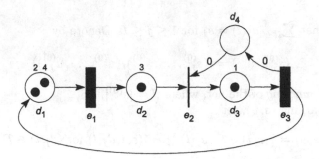

Figure 9.1. CSPN representation of machine repair system (four machines).

Whenever the (transition, color) pairs in a set $\mathcal{E}^*$ (with $|\mathcal{E}^*| > 1$) qualify to trigger a marking change, exactly one pair is selected for firing. For $(e, i) \in \mathcal{E}^*$, denote by $q(e, i; \mathcal{E}^*)$ the probability that transition e is selected to fire in color i. These *firing probabilities* satisfy $\sum_{(e,i)\in\mathcal{E}^*} q(e, i; \mathcal{E}^*) = 1$. For ease of exposition, we focus throughout on nets in which, with probability 1, no two clocks for timed transitions ever run down to 0 simultaneously—the firing probabilities are used to deal exclusively with situations in which two or more pairs in $\mathcal{E}'$ simultaneously qualify to trigger the next marking change.

clocks corresponding to immediate transitions are enabled simultaneously.

9.1.2 Modelling with CSPNs

This subsection contains several examples that illustrate the use of the CSPN building blocks for formal specification of discrete-event systems.

EXAMPLE 1.1 (Machine repair). Consider the system of N machines under the care of a single repairperson from Example 2.28 in Chapter 6. Suppose that the successive times (lifetimes) between end of repair and the next stoppage of machine j are i.i.d according to a random variable L_j with finite mean, and the successive times for the repairperson to repair (and restart) machine j are i.i.d. according to a random variable R_j with finite mean.

The machine repair model can be specified concisely as an N-bounded CSPN in which colors record the identity of the machines; see Figure 9.1 for $N = 4$. The set of colors is $U = \{0, 1, 2, \ldots, N\}$, and the color domains are given by $U_D(d_j) = \{1, 2, \ldots, N\}$ for $1 \le j \le 3$, $U_D(d_4) = \{0\}$, and

$U(e_k) = \{1, 2, \ldots, N\}$ for $1 \leq k \leq 3$. Place d_1 contains a token of color i if and only if machine i is running, place d_2 contains a token of color i if and only if machine i is awaiting repair, and place d_3 contains a token of color i if and only if machine i is under repair. Place d_4 contains a token of color 0 if and only if the repairperson is idle; otherwise, place d_4 contains no tokens.

The input incidence function is given by

$$w^-(e_1, i, d_1, l) = w^-(e_2, i, d_2, l) = w^-(e_3, i, d_3, l) = 1_{\{i\}}(l)$$

for $1 \leq i, l \leq N$,

$$w^-(e_2, i, d_4, 0) = 1$$

for $1 \leq i \leq N$, and $w^-(e, i, d, l) = 0$ otherwise. The output incidence function is given by

$$w^+(e_1, i, d_2, l) = w^+(e_2, i, d_3, l) = w^+(e_3, i, d_1, l) = 1_{\{i\}}(l)$$

for $1 \leq i, l \leq N$,

$$w^+(e_3, i, d_4, 0) = 1$$

for $1 \leq i, l \leq N$, and $w^+(e, i, d, l) = 0$ otherwise. According to this specification, transition e_1 is enabled in color i $(1 \leq i \leq N)$ if and only if place d_1 contains at least one token of color i; when transition e_1 fires in color i, it removes one token of color i from place d_1 and deposits one token of color i in place d_2. Transition e_2 is enabled in color i if and only place d_2 contains at least one token of color i and place d_4 contains at least one token of color 0; when transition e_2 fires in color i, it removes one token of color i from place d_2, removes one token of color 0 from place d_4, and deposits one token of color i in place d_3. Transition e_3 is enabled in color i if and only if place d_3 contains at least one token of color i; when transition e_3 fires in color i, it removes one token of color i from place d_3, deposits one token of color i in place d_1, and deposits one token of color 0 in place d_4.

The firing probabilities are defined for $\mathcal{E}^* \subseteq \{(e_2, 1), (e_2, 2), \ldots, (e_2, N)\}$ by

$$q(e_2, i; \mathcal{E}^*) = \begin{cases} 1 & \text{if } i = i^*; \\ 0 & \text{if } i \neq i^*, \end{cases}$$

where $i^* = \min\{i : e_{2,i} \in \mathcal{E}^*\}$. This definition reflects the fact that the repairperson always selects the lowest-numbered stopped machine for service. The clock-setting distribution functions are given by $F(x; e_1, i) = P\{L_i \leq x\}$ and $F(x; e_3, i) = P\{R_i \leq x\}$ for $1 \leq i \leq N$. We assume that, with probability 1, no two clocks corresponding to timed transitions ever run down to 0 simultaneously.

It is instructive to compare the CSPN representation of the machine repair system in Figure 9.1 to the SPN representation in Figure 6.1. Clearly, representing the system as a CSPN results in a more concise graph of places and transitions.

(e_1, i) = arrival of packet for transmission by port i

(e_2, i) = end of transmission by port i

(e_3, i) = observation of ring token by port $i + 1$

(e_4, i) = start of transmission by port i

(e_5, i) = start of propagation from port i

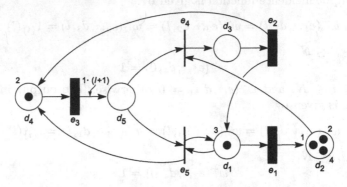

Figure 9.2. CSPN representation of token ring (four ports).

EXAMPLE 1.2 (Token ring). The ring network of Example 2.6 in Chapter 2 can be specified as a 1-bounded CSPN in which colors record both the port at which a packet arrives and the location of the ring token; see Figure 9.2 for $N = 4$ ports. The set of colors is $U = \{1, 2, \ldots, N\}$. Moreover, $U_D(d) = \{1, 2, \ldots, N\}$ for $d \in D$ and $U_E(e) = \{1, 2, \ldots, N\}$ for $e \in E$. Place d_1 contains a token of color i if and only if port i is not transmitting a packet and there is no packet awaiting transmission by port i. Place d_2 contains a token of color i if and only if there is a packet awaiting transmission by port i. Places d_3, d_4, and d_5 each contain at most one token. Place d_3 contains a token of color i if and only if port i is transmitting a packet. Place d_4 contains a token of color i if and only if the ring token is propagating to port $i + 1$. Place d_5 contains a token of color i whenever port i observes the ring token. Thus, in Figure 9.2, ports 1, 2, and 4 each have a packet awaiting transmission and the ring token is propagating to port 3.

Observe that the arc from transition e_3 to place d_5 has the label "1 · $(I + 1)$." This label indicates that whenever transition e_3 fires in color i $(1 \leq i < N)$, it deposits one token of color $i + 1$ in place d_5, and whenever transition e_3 fires in color N, it deposits one token of color 1 in place d_5. Also observe that immediate transitions e_4 and e_5 are never enabled simultaneously, so that firing probabilities need not be explicitly defined.

In the next example colors are used in a queueing system model to distinguish between jobs having different service requirements.

(e_1, i) = start of service to job i at center 1

(e_2, i) = end of service to job i at center 1

(e_3, i) = return of job i to queue at center 1

(e_4, i) = arrival of job i at center 2

(e_5, i) = start of service to job i at center 2

(e_6, i) = end of service to job i at center 2

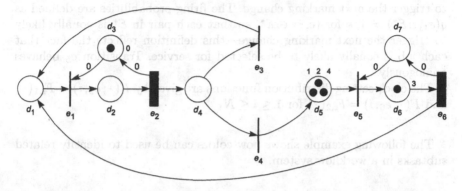

Figure 9.3. CSPN representation of cyclic queues with feedback and four stochastically nonidentical jobs.

EXAMPLE 1.3 (Cyclic queues with feedback and stochastically nonidentical jobs). Consider the queueing system of Example 1.4 in Chapter 2, but now suppose that the N jobs have nonidentical service requirements. In particular, suppose that successive service times for job i at center j are i.i.d. according to a continuous distribution function $F_{i,j}$. Also suppose that whenever there is a service completion at a center with one or more jobs waiting in queue, each of these jobs is equally likely to be selected for service.

This system can be specified as an N-bounded CSPN in which colors record the identity of jobs at each center; see Figure 9.3 for $N = 4$. The set of colors is $U = \{0, 1, \ldots, N\}$, and we have $U_D(d) = \{1, 2, \ldots, N\}$ for $d \in D - \{d_3, d_7\}$, $U_D(d) = \{0\}$ for $d \in \{d_3, d_7\}$, and $U_E(e) = \{1, 2, \ldots, N\}$ for $e \in E$.

Except for places d_1 and d_5, each place contains at most one token. Place d_1 contains a token of color i if and only if job i is waiting in queue at center 1, and place d_2 contains a token of color i if and only if job i is undergoing service at center 1. Place d_3 contains a token of color 0 if and only if the server at center 1 is idle. Place d_4 contains a token of color i if and only if service has just ended for job i at center 1. In this case, immediate transitions e_3 and e_4 are enabled simultaneously in color i. With probability $1 - p$, transition e_3 fires in color i, and job i joins the tail of the queue at center 1; with probability p, transition e_4 fires in color i, and job i moves to center 2. That is, $q(e_3, i; \mathcal{E}^*) = 1 - p$ and $q(e_4, i; \mathcal{E}^*) = p$ for $1 \le i \le N$, where $\mathcal{E}^* = \{(e_3, i), (e_4, i)\}$. The interpretations of places d_5, d_6, and d_7 are similar to the interpretations of places d_1, d_2, and d_3, respectively, but for center 2 rather than center 1.

Whenever there are $k \ge 1$ tokens of respective (distinct) colors $i_1, i_2, \dots, i_k$ in place d_1, no tokens in place d_2, and a single token of color 0 in place d_3, the pairs in the set $\mathcal{E}^* = \{(e_1, i_1), (e_1, i_2), \dots, (e_1, i_k)\}$ qualify to trigger the next marking change. The firing probabilities are defined as $q(e_1, i; \mathcal{E}^*) = 1/k$ for $(e, i) \in \mathcal{E}^*$, so that each pair in $\mathcal{E}^*$ is equally likely to trigger the next marking change—this definition reflects the fact that each job is equally likely to be selected for service. Transition e_5 behaves analogously.

The clock-setting distribution functions are given by $F(\,\cdot\,; e_2, i) = F_{i,1}(\,\cdot\,)$ and $F(\,\cdot\,; e_6, i) = F_{i,2}(\,\cdot\,)$ for $1 \le i \le N$.

The following example shows how colors can be used to identify related subtasks in a workflow system.

EXAMPLE 1.4 (Complaint processing). Consider a system for the processing of customer complaints. The complaints arrive one at a time and are processed asynchronously and in parallel. An arriving complaint is registered and evaluated, and then a questionnaire is sent to the complainant. A complaint is evaluated as "process" with probability $p_1 \in (0, 1)$ and as "archive" with probability $1 - p_1$. If the complaint is evaluated as "process" (resp., "archive"), then the complaint is processed (resp., archived) when the questionnaire is returned or when a timeout period of deterministic length L (> 0) elapses, whichever occurs first. After a complaint is processed it undergoes a quality inspection, which it passes with probability $p_2 \in (0, 1)$. If the complaint passes the inspection, then it is archived; otherwise, it undergoes processing again. The system can handle at most N (> 1) complaints; when the system is processing $N - 1$ complaints and another complaint arrives, the arrival process for complaints shuts down (i.e., any further complaints are routed to a different processing center). This process remains shut down until the first subsequent completion of archiving for a complaint. The times between successive arrivals of complaints are i.i.d. as a positive random variable, as are the successive times to register and evaluate a complaint, the successive times to return a questionnaire, the successive times to process and inspect a complaint, and the successive times to archive a complaint.

The complaint processing system can be specified as an N-bounded CSPN; see Figure 9.4 for $N = 7$. Places d_1, d_3, d_5, and d_{10} each contain at most one token; the remaining places contain between 0 and N tokens. The set of colors is $U = \{0, 1, 2, \dots, N\}$. Moreover, $U_D(d_1) = \{0\}$,

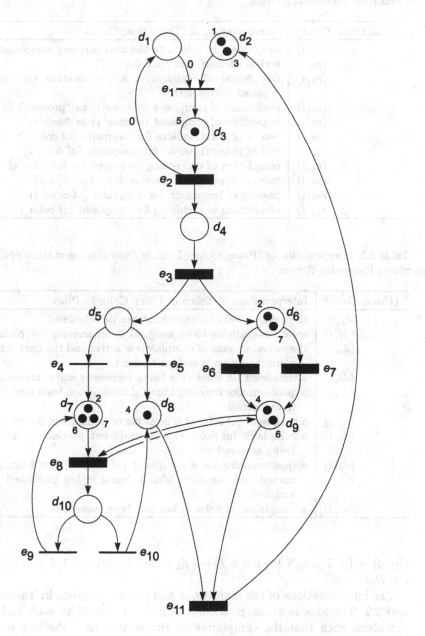

Figure 9.4. CSPN representation of complaint processing system.

Table 9.1. Interpretation of (Transition, Color) Pairs in CSPN Representation of Complaint Processing System

(Transition, Color)	Interpretation of (Transition, Color)
(e_1, i)	assignment of color i to the next arriving complaint
(e_2, i)	arrival of complaint (of color i)
(e_3, i)	completion of registration and evaluation for complaint (of color i)
(e_4, i)	evaluation of complaint (of color i) as "process"
(e_5, i)	evaluation of complaint (of color i) as "archive"
(e_6, i)	return of questionnaire for complaint (of color i)
(e_7, i)	end of timeout period for complaint (of color i)
(e_8, i)	completion of processing for complaint (of color i)
(e_9, i)	failure of inspection for complaint (of color i)
(e_{10}, i)	passing of inspection for complaint (of color i)
(e_{11}, i)	completion of archiving for complaint (of color i)

Table 9.2. Interpretation of (Place, Color) Pairs in CSPN Representation of Complaint Processing System

(Place, Color)	Interpretation of Token of Given Color in Place
$(d_1, 0)$	the next arrival of a complaint can be scheduled
(d_2, i)	color i is available to be assigned to an arriving complaint
(d_3, i)	the arrival process of complaints is active and the next complaint to arrive is assigned color i
(d_4, i)	a complaint (of color i) is being registered and evaluated
(d_6, i)	a questionnaire for complaint (of color i) has been sent but not returned
(d_7, i)	a complaint (of color i) is awaiting or undergoing processing
(d_8, i)	a complaint (of color i) has completed processing and is being archived
(d_9, i)	a questionnaire for a complaint (of color i) has been returned and the complaint is being either processed or archived
(d_{10}, i)	a complaint (of color i) has just been inspected

$U_D(d) = \{1, 2, \ldots, N\}$ for $d \in D - \{d_1\}$, and $U_E(e) = \{1, 2, \ldots, N\}$ for $e \in E$.

The interpretations of the transitions and places are given in Tables 9.1 and 9.2. The idea is to assign a color between 1 and N to each arriving complaint such that the complaints in the system have distinct colors; colors are "recycled" as necessary. In more detail, a token of color i in place d_2 means that color i is available to be assigned to the next arriving complaint. Whenever place d_2 contains one or more tokens and place d_1 contains a token of color 0—so that it is time to schedule the next arrival of a complaint by depositing a token in place d_3—each pair (e_1, i) with $u_2(i) = 1$ becomes qualified to trigger the next marking change, and exactly one of these pairs is selected (randomly and uniformly) for firing. If immediate transition e_1 fires in color i (where $1 \leq i \leq N$), then a token of color i is deposited in place d_3 and the color i is assigned to the next arriving complaint. This color-assignment mechanism is formally specified by setting

$$q(e_1, i_1; \mathcal{E}^*) = q(e_1, i_2; \mathcal{E}^*) = \cdots = q(e_1, i_k; \mathcal{E}^*) = 1/k$$

for $\mathcal{E}^* = \{ (e_1, i_1), (e_1, i_2), \ldots, (e_1, i_k) \}$ (where $1 \leq k \leq N$ and the i_j's are distinct). The color assigned to a complaint is recycled when the complaint is archived. Specifically, whenever $e_{11} = $ "completion of archiving for complaint" fires in a color i, it removes a token of color i from each of places d_8 and d_9 and deposits a token of color i in place d_2, so that color i is available to be assigned to the next arriving complaint. Observe that the role of place d_1 and immediate transition e_1 is to ensure that at most one arrival of a complaint is scheduled at any time.

The remaining firing probabilities are defined in a straightforward way:

$$q(e_4, i; \mathcal{E}^*) = 1 - q(e_5, i; \mathcal{E}^*) = p_1$$

for $\mathcal{E}^* = \{ (e_4, i), (e_5, i) \}$, and

$$q(e_{10}, i; \mathcal{E}^*) = 1 - q(e_9, i; \mathcal{E}^*) = p_2$$

for $\mathcal{E}^* = \{ (e_9, i), (e_{10}, i) \}$.

9.1.3 The Marking Process

The marking process of a CSPN records the marking as it evolves over continuous time. As with an ordinary SPN, formal definition of the marking process is in terms of a general state-space Markov chain that describes the net at successive marking changes.

Define *new-marking probabilities* as follows. For a marking (s, u) and a pair $(e, i) \in \mathcal{E}(s, u)$, set

$$p(s', u'; s, u, e, i) = \begin{cases} 1 & \text{if } (s', u') = g(s, u, e, i); \\ 0 & \text{otherwise,} \end{cases}$$

and for a subset $\mathcal{E}^* \subseteq \mathcal{E}(s, u)$ set

$$p(s', u'; s, u, \mathcal{E}^*) = \sum_{(e,i) \in \mathcal{E}^*} q(e, i; \mathcal{E}^*) p(s', u'; s, u, e, i).$$

For a marking (s, u) such that $\mathcal{E}(s, u) \cap \mathcal{E}' = \varnothing$, write $(s, u) \rightarrow (s', u')$ if $p(s', u'; s, u, e, i) > 0$ for some $(e, i) \in \mathcal{E}(s, u)$. Similarly, for (s, u) such that

$\mathcal{E}(s, u) \cap \mathcal{E}' \neq \varnothing$, write $(s, u) \to (s', u')$ if $p(s', u'; s, u, \mathcal{E}(s, u) \cap \mathcal{E}') > 0$. Finally, write $(s, u) \rightsquigarrow (s', u')$ if either $(s, u) \to (s', u')$ or there exist $n \geq 1$ and markings $(s^{(1)}, u^{(1)}), (s^{(2)}, u^{(2)}), \ldots, (s^{(n)}, u^{(n)})$ such that

$$(s, u) \to (s^{(1)}, u^{(1)}) \to \cdots \to (s^{(n)}, u^{(n)}) \to (s', u').$$

The *marking set H* of the CSPN is defined as the set

$$H = \{ (s, u) : (s^{(0)}, u^{(0)}) \rightsquigarrow (s, u) \},$$

where $(s^{(0)}, u^{(0)})$ is the initial marking. To avoid trivialities, we always assume without comment that $\mathcal{E}(s, u) \neq \varnothing$ for $(s, u) \in H$ and that, for each pair $(e, i) \in \mathcal{E}$, there exists a marking $(s, u) \in H$ such that $(e, i) \in \mathcal{E}(s, u)$. Define the set of *immediate markings* by

$$H' = \{ (s, u) \in H : \mathcal{E}(s, u) \cap \mathcal{E}' \neq \varnothing \}$$

and the set of *timed markings* by

$$H - H' = \{ (s, u) \in H : \mathcal{E}(s, u) \cap \mathcal{E}' = \varnothing \}.$$

EXAMPLE 1.5 (Machine repair). Suppose that all machines are running at time 0. Then the initial marking is $(s^{(0)}, u^{(0)})$, where $s^{(0)} = (N, 0, 0, 1)$, $u_4^{(0)}(0) = 1$, $u_1^{(0)}(i) = 1$ for $1 \leq i \leq N$, and $u_j^{(0)}(i) = 0$ otherwise. The marking set H is the set of all markings $(s, u) = ((s_1, s_2, s_3, s_4), u)$ such that

(i) $\sum_{j=1}^{3} u_j(i) = 1$ for $1 \leq i \leq N$,

(ii) $s_3 + s_4 = 1$, and

(iii) $u_4(0) = s_4$.

The set H' of immediate markings is given by $H' = \{ (s, u) \in H : s_4 = 1 \text{ and } s_2 > 0 \}$.

Denote by $C(s, u)$ the set of possible *clock-reading vectors* when the marking is (s, u). In the CSPN setting, a clock-reading vector c is a nonnegative real-valued vector of length $r = \sum_{e \in E} |U_E(e)|$. The clock readings for transition e_j appear to the left of the clock readings for transition e_k whenever $j < k$. For each transition e the clock readings for colors $\{ i : i \in U_E(e) \}$ appear from left to right in the order of the fixed enumeration of the set U of colors. Given a clock-reading vector c, we denote by $c_k(i)$ the clock reading for transition e_k and color i—that is, $c_k(i) = c_m$, where $m = \sum_{j=1}^{k-1} |U_E(e_j)| + \sum_{l \leq i} 1_{U_E(e_k)}(l)$. Thus

$$C(s, u) = \{ c = (c_1, \ldots, c_r) : c_k(i) \geq 0$$
$$\text{and } c_k(i) > 0 \text{ only if } (e_k, i) \in \mathcal{E}(s, u) - \mathcal{E}' \}.$$

Beginning in marking (s, u) with clock-reading vector c, the time $t^*(s, u, c)$ to the next marking change is given by

$$t^*(s, u, c) = \min_{\{i,k:\, (e_k,i) \in \mathcal{E}(s,u)\}} c_k(i),$$

and the set of (transition, color) pairs that *qualify* to trigger the next marking change is given by

$$\mathcal{E}^*(s, u, c) = \{(e_k, i) \in \mathcal{E}(s,u) : c_k^*(i; s, u, c) = 0\},$$

where $c_k^*(i; s, u, c) = c_k(i) - t^*(s, u, c)$. As mentioned previously, we assume that with probability 1 timed transitions never fire simultaneously, and so we can restrict attention to the following two cases:

1. $(s, u) \in H - H'$ and $\mathcal{E}^*(s, u, c) = \{(e, i)\}$ for some $(e, i) \in \mathcal{E} - \mathcal{E}'$.

2. $(s, u) \in H'$ and $\mathcal{E}^*(s, u, c) = \mathcal{E}(s, u) \cap \mathcal{E}'$.

Next consider a discrete-time Markov chain $\{(S_n, U_n, C_n) : n \geq 0\}$ taking values in the set

$$\Upsilon = \bigcup_{(s,u) \in H} (\{s\} \times \{u\} \times C(s,u)),$$

where (S_n, U_n) represents the marking and C_n represents the clock-reading vector just after the nth marking change. We denote by

- $S_{n,j}$ the number of tokens in place d_j

- $U_{n,j}(i)$ the number of tokens of color i in place d_j

- $C_{n,k}(i)$ the clock reading for transition e_k and color i

just after the nth marking change. The transition kernel of the chain is given by

$$P((s, u, c), A) = \sum_{(e_k, i)} q(e_k, i; \mathcal{E}^*) \prod_{(e_j, l) \in N} F(a_{j,l}; e_j, l) \prod_{(e_j, l) \in O} 1_{[0, a_{j,l}]}(c_j^*(l))$$

for all sets

$$A = \{s'\} \times \{u'\} \times \{c' \in C(s', u') : 0 \leq c_j'(l) \leq a_{j,l} \text{ for } (e_j, l) \in \mathcal{E}\},$$

where $c_j^*(l) = c_j^*(l; s, u, c)$, $\mathcal{E}^* = \mathcal{E}^*(s, u, c)$, $N = N(s, u, e_k, i)$, $O = O(s, u, e_k, i)$, and the sum is taken over all $(e_k, i) \in \mathcal{E}^*$ such that $g(s, u, e_k, i) = (s', u')$.

The initial distribution μ of the chain is given by

$$\mu(A) = 1_{\{(s^{(0)}, u^{(0)})\}}(s, u) \prod_{(e_j, l) \in \mathcal{E}(s^{(0)}, u^{(0)})} F(a_{j,l}; e_j, l)$$

for all sets

$$A = \{\,s\,\} \times \{\,u\,\} \times \{\,c \in C(s,u) \colon 0 \leq c_j(l) \leq a_{j,l} \text{ for } (e_j,l) \in \mathcal{E}\,\}.$$

That is, $(s^{(0)}, u^{(0)})$ is selected as the initial marking and then, for each transition e_j and color l such that $(e_j, l) \in \mathcal{E}(s^{(0)}, u^{(0)})$, an initial clock reading is selected according to the clock-setting distribution function $F(\cdot\,; e_j, l)$. Denote by P_μ the probability law of the chain when the initial distribution is μ.

Finally, construct a continuous-time process $\{\,Z(t) \colon t \geq 0\,\}$ from the chain $\{\,(S_n, U_n, C_n) \colon n \geq 0\,\}$ in a manner similar to the construction of the marking process for an ordinary SPN. Let ζ_n be the (nonnegative, real-valued) time of the nth marking change: $\zeta_0 = 0$ and

$$\zeta_n = \sum_{k=0}^{n-1} t^*(S_k, U_k, C_k)$$

for $n \geq 1$. Let $\Delta = (\Delta_1, \Delta_2) \notin H$ and set $Z(t) = \big(X(t), Y(t)\big)$, where

$$X(t) = \begin{cases} S_{N(t)} & \text{if } N(t) < \infty; \\ \Delta_1 & \text{if } N(t) = \infty, \end{cases}$$

$$Y(t) = \begin{cases} U_{N(t)} & \text{if } N(t) < \infty; \\ \Delta_2 & \text{if } N(t) = \infty, \end{cases}$$

and

$$N(t) = \sup\,\{\,n \geq 0 \colon \zeta_n \leq t\,\}.$$

The stochastic process $\{\,Z(t) \colon t \geq 0\,\}$ is the *marking process* of the CSPN.

By construction, the marking process takes values in the set $(H - H') \cup \{\Delta\}$ and has piecewise-constant, right-continuous sample paths. Observe that $Z(t) = \Delta$ for at least one finite time t if and only if the *lifetime*

$$\tau_\Delta = \sup_{n \geq 0} \zeta_n$$

is finite. We assume throughout that the lifetime is a.s. infinite. Theorem 1.6 below can be used to verify for specific models that this assumption holds—the theorem can be established using arguments similar to those in Section 3.3. We write $H' \rightsquigarrow H - H'$ if for each $(s', u') \in H'$ there exists $(s, u) \in H - H'$ such that $(s', u') \rightsquigarrow (s, u)$.

Theorem 1.6. *Suppose that*

$$P_\mu\{\,(S_n, U_n) \in H - H' \text{ i.o.}\,\} = 1. \tag{1.7}$$

Then $P_\mu\{\,\tau_\Delta = \infty\,\} = 1$. *If* H' *is finite, then the condition in* (1.7) *holds for any initial distribution* μ *if and only if* $H' \rightsquigarrow H - H'$.

As with ordinary SPNs, the marking process of a CSPN is a CTMC under appropriate assumptions on the clock-setting distribution functions. The precise result is given by Theorem 1.8 below. Let $\{\gamma(n): n \geq 0\}$ be the indices of the successive marking changes at which the new marking is timed: $\gamma(-1) = -1$ and

$$\gamma(n) = \inf\{j > \gamma(n-1): (S_j, U_j) \in H - H'\}$$

for $n \geq 0$. For timed markings $(s, u), (s', u') \in H - H'$, let $p^+(s', u'; s, u, e, i)$ be the probability that the next timed marking is (s', u') when the current marking is (s, u) and transition e fires in color i:

$$p^+(s', u'; s, u, e, i) = \sum \left(p(s_1, u_1; s, u, e, i) \prod_{j=2}^{k} p(s_j, u_j; s_{j-1}, u_{j-1}, \mathcal{E}_{j-1}^*) \right),$$

where $\mathcal{E}_{j-1}^* = \mathcal{E}' \cap \mathcal{E}(s_{j-1}, u_{j-1})$ and the summation is over all finite sequences $(s_1, u_1), \ldots, (s_k, u_k)$ $(k \geq 1)$ such that $(s_k, u_k) = (s', u')$ and $(s_j, u_j) \in H'$ for $1 \leq j < k$.

Theorem 1.8. *Suppose that for each $(e, i) \in \mathcal{E} - \mathcal{E}'$, the clock-setting distribution function has the form $F(x; e, i) = 1 - \exp(-v(e, i)x)$ for some positive finite constant $v(e, i)$. Also suppose that*

$$P_\mu\{(S_n, U_n) \in H - H' \ i.o.\} = 1.$$

Then the marking process $\{Z(t): t \geq 0\}$ is a nonexplosive time-homogeneous CTMC. The initial distribution is given by

$$\nu(s, u) = P_\mu\{(S_{\gamma(0)}, U_{\gamma(0)}) = (s, u)\}$$

for $(s, u) \in H - H'$, the intensity vector is given by

$$q(s, u) = \sum_{(e, i) \in \mathcal{E}(s, u)} (1 - p^+(s, u; s, u, e, i))v(e, i)$$

for $(s, u) \in H - H'$, and the transition matrix for the embedded jump chain is given by

$$W((s, u), (s', u'))$$
$$= \begin{cases} \sum_{(e, i) \in \mathcal{E}(s, u)} \frac{v(e, i)}{q(s, u)} p^+(s', u'; s, u, e, i) & \text{if } (s', u') \neq (s, u); \\ 0 & \text{if } (s', u') = (s, u) \end{cases}$$

for $(s', u'), (s, u) \in H - H'$.

Because tokens are removed and deposited in essentially a deterministic manner, CSPNs appear to have less modelling power than ordinary SPNs. For example, it appears impossible to model the queue with batch arrivals as a CSPN, but this system can be modelled as an SPN—see Example 2.4 in Chapter 2. On the other hand, a broad range of interesting models can be specified as CSPNs. In particular, it can be shown that CSPNs have at least the modelling power of GSMPs with finite state space and unit speeds. That is, for any GSMP with finite state space and unit speeds there exists a CSPN having a marking process that strongly mimics the GSMP in a sense analogous to that defined in Chapter 4. The proof of this result is similar to the proof of Theorem 3.3 in Chapter 4.

9.2   Stability and Simulation

In this section we give sufficient conditions for stability of a CSPN, as well as conditions under which various simulation methods are applicable. We simply state the relevant results—in all cases the proofs are similar to those given in the setting of ordinary SPNs.

9.2.1   Recurrence

As with ordinary SPNs, recurrence arguments can be based on drift criteria or geometric trials criteria.

Drift Criteria

We first give a "colored" version of Assumption PD. Recall from Section 5.1.2 that $\mathcal{G}^+$ is the set of distribution functions on $[0, \infty)$ that have a convergent LaPlace–Stieltjes transform in a neighborhood of the origin. As with an ordinary SPN, a CSPN is said to be *irreducible* if $(s, u) \rightsquigarrow (s', u')$ for all $(s, u), (s', u') \in H$.

Definition 2.1. *Assumption CPD is said to hold for a specified CSPN if*

(i) the marking set H is finite,

(ii) the CSPN is irreducible, and

(iii) there exists $0 < \bar{x} < \infty$ such that each clock-setting distribution function $F(\,\cdot\,; e, i)$ with $(e, i) \in \mathcal{E} - \mathcal{E}'$ belongs to $\mathcal{G}^+$ and has a density component that is positive and continuous on $(0, \bar{x})$.

Whenever Assumption CPD holds, we can find a real number $q > 0$ such that
$$\int_0^\infty e^{qx}\, dF(x; e, i) < \infty, \qquad (e, i) \in \mathcal{E}. \tag{2.2}$$

Recall that the random indices $\{\gamma(n) : n \geq 0\}$ correspond to the successive marking changes at which the new marking is timed. Define the *embedded chain* $\{(S_n^+, U_n^+, C_n^+) : n \geq 0\}$ by setting $(S_n^+, U_n^+, C_n^+) = (S_{\gamma(n)}, U_{\gamma(n)},$

$C_{\gamma(n)})$ for $n \geq 0$, and denote by Υ^+ the state space of the embedded chain. Theorem 2.5 below asserts that the embedded chain of a CSPN satisfies a drift criterion for stability provided that Assumption CPD holds. When Assumption CPD holds, let $\bar{\phi}$ be the unique measure on subsets of Υ^+ such that

$$\bar{\phi}(A) = \prod_{\{(j,l)\,:\,(e_j,l)\in\mathcal{E}(s,u)\}} \min(x_{j,l},\bar{x}) \tag{2.3}$$

for all sets

$$A = \{s\} \times \{u\} \times \{c \in C(s,u)\colon 0 \leq c_j(l) \leq x_{j,l} \text{ for } (e_j,l) \in \mathcal{E}(s,u)\}.$$

Set

$$h_q(s,u,c) = \exp\Big(q \max_{(e_j,i)\in\mathcal{E}(s,u)} c_j(i)\Big)$$

and

$$H_b = \Big\{(s,u,c) \in \Upsilon^+\colon \max_{(e_j,i)\in\mathcal{E}(s,u)} c_j(i) \leq b\Big\}. \tag{2.4}$$

Theorem 2.5. *If Assumption CPD holds, then*

(i) *the embedded chain* $\{(S_n^+,U_n^+,C_n^+)\colon n \geq 0\}$ *is $\bar{\phi}$-irreducible, where $\bar{\phi}$ is defined by (2.3), and*

(ii) *for each $b > 0$ the set H_b defined by (2.4) is petite with respect to* $\{(S_n^+,U_n^+,C_n^+)\colon n \geq 0\}$.

Moreover, for some $m \geq 1$, all q satisfying (2.2), and all sufficiently large b,

(iii) $\sup_{(s,u,c)\in H_b} E_{(s,u,c)}\big[h_q(S_m^+,U_m^+,C_m^+) - h_q(S_0^+,U_0^+,C_0^+)\big] < \infty$, *and*

(iv) *there exists $\beta \in (0,1)$ such that*

$$E_{(s,u,c)}\big[h_q(S_m^+,U_m^+,C_m^+) - h_q(S_0^+,U_0^+,C_0^+)\big] \leq -\beta h_q(s,u,c)$$

for $(s,u,c) \in \Upsilon^+ - H_b$.

Corollary 2.6. *Suppose that Assumption CPD holds for a CSPN. Then the embedded chain of the marking process is positive Harris recurrent with recurrence measure $\bar{\phi}$ given by (2.3) and hence admits a stationary distribution π. Moreover, if q satisfies (2.2), then $\pi(|f|) < \infty$ for any function f such that $f = O(h_q)$.*

Geometric Trials Criteria

Recurrence properties can also be established by means of geometric trials arguments. Set $\mathcal{E}_n^* = \mathcal{E}^*(S_n,U_n,C_n)$ and $t_n^* = t^*(S_n,U_n,C_n)$ for $n \geq 0$, and define the *partial history* $\mathcal{F}_n$ of the underlying chain up to the nth marking change by setting $\mathcal{F}_0 = \{S_0,U_0\}$ and

$$\mathcal{F}_n = \{S_0,U_0,\mathcal{E}_0^*,t_0^*,S_1,U_1,\mathcal{E}_1^*,t_1^*,\ldots,S_{n-1},U_{n-1},\mathcal{E}_{n-1}^*,t_{n-1}^*,S_n,U_n\}$$

for $n \geq 1$.

Lemma 2.7. *Let $\{\,\beta(n)\colon n \geq 1\,\}$ and $\{\,\alpha(n)\colon n \geq 1\,\}$ be increasing sequences of a.s. finite random indices such that each $\alpha(n)$ and $\beta(n)$ is a stopping time with respect to $\{\,\mathcal{F}_n\colon n \geq 0\,\}$ and, moreover, $\beta(n-1) \leq \alpha(n) < \beta(n)$ for $n \geq 1$. [Take $\beta(0) = 0$.] Suppose that*

$$P_\mu\big\{\,(S_{\beta(n)}, U_{\beta(n)}) \in \bar{H} \mid \mathcal{F}_{\alpha(n)}\,\big\} \geq \delta \ \ a.s.$$

for some $\delta > 0$ and all $n \geq 1$. Then $P_\mu\big\{\,(S_{\beta(n)}, U_{\beta(n)}) \in \bar{H} \ i.o.\,\big\} = 1$.

As with ordinary SPNs, representations of conditional clock-reading distributions play a key role when establishing the foregoing geometric trials criterion. The basic result in the CSPN setting is as follows. For $n \geq 0$ and a pair $(e_i, l) \in \mathcal{E}(S_n, U_n)$, denote by $Q_{n,i}(l)$ the amount of time elapsed on the clock for (e_i, l) between the most recent time prior to ζ_n at which this clock was set and time ζ_n itself. Observe that the value of $Q_{n,i}(l)$ is determined by $\mathcal{F}_n$ for $n \geq 0$.

Lemma 2.8. *Let γ be an a.s. finite stopping time with respect to $\{\,\mathcal{F}_n\colon n \geq 0\,\}$. Then*

$$P_\mu\big\{\,C_{\gamma,i}(l) > x_{i,l} \ for \ (e_i, l) \in \mathcal{H} \mid \mathcal{F}_\gamma\,\big\}$$

$$= \begin{cases} \prod_{(e_i,l)\in\mathcal{H}} \overline{F}(x_{i,l} + Q_{\gamma,i}(l); e_i, l)/\overline{F}(Q_{\gamma,i}(l); e_i, l) & if \ \mathcal{H} \subseteq \mathcal{E}(S_\gamma, U_\gamma); \\ 0 & otherwise \end{cases}$$

with probability 1 for any subset $\mathcal{H} \subseteq \mathcal{E} - \mathcal{E}'$ and nonnegative numbers $\{\,x_{i,l}\colon (e_i, l) \in \mathcal{H}\,\}$.

Using the foregoing result and its extensions, we can establish the following CSPN analogs of Theorems 2.21 and 2.29 in Chapter 5. For ease of exposition, we assume that $F(\,\cdot\,; e, i) \neq F(\,\cdot\,; e', i')$ whenever $(e, i) \neq (e', i')$ with $(e, i), (e', i') \in \mathcal{E} - \mathcal{E}'$, and we enumerate the clock-setting distributions corresponding to the timed transitions as $F_1, F_2, \ldots, F_J$, where $J = |\mathcal{E} - \mathcal{E}'|$. For a sequence $\{\,\alpha(n)\colon n \geq 0\,\}$ as in Lemma 2.7, we define H_α to be the state space of the process $\{\,(S_{\alpha(n)}, U_{\alpha(n)})\colon n \geq 1\,\}$. In addition, $\{\,k(i, j, s, u)\colon (s, u) \in H_\alpha, \ 1 \leq i, j \leq J\,\}$ is a collection of finite nonnegative integers such that

$$k(i, j) \overset{\text{def}}{=} \sup_{(s,u)\in H_\alpha} k(i, j, s, u) < \infty \tag{2.9}$$

for each i and j. Denote by $\alpha(n, j, l)$ the random index of the lth marking change after $\alpha(n)$ at which a new clock reading is generated from F_j and by $A_{n,j,l}$ the value of this new clock reading. For $(s, u) \in H_\alpha$, denote by $I(s, u)$ the unique subset of $\{\,1, 2, \ldots, J\,\}$ such that $i \in I(s, u)$ if and only if $(e, l) \in \mathcal{E}(s, u)$ and $F(\,\cdot\,; e, l) = F_i(\,\cdot\,)$. Finally, for $i \in I(S_{\alpha(n)}, U_{\alpha(n)})$, denote by $B_{n,i}$ the clock reading at time $\alpha(n)$ corresponding to the unique pair (e, l) such that $F(\,\cdot\,; e, l) = F_i(\,\cdot\,)$.

Theorem 2.10. *Let* $\tilde{I} \subseteq \{1, 2, \ldots, J\}$, $q \in \{1, 2, \ldots, J\}$, *and* $\bar{H} \subseteq H$, *and let* $\{x_i^*: i \in \tilde{I}\}$ *be a collection of nonnegative numbers. Also let* $\{\beta(n): n \geq 1\}$ *and* $\{\alpha(n): n \geq 0\}$ *be as in Lemma 2.7 and* $\{k(i, j, s, u): (s, u) \in H_\alpha, 1 \leq i, j \leq J\}$ *be nonnegative integers satisfying* (2.9). *Set* $\tilde{I}_n = \tilde{I} \cap I(S_{\alpha(n)}, U_{\alpha(n)})$ *and* $K_n(i, j) = k(i, j, S_{\alpha(n)}, U_{\alpha(n)})$, *and suppose that*

(i) *for each* $i \in \tilde{I}$, *the clock-setting distribution function* F_i *is* GNBU *with lower bound* x_i^*,

(ii) *a new clock reading is generated from* F_q *at the* $\alpha(n)$*th marking change for* $n \geq 0$ *and*

$$P_\mu \left\{ (S_{\beta(n)}, U_{\beta(n)}) \in \bar{H} \mid \mathcal{F}_{\alpha(n)} \right\}$$

$$\geq P_\mu \left\{ B_{n,i} + \sum_{j=1}^{J} \sum_{l=1}^{K_n(i,j)} A_{n,j,l} < B_{n,q}, \ i \in \tilde{I}_n \ \Big| \ \mathcal{F}_{\alpha(n)} \right\} \ a.s.$$

for $n \geq 0$, *and*

(iii) *the positivity condition*

$$x_i^* + \sum_{j=1}^{J} k(i, j) y_j < z \qquad for \ i \in \tilde{I}$$

holds, where $z = \operatorname{ess\,sup} F_q$ *and* $y_j = \operatorname{ess\,inf} F_j$ *for* $1 \leq j \leq J$.
Then $P_\mu \{ (S_{\beta(n)}, (U_{\beta(n)}) \in \bar{H} \ i.o. \} = 1$.

In Theorem 2.12 below $\{k(j, s, u): (s, u) \in H_\alpha, 1 \leq j \leq J\}$ is a collection of finite nonnegative integers such that, for each j,

$$k(j) \stackrel{\text{def}}{=} \sup_{(s,u) \in H_\alpha} k(j, s, u) < \infty. \tag{2.11}$$

Theorem 2.12. *Let* $\tilde{I}$, q, $\bar{H}$, *and* $\{x_i^*: i \in \tilde{I}\}$ *be as in Theorem 2.10. Also let* $\{\beta(n): n \geq 1\}$ *and* $\{\alpha(n): n \geq 0\}$ *be sequences of random indices as in Lemma 2.7 and* $\{k(j, s, u): (s, u) \in H_\alpha, 1 \leq j \leq J\}$ *be nonnegative integers satisfying* (2.11). *Set* $\tilde{I}_n = \tilde{I} \cap I(S_{\alpha(n)}, U_{\alpha(n)})$ *and* $K_n(j) = k(j, S_{\alpha(n)}, U_{\alpha(n)})$, *and suppose that*

(i) *for each* $i \in \tilde{I}$, *the clock-setting distribution function* F_i *is* GNBU *with lower bound* x_i^*,

(ii) *a new clock reading is generated from* F_q *at the* $\alpha(n)$*th marking change for* $n \geq 0$ *and*

$$P_\mu \left\{ (S_{\beta(n)}, U_{\beta(n)}) \in \bar{H} \mid \mathcal{F}_{\alpha(n)} \right\}$$

$$\geq P_\mu \left\{ \sum_{i \in \tilde{I}_n} B_{n,i} + \sum_{j=1}^{J} \sum_{l=1}^{K_n(j)} A_{n,j,l} < B_{n,q} \ \Big| \ \mathcal{F}_{\alpha(n)} \right\} \ a.s.$$

for $n \geq 0$, *and*

(iii) the positivity condition

$$\sum_{i \in \bar{I}} x_i^* + \sum_{j=1}^{J} k(j) y_j < z$$

holds, where $z = \operatorname{ess\,sup} F_q$ and $y_j = \operatorname{ess\,inf} F_j$ for $1 \le j \le J$.

Then $P_\mu \{ (S_{\beta(n)}, U_{\beta(n)}) \in \bar{H} \ i.o. \} = 1$.

9.2.2 CSPNs and Regeneration

In this section we give conditions on the building blocks of a CSPN under which there exists a sequence of regeneration points for the marking process or underlying chain or both, and under which the integral or sum of the output process over a cycle has finite moments. Under these conditions, estimation methods as in Section 6.3 can be used to obtain strongly consistent point estimates and asymptotic confidence intervals for time-average limits. As with ordinary SPNs, we first give general sufficient conditions for regenerative structure and then refine these conditions for CSPNs in which either Assumption CPD or a geometric trials recurrence criterion holds.

General Conditions for Regenerative Structure

For a marking $(\bar{s}, \bar{u}) \in H$ and set $\bar{\mathcal{E}} \subseteq \mathcal{E}(\bar{s}, \bar{u})$, denote by $\{ \theta(k) \colon k \ge 0 \}$ the indices of the successive marking changes at which the marking is $(\bar{s}, \bar{u})$ and the clocks corresponding to the (transition, color) pairs in $\bar{\mathcal{E}}$ run down to 0 simultaneously: $\theta(-1) = 0$ and

$$\theta(k) = \inf \left\{ n > \theta(k-1) \colon (S_{n-1}, U_{n-1}) = (\bar{s}, \bar{u}) \text{ and } \mathcal{E}_{n-1}^* = \bar{\mathcal{E}} \right\}. \quad (2.13)$$

Theorem 2.14. *Let $(\bar{s}, \bar{u}) \in H$ and $\bar{\mathcal{E}} \subseteq \mathcal{E}(\bar{s}, \bar{u})$, and suppose that*

$$P_\mu \left\{ (S_n, U_n, \mathcal{E}_n^*) = (\bar{s}, \bar{u}, \bar{\mathcal{E}}) \ i.o. \right\} = 1.$$

Also suppose that for all $(\bar{e}, \bar{\imath}) \in \bar{\mathcal{E}}$ with $q(\bar{e}, \bar{\imath}; \bar{\mathcal{E}}) > 0$ either

(a) $O(\bar{s}, \bar{u}, \bar{e}, \bar{\imath}) = \varnothing$, *or*

(b) $O(\bar{s}, \bar{u}, \bar{e}, \bar{\imath}) \ne \varnothing$ *and* $F(x; e, i) = 1 - \exp(-v(e, i) x)$ *for each pair* $(e, i) \in O(\bar{s}, \bar{u}, \bar{e}, \bar{\imath})$, *where* $v(e, i)$ *is a positive finite constant.*

Then the random times $\{ \zeta_{\theta(k)} \colon k \ge 0 \}$ defined via (2.13) form a sequence of regeneration points for $\{ Z(t) \colon t \ge 0 \}$. If, in particular, the condition in (a) holds for all $(\bar{e}, \bar{\imath}) \in \bar{\mathcal{E}}$ with $q(\bar{e}, \bar{\imath}; \bar{\mathcal{E}}) > 0$, then the random indices $\{ \theta(k) \colon k \ge 0 \}$ form a sequence of regeneration points (in discrete time) for $\{ (S_n, U_n, C_n) \colon n \ge 0 \}$.

As with ordinary SPNs, the condition in (a) holds if $(\bar{s}, \bar{u})$ is a *single state*—that is, if $\mathcal{E}(\bar{s}, \bar{u}) = \{ (\bar{e}, \bar{\imath}) \}$ for some $(\bar{e}, \bar{\imath}) \in \mathcal{E}$.

CSPNs with Positive Clock-Setting Densities

We now refine the foregoing result when Assumption CPD holds. To this end, we first define the notion of a polynomially dominated function in the setting of CSPNs. Set

$$\tilde{g}_q(s,u,c) = \begin{cases} 1 + \max_{(e_j,i)\in\mathcal{E}} c_j^q(i) & \text{if } (s,u,c) \in \Upsilon^+; \\ 1 & \text{if } (s,u,c) \in \Upsilon - \Upsilon^+ \end{cases}$$

for $(s,u,c) \in \Upsilon$ and $q \geq 0$.

Definition 2.15. A real-valued function $\tilde{f}$ defined on Υ is *polynomially dominated* if $\tilde{f} = O(\tilde{g}_q)$ for some $q \geq 0$.

For a sequence of random indices $\{\theta(k): k \geq 0\}$ defined as in (2.13), set

$$Y_k(f) = \int_{\zeta_{\theta(k-1)}}^{\zeta_{\theta(k)}} f\big(Z(u)\big)\, du \tag{2.16}$$

for each real-valued function f defined on $H - H'$ and

$$\tilde{Y}_k(\tilde{f}) = \sum_{j=\theta(k-1)}^{\theta(k)-1} \tilde{f}(S_j, U_j, C_j). \tag{2.17}$$

for each real-valued function $\tilde{f}$ defined on Υ.

Theorem 2.18. *Let $(\bar{s},\bar{u}) \in H - H'$ and $(\bar{e},\bar{\imath}) \in \mathcal{E}(\bar{s},\bar{u})$. Suppose that Assumption CPD holds and that either*

(a) $O(\bar{s},\bar{u},\bar{e},\bar{\imath}) = \varnothing$, *or*

(b) $O(\bar{s},\bar{u},\bar{e},\bar{\imath}) \neq \varnothing$ *and* $F(x; e, i) = 1 - \exp(-v(e,i)x)$ *for each pair* $(e,i) \in O(\bar{s},\bar{u},\bar{e},\bar{\imath})$, *where $v(e,i)$ is a positive finite constant.*

Then

(i) *The random times $\{\zeta_{\theta(k)}: k \geq 0\}$ defined via (2.13) with $\bar{\mathcal{E}} = \{(\bar{e},\bar{\imath})\}$ form a sequence of regeneration points for the marking process $\{Z(t): t \geq 0\}$.*

(ii) $E_\mu[Y_1^r(|f|)] < \infty$ *for $r \geq 0$ and any real-valued function f defined on $H - H'$, where $Y_1(f)$ is defined by (2.16).*

(iii) $E_\mu[\tilde{Y}_1^r(|\tilde{f}|)] < \infty$ *for $r \geq 0$ and any polynomially dominated function $\tilde{f}$ defined on Υ, where $\tilde{Y}_1(\tilde{f})$ is defined by (2.17).*

If, in particular, the condition in (a) holds, then also

(iv) *The random indices $\{\theta(k): k \geq 0\}$ form a sequence of regeneration points for $\{(S_n, U_n, C_n): n \geq 0\}$.*

The assumption that $(\bar{s},\bar{u})$ is a timed marking can be relaxed, similarly to the way in which Theorem 2.31 in Chapter 6 is obtained from Theorem 2.24 in the same chapter.

CSPNs Satisfying Geometric Trials Criteria

We now refine Theorem 2.14 when a geometric trials recurrence criterion holds. For a fixed set of pairs $\bar{\mathcal{E}} \subseteq \mathcal{E}$, set $\beta(-1) = -1$ and

$$\beta(n) = \inf\left\{ k > \beta(n-1) \colon \mathcal{E}^*(S_k, U_k, C_k) = \bar{\mathcal{E}} \right\} \qquad (2.19)$$

for $n \geq 0$. For a marking $(\bar{s}, \bar{u}) \in H$ with $\bar{\mathcal{E}} \subseteq \mathcal{E}(\bar{s}, \bar{u})$, define $\{\, \theta(k) \colon k \geq 0 \,\}$ as in (2.13) to be the random indices of the successive marking changes at which the marking is $(\bar{s}, \bar{u})$ and the clocks corresponding to the (transition, color) pairs in $\bar{\mathcal{E}}$ run down to 0 simultaneously. Thus $\{\, \theta(k) \colon k \geq 0 \,\}$ is a random subsequence of $\{\, \beta(n) + 1 \colon n \geq 0 \,\}$. In the following, $Y_1(f)$ is defined as in (2.16) and, as before, $\mathcal{F}_n$ denotes the partial history of the underlying chain up to the nth marking change.

Theorem 2.20. *Let $(\bar{s}, \bar{u}) \in H$ and $\bar{\mathcal{E}} \subseteq \mathcal{E}(\bar{s}, \bar{u})$. Suppose that each random index $\beta(n)$ defined in (2.19) is a.s. finite. Let $\{\, \alpha(n) \colon n \geq 1 \,\}$ be an increasing sequence of random indices such that each $\alpha(n)$ is a stopping time with respect to $\{\, \mathcal{F}_k \colon k \geq 0 \,\}$ and $\beta(n-1) \leq \alpha(n) < \beta(n)$. Suppose that*

$$P_\mu\left\{ (S_{\beta(n)}, U_{\beta(n)}) = (\bar{s}, \bar{u}) \mid \mathcal{F}_{\alpha(n)} \right\} > \delta \ a.s.$$

for some $\delta > 0$ and all $n \geq 0$. Also suppose that for all $(\bar{e}, \bar{\imath}) \in \bar{\mathcal{E}}$ with $q(\bar{e}, \bar{\imath}; \bar{\mathcal{E}}) > 0$ either

(a) $O(\bar{s}, \bar{u}, \bar{e}, \bar{\imath}) = \varnothing$, or

(b) $O(\bar{s}, \bar{u}, \bar{e}, \bar{\imath}) \neq \varnothing$ and $F(x; e, i) = 1 - \exp(-v(e, i)x)$ for each pair $(e, i) \in O(\bar{s}, \bar{u}, \bar{e}, \bar{\imath})$, where $v(e, i)$ is a positive finite constant.

Then the random times $\{\, \zeta_{\theta(k)} \colon k \geq 0 \,\}$ defined via (2.13) form a sequence of regeneration points for the marking process $\{\, Z(t) \colon t \geq 0 \,\}$. Moreover, for any bounded real-valued function f defined on $H - H'$, the cycle sum $Y_1(|f|)$ has finite mean if

$$\liminf_{n \geq 0} E_\mu\left[\zeta_{\beta(n+1)+1} - \zeta_{\beta(n)+1} \right] < \infty$$

and finite rth moment $(r > 1)$ if

$$\liminf_{n \geq 0} E_\mu\left[(\zeta_{\beta(n+1)+1} - \zeta_{\beta(n)+1})^{r+\epsilon} \right] < \infty \qquad (2.21)$$

for some $\epsilon > 0$.

The following analog to Lemma 2.39 in Chapter 6 can be useful when verifying that (2.21) holds.

Lemma 2.22. *Let $(e_i, l) \in \mathcal{E} - \mathcal{E}'$ and β be an a.s. finite stopping time with respect to the sequence $\{\, \mathcal{F}_n \colon n \geq 0 \,\}$ of partial histories of the underlying chain. Suppose that the clock-setting distribution function $F(\,\cdot\,; e_i, l)$ is GNBU. Then $E_\mu[C^r_{\beta, i}(l)] < \infty$ for $r \geq 0$.*

As with ordinary SPNs, sometimes a discrete-time version of the condition in (2.21) is easier to verify than (2.21) itself.

Theorem 2.23. *Suppose that the conditions of Theorem 2.20 hold. Also suppose that the marking set H is finite. Then, for any real-valued function f defined on $H - H'$, the cycle sum $Y_1(|f|)$ has finite mean if each clock-setting distribution has finite mean and*

$$\sup_{n \geq 0} E_\mu[\beta(n+1) - \beta(n)] < \infty,$$

and $Y_1(|f|)$ has finite rth moment ($r > 1$) if each clock-setting distribution has finite rth moment and

$$\sup_{n \geq 0} E_\mu[(\beta(n+1) - \beta(n))^{r+\epsilon}] < \infty$$

for some $\epsilon > 0$.

We conclude this subsection by giving the discrete-time analog of Theorem 2.20.

Theorem 2.24. *Let $(\bar{s}, \bar{u}) \in H$ and $\bar{\mathcal{E}} \subseteq \mathcal{E}(\bar{s}, \bar{u})$. Suppose that each random index $\beta(n)$ defined in (2.19) is a.s. finite. Let $\{\alpha(n): n \geq 1\}$ be an increasing sequence of random indices such that each $\alpha(n)$ is a stopping time with respect to $\{\mathcal{F}_k: k \geq 0\}$ and $\beta(n-1) \leq \alpha(n) < \beta(n)$. Suppose that*

$$P_\mu\{(S_{\beta(n)}, U_{\beta(n)}) = (\bar{s}, \bar{u}) \mid \mathcal{F}_{\alpha(n)}\} > \delta \ a.s.$$

for some $\delta > 0$ and all $n \geq 0$. Also suppose that $O(\bar{s}, \bar{u}, \bar{e}, \bar{\imath}) = \varnothing$ for all $(\bar{e}, \bar{\imath}) \in \bar{\mathcal{E}}$ with $q(\bar{e}, \bar{\imath}; \bar{\mathcal{E}}) > 0$. Then the random indices $\{\theta(k): k \geq 0\}$ defined via (2.13) form a sequence of regeneration points for the underlying chain $\{(S_n, U_n, C_n): n \geq 0\}$. Moreover, for any bounded real-valued function $\tilde{f}$ defined on Υ, the cycle sum $\tilde{Y}_1(|\tilde{f}|)$ has finite mean if

$$\sup_{n \geq 0} E_\mu[\beta(n+1) - \beta(n)] < \infty$$

and finite rth moment ($r > 1$) if

$$\sup_{n \geq 0} E_\mu[(\beta(n+1) - \beta(n))^{r+\epsilon}] < \infty$$

for some $\epsilon > 0$.

9.2.3 CSPNs and STS Estimation Methods

In this section we provide SLLNs and FCLTs for the marking process and underlying chain of a CSPN. When such limit theorems hold, methods based on standardized time series—see Section 7.2—can be used to obtain point estimates and confidence intervals for time-average limits and functions of such limits.

Lemma 2.25 asserts that—in analogy to ordinary SPNs—the underlying chain for a CSPN is an od-regenerative process in discrete time under Assumption CPD, and a broad class of cycle sums have finite moments of all orders.

Lemma 2.25. *Suppose that Assumption CPD holds. Then there exists a sequence $\{\theta(k): k \geq 0\}$ of od-regeneration points for the underlying chain $\{(S_n, U_n, C_n): n \geq 0\}$. Moreover, the cycle sum*

$$\tilde{Y}_1(\tilde{f}) = \sum_{n=\theta(0)}^{\theta(1)-1} \tilde{f}(S_n, U_n, C_n) \qquad (2.26)$$

has finite moments of all orders for any polynomially dominated real-valued function $\tilde{f}$ defined on Υ.

It follows from Lemma 2.25 that the chain $\{(S_n, U_n, C_n): n \geq 0\}$ is positive Harris recurrent whenever Assumption CPD holds. Moreover, the desired SLLNs and FCLTs for the marking process and underlying chain can be obtained by applying the results in Section 7.2.1.

We first state an SLLN for the underlying chain. In the theorem an $\Re^l$-valued function $\tilde{f} = (\tilde{f}_1, \tilde{f}_2, \ldots, \tilde{f}_l)$ defined on Υ is said to be *polynomially dominated* if each $\tilde{f}_j$ is polynomially dominated in the sense of Definition 2.15. Given such a function together with a sequence $\{\theta(k): k \geq 0\}$ of od-regeneration points for the chain, set

$$\tilde{r}(\tilde{f}) = \frac{E_\mu[\tilde{Y}_1(\tilde{f})]}{E_\mu[\tilde{\tau}_1]}, \qquad (2.27)$$

where

$$\tilde{Y}_1(\tilde{f}) = \sum_{j=\theta(0)}^{\theta(1)-1} \tilde{f}(S_j, U_j, C_j)$$

and $\tilde{\tau}_1 = \theta(1) - \theta(0)$.

Theorem 2.28. *Suppose that Assumption CPD holds, so that there exists a sequence $\{\theta(k): k \geq 0\}$ of od-regeneration points for the underlying chain $\{(S_n, U_n, C_n): n \geq 0\}$. Then $\tilde{r}(|\tilde{f}|) < \infty$ and*

$$\lim_{n \to \infty} \frac{1}{n} \sum_{j=0}^{n-1} \tilde{f}(S_j, U_j, C_j) = \tilde{r}(\tilde{f}) \quad a.s.$$

for any polynomially dominated $\Re^l$-valued function $\tilde{f}$ defined on Σ, where $\tilde{r}(\tilde{f})$ is defined by (2.27).

We next give an FCLT for the underlying chain. Recall that $C^l[0,1]$ ($l \geq 1$) is the space of continuous $\Re^l$-valued functions on $[0,1]$. Whenever there exists a sequence $\{\theta(k): k \geq 0\}$ of od-regeneration points for the underlying chain and the quantity $\tilde{r}(\tilde{f})$ given by (2.27) is well defined and finite, we can define a sequence of $C^l[0,1]$-valued random functions $\tilde{R}_1(\tilde{f}), \tilde{R}_2(\tilde{f}), \dots$ by setting

$$\tilde{R}_n(\tilde{f})(t) = \frac{1}{\sqrt{n}} \int_0^{nt} \left(\tilde{f}(S_{\lfloor u \rfloor}, U_{\lfloor u \rfloor}, C_{\lfloor u \rfloor}) - \tilde{r}(\tilde{f}) \right) du$$

for $0 \leq t \leq 1$ and $n \geq 1$. In the following, denote by $\Rightarrow$ weak convergence on $C^l[0,1]$ and by $W^{(l)}$ a standard l-dimensional Brownian motion on $[0,1]$.

Theorem 2.29. *Suppose that Assumption CPD holds—so that there exists a sequence $\{\theta(k): k \geq 0\}$ of od-regeneration points for the underlying chain—and let $\tilde{f}$ be a polynomially dominated $\Re^l$-valued function defined on Υ. Then there exists an $l \times l$ matrix $Q(\tilde{f})$ such that $\tilde{R}_n(\tilde{f}) \Rightarrow Q(\tilde{f})W^{(l)}$ as $n \to \infty$ for any initial distribution μ.*

As usual,

$$\frac{1}{\sqrt{n}} \sum_{j=0}^{n-1} (\tilde{f}(S_j, U_j, C_j) - \tilde{r}(\tilde{f})) \Rightarrow \tilde{\sigma}(\tilde{f})N(0,1)$$

under the conditions of Theorem 2.29, where $\tilde{\sigma}(\tilde{f})$ is a nonnegative constant and $\Rightarrow$ denotes ordinary convergence in distribution. That is, the foregoing FCLT implies an ordinary CLT.

As with ordinary SPNs, both SLLNs and FCLTs for processes of the form $\{f(Z(t)): t \geq 0\}$ can be obtained from the corresponding results for the underlying chain. For a sequence $\{\theta(k): k \geq 0\}$ of od-regeneration points for the underlying chain, set

$$r(f) = \frac{E_\mu[\tilde{Y}_1(ft^*)]}{E_\mu[\tilde{Y}_1(t^*)]} . \tag{2.30}$$

for each $\Re^l$-valued function f defined on $H - H'$, where $(ft^*)(s,u,c) = f(s,u)t^*(s,u,c)$ for $(s,u,c) \in \Upsilon$ and $\tilde{Y}_1(\tilde{f})$ is defined in (2.26).

Theorem 2.31. *Suppose that Assumption CPD holds, so that there exists a sequence $\{\theta(k): k \geq 0\}$ of od-regeneration points for the underlying chain. Then $r(|f|) < \infty$ and*

$$\lim_{t \to \infty} \frac{1}{t} \int_0^t f(Z(u)) \, du = r(f) \quad a.s.$$

for any $\Re^l$-valued function f defined on $H - H'$, where $r(f)$ is defined by (2.30).

When there exists a sequence $\{\,\theta(k)\colon k \geq 0\,\}$ of od-regeneration points for the underlying chain and the quantity $r(f)$ given by (2.30) is well defined and finite, set

$$R_\nu(f)(t) = \frac{1}{\sqrt{\nu}} \int_0^{\nu t} \Big(f(Z(u)) - r(f)\Big)\, du$$

for $0 \leq t \leq 1$ and $\nu > 0$.

Theorem 2.32. *Suppose that Assumption CPD holds, and let f be an arbitrary $\Re^l$-valued function defined on $H - H'$. Then there exists an $l \times l$ matrix $Q(f)$ such that $R_\nu(f) \Rightarrow Q(f)W^{(l)}$ as $\nu \to \infty$ for any initial distribution μ.*

9.2.4   Consistent Estimation Methods

In this subsection we give conditions on the building blocks of a CSPN under which variable batch-means and spectral methods are valid. The development, which parallels that in Section 7.3 for ordinary SPNs, applies to other consistent estimation methods as well.

Consider a CSPN with an underlying chain $\{\,(S_n, U_n, C_n)\colon n \geq 0\,\}$ having state space Υ, together with a real-valued function $\tilde{f}$ defined on Υ, such that

$$\lim_{n \to \infty} \bar{r}(n; \tilde{f}) = \tilde{r}(\tilde{f}) \text{ a.s.}$$

for some finite constant $\tilde{r}(\tilde{f})$ and

$$\frac{\sqrt{n}\big(\bar{r}(n; \tilde{f}) - \tilde{r}(\tilde{f})\big)}{\tilde{\sigma}(\tilde{f})} \Rightarrow N(0, 1) \tag{2.33}$$

as $n \to \infty$ for some constant $\tilde{\sigma}(\tilde{f}) \in (0, \infty)$, where

$$\bar{r}(n; \tilde{f}) = \frac{1}{n} \sum_{j=0}^{n-1} \tilde{f}(S_j, U_j, C_j). \tag{2.34}$$

If we can find an estimator V_n that is consistent for $\tilde{\sigma}^2(\tilde{f})$ in (2.33), then the random interval

$$\left[\bar{r}(n; \tilde{f}) - \frac{z_p V_n^{1/2}}{\sqrt{n}}, \bar{r}(n; \tilde{f}) + \frac{z_p V_n^{1/2}}{\sqrt{n}}\right]$$

is an asymptotic $100p\%$ confidence interval for $\tilde{r}(\tilde{f})$, where z_p is the $(1+p)/2$ quantile of the standard normal distribution as before; cf. Section 7.3.

Aperiodicity and Harris Ergodicity

As with ordinary SPNs, the first step is to obtain conditions under which the underlying chain is Harris ergodic. A *d-cycle* of a CSPN is a finite collection $\{H_1, H_2, \ldots, H_d\}$ of disjoint subsets of H such that $(s', u') \in H_{i+1}$ whenever $(s, u) \in H_i$ and $(s, u) \to (s', u')$. (Take $H_{d+1} = H_1$.) The *period* of the CSPN is the largest d for which a d-cycle exists; the CSPN is called *aperiodic* if $d = 1$ and *periodic* if $d > 1$.

Theorem 2.35. *Let $\{(S_n, U_n, C_n): n \geq 0\}$ be the underlying chain of an aperiodic CSPN. If Assumption CPD holds, then $\{(S_n, U_n, C_n): n \geq 0\}$ is aperiodic.*

Corollary 2.36. *Let $\{(S_n, U_n, C_n): n \geq 0\}$ be the underlying chain of an aperiodic CSPN. If Assumption CPD holds, then $\{(S_n, U_n, C_n): n \geq 0\}$ is Harris ergodic.*

Consistent Estimation in Discrete Time

Let $\bar{r}(n; \tilde{f})$ be defined as in (2.34). As discussed previously, if Assumption CPD holds, then for any polynomially dominated function $\tilde{f}$ there exist constants $\tilde{r}(\tilde{f})$ and $\tilde{\sigma}(\tilde{f})$ such that $\lim_{n \to \infty} \bar{r}(n; \tilde{f}) = \tilde{r}(\tilde{f})$ a.s. and $\sqrt{n}(\bar{r}(n; \tilde{f}) - \tilde{r}(\tilde{f})) \Rightarrow \tilde{\sigma}(\tilde{f})N(0, 1)$ as $n \to \infty$. As with ordinary SPNs, we assume that $\tilde{\sigma}^2(\tilde{f}) > 0$, and consider *quadratic-form* estimators, that is, estimators of the form

$$V_n = V_n(\tilde{f}) = \sum_{i=0}^{n} \sum_{j=0}^{n} \tilde{f}(S_i, U_i, C_i)\tilde{f}(S_j, U_j, C_j)q_{i,j}^{(n)},$$

where each $q_{i,j}^{(n)}$ is a finite constant and $q_{i,j}^{(n)} = q_{j,i}^{(n)}$ for all i, j.

When Assumption CPD holds, there exists an invariant probability measure π for the underlying chain $\{(S_n, U_n, C_n): n \geq 0\}$. By applying general results on consistent variance estimation for stationary processes, it sometimes can be established that $V_n(\tilde{f}) \Rightarrow \tilde{\sigma}^2(\tilde{f})$ for a specified estimator $V_n(\tilde{f})$ when the initial distribution of the underlying chain is π. To this end, we have the following analogs of Propositions 3.10 and 3.12 in Chapter 7.

Proposition 2.37. *Let $\{(S_n, U_n, C_n): n \geq 0\}$ be the underlying chain of an aperiodic CSPN, and let $\tilde{f}$ be a polynomially dominated real-valued function defined on Υ. Suppose that Assumption CPD holds, so that there exists an invariant distribution π for the chain and $\{\tilde{f}(S_n, U_n, C_n): n \geq 0\}$ obeys a CLT with variance constant $\tilde{\sigma}^2(\tilde{f})$. Then $\tilde{\sigma}^2(\tilde{f})$ has the representation*

$$\tilde{\sigma}^2(\tilde{f}) = \lim_{n \to \infty} n \mathrm{Var}_\pi \left[\frac{1}{n} \sum_{j=0}^{n-1} \tilde{f}(S_j, U_j, C_j)\right].$$

Proposition 2.38. *Suppose that Assumption CPD holds for an aperiodic CSPN. Then there exist $\rho \in (0,1)$ and $c \in [0,\infty)$ such that*

$$\left| \mathrm{Cov}_\pi \left[\tilde{f}_1(S_0, U_0, C_0), \tilde{f}_2(S_k, U_k, C_k) \right] \right| \leq c\rho^k$$

for $k \geq 0$ and any polynomially dominated functions $\tilde{f}_1$ and $\tilde{f}_2$.

Theorem 2.40 below can be used to extend consistency results from the stationary to the nonstationary setting. Recall from Definition 3.13 in Chapter 7 that a quadratic-form estimator is *localized* if there exist $a_1 \in (0,\infty)$ and sequences $\{a_2(n) : n \geq 0\}$ and $\{m(n) : n \geq 0\}$ of non-negative constants with $a_2(n) \to 0$ and $m(n)/n \to 0$ such that

$$|q_{i,j}^{(n)}| \leq \begin{cases} a_1/n & \text{if } |i-j| \leq m(n); \\ a_2(n)/n & \text{if } |i-j| > m(n). \end{cases} \tag{2.39}$$

Theorem 2.40. *Let $\{(S_n, U_n, C_n) : n \geq 0\}$ be the underlying chain of an aperiodic CSPN, and let $\tilde{f}$ be a polynomially dominated real-valued function defined on Υ. Suppose that Assumption CPD holds, so that there exists an invariant distribution π for the chain and $\{\tilde{f}(S_n, U_n, C_n) : n \geq 0\}$ obeys a CLT with variance constant $\tilde{\sigma}^2(\tilde{f})$. If a localized quadratic-form estimator $V_n(\tilde{f})$ satisfies $V_n(\tilde{f}) \Rightarrow \tilde{\sigma}^2(\tilde{f})$ when the initial distribution equals π, then $V_n(\tilde{f}) \Rightarrow \tilde{\sigma}^2(\tilde{f})$ for any initial distribution.*

Applications to Batch-Means and Spectral Methods

For a CSPN with underlying chain $\{(S_n, U_n, C_n) : n \geq 0\}$ and a specified function $\tilde{f}$, consider the batch-means estimator based on b batches of length m:

$$V_n^{(B)} = \frac{m}{b-1} \sum_{j=1}^{b} \left(\bar{X}_n(j) - \bar{X}_n \right)^2, \tag{2.41}$$

where $n = bm$,

$$\bar{X}_n(j) = \frac{1}{m} \sum_{i=(j-1)m}^{jm-1} \tilde{f}(S_n, U_n, C_n), \tag{2.42}$$

and $\bar{X}_n = (1/b) \sum_{j=1}^{b} \bar{X}_n(j)$. Also consider the spectral estimator

$$V_n^{(S)} = \frac{1}{n} \sum_{h=-(m-1)}^{m-1} \lambda(h/m)\hat{R}_h, \tag{2.43}$$

where

$$\hat{R}_h = \frac{1}{n} \sum_{i=0}^{n-|h|-1} (Z_i - \bar{Z}_n)(Z_{i+|h|} - \bar{Z}_n), \tag{2.44}$$

with $Z_i = \tilde{f}(S_i, U_i, C_i)$ for $0 \leq i \leq n$ and $\bar{Z}_n = (1/n)\sum_{i=0}^{n-1} Z_i$. We assume throughout that the lag window λ belongs to the class Λ defined in Section 7.3.3. Arguing as in Section 7.3.3, we can establish the following two results.

Theorem 2.45. *Let* $\{(S_n, U_n, C_n): n \geq 0\}$ *be the underlying chain of an aperiodic* CSPN, *and let* $V_n^{(B)}$ *be given by* (2.41) *and* (2.42)*, where* $\tilde{f}$ *is a polynomially dominated real-valued function defined on* Υ. *Suppose that Assumption* CPD *holds, so that* $\{\tilde{f}(S_n, U_n, C_n): n \geq 0\}$ *obeys a* CLT *with variance constant* $\tilde{\sigma}^2(\tilde{f})$. *Also suppose that the batch size* $b = b(n)$ *and batch length* $m = m(n)$ *satisfy* $b(n) \rightarrow \infty$ *and* $m(n) \rightarrow \infty$ *as* $n \rightarrow \infty$. *Then* $V_n^{(B)} \Rightarrow \tilde{\sigma}^2(\tilde{f})$ *as* $n \rightarrow \infty$.

Theorem 2.46. *Let* $\{(S_n, U_n, C_n): n \geq 0\}$ *be the underlying chain of an aperiodic* CSPN. *Also let* $V_n^{(S)}$ *be defined by* (2.43) *and* (2.44) *with* $\lambda \in \Lambda$ *and* $Z_n = \tilde{f}(S_n, U_n, C_n)$, *where* $\tilde{f}$ *is a polynomially dominated real-valued function defined on* Υ. *Suppose that Assumption* CPD *holds, so that* $\{\tilde{f}(S_n, U_n, C_n): n \geq 0\}$ *obeys a* CLT *with variance constant* $\tilde{\sigma}^2(\tilde{f})$. *Also suppose that the spectral window length* $m = m(n)$ *satisfies* $m(n) \rightarrow \infty$ *and* $m^2(n)/n \rightarrow 0$. *Then* $V_n^{(S)} \Rightarrow \tilde{\sigma}^2(\tilde{f})$ *as* $n \rightarrow \infty$.

Functions of Time-Average Limits

Fix $l \geq 1$ and let $\tilde{f} = (\tilde{f}_1, \tilde{f}_2, \ldots, \tilde{f}_l)$ be a polynomially dominated $\Re^l$-valued function defined on Υ. If Assumption CPD holds, then there exists an l-vector $\tilde{r}(\tilde{f}) = (\tilde{r}(\tilde{f}_1), \tilde{r}(\tilde{f}_2), \ldots, \tilde{r}(\tilde{f}_l))$ such that $\bar{r}(n; \tilde{f}) \rightarrow \tilde{r}(\tilde{f})$ a.s., where

$$\bar{r}(n; \tilde{f}) = \frac{1}{n} \sum_{j=0}^{n-1} \tilde{f}(S_n, U_n, C_n).$$

We now consider estimation methods for quantities of the form

$$r = g(\tilde{r}(\tilde{f})) = g(\tilde{r}(\tilde{f}_1), \tilde{r}(\tilde{f}_2), \ldots, \tilde{r}(\tilde{f}_l)), \tag{2.47}$$

where $g: \Re^l \mapsto \Re$ is differentiable in a neighborhood of $\tilde{r}(\tilde{f})$. The quantity $\tilde{r}(\tilde{f})$ exists whenever Assumption CPD holds and the CSPN is aperiodic. As with ordinary SPNs, we define a point estimator of r by $r_n = g(\bar{r}(n; \tilde{f}))$. Letting the matrix $W = \|w_{s,t}\|$ given by

$$w_{s,t} = \lim_{n \rightarrow \infty} n\text{Cov}_\pi[\bar{r}(n; \tilde{f}_s), \bar{r}(n; \tilde{f}_t)], \tag{2.48}$$

we have the following result.

Theorem 2.49. *Let* $\{(S_n, U_n, C_n): n \geq 0\}$ *be the underlying chain of an aperiodic* CSPN, *and let* $\tilde{f} = (\tilde{f}_1, \tilde{f}_2, \ldots, \tilde{f}_l)$ *be polynomially dominated. Suppose that Assumption* CPD *holds, so that there exists an invariant measure* π *for the chain and* $\bar{r}(n; \tilde{f}) \to \tilde{r}(\tilde{f})$ *a.s. for some finite l-vector* $\tilde{r}(\tilde{f})$. *Then* $r_n \to r$ *a.s. and*

$$\sqrt{n}(r_n - r) \Rightarrow \sigma N(0,1)$$

as $n \to \infty$, *where*

$$\sigma^2 = \nabla g(\tilde{r}(\tilde{f}))^t \, W \, \nabla g(\tilde{r}(\tilde{f})).$$

We now consider the problem of consistently estimating σ^2.

As in Section 7.3.4, define an $l \times l$ matrix $W_n = \|V_n(s,t)\|$, where

$$V_n(s,t) = \sum_{i=1}^{n} \sum_{j=1}^{n} \tilde{f}_s(S_i, U_i, C_i) \tilde{f}_t(S_j, U_j, C_j) q_{i,j}^{(n)}$$

for $s, t \in \{1, 2, \ldots, n\}$ and the $q_{i,j}^{(n)}$ are coefficients of a quadratic-form estimator. For the batch-means and spectral estimators described previously, we can show that $W_n \Rightarrow W$ when the initial distribution is π and the conditions of Theorems 2.45 and 2.46 hold, respectively.

As for ordinary SPNs, the coupling argument used to establish Theorem 2.40 can be extended to obtain a multidimensional limit result. In the following theorem the matrix W_n is said to be a *localized* estimator of W if and only if each $q_{i,j}^{(n)}$ satisfies (2.39).

Theorem 2.50. *Let* $\{(S_n, U_n, C_n): n \geq 0\}$ *be the underlying chain of an aperiodic* CSPN, *and let* $\tilde{f} = (\tilde{f}_1, \tilde{f}_2, \ldots, \tilde{f}_l)$ *be a polynomially dominated* $\Re^l$*-valued function defined on* Υ. *Suppose that Assumption* CPD *holds, so that there exists an invariant distribution* π *for the chain and the process* $\{\tilde{f}(S_n, U_n, C_n): n \geq 0\}$ *obeys a* CLT *with covariance matrix* W. *If a localized estimator* W_n *satisfies* $W_n \Rightarrow W$ *when the initial distribution equals* π, *then* $W_n \Rightarrow W$ *for any initial distribution.*

The foregoing results can be combined to yield confidence intervals for $r = g(\tilde{r}(\tilde{f}))$. Suppose, for example, that W_n is the batch-means estimator of W and the conditions of Theorem 2.45 hold, or that W_n is a spectral estimator of W and the conditions of Theorem 2.46 hold. Set

$$\sigma_n^2 = \nabla g(\bar{r}(n; \tilde{f}))^t \, W_n \, \nabla g(\bar{r}(n; \tilde{f}))$$

for $n \geq 1$. Arguing as in Section 7.3.4, we find that $\sigma_n^2 \Rightarrow \sigma^2$, which implies that

$$\left[r_n - \frac{z_p \sigma_n}{\sqrt{n}}, r_n + \frac{z_p \sigma_n}{\sqrt{n}} \right]$$

is an asymptotic $100p\%$ confidence interval for r, where z_p is the $(1+p)/2$ quantile of the standard normal distribution.

Consistent Estimation in Continuous Time

Consider an aperiodic SPN and suppose that Assumption CPD holds. It follows that $\bar{r}(t; f) \to r(f)$ a.s. for some finite constant $r(f)$ and any real-valued function f defined on H; here

$$\bar{r}(t; f) = \frac{1}{t} \int_0^t f(Z(u))\, du.$$

As with ordinary SPNs, the foregoing methodology for functions of time-average limits in discrete time leads to confidence-interval procedures for $r(f)$. Fix f and recall that $(ft^*)(s, u, c) = f(s, u)t^*(s, u, c)$, where t^* is the holding-time function. The idea, as before, is to express $r(f)$ in the form (2.47) with $\tilde{f} = (ft^*, t^*)$ and $g(x, y) = x/y$.

Let $\|q_{i,j}^{(n)}\|$ be a set of coefficients such that, for any polynomially dominated functions $\tilde{f}_s$ and $\tilde{f}_t$ defined on Υ, the quadratic-form estimator

$$V_n(\tilde{f}_s, \tilde{f}_t) = \sum_{i=0}^n \sum_{j=0}^n \tilde{f}_s(S_i, U_i, C_i)\tilde{f}_t(S_j, U_j, C_j)q_{i,j}^{(n)}$$

is consistent for $w_{s,t}$, where $w_{s,t}$ is given by (2.48). Then

$$\left[\hat{r}_n - \frac{z_p\,\sigma_n}{\sqrt{n}}, \hat{r}_n + \frac{z_p\,\sigma_n}{\sqrt{n}}\right]$$

is an asymptotic $100p\%$ confidence interval for $r(f)$, where

$$\hat{r}_n = \frac{\bar{r}(n; ft^*)}{\bar{r}(n; t^*)}$$

and

$$\sigma_n^2 = \frac{1}{\bar{r}^2(n; t^*)}\left(V_n(1, 1) - 2\hat{r}_n V_n(1, 2) + \hat{r}_n^2 V_n(2, 2)\right).$$

Here $\bar{r}(n; \tilde{f})$ is defined as in (2.34),

$$V_n(1, 1) = \sum_{i=0}^n \sum_{j=0}^n (ft^*)(S_i, U_i, C_i)\,(ft^*)(S_j, U_j, C_j)\,q_{i,j}^{(n)},$$

$$V_n(1, 2) = \sum_{i=0}^n \sum_{j=0}^n (ft^*)(S_i, U_i, C_i)\,t^*(S_j, U_j, C_j)\,q_{i,j}^{(n)},$$

and

$$V_n(2, 2) = \sum_{i=0}^n \sum_{j=0}^n t^*(S_i, U_i, C_i)\,t^*(S_j, U_j, C_j)\,q_{i,j}^{(n)}.$$

9.2.5   Delays

Delays in CSPNs are specified similarly to delays in ordinary SPNs. A sequence $\{D_j: j \geq 0\}$ of delays in a CSPN is specified in terms of starts $\{A_j: j \geq 0\}$ and terminations $\{B_j: j \geq 0\}$ defined on the same probability space as the underlying chain $\{(S_n, U_n, C_n): n \geq 0\}$ (via the relation $D_j = B_j - A_j$). As always, we restrict attention to delays that start and terminate only at marking changes—thus we have $A_j = \zeta_{\alpha(j)}$ and $B_j = \zeta_{\beta(j)}$ for $j \geq 0$, where $\alpha(j)$ and $\beta(j)$ are a.s. finite random indices. We also focus on sequences for which the $\alpha(j)$'s are nondecreasing, so that delays are enumerated in start order.

Start Vectors

As in Chapter 8, a recursively generated sequence of start vectors provides the link between the starts and terminations of individual delay intervals. For CSPNs, the sequence $\{V_n: n \geq 0\}$ of start vectors is determined by the sample paths of the chain $\{(S_n, U_n, C_n): n \geq 0\}$. In particular, we assume that the current marking determines the length of the start vector and denote this length by $\psi(s, u)$ when the current marking is (s, u). The nth start vector V_n records the starts of delay intervals for all ongoing and newly started delays (of positive duration) at time ζ_n. Some components of V_n may be equal to -1; as before, lengths are never computed for delay intervals with negative starts. The initial start vector is a specified vector, denoted $v_0(S_0, U_0)$, that is determined by the initial marking (S_0, U_0) and has components equal to 0 or -1. Take $v_0(S_0, U_0)$ to be the empty vector $\varnothing$ when $\psi(S_0, U_0) = 0$.

Whenever the clocks corresponding to the (transition, color) pairs in a set $\mathcal{E}^*$ run down to 0 simultaneously and trigger a marking change from (s, u) to (s', u'), a new start vector is obtained from the current start vector by

1. Inserting the current time at zero or more positions specified by an index vector $i_\alpha(s', u'; s, u, \mathcal{E}^*)$

2. Deleting components at zero or more positions specified by an index vector $i_\beta(s', u'; s, u, \mathcal{E}^*)$

3. Permuting the components according to an index vector $i_\pi(s', u'; s, u, \mathcal{E}^*)$

Components are deleted one at a time in the order in which the indices appear in the vector $i_\beta(s', u'; s, u, \mathcal{E}^*)$. For each nonnegative component that is deleted, the length of a delay interval is computed by subtracting the deleted component from the current time.

The formal definitions of the sequences $\{V_n: n \geq 0\}$, $\{A_n: n \geq 0\}$, $\{B_n: n \geq 0\}$, and $\{D_n: n \geq 0\}$ are completely analogous to those in

Section 8.1.2. Denote by $V_{n,i}$ the ith component of the vector V_n for $1 \leq i \leq \psi(S_n, U_n)$, and set

$$K = \inf \{ n \geq 0 : V_{n,i} \neq -1 \text{ for } 0 \leq i \leq \psi(S_n, U_n) \}. \qquad (2.51)$$

When $\mathcal{E}^* = \{ (e^*, i^*) \}$ for some $(e^*, i^*) \in \mathcal{E}(s, u)$, we often write $i_\alpha(s', u')$; $s, u, e^*, i^*)$ for $i_\alpha(s', u'; s, u, \mathcal{E}^*)$, and so forth.

Regenerative Methods for Delays

We now consider methods for estimating general time-average limits of the form $\lim_{n \to \infty} (1/n) \sum_{j=0}^{n-1} f(D_j)$, where the sequence of delays $\{ D_j : j \geq 0 \}$ is determined from the marking changes of a CSPN by means of start vectors.

We first suppose that there exists a recurrent single state $(\bar{s}, \bar{u})$, so that $\mathcal{E}(\bar{s}, \bar{u}) = \{ (\bar{e}, \bar{\imath}) \}$ for some $(\bar{e}, \bar{\imath}) \in \mathcal{E}$ and $P_\mu \{ (S_n, U_n) = (\bar{s}, \bar{u}) \text{ i.o.} \} = 1$. Then the successive times at which the marking is $(\bar{s}, \bar{u})$ and transition $\bar{e}$ fires in color $\bar{\imath}$ form a sequence of regeneration points. More specifically, if we set $\theta(0) = 0$ and

$$\theta(k) = \inf \{ n > \theta(k-1) : (S_{n-1}, U_{n-1}) = (\bar{s}, \bar{u}) \}$$

for $k \geq 1$, then the random indices $\{ \theta(k) : k \geq 0 \}$ form a sequence of regeneration points for $\{ (S_n, U_n, C_n) : n \geq 0 \}$ and the random times $\{ \zeta_{\theta(k)} : k \geq 0 \}$ form a sequence of regeneration points for $\{ Z(t) : t \geq 0 \}$.

We assume throughout that the system behaves as if a regeneration occurs at time 0—cf. the discussion at the beginning of Section 8.2—and that the starts $\{ A_j : j \geq 0 \}$, the terminations $\{ B_j : j \geq 0 \}$, and the random index K defined by (2.51) satisfy

$$P_\mu \{ K < \infty \} = 1 \qquad (2.52)$$

$$P_\mu \{ A_j < \infty \} = P_\mu \{ B_j < \infty \} = 1, \qquad (2.53)$$

for $j \geq 0$ and

$$P_\mu \{ \lim_{j \to \infty} A_j = \infty \} = 1. \qquad (2.54)$$

As in Section 8.2.1, we can construct a sequence $\{ \breve{\gamma}(k) : k \geq 0 \}$ of random indices that decomposes sample paths of $\{ D_j : j \geq 0 \}$ into one-dependent stationary cycles. Start with the sequence $\{ \zeta_{\theta(k)} : k \geq 0 \}$ of regeneration points for the marking process and recursively construct a subsequence $\{ \zeta_{\breve{\theta}(k)} : k \geq 0 \}$. To do this, take $\breve{\theta}(0) = \theta(0) = 0$ and then, given $\breve{\theta}(k)$, wait until the first marking change $\breve{\nu}(k)$ at which all the ongoing delays at the $\breve{\theta}(k)$th marking change have terminated, and take as $\breve{\theta}(k+1)$ the smallest $\theta(l)$ such that $\theta(l) \geq \breve{\nu}(k)$. If there are no ongoing delays at the $\breve{\theta}(k)$th marking change, take as $\breve{\theta}(k+1)$ the smallest $\theta(l)$ such that $\theta(l) > \breve{\theta}(k)$. For $k = 0$, take as $\breve{\theta}(1)$ the smallest $\theta(l)$ such that $\theta(l) \geq K$. To complete the construction, set $\breve{\gamma}(0) = 0$ and

$$\breve{\gamma}(k) = \inf \{ j > \breve{\gamma}(k-1) : \alpha(j-1) < \breve{\theta}(m) \leq \alpha(j) \text{ for some } m \geq 0 \} \quad (2.55)$$

for $k \geq 1$. Denote by δ_k ($k \geq 1$) the number of delays that start during the interval $[\zeta_{\theta(k-1)}, \zeta_{\theta(k)})$ and set $\tau_k = \zeta_{\theta(k)} - \zeta_{\theta(k-1)}$.

Theorem 2.56. *Let $\{\, D_j \colon j \geq 0 \,\}$ be a sequence of delays determined from the underlying chain of a marking process using the method of start vectors. Suppose that there exists a recurrent single state $(\bar{s}, \bar{u})$ and that the conditions in (2.52)–(2.54) hold. Then*

 (i) *The random indices $\{\, \check{\gamma}(k) \colon k \geq 0 \,\}$ defined by (2.55) form a sequence of od-equilibrium points for $\{\, D_j \colon j \geq 0 \,\}$.*

 (ii) *The random indices $\{\, \check{\gamma}(k) \colon k \geq 0 \,\}$ also form a sequence of regeneration points for $\{\, D_j \colon j \geq 0 \,\}$, provided that there are no ongoing delays at the $\theta(k)$th marking change for $k \geq 0$.*

 (iii) *The cycle sum $\check{Y}_1(|f|) = \sum_{j=\check{\gamma}(0)}^{\check{\gamma}(1)-1} |f(D_j)|$ has finite rth moment for any real-valued function f that is polynomially dominated to degree b (where $r, b \geq 1$), provided that $E_\mu[\delta_1^{rp}] < \infty$ and $E_\mu[\tau_1^{rbq}] < \infty$ for nonnegative real numbers p and q with $p^{-1} + q^{-1} = 1$.*

Under the conditions of Theorem 2.56, both the extended regenerative method for delays (as in Section 8.2.2) and the multiple-runs method (as in Section 8.2.3) can be used to obtain strongly consistent point estimates and asymptotic confidence intervals.

Remark 2.57. Observe that there is no need for an analog of the condition in (2.6) of Chapter 8, because there is only one possible new marking whenever the current marking is the single state $(\bar{s}, \bar{u})$ and transition $\bar{e}$ fires in color $\bar{\imath}$.

Limiting Average Delays

Under appropriate conditions, the limiting average delay

$$r = \lim_{n \to \infty} \frac{1}{n} \sum_{j=0}^{n-1} D_j$$

exists a.s., and estimates for r can be obtained without measuring the lengths of individual delay intervals. Recall that $\psi(s, u)$ is the length of the start vector when the marking is s. Denote by $n_\alpha(s', u'; s, u, \mathcal{E}^*)$ the length of the vector $i_\alpha(s', u'; s, u, \mathcal{E}^*)$. Thus $n_\alpha(s', u'; s, u, \mathcal{E}^*)$ is the number of newly started delays whenever the clocks corresponding to the (transition, color) pairs in a set $\mathcal{E}^*$ run down to 0 simultaneously and trigger a marking change from (s, u) to (s', u'). When $\mathcal{E}^* = \{\, (e^*, i^*) \,\}$ for some $(e^*, i^*) \in \mathcal{E}(s, u)$, we often write $n_\alpha(s', u'; s, u, e^*, i^*)$ for $n_\alpha(s', u'; s, u, \mathcal{E}^*)$. As before, we assume that clocks for timed transitions never run down to

0 simultaneously. Set

$$Z_k = \int_{\zeta_{\theta(k-1)}}^{\zeta_{\theta(k)}} \psi\big(X(t), Y(t)\big)\, dt = \sum_{n=\theta(k-1)}^{\theta(k)-1} \psi(S_n, U_n) t^*(S_n, U_n, C_n),$$

$$\tau_k = \zeta_{\theta(k)} - \zeta_{\theta(k-1)},$$

and

$$\delta_k = \sum_{n=\theta(k-1)}^{\theta(k)-1} n_\alpha(S_n, U_n; S_{n-1}, U_{n-1}, E_{n-1}^*)$$

for $k \geq 1$. Since Z_k, τ_k, and δ_k are determined by $\{(S_n, U_n, C_n) : \theta(k-1) \leq n < \theta(k)\}$ for $k \geq 1$, it follows that the sequence $\{(Z_k, \tau_k, \delta_k) : k \geq 1\}$ consists of i.i.d. random vectors.

Theorem 2.58. *Suppose that $E_\mu[Z_1] < \infty$ and $E_\mu[\delta_1] < \infty$ and that (2.52)–(2.54) hold. Then*

$$\lim_{n\to\infty} \frac{1}{n} \sum_{j=0}^{n-1} D_j = \frac{E_\mu[Z_1]}{E_\mu[\delta_1]} \quad a.s..$$

It follows from the foregoing result that a version of the standard regenerative method can be used to obtain strongly consistent point estimates and asymptotic confidence intervals for the limiting average delay. This algorithm is almost identical to Algorithm 2.34 in Chapter 8.

STS Methods for Delays

We now consider STS methods for estimating time-average limits of a sequence of delays $\{D_j : j \geq 0\}$ determined from the underlying chain $\{(S_n, U_n, C_n) : n \geq 0\}$ of a CSPN using the method of start vectors. As with ordinary SPNs, we give conditions on the CSPN building blocks, start-vector mechanism, and function f under which the output process $\{f(D_j) : j \geq 0\}$ obeys an FCLT. It follows that STS methods such as the method of batch means can be used to obtain strongly consistent point estimates and asymptotic confidence intervals for time-average limits and functions of time-average limits.

Denote by $\mathcal{X}$ the set of all infinite-length sequences $(s^{(0)}, u^{(0)}, \mathcal{E}^{(0)}, s^{(1)}, u^{(1)}, \mathcal{E}^{(1)}, \ldots)$ such that $p(s^{(k+1)}, u^{(k+1)}; s^{(k)}, u^{(k)}, \mathcal{E}^{(k)}) > 0$ for $k \geq 0$. For an element $x = (s^{(0)}, u^{(0)}, \mathcal{E}^{(0)}, s^{(1)}, u^{(1)}, \mathcal{E}^{(1)}, \ldots) \in \mathcal{X}$, recursively define a sequence of vectors $v_0, v_1, \ldots$ by setting $v_0(x)$ equal to the vector of length $\psi(s^{(0)}, u^{(0)})$ whose components are all equal to -1, and then setting

$$v_n'(x) = \text{Ins}\big(v_{n-1}(x), i_\alpha(s^{(n)}, u^{(n)}; s^{(n-1)}, u^{(n-1)}, \mathcal{E}^{(n-1)}), 0\big),$$
$$v_n''(x) = \text{Del}\big(v_n'(x), i_\beta(s^{(n)}, u^{(n)}; s^{(n-1)}, u^{(n-1)}, \mathcal{E}^{(n-1)})\big),$$

and

$$v_n(x) = \text{Per}\big(v_n''(x), i_\pi(s^{(n)}, u^{(n)}; s^{(n-1)}, u^{(n-1)}, \mathcal{E}^{(n-1)})\big)$$

for $n \geq 1$. Denote by $\iota(x)$ the smallest integer n such that $v_n(x) = (0, 0, \dots, 0)$; if such an integer n does not exist, then set $\iota(x) = \infty$.

Definition 2.59. A start-vector mechanism for a specified CSPN is said to be *regular* if

(i) there exist $(s, u) \in H - H'$ and $(e^*, i^*) \in \mathcal{E}(s, u)$ such that $n_\alpha(s', u'; s, u, e^*, i^*) > 0$, where $(s', u') = g(s, u, e^*, i^*)$, and

(ii) there exists $x = (s^{(0)}, u^{(0)}, \mathcal{E}^{(0)}, s^{(1)}, u^{(1)}, E^{(1)}, \dots) \in \mathcal{X}$ such that $\iota(x) < \infty$.

Theorem 2.60. *Let $\{D_j : j \geq 0\}$ be a sequence of delays determined from the underlying chain of a marking process by means of a regular start-vector mechanism, and let f be a polynomially dominated $\Re^l$-valued function defined on $\Re_+$ ($l \geq 1$). Suppose that Assumption CPD holds. Then*

(i) *There exists a finite constant $r(f) \in \Re^l$ such that*

$$\lim_{n \to \infty} \frac{1}{n} \sum_{j=0}^{n-1} f(D_j) = r(f) \quad a.s..$$

(ii) *There exists an $l \times l$ matrix $Q(f)$ such that $R_n(f) \Rightarrow Q(f)W^{(l)}$ as $n \to \infty$ for any initial distribution μ, where $\Rightarrow$ denotes weak convergence on $C^l[0, 1]$, $W^{(l)}$ is a standard l-dimensional Brownian motion, and*

$$R_n(f)(t) = \frac{1}{\sqrt{n}} \int_0^{nt} \Big(f(D_{\lfloor u \rfloor}) - r(f)\Big) \, du \qquad (2.61)$$

for $0 \leq t \leq 1$ and $n \geq 0$.

The proof of Theorem 2.60 is similar to that of Corollary 3.9 in Chapter 8. The idea is to show that there exists a sequence of od-regeneration points that decompose the underlying chain $\{(S_n, U_n, C_n) : n \geq 0\}$ into o.d.s. cycles. Under Assumption CPD, and the regularity condition on the start-vector mechanism, the output process $\{f(D_j) : j \geq 0\}$ inherits the od-regenerative property. If, moreover, the function f is polynomially dominated, then the sum of the process over a cycle has finite moments of all orders. The conclusion of the theorem then follows from the limit theorems for od-regenerative processes given in Section 7.2.1.

Thus, when the start-vector mechanism is regular and Assumption CPD holds, the time-average limit $r(f)$ of the output process $\{f(D_j) : j \geq 0\}$ is well defined and finite for any polynomially dominated real-valued function f. Moreover, we can obtain point estimates and confidence intervals for $r(f)$ using STS methods. Fix f and set

$$\bar{Y}_n(t) = \frac{1}{n} \int_0^{nt} f(D_{\lfloor u \rfloor}) \, du$$

for $0 \le t \le 1$ and $n \ge 1$. Also set

$$\hat{r}_n = \bar{Y}_n(1) = \frac{1}{n} \sum_{j=0}^{n-1} f(D_j).$$

By Theorem 2.60(i), the point estimator $\hat{r}_n$ is strongly consistent for $r(f)$. To obtain asymptotic confidence intervals for $r(f)$, we proceed as in Section 8.3.2. Define the set Ξ of mappings from $C[0,1]$ to $\Re$ as in Section 7.2.2 and fix $\xi \in \Xi$. Let $\sigma(f)$ be a nonnegative constant such that $R_n(f) \Rightarrow \sigma(f)W$ as $n \to \infty$, where $R_n(f)$ is given by (2.61) and W is a standard one-dimensional Brownian motion. The existence of $\sigma(f)$ follows from Theorem 3.9(ii) with $l = 1$, and we assume throughout that $\sigma(f) > 0$. Finally, set $\xi_n = \xi(\bar{Y}_n)$. Arguing as in Section 7.2.2, we find that

$$[\hat{r}_n - \xi_n z_p, \hat{r}_n + \xi_n z_p]$$

is an asymptotic $100p\%$ confidence interval for $r(f)$, where $p \in (0,1)$ and z_p is a positive constant such that $P\{ -z_p \le W(1)/\xi(W) \le z_p \} = p$. As discussed previously, the foregoing estimation procedure reduces to the method of batch means for an appropriate choice of the mapping ξ. We can use a "jackknifed" extension of the batch-means method to estimate functions of time-average limits of a sequence of delays—the development is identical to that in Section 8.3.2. Also as discussed in Section 8.3.2, the task of measuring individual delays can be avoided when the performance measure of interest is the limiting average delay $\lim_{n \to \infty} (1/n) \sum_{j=0}^{n-1} D_j$.

9.3 Symmetric CSPNs

In this section we introduce the class of symmetric CSPNs and illustrate two ways in which symmetry can be exploited in the context of regenerative simulation.[1]

9.3.1 The Symmetry Conditions

Heuristically, a CSPN is symmetric if its building blocks are invariant under certain permutations of the colors. We formalize this notion as follows. For

---

[1]A third technique that involves "permuted regenerative estimators" is briefly discussed in the notes at the end of the chapter.

a permutation λ of the set U of colors and a subset $\tilde{U} \subseteq U$, write $\lambda(\tilde{U}) = \{\lambda(i): i \in U\}$. Also write $\mathcal{E}_\lambda^* = \{(e, \lambda(i)): (e, i) \in \mathcal{E}^*\}$ for $\mathcal{E}^* \subseteq \mathcal{E}$. For $(s, u) \in H$, denote by $\xi_\lambda(s, u)$ the marking (s, u') such that $u_j'(\lambda(i)) = u_j(i)$ for $i \in U$ and $1 \leq j \leq L$.

Definition 3.1. A set Λ of permutations of the set U is *complete* if

(i) $\xi_\lambda(s, u) \in H$ for $\lambda \in \Lambda$ and $(s, u) \in H$, and

(ii) for each pair of markings $(s, u), (s, u') \in H$, there exists a permutation $\lambda \in \Lambda$ such that $(s, u') = \xi_\lambda(s, u)$.

Definition 3.2. The *symmetry conditions* are said to hold for a CSPN if there exists a complete set Λ of permutations of the set U of colors such that

(i) $\lambda(U_D(d)) = U_D(d)$ and $\lambda(U_E(e)) = U_E(e)$,

(ii) $w^-(e, i, d, l) = w^-(e, \lambda(i), d, \lambda(l))$ and $w^+(e, i, d, l) = w^+(e, \lambda(i), d, \lambda(l))$,

(iii) $q(e', i'; \mathcal{E}^*) = q(e', \lambda(i'); \mathcal{E}_\lambda^*)$, and

(iv) $F(\,\cdot\,; e, i) = F(\,\cdot\,; e, \lambda(i))$

for $\lambda \in \Lambda$, $(e, i), (e', i') \in \mathcal{E}$, $(d, l) \in \mathcal{D}$, and $\mathcal{E}^* \subseteq \mathcal{E}$.

EXAMPLE 3.3 (Symmetric machine repair). Consider the system of Example 1.1, but now suppose that, at the start of each repair, each of the stopped machines is equally likely to be selected for repair. Also suppose that the repair times for the N machines are stochastically identical in that $P\{R_1 \leq x\} = P\{R_2 \leq x\} = \cdots = P\{R_N \leq x\}$ for $x \geq 0$. Similarly, suppose that the lifetimes for the N machines are stochastically identical. The CSPN specification for this system is as in Figure 9.1, but now the firing probabilities are given by

$$q(e_2, i; \mathcal{E}^*) = \frac{1}{|\mathcal{E}^*|}$$

for $\mathcal{E}^* \subseteq \{(e_2, 1), (e_2, 2), \ldots, (e_2, N)\}$ and $(e_2, i) \in \mathcal{E}^*$. The symmetry conditions hold for this CSPN, where Λ is the set of all permutations of $U = \{0, 1, \ldots, N\}$ such that $\lambda(0) = 0$.

EXAMPLE 3.4 (Cyclic queues with feedback). Consider the cyclic queues of Example 1.3, but now suppose that the jobs are stochastically identical: $F_{1,j} = F_{2,j} = \cdots = F_{N,j}$ for $j = 1, 2$. The CSPN specification for this system is as in Figure 9.3. The symmetry conditions hold for this CSPN, where Λ is the set of all permutations of $U = \{0, 1, \ldots, N\}$ such that $\lambda(0) = 0$.

EXAMPLE 3.5 (Symmetric token ring). Consider the token ring of Example 1.2, but now suppose that there exist a positive constant R and positive random variables A and L such that, for each port j,

- The time for the ring token to propagate to the next port is equal to R.

- The successive times from the end of transmission until the arrival of the next packet for transmission are i.i.d. as the random variable A.

- The successive times to transmit a packet are i.i.d. as the random variable L.

The CSPN specification for this system is as in Figure 9.2. The symmetry conditions hold for this CSPN, where Λ is the set of all cyclic permutations of $U = \{1, 2, \ldots, N\}$; that is, $\Lambda = \{\lambda_1, \lambda_2, \ldots, \lambda_N\}$, where

$$\lambda_j(i) = ((i + j - 1) \bmod N) + 1.$$

9.3.2 Exploiting Symmetry: Shorter Cycle Lengths

It can be difficult in practice to apply the standard regenerative method to a specified CSPN when the number of marking changes between successive regeneration points is large—see Section 7.1 for a discussion in the context of ordinary SPNs. Theorem 3.8 below can be used to alleviate this situation when the CSPN satisfies the symmetry conditions. The idea is to simulate the marking process in independent, nonidentically distributed blocks; the blocks are in general much shorter than the original regenerative cycles. A similar idea can be applied when estimating long-run averages for a sequence of delays. In this latter setting we can exploit symmetry in order to obtain od-equilibrium cycles having shorter lengths than the usual cycles.

Simulation of the Marking Process

Recall from Section 9.2.2 that a marking $(\bar{s}, \bar{u})$ is said to be a single state if $|\mathcal{E}(\bar{s}, \bar{u})| = 1$. Consider a CSPN that has a recurrent single state $(\bar{s}, \bar{u})$, and suppose that the net behaves as if the marking is $(\bar{s}, \bar{u})$ just before time 0. Set $\theta'(0) = 0$ and

$$\theta'(k) = \{n > \theta'(k-1) \colon (S_{n-1}, U_{n-1}) = (\bar{s}, \bar{u})\} \qquad (3.6)$$

for $k \geq 1$. It then follows from Theorem 2.14 that the random indices $\{\theta'(k) \colon k \geq 0\}$ form a sequence of regeneration points for the process $\{(S_n, U_n, C_n) \colon n \geq 0\}$. Moreover, the random times $\{\zeta_{\theta'(k)} \colon k \geq 0\}$ form a sequence of regeneration points for the process $\{Z(t) \colon t \geq 0\}$. Under appropriate regularity conditions, the standard regenerative method can therefore be used to obtain strongly consistent point estimates and asymptotic confidence intervals for time-average limits of the form

$$\tilde{r}(\tilde{f}) = \lim_{n\to\infty} \frac{1}{n} \sum_{j=0}^{n-1} \tilde{f}(S_j, U_j, C_j)$$

or

$$r(f) = \lim_{t\to\infty} \frac{1}{t} \int_0^t f(Z(t))\, dt,$$

where $\tilde{f}$ and f are real-valued functions defined on Υ and $H - H'$.

In the presence of symmetry, an alternative simulation method is available, based on Theorem 3.8 below. A vector of token counts $\bar{s}$ is a *single state* if $|\mathcal{E}(\bar{s}, u)| = 1$ for all u such that $(\bar{s}, u) \in H$—we require that $(\bar{s}, u) \in H$ for at least one u. Observe that if $(\bar{s}, \bar{u})$ is a single state (in the original sense) and the symmetry conditions hold, then $\bar{s}$ is a single state (in the modified sense). Denote by $\tilde{\xi}_\lambda(s, u, c)$ the state $(s, u', c') \in \Upsilon$ such that $(s, u') = \xi_\lambda(s, u)$ and $c'_j(\lambda(i)) = c_j(i)$ for $i \in U$ and $1 \le j \le M$. A real-valued function $\tilde{f}$ defined on Υ is said to be *symmetric* with respect to a complete set Λ of permutations if $\tilde{f}(s, u, c) = \tilde{f}(\tilde{\xi}_\lambda(s, u, c))$ for $(s, u, c) \in \Upsilon$ and $\lambda \in \Lambda$. Similarly, a real-valued function f defined on $H - H'$ is said to be *symmetric* with respect to Λ if $f(s, u) = f(\xi_\lambda(s, u))$ for $(s, u) \in H - H'$ and $\lambda \in \Lambda$.

As before, consider a CSPN that behaves as if the marking just before time 0 is of the form $(\bar{s}, \bar{u})$, where $(\bar{s}, \bar{u})$ is a recurrent single state. If the symmetry conditions hold, then $\bar{s}$ also is a recurrent single state. Set $\theta(0) = 0$ and

$$\theta(k) = \{\, n > \theta(k-1)\colon S_{n-1} = \bar{s}\,\} \tag{3.7}$$

for $k \ge 1$.

Theorem 3.8. *Suppose that there exists a single state $(\bar{s}, \bar{u})$ such that $P_\mu\{\,(S_n, U_n) = (\bar{s}, \bar{u})\ i.o.\,\} = 1$. Also suppose that the symmetry conditions hold with permutation set Λ. Then*

(i) *the random indices $\{\,\theta(k)\colon k \ge 0\,\}$ form a sequence of regeneration points for the process $\{\,\tilde{f}(S_n, U_n, C_n)\colon n \ge 0\,\}$, where $\tilde{f}$ is any real-valued function defined on Υ that is symmetric with respect to Λ, and*

(ii) *the random times $\{\,\zeta_{\theta(k)}\colon k \ge 0\,\}$ form a sequence of regeneration points for the process $\{\,f(Z(t))\colon t \ge 0\,\}$, where f is any real-valued function defined on $H - H'$ that is symmetric with respect to Λ.*

PROOF. Fix a symmetric function $\tilde{f}\colon \Upsilon \to \Re$. An argument similar to the proof of Theorem 2.2 in Chapter 6 shows that the random indices $\{\,\theta(k)\colon k \ge 0\,\}$ decompose sample paths of the chain $\{\,(S_n, U_n, C_n)\colon n \ge$

$0\}$, and hence of the output process $\{\tilde{f}(S_n, U_n, C_n): n \geq 0\}$, into independent cycles. It remains only to show that the cycles of the output process are identically distributed. To this end, set $\mu_0(\cdot) = P((\bar{s}, \bar{u}, \bar{c}), \cdot)$, where P is the transition kernel for the underlying chain and $\bar{c}$ is an arbitrary but fixed clock-reading vector such that $(\bar{s}, \bar{u}, \bar{c}) \in \Upsilon$—the probability measure μ_0 is well defined because $(\bar{s}, \bar{u})$ is a single state. Next, consider an arbitrary element $u \in U^L$ such that $(\bar{s}, u) \in H$, and set $\mu(\cdot) = P((\bar{s}, u, c), \cdot)$ for an arbitrary but fixed clock-reading vector c. It suffices to show that

$$P_{\mu_0}\left\{ \left(\theta(1), \tilde{f}(S_0, U_0, C_0), \ldots, \tilde{f}(S_{\theta(1)}, U_{\theta(1)}, C_{\theta(1)})\right) \in A \right\}$$
$$= P_\mu\left\{ \left(\theta(1), \tilde{f}(S_0, U_0, C_0), \ldots, \tilde{f}(S_{\theta(1)}, U_{\theta(1)}, C_{\theta(1)})\right) \in A \right\}$$

for $A \subseteq \bigcup_{k=1}^{\infty} \{k\} \times \Re^k$. Let $\lambda \in \Lambda$ be the unique permutation such that $\xi_\lambda(\bar{s}, u) = (\bar{s}, \bar{u})$—such a permutation exists because Λ is complete. It follows from the symmetry conditions that $\mu_0(A) = \mu(\tilde{\xi}_\lambda^{-1} A)$ and

$$P(\tilde{\xi}_\lambda(\bar{s}, \tilde{u}, \tilde{c}), A) = P((\bar{s}, \tilde{u}, \tilde{c}), \tilde{\xi}_\lambda^{-1} A)$$

for all $A \subseteq \Upsilon$ and $(\bar{s}, \tilde{u}, \tilde{c}) \in \Upsilon$, where $\tilde{\xi}_\lambda^{-1} A = \{(s, u, c): \tilde{\xi}_\lambda(s, u, c) \in A\}$. An argument similar to the proof of Theorem 2.10 in Chapter 4 then shows that $\{(S_n, U_n, C_n): n \geq 0\}$ and $\{\tilde{\xi}_\lambda(S_n, U_n, C_n): n \geq 0\}$ have the same finite-dimensional distributions when the initial distributions of the processes are μ_0 and μ, respectively. Since each $\theta(k)$ is a stopping time with respect to the underlying chain and is invariant under the permutation λ, it follows that the cycles $\left(\theta(1), (S_0, U_0, C_0), \ldots, (S_{\theta(1)}, U_{\theta(1)}, C_{\theta(1)})\right)$ and $\left(\theta(1), \tilde{\xi}_\lambda(S_0, U_0, C_0), \ldots, \tilde{\xi}_\lambda(S_{\theta(1)}, U_{\theta(1)}, C_{\theta(1)})\right)$ are identically distributed under respective initial distributions μ_0 and μ. The first assertion of the theorem then follows from the symmetry of $\tilde{f}$. The second assertion follows from the first by considering the symmetric function $\tilde{f}(s, u, c) = f(s, u)t^*(s, u, c)$. □

Under the conditions of Theorem 3.8, we can use the standard regenerative method—but with regeneration points replaced by the random indices defined by (3.7)—to obtain point estimates and confidence intervals for time-average limits $\tilde{r}(\tilde{f})$ and $r(f)$ as given previously. The key point is that the random indices $\{\theta(k): k \geq 0\}$ defined by (3.7) are typically much more frequent than the original indices $\{\theta'(k): k \geq 0\}$ defined by (3.6). Moreover, a.s.-finiteness can often be more easily established for the former random indices. We emphasize that the random indices $\{\theta(k): k \geq 0\}$ do not form a sequence of regeneration points for the chain $\{(S_n, U_n, C_n): n \geq 0\}$, and the random times $\{\zeta_{\theta(k)}: k \geq 0\}$ do not form a sequence of regeneration points for the process $\{Z(t): t \geq 0\}$. The random indices $\{\theta(k): k \geq 0\}$ do, however, decompose the sample paths of $\{(S_n, U_n, C_n): n \geq 0\}$ into independent, nonidentically distributed blocks, and similarly for the random times $\{\zeta_{\theta(k)}: k \geq 0\}$.

EXAMPLE 3.9 (Symmetric token ring). For the CSPN of Example 3.5, suppose that we wish to estimate the long-run average utilization—that is, we wish to estimate the time-average limit $r(f)$ with $f(s, u) = s_3$. Take $\bar{s} = (0, N, 0, 1, 0)$ and let $\bar{u}$ be the unique element of $U^{|\bar{s}|}$ such that $(\bar{s}, \bar{u}) \in H$ and $u_4(N) = 1$. Thus, whenever the marking is $(\bar{s}, \bar{u})$, the ring token is propagating to port 1 and each port has a packet awaiting transmission. This marking is a recurrent single state and, under appropriate moment conditions, the standard regenerative method can be used to obtain point estimates and confidence intervals for $r(f)$. The regeneration points $\{\zeta_{\theta'(k)} : k \geq 0\}$ correspond to the successive times at which port 1 observes the ring token with each port having a packet awaiting transmission. Observe that the function f is symmetric, so that, by Theorem 3.8, we can also estimate $r(f)$ based on simulation of the marking process in independent, nonidentically distributed blocks. The random times $\{\zeta_{\theta(k)} : k \geq 0\}$ that demarcate these blocks correspond to the successive times at which *some* port observes the ring token with each port having a packet awaiting transmission. These random times are about N times as frequent as the original regeneration points.

Remark 3.10. The foregoing results can be extended to more general types of regenerative structure as in Theorem 2.14, for example.

Simulation of Delays

Time-average limits for a sequence of delays in a symmetric CSPN can also be estimated by decomposing the sample paths of the underlying chain into independent, nonidentically distributed blocks. This decomposition leads to od-equilibrium points for the output process $\{f(D_j): j \geq 0\}$ that are more frequent than the usual od-equilibrium points. The idea is to impose symmetry conditions not only on the CSPN building blocks, but also on the associated start-vector mechanism. In the following definition we set $\mathcal{E}_\lambda^* = \{(e, \lambda(i)): (e, i) \in \mathcal{E}^*\}$ for $\mathcal{E}^* \subseteq \mathcal{E}$ as before.

Definition 3.11. The *extended symmetry conditions* are said to hold for a CSPN if there exists a complete set Λ of permutations of the set U of colors such that the symmetry conditions of Definition 3.2 hold and, moreover,

(i) $\psi(s, u) = \psi(\xi_\lambda(s, u))$,

(ii) $i_\alpha(s', u'; s, u, \mathcal{E}^*) = i_\alpha(\xi_\lambda(s', u'); \xi_\lambda(s, u), \mathcal{E}_\lambda^*)$,

(iii) $i_\beta(s', u'; s, u, \mathcal{E}^*) = i_\beta(\xi_\lambda(s', u'); \xi_\lambda(s, u), \mathcal{E}_\lambda^*)$, and

(iv) $i_\pi(s', u'; s, u, \mathcal{E}^*) = i_\pi(\xi_\lambda(s', u'); \xi_\lambda(s, u), \mathcal{E}_\lambda^*)$

for $\lambda \in \Lambda$, $\mathcal{E}^* \subseteq \mathcal{E}(s, u)$, and $(s, u), (s', u') \in H$.

We consider a sequence of delays $\{D_j : j \geq 0\}$ defined for a specified CSPN using the method of start vectors and satisfying the regularity conditions in (2.52)–(2.54). Suppose that there exists a recurrent single state $(\bar{s}, \bar{u})$, and define a sequence of regeneration points $\{\theta'(k): k \geq 0\}$ for the marking process as in (3.6). As in Section 9.2.5, define a subsequence $\{\check{\theta}'(k): k \geq 0\}$ of the regeneration points so that a delay that starts in one of the resulting longer cycles terminates before the end of the next such cycle. Then define a sequence of random indices $\{\tilde{\gamma}'(k): k \geq 0\}$ as in (2.55): $\tilde{\gamma}'(0) = 0$ and

$$\tilde{\gamma}'(k) = \inf\big\{ j > \tilde{\gamma}'(k-1): \alpha(j-1) < \check{\theta}'(m) \leq \alpha(j) \text{ for some } m \geq 0 \big\}.$$

It follows from Theorem 2.56 that the random indices $\{\tilde{\gamma}'(k): k \geq 0\}$ form a sequence of od-equilibrium points for the process $\{D_j: j \geq 0\}$ and, moreover, form a sequence of regeneration points for $\{D_j: j \geq 0\}$ if there are no ongoing delays at the $\theta'(k)$th marking change for $k \geq 0$. We can then use the extended regenerative method for delays or the multiple-runs method to obtain strongly consistent point estimates and asymptotic confidence intervals for time-average limits.

When the CSPN and start-vector mechanism are symmetric, we can apply the foregoing estimation techniques using od-equilibrium points $\{\tilde{\gamma}(k): k \geq 0\}$ that are more frequent than the points $\{\tilde{\gamma}'(k): k \geq 0\}$. The idea is to define a sequence of blocking points $\{\theta(k): k \geq 0\}$ for the underlying chain as in (3.7). Then, in the usual manner, define a subsequence $\{\check{\theta}(k): k \geq 0\}$ of the blocking points so that a delay that starts in one of the cycles demarcated by the points $\{\zeta_{\check{\theta}(k)}: k \geq 0\}$ always terminates before the end of the next such cycle. Finally, define a sequence of random indices $\{\tilde{\gamma}(k): k \geq 0\}$ as in (2.55): $\tilde{\gamma}(0) = 0$ and

$$\tilde{\gamma}(k) = \inf\big\{ j > \tilde{\gamma}(k-1): \alpha(j-1) < \check{\theta}(m) \leq \alpha(j) \text{ for some } m \geq 0 \big\} \quad (3.12)$$

for $k \geq 1$. This procedure is justified by the following result.

Theorem 3.13. *Let $\{D_j: j \geq 0\}$ be a sequence of delays determined from the underlying chain of a marking process using the method of start vectors. Suppose that there exists a recurrent single state $\bar{s}$ and that the conditions in (2.52)–(2.54) hold. Also suppose that the extended symmetry conditions hold with permutation set Λ. Then*

    (i) *the random indices $\{\tilde{\gamma}(k): k \geq 0\}$ defined by (3.12) form a sequence of od-equilibrium points for $\{D_j: j \geq 0\}$, and*

    (ii) *the random indices $\{\tilde{\gamma}(k): k \geq 0\}$ also form a sequence of regeneration points for $\{D_j: j \geq 0\}$, provided that there are no ongoing delays at the $\theta(k)$th marking change for $k \geq 0$.*

The proof of Theorem 3.13 is similar to the proof of Theorem 3.8. The idea is to mimic the proof of Theorem 2.8 in Chapter 8 to show that the random indices $\{\hat\gamma(k)\colon k \geq 0\}$ decompose sample paths $\{D_j\colon j \geq 0\}$ into one-dependent cycles. We then use an argument as in the proof of Theorem 3.8 to show that the cycles are identically distributed under the extended symmetry conditions.

EXAMPLE 3.14 (Symmetric token ring). For the CSPN of Example 3.5, consider the delay intervals from whenever a packet arrives at a port for transmission until the end of transmission of the packet. Suppose that we wish to estimate time-average limits of the sequence of delays for all ports. The method of start vectors can be used to specify and measure individual delays in the CSPN of Figure 9.2. The start vector V_n records the arrival time for each packet awaiting or under transmission at time ζ_n. A start corresponding to a packet at port i appears to the left of a start corresponding to a packet at port j if and only if, at time ζ_n, the next observation of the ring token by port i will occur before the next observation by port j. It follows that if a transmission by port i is underway at time ζ_n, then the start corresponding to the packet at port i appears as the rightmost component of the start vector V_n.

To formally specify the start-vector mechanism, we need to introduce some notation. First, let $d(i,j)$ be the clockwise "distance" from port i to port j:

$$d(i,j) = \begin{cases} j - i & \text{if } i \leq j; \\ N - (i - j) & \text{if } i > j. \end{cases}$$

Next, denote by $n(s,u)$ the next port to observe the ring token when the marking is (s,u):

$$n(s,u) = \begin{cases} i \text{ such that } u_3(i-1) = 1 & \text{if } s_3 = 1; \\ i \text{ such that } u_4(i-1) = 1 & \text{if } s_4 = 1. \end{cases}$$

Also let $I(s,u) = u_2\big(n(s,u)\big)$ be the indicator variable that equals 1 if the next port to observe the ring token has a packet awaiting transmission and equals 0 otherwise. Finally, denote by $p(j) = p(j; s, u)$ the index of the port that corresponds to the jth component of the start vector, and set

$$j(s, u, i^*) = \max\big\{ j\colon d\big(n(s,u), p(j)\big) < d\big(n(s,u), i^*\big) \big\},$$

where we take $j(s, u, i^*) = 0$ if the maximization is over an empty set. In terms of this notation, set $\psi(s,u) = s_2 + s_3$,

$$i_\alpha(s', u'; s, u, e^*, i^*) = \begin{cases} \big(j(s, u, i^*)\big) & \text{if } e^* = e_1; \\ \varnothing & \text{otherwise,} \end{cases}$$

$$i_\beta(s', u'; s, u, e^*, i^*) = \begin{cases} \big(\psi(s, u)\big) & \text{if } e^* = e_2; \\ \varnothing & \text{otherwise,} \end{cases}$$

and

$$i_\pi(s', u'; s, u, e^*, i^*) = \begin{cases} (2, 3, \ldots, \psi(s, u), 1) & \text{if } e^* = e_3 \text{ and } I(s, u) = 1; \\ \varnothing & \text{otherwise} \end{cases}$$

for $(s, u), (s', u') \in H$ and $(e^*, i^*) \in \mathcal{E}(s, u)$.

This CSPN satisfies the extended symmetry conditions, so that a sequence of od-equilibrium points $\{ \tilde{\gamma}(k) \colon k \geq 0 \}$ for the process $\{ D_j \colon j \geq 0 \}$ can be constructed based on the random indices $\{ \theta(k) \colon k \geq 0 \}$ defined in Example 3.9. Indeed, the random indices $\{ \tilde{\gamma}(k) \colon k \geq 0 \}$ form a sequence of regeneration points for $\{ D_j \colon j \geq 0 \}$ because there are no ongoing delays at any point $\zeta_{\theta(k)}$. The regenerative cycles are shorter by roughly a factor of N than the "naive" cycles based on the random indices $\{ \theta'(k) \colon k \geq 0 \}$ defined in Example 3.9.

Observe that the start-vector mechanism given here differs from that given in Example 1.7 in Chapter 8. If we were to use the latter start-vector mechanism (adapted to the CSPN setting), the extended symmetry conditions would not hold, because, for example,

$$i_\alpha(s', u'; s, u, \mathcal{E}^*) \neq i_\alpha(\xi_\lambda(s', u'); \xi_\lambda(s, u), \mathcal{E}_\lambda^*).$$

9.3.3 Exploiting Symmetry: Increased Efficiency

Suppose that, for the token ring model of Example 3.14, we wish to estimate the long-run average delay between the arrival of a packet at port 1 for transmission and the end of transmission of the packet. We can observe the successive delays for port 1 and use the regenerative method to obtain point estimates and confidence intervals. If the CSPN and start-vector mechanism are symmetric, however, it is intuitively plausible that a valid estimate can also be obtained from observation of the combined delays for all the ports. Indeed, one might expect the resulting estimation procedure to be statistically more efficient than the first approach, because more delays are observed for a fixed simulation run length. In this subsection we give conditions under which symmetry in a CSPN can be exploited in this manner to yield more efficient estimates.

Associating Colors with Delays

We associate a color with each simulated delay by slightly modifying the previously described start-vector mechanism for CSPNs. Denote by $\tilde{U}$ ($\subseteq U$) the set of possible colors for the delays. Each component of the start vector is no longer simply a start ζ, but a pair (ζ, i), where $i \in \tilde{U}$. The color i is assigned when the corresponding delay starts and remains fixed until the delay terminates; we denote by $\mathcal{C}_j$ the color associated with delay D_j. To effect this modification, we redefine $i_\alpha(s', u'; s, u, \mathcal{E}^*)$ to be a vector of (index, color) pairs, that is, a vector of elements of $\{0, 1, \ldots, \psi(s, u)\} \times \tilde{U}$. For example, suppose that the clocks associated with the (transition, color) pairs in set $\mathcal{E}^*$ simultaneously run down to 0 and trigger a marking change from (s, u) to (s', u') and that this is the nth marking change. Also suppose that $\zeta_n = 5.3$,

$$V_{n-1} = ((3.2, 6), (1.2, 4), (4.8, 2)),$$
$$i_\alpha(s', u'; s, u, \mathcal{E}^*) = ((0, 7), (2, 9)),$$
$$i_\beta(s', u'; s, u, \mathcal{E}^*) = (5),$$

and $i_\pi(s', u'; s, u, \mathcal{E}^*) = \varnothing$. Finally, suppose that no delays have terminated so far. Then, using notation as in Section 8.1.2, we have

$$V_n' = ((5.3, 7), (3.2, 6), (1.2, 4), (5.3, 9), (4.8, 2)),$$
$$V_n'' = ((5.3, 7), (3.2, 6), (1.2, 4), (5.3, 9)),$$

and $V_n = V_n''$; the latter equality follows because the start-vector components are not permuted. This marking change corresponds to the termination B_2—recall that we enumerate delays $D_0, D_1, \ldots$ in start order—so that $D_2 = 5.3 - 4.8 = 0.5$, and the associated color is $C_2 = 2$.

We now define a strengthened version of the extended symmetry conditions. For a vector $i_\alpha = ((i_1, j_1), (i_2, j_2), \ldots, (i_m, j_m))$, and a permutation λ of the colors in U, set

$$i_\alpha^\lambda = \Big((i_1, \lambda(j_1)), (i_2, \lambda(j_2)), \ldots, (i_m, \lambda(j_m))\Big).$$

Definition 3.15. The *expanded symmetry conditions* are said to hold for a CSPN if there exists a complete set Λ of permutations of the set U of colors such that (a) for each $i, j \in \tilde{U}$, there exists $\lambda \in \Lambda$ such that $\lambda(i) = j$ and (b) the extended symmetry conditions in Definition 3.11 hold with the condition in (ii) modified to read

(ii′) $i_\alpha^\lambda(s', u'; s, u, \mathcal{E}^*) = i_\alpha\big(\xi_\lambda(s', u'); \xi_\lambda(s, u), \mathcal{E}_\lambda^*\big).$

An Estimation Procedure That Ignores Symmetry

Consider a sequence $\{(D_j, C_j) : j \geq 0\}$ defined for a specified CSPN using the method of start vectors and satisfying the regularity conditions in (2.52)–(2.54). For ease of exposition, we assume that the color set $U = \{1, 2, \ldots, N\}$ is enumerated such that the set of colors $\tilde{U}$ associated with the delays is $\tilde{U} = \{1, 2, \ldots, N_0\}$ for some $N_0 \leq N$. For $1 \leq q \leq N_0$, let $D_1^q, D_2^q, \ldots$ be an enumeration, in start order, of those delays in the sequence $\{D_j : j \geq 0\}$ that have an associated color equal to q. Suppose that there are infinitely many delays of each color and that we are interested in

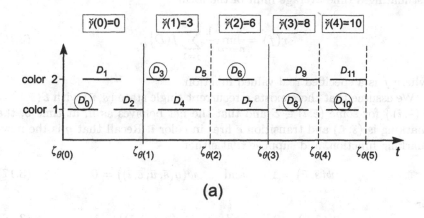

(a)

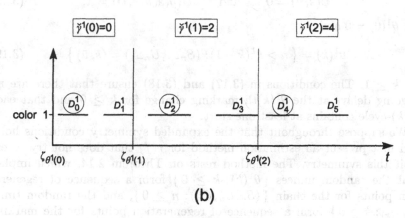

(b)

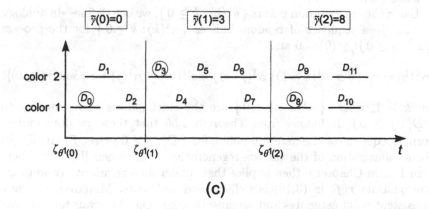

(c)

Figure 9.5. Cycles for delays in a CSPN (two colors).

estimating a time-average limit of the form

$$r(f) = \lim_{n \to \infty} \frac{1}{n} \sum_{j=1}^{n-1} f(D_j^1), \tag{3.16}$$

where f is a specified real-valued function.

We assume that there exists a recurrent single state $(\bar{s}, \bar{u})$ with $\mathcal{E}(\bar{s}, \bar{u}) = \{(\bar{e}, \bar{\imath})\}$ for some $(\bar{e}, \bar{\imath}) \in \mathcal{E}$ and that the net behaves as if, at time 0, the marking is $(\bar{s}, \bar{u})$ and transition $\bar{e}$ fires in color $\bar{\imath}$. Recall that g is the new-marking function, and suppose that either

$$\psi(\bar{s}, \bar{u}) = 1 \quad \text{and} \quad \psi\big(g(\bar{s}, \bar{u}, \bar{e}, \bar{\imath})\big) = 0 \tag{3.17}$$

or

$$\psi(\bar{s}, \bar{u}) = 0 \quad \text{and} \quad \psi\big(g(\bar{s}, \bar{u}, \bar{e}, \bar{\imath})\big) = 1. \tag{3.18}$$

Set $\theta^1(0) = 0$ and

$$\theta^1(k) = \{ n > \theta^1(k-1) \colon (S_{n-1}, U_{n-1}) = (\bar{s}, \bar{u}) \} \tag{3.19}$$

for $k \geq 1$. The conditions in (3.17) and (3.18) ensure that there are no ongoing delays at the $\theta^1(k)$th marking change for $k \geq 0$ and that each $\theta^1(k)$-cycle contains at least one start.

We suppose throughout that the expanded symmetry conditions hold, and first present an estimation method for $r(f)$ that does not try to exploit this symmetry. The method rests on Theorem 2.14, which implies that the random indices $\{\theta^1(k) \colon k \geq 0\}$ form a sequence of regeneration points for the chain $\{(S_n, U_n, C_n) \colon n \geq 0\}$, and the random times $\{\zeta_{\theta^1(k)} \colon k \geq 0\}$ form a sequence of regeneration points for the marking process $\{Z(t) \colon t \geq 0\}$.

Using the regeneration points $\{\theta^1(k) \colon k \geq 0\}$, we can define—in analogy to (2.55)—a sequence of random indices $\{\gamma^1(k) \colon k \geq 0\}$ for the process $\{D_j^1 \colon j \geq 0\}$: $\gamma^1(0) = 0$ and

$$\gamma^1(k) = \inf\{ j > \gamma^1(k-1) \colon \alpha^1(j-1) < \theta^1(m) \leq \alpha^1(j) \text{ for some } m \geq 0 \}$$

for $k \geq 1$, where $\{\alpha^1(k) \colon k \geq 0\}$ are the start indices corresponding to $\{D_j^1 \colon j \geq 0\}$. It follows from Theorem 2.56 that these random indices form a sequence of regeneration points for $\{D_j^1 \colon j \geq 0\}$—see Figure 9.5(b) for an illustration of the various regenerative cycles when $|\tilde{U}| = 2$. Theorem 1.12 in Chapter 6 then implies that, under mild regularity conditions, the quantity $r(f)$ in (3.16) is well defined and finite. Moreover, strongly consistent point estimates and asymptotic confidence intervals for $r(f)$ can be based on simulation of the CSPN over a fixed time interval $[0, t]$—we focus on simulation until a fixed time to facilitate comparison with other estimation methods.

Denote by $m^1(t)$ the number of delays $\{ D_j^1 : j \geq 0 \}$ that terminate in the interval $[0, t]$, and set

$$\bar{r}^1(t) = \frac{1}{m^1(t)} \sum_{j=1}^{m^1(t)} f(D_j^1).$$

Also set

$$Y_k^1(f) = \sum_{j=\gamma^1(k-1)}^{\gamma^1(k)-1} f(D_j^1),$$

$$\delta_k^1 = \gamma^1(k) - \gamma^1(k-1),$$

and

$$\tau_k^1 = \zeta_{\theta^1(k)} - \zeta_{\theta^1(k-1)}$$

for $k \geq 1$. Finally, set

$$\sigma^1(f) = \text{Var}_\mu^{1/2} \left[Y_1^1(f) - r(f)\delta_1^1 \right].$$

Estimation of $r(f)$ is based on the following result.

Theorem 3.20. *Suppose that $E_\mu[\tau_1^1] < \infty$, that δ_1^1 and $Y_k^1(|f|)$ each have finite second moment, and that* (2.52)–(2.54) *hold. Then $\lim_{t\to\infty} \bar{r}^1(t) = r(f)$ a.s. and*

$$\frac{t^{1/2} \left(\bar{r}^1(t) - r(f) \right)}{e^1(f)} \Rightarrow N(0,1)$$

at $t \to \infty$, where

$$e^1(f) = \frac{E_\mu^{1/2}[\tau_1^1] \sigma^1(f)}{E_\mu[\delta_1^1]}.$$

PROOF. The first assertion of the theorem follows from Theorem 1.12 in Chapter 6 and the fact that $\lim_{t\to\infty} m^1(t) = \infty$ a.s.; the latter convergence is a consequence of (2.53) and the expanded symmetry conditions. To prove the second assertion, use essentially a discrete-time analog of the proof of Theorem 3.51 in Chapter 6 to show that

$$\frac{m^{1/2} \left(m^{-1} \sum_{j=1}^m f(D_j^1) - r(f) \right)}{\sigma^1(f)/E_\mu^{1/2}[\delta_1^1]} \Rightarrow N(0,1) \tag{3.21}$$

at $m \to \infty$. It is straightforward to show that

$$\lim_{t\to\infty} \frac{m^1(t)}{t} = \frac{E_\mu[\delta_1^1]}{E_\mu[\tau_1^1]} \quad \text{a.s.}$$

using, for example, the regenerative structure of the marking process together with Theorem 2.14 in Chapter 3. An argument similar to the proof of Theorem 3.18 in Chapter 6 then shows that m can be replaced by $m^1(t)$ in (3.21), so that

$$\frac{\left(m^1(t)\right)^{1/2}\left(\bar{r}^1(t) - r(f)\right)}{\sigma^1(f)/E_\mu^{1/2}[\delta_1^1]} \Rightarrow N(0,1)$$

at $t \to \infty$. To complete the proof, write

$$\frac{t^{1/2}\left(\bar{r}^1(t) - r(f)\right)}{e^1(f)}$$

$$= \left(\frac{t}{m^1(t)} \cdot \frac{E_\mu[\delta_1^1]}{E_\mu[\tau_1^1]}\right)^{1/2} \frac{\left(m^1(t)\right)^{1/2}\left(\bar{r}^1(t) - r(f)\right)}{\sigma^1(f)/E_\mu^{1/2}[\delta_1^1]}$$

and apply Slutsky's theorem.   □

Thus the estimator $\bar{r}^1(t)$ is strongly consistent for $r(f)$. Moreover, we can obtain asymptotic confidence intervals in the usual manner. Specifically, denote by $K^1(t)$ the number of regeneration points $\left\{ \zeta_{\theta^1(k)} : k \geq 1 \right\}$ that lie in the interval $(0, t]$, and set

$$\bar{\tau}^1(t) = \frac{1}{K^1(t)} \sum_{k=1}^{K^1(t)} \tau_k^1,$$

$$\bar{\delta}^1(t) = \frac{1}{K^1(t)} \sum_{k=1}^{K^1(t)} \delta_k^1,$$

and

$$s^1(t) = \left(\frac{1}{K^1(t) - 1} \sum_{k=1}^{K^1(t)} \left(Y_k^1(f) - \bar{r}(t)\tau_k^1\right)^2\right)^{1/2}.$$

The estimator

$$\hat{e}^1(t) = \frac{\left(\bar{\tau}^1(t)\right)^{1/2} s^1(t)}{\bar{\delta}^1(t)}$$

is strongly consistent for $e^1(f)$, and

$$J^1(t; p) = \left[\bar{r}^1(t) - \frac{z_p\, \hat{e}^1(t)}{\sqrt{t}}, \bar{r}^1(t) + \frac{z_p\, \hat{e}^1(t)}{\sqrt{t}}\right]$$

is an asymptotic $100p\%$ confidence interval for $r(f)$, where z_p is the $(1+p)/2$ quantile of the standard normal distribution.

An Estimation Procedure That Exploits Symmetry

We now give an alternative estimation procedure that exploits symmetry (in the foregoing expanded sense) of a CSPN model. This procedure not only uses shorter cycles than the estimation procedure outlined above, but also is statistically more efficient. The idea is to exploit the fact—which is clear from symmetry considerations and not hard to prove rigorously—that

$$\lim_{n\to\infty} \frac{1}{n} \sum_{j=0}^{n-1} f(D_j) = r(f) \text{ a.s.,}$$

where $r(f)$ is defined by (3.16), so that the time-average limit for the combined sequence of delays of all colors coincides with the time-average limit for the sequence of delays of color 1.

First, define a sequence $\{\theta(k): k \geq 0\}$ as in (3.7). Then, using these blocking points, define a sequence of random indices $\{\gamma(k): k \geq 0\}$ for the combined sequence $\{D_j: j \geq 0\}$ of delays of all colors: $\gamma(0) = 0$ and

$$\gamma(k) = \inf\{j > \gamma(k-1): \alpha(j-1) < \theta(m) \leq \alpha(j) \text{ for some } m \geq 0\}$$

for $k \geq 1$. Since we assume that either (3.17) or (3.18) holds, it follows from Theorem 3.13 that the random indices $\{\gamma(k): k \geq 0\}$ form a sequence of regeneration points for the process $\{D_j: j \geq 0\}$—see Figure 9.5(a) for an illustration of the cycles when $|\tilde{U}| = 2$. In the figure the solid vertical lines correspond to time points in the set $\{\zeta_{\theta(k)}: k \geq 0\} \cap \{\zeta_{\theta^1(k)}: k \geq 0\}$ and the dashed vertical lines correspond to time points in the set $\{\zeta_{\theta(k)}: k \geq 0\} - \{\zeta_{\theta^1(k)}: k \geq 0\}$.

Theorem 3.22 below leads to an estimation procedure based on simulation of the CSPN over a fixed time interval $[0, t]$. Denote by $m(t)$ the number of delays $\{D_j: j \geq 0\}$ that terminate in the interval $(0, t]$, and set

$$\bar{r}(t) = \frac{1}{m(t)} \sum_{j=1}^{m(t)} f(D_j).$$

Also set

$$Y_k(f) = \sum_{j=\gamma(k-1)}^{\gamma(k)-1} f(D_j),$$

$$\delta_k = \gamma(k) - \gamma(k-1),$$

and

$$\tau_k = \zeta_{\theta(k)} - \zeta_{\theta(k-1)}$$

for $k \geq 1$. Finally, set

$$\sigma^2(f) = \text{Var}_\mu\left[Y_1(f) - r(f)\delta_1\right].$$

Theorem 3.22. *Suppose that $E_\mu[\tau_1] < \infty$, that δ_1 and $Y_k(|f|)$ each have finite second moment, and that (2.52)–(2.54) hold. Then $\lim_{t\to\infty} \bar{r}(t) = r(f)$ a.s. and*

$$\frac{t^{1/2}\big(\bar{r}(t) - r(f)\big)}{e(f)} \Rightarrow N(0,1) \tag{3.23}$$

at $t \to \infty$, where

$$e(f) = \frac{E_\mu^{1/2}[\tau_1]\sigma(f)}{E_\mu[\delta_1]}.$$

The proof of Theorem 3.22 is almost identical to that of Theorem 3.20.

Theorem 3.22 asserts that the estimator $\bar{r}(t)$ is strongly consistent for $r(f)$; the theorem also leads to a confidence interval for $r(f)$. Denote by $K(t)$ the number of regeneration points $\{\,\zeta_{\theta(k)} : k \geq 1\,\}$ that lie in the interval $(0,t]$, and set

$$\bar{\tau}(t) = \frac{1}{K(t)} \sum_{k=1}^{K(t)} \tau_k,$$

$$\bar{\delta}(t) = \frac{1}{K(t)} \sum_{k=1}^{K(t)} \delta_k,$$

and

$$s(t) = \left(\frac{1}{K(t)-1} \sum_{k=1}^{K(t)} \big(Y_k(f) - \bar{r}(t)\tau_k\big)^2 \right)^{1/2}.$$

The estimator

$$\hat{e}(t) = \frac{\big(\bar{\tau}(t)\big)^{1/2} s(t)}{\bar{\delta}(t)}$$

is strongly consistent for $e(f)$, and

$$J(t;p) = \left[\bar{r}(t) - \frac{z_p\, \hat{e}(t)}{\sqrt{t}}, \bar{r}(t) + \frac{z_p\, \hat{e}(t)}{\sqrt{t}} \right]$$

is an asymptotic $100p\%$ confidence interval for $r(f)$.

Comparison of Estimation Procedures

We now compare the two estimation procedures described above. Analogously to the discussion in Chapter 8, for a fixed value of $p \in (0,1)$ we take the asymptotic relative efficiency (ARE) of the two procedures to be the limiting ratio of the lengths of the $100p\%$ confidence intervals for $r(f)$ as the simulated time becomes large. Denote by $I^1(t;p)$ and $I(t;p)$ the lengths of the intervals $J^1(t;p)$ and $J(t;p)$, and observe that

$$\lim_{t\to\infty} \frac{I(t;p)}{I^1(t;p)} = \frac{e(f)}{e^1(f)} \quad \text{a.s.}$$

for all $p \in (0,1)$. The next result asserts that the ARE is less than or equal to 1, so that the second of the two estimation methods—which exploits symmetry in the CSPN model—is always at least as efficient as the first estimation method.

Theorem 3.24. *Under the assumptions of this section, $e(f) \le e^1(f)$ for all functions f such that both quantities are well defined.*

To prove Theorem 3.24, we need the following result for "cumulative processes." A real-valued stochastic process $\{Y(t): t \ge 0\}$ is a cumulative process if

1. There exists a sequence $T_0 = 0, T_1, T_2, \ldots$ of increasing a.s. finite random times such that the sequence $\{(Y_n, \tau_n): n \ge 1\}$ consists of i.i.d. random pairs, where $Y_n = Y(T_n) - Y(T_{n-1})$ and $\tau_n = T_n - T_{n-1}$ for $n \ge 1$.

2. The process $\{Y(t): t \ge 0\}$ is, with probability 1, of bounded variation over every finite interval—see Definition 1.6 in the Appendix.

If $\{X(t): t \ge 0\}$ is a regenerative process with piecewise-constant sample paths and $Y(t) = \int_0^t f(X(u))\, du$ for $t \ge 0$ and some real-valued function f, then $\{Y(t): t \ge 0\}$ is a cumulative process.

Lemma 3.25. *Let $\{Y(t): t \ge 0\}$ be a cumulative process and set $\sigma^2 = \mathrm{Var}\,[Y_1]$. Suppose that $E\,[Y_1] = 0$, $\sigma^2 < \infty$ and that $E\,[\tau_1] < \infty$. Then*

$$\lim_{t \to \infty} \frac{\mathrm{Var}\,[Y(t)]}{t} = \frac{\sigma^2}{E\,[\tau_1]}.$$

Although a complete proof of the lemma uses techniques from renewal theory (see Section A.2.3) and is beyond the scope of the current discussion, the following argument shows why the result is plausible. Let $N(t) = \sup\{n: T_n \le t\}$ for $t \ge 0$. Using Theorem 2.9(ii) in Chapter 3 together with the SLLN for i.i.d. random variables, it is easy to show that $\lim_{t \to \infty} N(t)/t = 1/E\,[\tau_1]$ a.s.. In light of this result, it is plausible that

$$\lim_{t \to \infty} \frac{E\,[N(t)]}{t} \to \frac{1}{E\,[\tau_1]}, \tag{3.26}$$

and the elementary renewal theorem (Proposition 2.13 in the Appendix) asserts that the convergence in (3.26) does indeed hold. Setting $S_n = \sum_{k=1}^n Y_k$, we can write

$$\frac{\mathrm{Var}\,[Y(t)]}{t} = \frac{\mathrm{Var}\,[S_{N(t)}]}{E\,[N(t)]} \cdot \frac{E\,[N(t)]}{t} + \frac{r(t)}{t} \tag{3.27}$$

for $t \ge 0$, where

$$r(t) = \mathrm{Var}\,[Y(t) - S_{N(t)}] + 2\,\mathrm{Cov}\,[S_{N(t)}, Y(t) - S_{N(t)}].$$

Since $\{Y_n : n \geq 1\}$ is an i.i.d. sequence, we have $\mathrm{Var}\,[S_n] = n\sigma^2$ for $n \geq 1$. By Wald's second moment identity [Proposition 1.19(ii) in the Appendix] we can replace the deterministic index n by the random index $N(t)$ to obtain $\mathrm{Var}\,[S_{N(t)}] = E\,[N(t)]\,\sigma^2$. Substituting this expression into (3.27), taking limits, and applying (3.26), we obtain

$$\lim_{t\to\infty} \frac{\mathrm{Var}\,[Y(t)]}{t} = \frac{\sigma^2}{E\,[\tau_1]} + \lim_{t\to\infty} \frac{r(t)}{t}.$$

Observe that, by the Cauchy–Schwarz inequality,

$$|r(t)| \leq \mathrm{Var}\,[Y(t) - S_{N(t)}] + 2\mathrm{Var}^{1/2}\,[Y(t) - S_{N(t)}]\,\mathrm{Var}^{1/2}\,[S_{N(t)}]\,.$$

Because the sample paths of $\{Y(t) : t \geq 0\}$ are well behaved, we do not expect the variance of $Y(t) - S_{N(t)}$ to increase systematically, and so it is plausible that $\mathrm{Var}\,[Y(t) - S_{N(t)}]\,/t \to 0$ as $t \to \infty$. This latter convergence can indeed be established, and it follows that $\lim_{t\to\infty} r(t)/t = 0$.

PROOF OF THEOREM 3.24. Fix a function f, and observe that both the numerator on the left side of (3.23) and the limit $N(0,1)$ are independent of the particular choice of cycle boundaries for the marking process, and hence of the choice of cycles for the process $\{D_j : j \geq 0\}$. It follows easily that $e(f)$ is independent of the choice of cycle boundaries. In particular, we have $e(f) = \tilde{e}(f)$, where the latter quantity is defined as follows. Recall the sequence of regeneration points $\{\theta^1(k) : k \geq 0\}$ for the underlying chain defined previously via (3.19). Using these points, define a sequence of random indices $\{\tilde{\gamma}(k) : k \geq 0\}$ for the process[2] $\{D_j : j \geq 0\}$ in what is, by now, the usual way: $\tilde{\gamma}(0) = 0$ and

$$\tilde{\gamma}(k) = \inf\{\,j > \tilde{\gamma}(k-1) : \alpha(j-1) < \theta^1(m) \leq \alpha(j) \text{ for some } m \geq 0\,\}$$

for $k \geq 1$—see Figure 9.5(c). Set

$$\tilde{Y}_k(f) = \sum_{j=\tilde{\gamma}(k-1)}^{\tilde{\gamma}(k)-1} f(D_j),$$

$$\tilde{\delta}_k = \tilde{\gamma}(k) - \tilde{\gamma}(k-1),$$

and

$$\tilde{\tau}_k = \tau_k^1 = \zeta_{\theta^1(k)} - \zeta_{\theta^1(k-1)}$$

---

[2]In contrast, the random indices $\{\gamma^1(k) : k \geq 0\}$, though also specified using the regeneration points $\{\theta^1(k) : k \geq 0\}$, are defined for the process $\{D_j^1 : j \geq 0\}$ as in Figure 9.5(b).

for $k \geq 1$, and then set

$$\tilde{\sigma}(f) = \left(\text{Var}_\mu\left[\tilde{Y}_1(f) - r(f)\tilde{\delta}_1\right]\right)^{1/2}$$

and

$$\tilde{e}(f) = \frac{E_\mu^{1/2}[\tilde{\tau}_1]\tilde{\sigma}(f)}{E_\mu[\tilde{\delta}_1]}.$$

Thus $e(f)$ is based on cycles as in Figure 9.5(a), whereas $\tilde{e}(f)$ is based on cycles as in Figure 9.5(c). Because $e(f) = \tilde{e}(f)$, it suffices to show that

$$\tilde{\sigma}(f) \leq N_0\,\sigma^1(f) \tag{3.28}$$

and

$$E_\mu[\tilde{\delta}_1] = N_0 E_\mu\left[\delta_1^1\right], \tag{3.29}$$

where, as before, N_0 is the number of possible colors for the delays.

To establish (3.28), set

$$W(t) = \sum_{j=1}^{m(t)}\left(f(D_j) - r(f)\right)$$

for $t \geq 0$. For $1 \leq q \leq N_0$, denote by $m^q(t)$ the number of delays $\{D_j^q : j \geq 0\}$ that terminate in the interval $[0, t]$, and set

$$W^q(t) = \sum_{j=1}^{m^q(t)}\left(f(D_j^q) - r(f)\right)$$

for $t \geq 0$. For each q, fix a permutation $\lambda \in \Lambda$ such that $\lambda(1) = q$; the existence of such a permutation follows from the expanded symmetry conditions. Next, set $(\bar{s}, \bar{u}^q) = \xi_\lambda(\bar{s}, \bar{u})$ and observe that, by the expanded symmetry conditions, $(\bar{s}, \bar{u}^q)$ is a single state. We can therefore define random indices $\{\theta^q(k) : k \geq 0\}$ and $\{\gamma^q(k) : k \geq 0\}$ in a manner completely analogous to the definitions of $\{\theta^1(k) : k \geq 0\}$ and $\{\gamma^1(k) : k \geq 0\}$. In Figure 9.5(a), for example, the vertical solid lines correspond to the time points $\{\zeta_{\theta^1(k)} : k \geq 0\}$ and the vertical dashed lines correspond to the time points $\{\zeta_{\theta^2(k)} : k \geq 0\}$. We can also define quantities $Y_k^q(f)$, δ_k^q, τ_k^q, $\sigma^q(f)$, and so forth. Using Lemma 3.25, we have

$$\lim_{t \to \infty}\left(\frac{\text{Var}_\mu[W(t)]}{t}\right)^{1/2} = \frac{\tilde{\sigma}(f)}{E_\mu^{1/2}[\tilde{\tau}_1]} = \frac{\tilde{\sigma}(f)}{E_\mu^{1/2}[\tau_1^1]}$$

and, for each q,

$$\lim_{t \to \infty}\left(\frac{\text{Var}_\mu[W^q(t)]}{t}\right)^{1/2} = \frac{\sigma^q(f)}{E_\mu^{1/2}[\tau_1^q]} = \frac{\sigma^1(f)}{E_\mu^{1/2}[\tau_1^1]},$$

where the second equality follows by symmetry. Observe that $W(t) = \sum_{q=1}^{N_0} W^q(t)$ and apply the Cauchy–Schwarz inequality to obtain

$$\text{Var}_\mu^{1/2}[W(t)] \le \left(\sum_{q=1}^{N_0} \text{Var}_\mu[W^q(t)] + \sum_{p \ne q} \text{Var}_\mu^{1/2}[W^p(t)]\, \text{Var}_\mu^{1/2}[W^q(t)] \right)^{1/2}$$

$$= \sum_{q=1}^{N_0} \text{Var}_\mu^{1/2}[W^q(t)].$$

Dividing the leftmost and rightmost terms in the above inequality by $t^{1/2}$ and then letting $t \to \infty$ yields (3.28).

The equality in (3.29) is established in a similar manner. As in the proof of Theorem 3.20, we have

$$\lim_{t \to \infty} \frac{m(t)}{t} = \frac{E_\mu[\tilde{\delta}_1]}{E_\mu[\tilde{\tau}_1]} = \frac{E_\mu[\tilde{\delta}_1]}{E_\mu[\tau_1^1]}$$

and, for each q,

$$\lim_{t \to \infty} \frac{m^q(t)}{t} = \frac{E_\mu[\delta_1^q]}{E_\mu[\tau_1^q]} = \frac{E_\mu[\delta_1^1]}{E_\mu[\tau_1^1]},$$

where we have again used symmetry. Observe that

$$m(t) = \sum_{q=1}^{N_0} m^q(t)$$

for $t \ge 0$. Dividing both sides of the above equality by t and then letting $t \to \infty$ yields (3.29).   □

The following example shows that the difference in efficiency between the symmetric and nonsymmetric estimation methods can be significant.

EXAMPLE 3.30 (Symmetric token ring with fixed-sized packets). Similarly to Example 3.17 in Chapter 6, suppose that, for each port, the time to transmit a package is a deterministic constant L and the time for the ring token to propagate to the next port is a deterministic constant R. Also suppose that the successive times from the end of transmission until the arrival of the next packet for transmission are i.i.d. as an exponential random variable with intensity q. Finally, suppose that we wish to estimate the *access time* for port 1, that is, the time from when a packet arrives at port 1 for transmission until the start of transmission of the packet.

Assuming that the system is modelled by a CSPN as in Figure 9.2, we can specify the sequence of access times for all ports using a start-vector mechanism very similar to that in Example 3.14. We modify this start-vector mechanism as described at the beginning of this subsection and associate a color with each delay—the color associated with a delay is the number of the

Table 9.3. Simulation Results for Symmetric Token Ring with Fixed-Sized Packets

Number of Jobs	Access Time	ARE
4	0.5819±0.0005	0.6020
8	1.9549±0.0006	0.4123
12	3.5140±0.0005	0.3115

port at which the corresponding packet arrived. It is easy to show that the expanded symmetry conditions hold, and we can estimate the access time for port 1 using either of the two estimation methods described previously. To this end, we define the sequences $\{\theta^1(k): k \geq 0\}$ and $\{\theta(k): k \geq 0\}$ using the single state $(\bar{s}, \bar{u})$ given in Example 3.9. Thus $\theta^1(k)$ is the random index of the kth marking change at which port 1 observes the ring token with each port having a packet awaiting transmission and $\theta(k)$ is the index of the kth marking change at which some port observes the ring token with each port having a packet awaiting transmission.

Table 9.3 displays point estimates and 95% confidence intervals for the port 1 access time; these estimates were computed using the symmetric estimation method. Also displayed is the estimate ARE—that is, the ratio $e(f)/e^1(f)$—for the symmetric and nonsymmetric estimation methods. The parameter values used in the simulation are $L = 0.3$, $R = 0.1$, and $q = 1$. Results are reported for several values of N, the number of ports. All estimates are based on 10^4 cycles demarcated by the random times $\{\zeta_{\theta^1(k)}: k \geq 0\}$. As the total number of ports increases, the access time for port 1 increases, which is to be expected. As can be seen from the right-most column, the asymptotic confidence-interval length for the symmetric estimation method is significantly shorter than the interval length for the nonsymmetric method. The difference in efficiency between the methods becomes increasingly pronounced as N increases. This trend also is to be expected, since the symmetric estimation method observes roughly N times as many delays as the nonsymmetric estimation method does.

Notes

The notion of associating colors with the tokens and transitions of an (ordinary) Petri net dates back to a paper by Jensen (1981). This work can be viewed as a specific variation on the general theme of assigning "inscriptions" to the various components of a net model, thereby obtaining a "high-level" Petri net; see, for example, the discussion in Genrich and Lautenbach (1981). Zenie (1985) proposes augmenting the SPN formalism of Molloy (1981) with colors, Chiola, Bruno, and Demaria (1988) makes a similar proposal in the setting of GSPNs, and Lin and Marinescu (1988) proposes "stochastic high-level Petri nets." Our definition of CSPNs follows Haas and Shedler (1993b).

With the exception of Haas and Shedler (1993b), the foregoing work emphasizes computation of reachability sets and exact computation of steady-state probabilities for the marking process when all clock-setting distribution functions for timed transitions are exponential. The authors of these papers provide techniques for aggregation of markings to reduce the complexity of the computations; these techniques exploit various symmetries in the net model. More recently, attention has focused on SPNs in which the color mechanism has additional structure that facilitates automatic detection and exploitation of model symmetry. A notable example is provided by the "stochastic well-formed colored nets" (SWNs) introduced by Chiola et al. (1993). Gaeta and Chiola (1995) discuss methods for exploiting model symmetry to efficiently generate sample paths of the marking process of an SWN; the authors provide techniques both for reducing the amount of work needed to schedule or cancel the firing of a transition and for reducing the length of the "event list" of transitions currently scheduled to fire. Chiola (1995) has initiated an extension of the "recursion equation" approach to the analysis and parallel simulation of nets—as in Baccelli et al. (1993)—to the setting of SWNs.

The complaint processing system of Example 1.4 is based on an example in van der Aalst (1998). The results in Section 9.3.3 are adapted from Prisgrove and Shedler (1986). Lemma 3.25 is contained in Theorem 8 of Smith (1955). As mentioned at the end of Chapter 6, discussions of renewal theory can be found, for example, in the books of Asmussen (1987a), Çinlar (1975), Karlin and Taylor (1975), and Ross (1983).

Besides the methods discussed in Section 9.3, an additional technique for exploiting symmetry can be found in the work of Calvin and Nakayama (2000). Their methodology is applicable when the underlying chain of a CSPN is regenerative and the performance measure of interest is one of the following:

- The variance of the reward earned over a regenerative cycle—for example, the variance of the busy period in a single-server queue

- The variance constant of the process $\{\tilde{f}(S_n, U_n, C_n) : n \geq 0\}$, where $\tilde{f}$ is a suitable reward function

- The reward earned before the chain hits a specified set—for example, the mean time to failure

The idea is as follows. If there exists a sequence of regeneration points for the underlying chain of the marking process and the symmetry conditions hold, then there typically exist multiple sequences of regeneration points—each sequence corresponds to some permutation of the colors. Suppose that there are m such sequences, denoted $\Theta_1, \Theta_2, \ldots, \Theta_m$, with

$\Theta_i = (\theta_i(0), \theta_i(1), \ldots)$ for $1 \leq i \leq m$. Simulate a finite sample path that corresponds to a fixed number of regenerative cycles demarcated by the regeneration points in Θ_1. Starting with this sample path, generate a "permuted" sample path by first permuting the segments of the sample path demarcated by the Θ_1-regeneration points, then permuting the segments of the resulting path demarcated by the Θ_2-regeneration points, and so forth, for a total of m permutation steps. Conceptually, we can obtain an estimator of the performance measure of interest by generating all possible permuted paths, computing the value of the standard regenerative estimator of the specified performance measure over each such path, and then averaging these values. Calvin and Nakayama (2000) show how to compute such a "permuted regenerative estimator" estimator efficiently—without actually materializing all the permuted paths. The authors also show that for any finite-length simulation the mean-square error (MSE) of the permuted regenerative estimator is less than or equal to the MSE of the standard regenerative estimator (applied to the original sample path). Finally, the authors also show that the permuted regenerative estimator is strongly consistent, and they obtain a CLT that permits construction of asymptotic confidence intervals.

Appendix A
Selected Background

In the following sections we summarize various results on probability and stochastic processes that are used in the text. The notes at the end of the Appendix give references for further reading.

A.1 Probability, Random Variables, Expectation

A.1.1 Probability Spaces

We start with a set Ω, called the *sample space*, that represents the possible elementary outcomes of a probabilistic experiment. A subset $A \subseteq \Omega$ is called an *event*—if the outcome of the experiment is an element of A, then we say that "event A has occurred." Associated with Ω is a σ-*field* $\mathcal{F}$ of events, that is, a collection $\mathcal{F}$ of events such that

1. $\varnothing \in \mathcal{F}$ and $\Omega \in \mathcal{F}$.

2. $A^c \in \mathcal{F}$ whenever $A \in \mathcal{F}$, where $A^c = \Omega - A = \{\omega \in \Omega : \omega \notin A\}$.

3. $\bigcup_{i=1}^{\infty} A_i \in \mathcal{F}$ whenever $A_1, A_2, \ldots \in \mathcal{F}$.

A *probability measure* P is a nonnegative real-valued function on $\mathcal{F}$ that satisfies $P(\varnothing) = 0$, $P(\Omega) = 1$, and

$$P\left(\bigcup_n A_n\right) = \sum_n P(A_n)$$

whenever $A_1, A_2, \ldots$ form a finite or countably infinite collection of disjoint sets in $\mathcal{F}$. The triple $(\Omega, \mathcal{F}, P)$ is called a *probability space*.[1] The pair $(\Omega, \mathcal{F})$ is called a *measurable space*, and the elements of $\mathcal{F}$ are called *measurable sets* or *measurable events*. In general, a σ-field $\mathcal{G}$ ($\subseteq \mathcal{F}$) can be interpreted as a specified "body of information" about the underlying probabilistic experiment—being "given" the information in $\mathcal{G}$ means being told for each $A \in \mathcal{G}$ whether or not event A has occurred. Thus $\mathcal{F}$ represents the "maximal" body of information available about the experiment. The σ-field *generated* by a collection $\mathcal{A}$ ($\subseteq \mathcal{F}$) of events is defined as the smallest σ-field containing $\mathcal{A}$ or, equivalently, the intersection of all σ-fields containing $\mathcal{A}$.

Some elementary properties of probability measures are given in Proposition 1.1—all sums, limits, unions, and intersections are taken over a finite or countably infinite collection of events.

Proposition 1.1. *Let* $(\Omega, \mathcal{F}, P)$ *be a probability space. Then*

(i) $0 \leq P(A) \leq 1$,

(ii) $P(A^c) = 1 - P(A)$,

(iii) $P(A) \leq P(B)$ *whenever* $A \subseteq B$,

(iv) *(Continuity) if* $A_n \uparrow A$ *or* $A_n \downarrow A$ *then* $P(A_n) \to P(A)$,

(v) *(Boole's inequality)* $P(\bigcup_n A_n) \leq \sum_n P(A_n)$, *and*

(vi) *(Bonferroni's inequality)* $P(\bigcap_n A_n) \geq 1 - \sum_n P(A_n^c)$,

where all sets are elements of $\mathcal{F}$.

Events $A_1, A_2, \ldots, A_n$ are *mutually independent* if

$$P(A_{n_1} \cap A_{n_2} \cap \cdots \cap A_{n_k}) = P(A_{n_1})P(A_{n_2}) \cdots P(A_{n_k})$$

for $2 \leq k \leq n$ and $1 \leq n_1 < n_2 < \cdots < n_k \leq n$. The events in a countably infinite collection are said to be mutually independent if the events in every finite subcollection are mutually independent. The σ-fields $\{\mathcal{G}_n : n \in N\}$—where N is finite or countably infinite—are said to be mutually independent if the events $\{A_n : n \in N\}$ are mutually independent for each possible choice of A_n from $\mathcal{G}_n$.

For a countably infinite sequence of events $A_1, A_2, \ldots$, set

$$\limsup_n A_n = \bigcap_{n=1}^{\infty} \bigcup_{k=n}^{\infty} A_k.$$

---

[1]The reader may wonder why we do not simply define a probability measure on all of the subsets of Ω—it can be shown that, in general, such a definition is not possible.

Intuitively, the event $\limsup_n A_n$ occurs if and only if events $A_1, A_2, \ldots$ occur infinitely often, and we write "A_n i.o." for $\limsup_n A_n$. The events $A_1, A_2, \ldots$ are then said to be *recurrent*.

Proposition 1.2. *Let $A_1, A_2, \ldots$ be a countably infinite sequence of (arbitrary) events. If $\sum_{n=1}^{\infty} P(A_n) < \infty$, then $P\{A_n \text{ i.o.}\} = 0$.*

Proposition 1.3. *Let $A_1, A_2, \ldots$ be a countably infinite sequence of mutually independent events. If $\sum_{n=1}^{\infty} P(A_n) = \infty$, then $P\{A_n \text{ i.o.}\} = 1$.*

Propositions 1.2 and 1.3 are known as the first and second *Borel–Cantelli lemmas*. A reference to "the" Borel–Cantelli lemma is a reference to the second result, Proposition 1.3.

A.1.2 General Measures

A general measure μ defined on a σ-field $\mathcal{F}$ satisfies the defining requirements of a probability measure, except that we do not require that $\mu(\Omega) = 1$, or even that $\mu(\Omega) < \infty$. If $\mu(A) > 0$ for some $A \in \mathcal{F}$, then μ is called *nontrivial*. A triple $(\Omega, \mathcal{F}, \mu)$ is called a *measure space*.

The most well-known example of an unbounded measure is *Lebesgue measure* μ^{Leb}. Let $\mathcal{B}$ be the σ-field generated by the class of finite open intervals of the form (a, b)—the elements of $\mathcal{B}$ are called the *Borel sets*. Then μ^{Leb} is the unique measure on $(\mathfrak{R}, \mathcal{B})$ such that

$$\mu^{\text{Leb}}((a, b)) = b - a$$

for each interval (a, b). Thus μ^{Leb} corresponds to the usual intuitive notion of length, but is well defined for a much wider class of sets than just the intervals. As might be expected, μ^{Leb} is translation invariant:

$$\mu^{\text{Leb}}(A + x) = \mu^{\text{Leb}}(A)$$

for all $A \in \mathcal{B}$ and $x \in \mathfrak{R}$, where $A + x = \{a + x : a \in A\}$.

We also define k-dimensional Lebesgue measure as the unique measure μ^{Leb} on $(\mathfrak{R}^k, \mathcal{B}^k)$ such that

$$\mu^{\text{Leb}}(A) = \prod_{i=1}^{k} (b_i - a_i)$$

for all sets of the form

$$A = (a_1, b_1) \times (a_2, b_2) \times \cdots \times (a_k, b_k), \tag{1.4}$$

with $a_1, b_1, \ldots, a_k, b_k \in \mathfrak{R}$. Here $\mathcal{B}^k$ is the "product" σ-field of k-dimensional Borel sets—$\mathcal{B}^k$ is generated by the "rectangle sets" as in (1.4). Equivalently, $\mathcal{B}^k$ is generated by the collection of sets of the form $A_1 \times A_2 \times \cdots \times A_k$, where $A_i \in \mathcal{B}$ for $1 \leq i \leq k$. Just as Lebesgue measure in $\mathfrak{R}$ extends the usual notion of length, Lebesgue measure in $\mathfrak{R}^2$ and $\mathfrak{R}^3$ extends the usual notions of area and volume.

A.1.3 Random Variables

A *random variable* X on a probability space $(\Omega, \mathcal{F}, P)$ is a real-valued function defined on Ω and *measurable* with respect to $\mathcal{F}$ in the sense that $\{\omega\colon X(\omega) \le x\} \in \mathcal{F}$ for all $x \in \Re$. That is, X is a variable whose value is determined by the outcome ω of the probabilistic experiment. The σ-field *generated* by X, denoted by $\sigma\langle X\rangle$, is the smallest σ-field that contains all sets of the form $\{\omega\colon X(\omega) \le x\}$. In light of our previous interpretation of a σ-field, $\sigma\langle X\rangle$ can be viewed as "complete information about X"—that is, if we know for each $A \in \sigma\langle X\rangle$ whether or not event A has occurred, we can precisely determine the value of X. Note that $\sigma\langle X\rangle \subseteq \mathcal{F}$: the σ-field $\mathcal{F}$ may contain "more information" about the probabilistic experiment than merely information about X. In general, we say that a random variable X is measurable with respect to a σ-field $\mathcal{G}$ if the value of X is determined by $\mathcal{G}$ in the foregoing manner, that is, if $\{\omega\colon X(\omega) \le x\} \in \mathcal{G}$ for $x \in \Re$. For a (possibly uncountably infinite) collection of random variables $\mathcal{H} = \{X_\theta\colon \theta \in \Theta\}$, the σ-field generated by the random variables in $\mathcal{H}$ is the smallest σ-field that contains $\bigcup_{\theta \in \Theta} \sigma\langle X_\theta\rangle$.

The *distribution* of a random variable X is the unique probability measure μ defined on $(\Re, \mathcal{B})$ by

$$\mu(A) = P\{X \in A\} = P(\{\omega\colon X(\omega) \in A\})$$

for $A \in \mathcal{B}$. The right-continuous function F defined by

$$F(x) = P\{X \le x\} = \mu((-\infty, x])$$

for $x \in \Re$ is the *(cumulative) distribution function* of X. Provided that $P\{X < \infty\} = 1$, the distribution function F is *proper* in that

$$\lim_{x \to \infty} F(x) = 1.$$

As indicated above, μ determines F. Conversely, F determines μ as the unique measure on $(\Re, \mathcal{B})$ that satisfies $\mu((a, b]) = F(b) - F(a)$ for each interval $(a, b]$. A distribution function F is *discrete* if it is constant except at a finite or countably infinite sequence of jump points. If F is discrete, then X takes on some set of values $\{a_n\colon n \in N\}$ with respective probabilities $\{p_n\colon n \in N\}$, where N is finite or countably infinite and $p_n = F(a_n) - F(a_n-)$ for $n \in N$. In this case X is also said to be discrete. We say that F is *continuous* if $\lim_{y \to x} F(y) = F(x)$ at each point x and is *absolutely continuous* if it has the following property: for every $\epsilon > 0$, there exists $\delta > 0$ such that for each collection $\{[a_i, b_i]\colon 1 \le i \le k\}$ of nonoverlapping intervals

$$\sum_{i=1}^{k} |F(b_i) - F(a_i)| < \epsilon \quad \text{if} \quad \sum_{i=1}^{k} (b_i - a_i) < \delta.$$

It can be shown that F is absolutely continuous if and only if $\mu(A) = 0$ whenever $\mu^{\text{Leb}}(A) = 0$. An absolutely continuous distribution function is continuous, but the converse assertion is not true in general. The notion of absolute continuity is important because of the result in Proposition 1.5 below. We say that f is a *density function* of the distribution function F if $F(x) = \int_{-\infty}^{x} f(x)\, dx$ for all $x \in \Re$. In the following, set $F'(x) = dF(x)/dx$ for all x such that the derivative is defined.

Proposition 1.5. *A distribution function F possesses a density f if and only if F is absolutely continuous, in which case $f = F'$ except on a set of Lebesgue measure 0.*

A distribution function F is *singular* if the derivative $F'(x)$ exists and is equal to 0 except on a set of Lebesgue measure 0. We say that F is a *convex combination* of distribution functions $G_1, G_2, \ldots, G_k$ if $F = p_1 G_1 + p_2 G_2 + \cdots + p_k G_k$ for some nonnegative numbers $p_1, p_2, \ldots, p_k$ with $p_1 + p_2 + \cdots + p_k = 1$. An arbitrary distribution function F can be represented as a convex combination of distribution functions F_d, F_{ac}, and F_s, where F_d is discrete, F_{ac} is absolutely continuous, and F_s is singular and continuous. For most distribution functions encountered in practice, the component F_s is not actually present (i.e., has a weight of 0).[2]

Proposition 1.5 can easily be extended to general unbounded functions and can be further extended to functions that are not necessarily nondecreasing by using the notion of *bounded variation*. For a partition

$$\Delta : a = a_0 < a_1 < \cdots < a_k = b$$

of a finite interval $[a, b]$ and a real-valued function F, set

$$\|F\|_{\Delta} = \sum_{i=1}^{k} |F(a_i) - F(a_{i-1})|.$$

Definition 1.6. A real-valued function F is of *bounded variation* on $[a, b]$ if $\sup_{\Delta} \|F\|_{\Delta} < \infty$, where the supremum runs over all partitions of $[a, b]$.

It can be shown that (1) an absolutely continuous function is of bounded variation and (2) a function of bounded variation is expressible as the difference between two nondecreasing functions. Thus, if a function F (not necessarily nondecreasing) is absolutely continuous, then it can be written as $F = G - H$, where G and H are nondecreasing. The argument leading to Proposition 1.5 can be applied to G and H separately to show that F

---

[2]It is perhaps counterintuitive that a continuous function can increase from 0 to 1 while having a derivative that equals 0 except on a set of Lebesgue measure 0. Such functions exist, however. An example is given by the distribution function of the random variable $X = \sum_{n=1}^{\infty} X_n 2^{-n}$, where $\{X_n : n \geq 1\}$ is a sequence of i.i.d. random variables with $P\{X_n = 1\} = 1 - P\{X_n = 0\} = p$ for some $p \in (0, \infty)$ with $p \neq 1/2$.

can be expressed as the integral of a function f. The converse assertion can also be established: F can be expressed as the integral of a function f only if F is absolutely continuous.

Definition 1.7. A random variable X is *dominated* by a random variable Y if $P\{X \leq Y\} = 1$. A random variable X is *stochastically dominated* by a random variable Y (or is *stochastically smaller* than Y) if $F_X(u) \geq F_Y(u)$ for $u \in \Re$, where F_X and F_Y are the distribution functions of X and Y.

Observe that for X to be dominated by Y, the random variables X and Y must be defined on the same probability space—this requirement can be dropped in the case of stochastic domination. If X is stochastically dominated by Y, then we can construct random variables X' and Y' on a common probability space $(\Omega, \mathcal{F}, P)$ such that X' is distributed as X, Y' is distributed as Y, and X' is dominated by Y'. The idea is as follows. For a distribution function F, set

$$F^{-1}(u) = \inf\{x\colon F(x) \geq u\}$$

for $u \in [0,1]$. A simple computation shows that if U is a random variable uniformly distributed on $[0,1]$, then the random variable $Z = F^{-1}(U)$ has distribution function F. Thus X' and Y' can be obtained by constructing a probability space $(\Omega, \mathcal{F}, P)$ together with a uniform random variable U defined on $(\Omega, \mathcal{F}, P)$, and then setting $X' = F_X^{-1}(U)$ and $Y' = F_Y^{-1}(U)$.

The real-valued random variables $X_1, X_2, \ldots, X_n$ are *mutually independent* if the σ-fields generated by these random variables are mutually independent. To establish mutual independence, it suffices to show that

$$P\{X_1 \in A_1, X_2 \in A_2, \ldots, X_n \in A_n\}$$
$$= P\{X_1 \in A_1\} P\{X_2 \in A_2\} \cdots P\{X_n \in A_n\}$$

for all Borel sets $A_1, A_2, \ldots, A_n \in \mathcal{B}$, or that

$$P\{X_1 \leq x_1, X_2 \leq x_2, \ldots, X_n \leq x_n\}$$
$$= P\{X_1 \leq x_1\} P\{X_2 \leq x_2\} \cdots P\{X_n \leq x_n\}$$

for all $x_1, x_2, \ldots, x_n \in \Re$. A countably infinite collection of random variables is said to be mutually independent if the random variables in each finite subcollection are independent. A random variable X is said to be independent of a σ-field $\mathcal{G}$ if $\sigma\langle X\rangle$ and $\mathcal{G}$ are independent.

A.1.4 Expectation

The expectation of a random variable X on a probability space $(\Omega, \mathcal{F}, P)$ is most generally defined as the "Lebesgue integral of X with respect to P." We first define such an integral and relate our general definition to various elementary definitions of expectation. First consider a random variable $X \geq 0$, that is, $X(\omega) \geq 0$ for $\omega \in \Omega$. Then

$$\int X \, dP = \sup_{\mathcal{P}} \sum_{A \in \mathcal{P}} \left(\inf_{\omega \in A} X(\omega) \right) P(A),$$

where $\mathcal{P} \, (\subseteq \mathcal{F})$ is a finite partition of Ω into disjoint subsets and the supremum is taken over all finite partitions. In general, this integral can equal $+\infty$. For an arbitrary random variable X, we set $X^+ = \max(X, 0)$ and $X^- = \max(-X, 0)$, so that $X = X^+ - X^-$, and define

$$\int X \, dP = \int X^+ \, dP - \int X^- \, dP.$$

This integral is well defined (though perhaps equal to $+\infty$ or $-\infty$) provided that at least one of the two integrals on the right is finite. If both integrals on the right are finite, then $\int X \, dP$ is both well defined and finite, and X is said to be *integrable* (with respect to P). We sometimes write $\int X(\omega) \, P(d\omega)$ for $\int X \, dP$. We usually denote the expectation of X by $E[X]$, that is,

$$E[X] = \int X \, dP.$$

The expectation operator also has a representation as an integral with respect to the distribution μ of X:

$$E[X] = \int x \, \mu(dx). \tag{1.8}$$

The right-hand integral is sometimes written as $\int x \, dF(x)$, where F is the distribution function of X; in this form it is called the Stieltjes integral with respect to F. If F is absolutely continuous with density function f, then we have the representation

$$E[X] = \int_{-\infty}^{\infty} x f(x) \, dx,$$

which is familiar from elementary probability. If X is discrete, taking on values $\{ a_n : n \in N \}$ with respective probabilities $\{ p_n : n \in N \}$, then

$$E[X] = \sum_{n \in N} a_n p_n.$$

The next results summarize basic properties of the expectation operator that are used in the text.

Proposition 1.9. *Let X and Y be random variables and a and b be real-valued constants.*

(i) X is integrable if and only if $E\left[|X|\right] < \infty$.

(ii) If $X = 0$ a.s., then $E\left[X\right] = 0$.

(iii) If X and Y are integrable with $X \leq Y$ a.s., then $E\left[X\right] \leq E\left[Y\right]$.

(iv) If X and Y are integrable, then $E\left[aX + bY\right] = aE\left[X\right] + bE\left[Y\right]$.

(v) $\left|E\left[X\right]\right| \leq E\left[|X|\right]$.

(vi) If X and Y are independent, then $E\left[XY\right] = E\left[X\right]E\left[Y\right]$.

Note that, in general, the converse to Proposition 1.9(vi) does not hold.

To state our next result, we recall the notation

$$\liminf_n x_n = \lim_{n \to \infty} \left(\inf_{m \geq n} x_m\right),$$

where $x_1, x_2, \ldots$ is a sequence of real numbers.

Proposition 1.10 (Fatou's lemma). *Let $\{X_n : n \geq 0\}$ be a sequence of nonnegative random variables. Then*

$$E\left[\liminf_n X_n\right] \leq \liminf_n E\left[X_n\right].$$

The rth *moment* of a random variable X is $E\left[X^r\right]$ and the rth *central moment* is $E\left[(X - \mu)^r\right]$, where $\mu = E\left[X\right]$—the second central moment is called the *variance* of X and denoted $\mathrm{Var}\left[X\right]$. There are many identities and inequalities for moments of random variables—a very useful inequality for our purposes is *Hölder's inequality*: let X and Y be random variables, and let p and q be constants such that $1 < p < \infty$ and $1/p + 1/q = 1$. Then

$$E\left[|XY|\right] \leq E^{1/p}\left[|X|^p\right]E^{1/q}\left[|Y|^q\right]. \tag{1.11}$$

Take $p = q = 2$ to obtain the *Cauchy–Schwarz inequality*:

$$E\left[|XY|\right] \leq E^{1/2}\left[X^2\right]E^{1/2}\left[Y^2\right].$$

In particular, $E^2\left[X\right] \leq E\left[X^2\right]$—take $Y \equiv 1$ and use the fact that $E\left[X\right] \leq E\left[|X|\right]$. Next, fix $0 < \alpha \leq \beta$ and take $X = |Z|^\alpha$, $Y \equiv 1$, and $p = \beta/\alpha$ in (1.11) to obtain

$$E^{1/\alpha}\left[|Z|^\alpha\right] \leq E^{1/\beta}\left[|Z|^\beta\right],$$

which is *Lyapunov's inequality*. Observe that if a nonnegative random variable X has a finite rth moment for some $r > 0$, then Lyapunov's inequality implies that X has a finite qth moment for $q \in (0, r]$. Now fix $r \geq 1$ and let $a, b \geq 0$. Applying Lyapunov's inequality with $\alpha = 1$ and $\beta = r$ to the random variable that equals a or b with probability $1/2$ each, we find that $(a+b)^r \leq 2^{r-1}(a^r + b^r)$. An easy argument shows that $(a+b)^r \leq a^r + b^r$ for $0 < r < 1$. Thus, if X and Y are nonnegative random variables with finite rth moments $(r > 0)$, then

$$E\left[(X+Y)^r\right] \leq c_r E\left[X^r + Y^r\right] = c_r\left(E\left[X^r\right] + E\left[Y^r\right]\right), \qquad (1.12)$$

where c_r equals 2^{r-1} or 1, depending on the value of r. The inequality in (1.12) is sometimes called the c_r-*inequality*. This result can easily be generalized to $n > 2$ nonnegative random variables:

$$E\left[(X_1 + X_2 + \cdots + X_n)^r\right] \leq c_r E\left[X_1^r + X_2^r + \cdots + X_n^r\right],$$

where $c_r = n^{r-1}$ if $r \geq 1$ and $c_r = 1$ if $r \leq 1$. We also refer to this extension as the c_r-inequality.

A useful representation of the rth moment $(r \geq 1)$ of a nonnegative random variable X is as follows:

$$E\left[X^r\right] = \int_0^\infty rx^{r-1}\overline{F}(x)\,dx, \qquad (1.13)$$

where F is the distribution function of X and $\overline{F} = 1 - F$. Taking $r = 1$, we find that

$$E\left[X\right] = \int_0^\infty \overline{F}(x)\,dx. \qquad (1.14)$$

Proposition 1.15. X *is stochastically dominated by* Y *if and only if*

$$E\left[h(X)\right] \leq E\left[h(Y)\right]$$

for every nondecreasing function h.

Proving the "if" direction of Proposition 1.15 is easy—let[3] $h(u) = 1_{(x,\infty)}(u)$ for each $x \in \Re$. The "only if" direction can be proved either by using an argument based on (1.14) or by using the fact—discussed previously—that if X is stochastically dominated by Y, then there exist random variables X' and Y', defined on the same probability space, such that X' is distributed as X, Y' is distributed as Y, and X' is dominated by Y'. It follows from Proposition 1.15 that if X is stochastically dominated by Y and Y has finite rth moment $(r \geq 0)$, then X has finite rth moment.

Fix $r \geq 1$ and observe that

$$\int_0^\infty rx^{r-1}\overline{F}(x)\,dx = \sum_{n=1}^\infty \int_{n-1}^n rx^{r-1}\overline{F}(x)\,dx$$

---

[3]Here and elsewhere, 1_A denotes the *indicator function* of the set A, that is, $1_A(x) = 1$ if $x \in A$ and $1_A(x) = 0$ if $X \notin A$. In the special case where 1_A is defined on a probability space $(\Omega, \mathcal{F}, P)$, so that $1_A = 1_A(\omega)$, then 1_A is interpreted as a random variable that equals 1 if event A occurs and equals 0 otherwise.

and

$$r(n-1)^{r-1}\overline{F}(n) \le \int_{n-1}^{n} rx^{r-1}\overline{F}(x)\,dx \le rn^{r-1}\overline{F}(n-1)$$

for $n \ge 1$. Combining these results with (1.13), we find that

$$r\sum_{n=1}^{\infty}(n-1)^{r-1}P\{|X| > n\} \le E\left[|X|^r\right] \le r\sum_{n=1}^{\infty}n^{r-1}P\{|X| > n-1\}$$

$$(1.16)$$

for any random variable X.

A useful tool for analyzing a random variable X having distribution function F is the *LaPlace–Stieltjes transform* $\mathcal{L}_F$, defined by

$$\mathcal{L}_F(s) = \int e^{sx}\,dF(x) = E\left[e^{sX}\right]$$

for all s such that the expectation is finite—the domain of definition always takes the form of a (possibly degenerate) interval that contains the origin. We often write $\mathcal{L}_X$ instead of $\mathcal{L}_F$ when there is no ambiguity. If X is, for example, an exponential random variable with intensity $q > 0$, then $\mathcal{L}_X(s) = q/(q-s)$ for $s < q$. If X is a geometric random variable with parameter $p \in (0,1)$, so that $P\{X = k\} = p(1-p)^{k-1}$ for $k \ge 1$, then

$$\mathcal{L}_X(s) = \frac{pe^s}{1 - (1-p)e^s}$$

for $s < -\log(1-p)$. The LaPlace–Stieltjes transform uniquely determines the distribution of a random variable. If X has a finite LaPlace–Stieltjes transform in some neighborhood of the origin, then X has finite moments of all orders. The function $\mathcal{L}_X(s)$ is sometimes called the *moment generating function* of X, because $\mathcal{L}_X^{(k)}(0) = E\left[X^k\right]$ for $k \ge 0$, where $\mathcal{L}_X^{(k)}$ is the kth derivative of $\mathcal{L}_X$. The following result illustrates the usefulness of the transform.

Proposition 1.17. *Let $X, X_1, X_2, \ldots$ be a sequence of i.i.d. random variables and let N be a random variable that takes values in $\{1, 2, \ldots\}$ and is independent of $\{X_n : n \ge 1\}$. Suppose that the LaPlace–Stieltjes transforms $\mathcal{L}_X$ and $\mathcal{L}_N$ are both finite in some neighborhood of the origin, and set $S_N = X_1 + X_2 + \cdots + X_N$. Then $\mathcal{L}_{S_N}(s) = \mathcal{L}_N\left(\log \mathcal{L}_X(s)\right)$ whenever the left side is finite.*

PROOF. Set $p_n = P\{N = n\}$ for $n \ge 1$. Observe that, by the i.i.d. property, $\mathcal{L}_{S_n}(s) = \mathcal{L}_X^n(s)$ for $n \ge 1$, where $S_n = X_1 + X_2 + \cdots + X_n$. Since N is independent of $\{X_n : n \ge 1\}$, we have

$$\mathcal{L}_{S_N}(s) = E\left[e^{sS_N}\right] = \sum_{n \ge 1} E\left[e^{sS_n}\right]p_n = \sum_{n \ge 1}\mathcal{L}_X^n(s)p_n.$$

But
$$\sum_{n\geq 1} \mathcal{L}_X^n(s)p_n = \sum_{n\geq 1} e^{n\log\mathcal{L}_X(s)}p_n = \mathcal{L}_N\big(\log\mathcal{L}_X(s)\big). \qquad \square$$

For example, if X is an exponential random variable with intensity q and if N is a geometric random variable with parameter p, then Proposition 1.17 implies that S_N is exponential with intensity pq.

A.1.5   Moment Results for Random Sums

In this section we give some useful moment equalities and inequalities for sums of the form $S_N = X_1 + X_2 + \cdots + X_N$, where $X_1, X_2, \ldots$ are independent and identically distributed (i.i.d.) random variables and N is a random variable taking values in $\{1, 2, \ldots\}$. These results all concern an increasing sequence of σ-fields $\{\mathcal{F}_n \colon n \geq 1\}$ such that each X_n is measurable with respect to $\mathcal{F}_n$. Intuitively, the "information" embodied in $\mathcal{F}_n$ is sufficient to determine the values of $X_1, X_2, \ldots, X_n$, and $\mathcal{F}_n$ may contain some additional information (perhaps about some auxiliary random variables $Y_1, Y_2, \ldots, Y_n$). When $\mathcal{F}_n = \sigma\langle X_1, X_2, \ldots, X_n\rangle$, so that $\mathcal{F}_n$ is the σ-field generated by $X_1, X_2, \ldots, X_n$, then $\mathcal{F}_n$ consists precisely of information about $X_1, X_2, \ldots, X_n$ and does not contain any additional information.

Definition 1.18. An integer-valued random variable N is a *stopping time* with respect to an increasing sequence of σ-fields $\{\mathcal{F}_n \colon n \geq 1\}$ if and only if $\{N \leq n\} \in \mathcal{F}_n$ for $n \geq 1$.

Roughly speaking, the "information" in $\mathcal{F}_n$ is enough to determine whether or not the event $\{N \leq n\}$ has occurred. Equivalently, the information in $\mathcal{F}_n$ determines whether or not the event $\{N = n\}$ has occurred. Typically, N denotes the random time at which some event happens, and N is a stopping time if, at any time point, the occurrence or nonoccurrence of the event can be determined from observation of the past and present, without needing to look into the future. For example, the random index $N =$ "the first time at which the state of the system changes to s" is a stopping time, whereas the random index $N - 2 =$ "two state transitions before the first time at which the state of the system changes to s" is not. If $\mathcal{F}_n = \sigma\langle X_1, X_2, \ldots, X_n\rangle$ for $n \geq 1$, then N is said to be a stopping time with respect to $\{X_n \colon n \geq 1\}$. In this case the occurrence or nonoccurrence of the event $\{N = n\}$ is completely determined by the values of $X_1, X_2, \ldots, X_n$.

Part (i) of Proposition 1.19 below is known as *Wald's moment identity* and part (ii) as *Wald's second moment identity*. In the proposition denote by μ the common mean and by σ^2 the common variance of $X_1, X_2, \ldots$ whenever these quantities exist.

Proposition 1.19. *Let $S_N = \sum_{n=1}^{N} X_n$, where $\{X_n \colon n \geq 1\}$ is a sequence of i.i.d. random variables and N is a stopping time with respect* to an increasing sequence of σ-fields $\{\mathcal{F}_n : n \geq 1\}$ such that X_n is measurable with respect to $\mathcal{F}_n$ for $n \geq 1$ and independent of $\mathcal{F}_{n-1}$ for $n \geq 2$. Then

(i) $E[S_N] = \mu \cdot E[N]$ if either $E[|X_1|] < \infty$ and $E[N] < \infty$ or if $X_1 \geq 0$, and

(ii) $E[(S_N - N\mu)^2] = \sigma^2 \cdot E[N]$ if $\sigma^2 < \infty$ and $E[N] < \infty$.

The next result gives an inequality rather than an equality, but applies to moments of S_N higher than the second moment.

Proposition 1.20. Let $S_N = \sum_{n=1}^{N} X_n$, where $\{X_n : n \geq 1\}$ is a sequence of i.i.d. random variables and N is a stopping time with respect to an increasing sequence of σ-fields $\{\mathcal{F}_n : n \geq 1\}$ such that X_n is measurable with respect to $\mathcal{F}_n$ for $n \geq 1$ and independent of $\mathcal{F}_{n-1}$ for $n \geq 2$. Then for $r \geq 0$ there exists a constant b_r (depending only on r) such that

$$E[|S_N|^r] \leq b_r E[|X_1|^r] E[N^r].$$

Remark 1.21. When N is independent of $\{X_n : n \geq 1\}$, we can apply Proposition 1.20 by taking $\mathcal{F}_n = \sigma\langle X_1, Y_1, \ldots, X_n, Y_n \rangle$, where $Y_k = 1_{\{N \leq k\}}$ for $1 \leq k \leq n$.

A.1.6 General Integrals

The foregoing development of the integral can be generalized to an arbitrary measure space $(\Omega, \mathcal{F}, \mu)$; that is, μ need not be a probability measure. In this general setup, the integral $\int f \, d\mu$ or, equivalently, $\int f(\omega) \, \mu(d\omega)$ can be defined almost exactly as before for each measurable real-valued function f. We also define

$$\int_A f \, d\mu = \int f 1_A \, d\mu$$

for $A \in \mathcal{F}$.

Lemma 1.22. Let f be a nonnegative measurable function defined on a measure space $(\Omega, \mathcal{F}, \mu)$ and let $A \in \mathcal{F}$ satisfy $\mu(A) < \infty$. Suppose that $\int_A f \, d\mu > 0$. Then there exist a measurable set $Q \subseteq A$ and a number $\gamma > 0$ such that $\mu(Q) > 0$ and $f(\omega) > \gamma$ for $\omega \in Q$.

PROOF. Fix $\gamma > 0$ and set $Q_\gamma = \{\omega : f(\omega) > \gamma\}$. Suppose that $\mu(Q_\gamma) = 0$ for $\gamma > 0$. Then

$$\int_A f \, d\mu = \int_{A \cap Q_\gamma^c} f \, d\mu \leq \gamma \mu(A)$$

for $\gamma > 0$. Because $\mu(A) < \infty$ and γ is arbitrary, it follows that $\int_A f \, d\mu = 0$, a contradiction. Thus $\mu(Q_\gamma) > 0$ for at least one value of γ. $\qquad \square$

Lemma 1.23. *For a measure space $(\Omega, \mathcal{F}, \mu)$, let $A \in \mathcal{F}$ satisfy $\mu(A) > 0$ and let f be a nonnegative measurable function such that $f(\omega) > 0$ for $\omega \in A$. Then $\int_A f \, du > 0$.*

PROOF. Set $A_\gamma = \{\omega \in A: f(\omega) \geq \gamma\}$. Since $A_\gamma \uparrow A$ as $\gamma \downarrow 0$, it follows from Proposition 1.1(iv) that $P(A_\gamma) \to P(A) > 0$, and hence that $P(A_\gamma) > 0$ for some $\gamma > 0$. For this value of γ,

$$\int_A f \, d\mu \geq \int_{A_\gamma} f \, d\mu \geq \gamma \mu(A_\gamma) > 0. \qquad \square$$

To compare the modelling power of different formalisms for discrete-event systems in Chapter 4, we use the following "change of variable" result. Let $(\Omega, \mathcal{F})$ and $(\Omega', \mathcal{F}')$ be measurable spaces and let ϕ be a mapping from Ω to Ω'. We assume that ϕ is *measurable* in that $\phi^{-1}A' \in \mathcal{F}$ for all $A' \in \mathcal{F}'$, where $\phi^{-1}A' = \{x \in \Omega: \phi x \in A'\}$. For a measure μ on $\mathcal{F}$, we define a measure ν on $\mathcal{F}'$ by setting $\nu(A') = \mu(\phi^{-1}A')$ for $A' \in \mathcal{F}'$, and for a measurable real-valued function f on Ω', we define the function $f \circ \phi$ on Ω by setting $(f \circ \phi)(\omega) = f(\phi\omega)$ for $\omega \in \Omega$.

Proposition 1.24 (Change of variable). *Let f be a measurable real-valued function defined on $(\Omega', \mathcal{F}')$ and ϕ a measurable mapping from $(\Omega, \mathcal{F}, \mu)$ to $(\Omega', \mathcal{F}')$. Then f is integrable with respect to ν if and only if $f \circ \phi$ is integrable with respect to μ, in which case*

$$\int_{\phi^{-1}A'} f \circ \phi \, d\mu = \int_{A'} f \, d\nu.$$

We sometimes express the conclusion of Proposition 1.24 using the following notation:

$$\int_{\phi^{-1}A'} f(\phi\omega) \, \mu(d\omega) = \int_{A'} f(\omega) \, \mu(\phi^{-1}d\omega').$$

The representation of expected value as in (1.8) is a consequence of this result, where we take f as the identity function, $(\Omega', \mathcal{F}') = (\Re, \mathcal{B})$, $A' = \Omega'$, $\mu = P$, and $\phi\omega = X(\omega)$.

The next result concerns integrals of functions defined on a *product space* $(X \times Y, \mathcal{X} \times \mathcal{Y}, \pi)$ composed from the measure spaces $(X, \mathcal{X}, \mu)$ and $(Y, \mathcal{Y}, \nu)$. Here $X \times Y$ is the usual Cartesian product of X and Y, the σ-field $\mathcal{X} \times \mathcal{Y}$ is the σ-field generated by $\mathcal{A} = \{A \times B: A \in \mathcal{X} \text{ and } B \in \mathcal{Y}\}$, and π is the unique measure that satisfies $\pi(A \times B) = \mu(A)\nu(B)$ for $A \in \mathcal{X}$ and $B \in \mathcal{Y}$. The measure π is often called the *product measure* of μ and ν.

Proposition 1.25 (Fubini's theorem). *Let f be a function defined on the product space $(X \times Y, \mathcal{X} \times \mathcal{Y}, \pi)$, and suppose that either $f \geq 0$ or f is integrable with respect to π. Then*

$$\int_{X \times Y} f \, d\pi = \int_X \left(\int_Y f(x, y)\nu(dy) \right) \mu(dx) = \int_Y \left(\int_X f(x, y)\mu(dx) \right) \nu(dy).$$

$$(1.26)$$

Under the conditions of the proposition, the leftmost *double* integral is equal to each of the two *iterated* integrals. This result is usually used to justify the interchange of the order of integration in an iterated integral. The assertion that (1.26) holds when f is nonnegative is known as *Tonelli's theorem*. Proposition 1.25 can be used to obtain the identity in (1.13).

A.1.7   Conditional Expectation and Probability

Consider a probabilistic experiment that is described by a probability space $(\Omega, \mathcal{F}, P)$, along with an "observer" who has some degree of information about the experiment. First suppose that the observer has no information. If we ask the observer to assess the probability that event A has occurred— in other words, that the outcome ω is an element of A—then the answer is $P(A)$. Now suppose that the observer knows that event B has occurred, where $P(B) > 0$. Then the answer is $P(A \mid B) = P(A \cap B)/P(B)$, the "conditional probability" of event A, given that event B has occurred. Similarly, knowing that B has occurred, the observer computes the expected value of a random variable X defined on $(\Omega, \mathcal{F}, P)$ as $E[X \mid B] = E[X 1_B]/P(B)$. Before we conduct the experiment, and knowing that we will tell the observer whether or not B has occurred, we can view the observer's future assessment of the probability of A as a random variable Z, where

$$Z(\omega) = \begin{cases} P(A \cap B)/P(B) & \text{if } \omega \in B; \\ P(A \cap B^c)/P(B^c) & \text{if } \omega \in B^c. \end{cases}$$

Denote by $\mathcal{G} = \{\Omega, \varnothing, B, B^c\}$ the σ-field that represents "complete information about whether or not B has occurred." Then the random variable Z is said to be the "conditional probability of A, given $\mathcal{G}$," and we write $P(A \mid \mathcal{G})$ for Z—sometimes we write $P(A \mid \mathcal{G})_\omega$ to emphasize the dependence on ω. We can similarly define the conditional expectation $E[X \mid \mathcal{G}]$ as

$$E[X \mid \mathcal{G}] = E[X \mid \mathcal{G}]_\omega = \begin{cases} E[X 1_B]/P(B) & \text{if } \omega \in B; \\ E[X 1_{B^c}]/P(B^c) & \text{if } \omega \in B^c. \end{cases}$$

An easy calculation shows that

$$E[X] = E[E[X \mid \mathcal{G}]] = E[X \mid B] P(B) + E[X \mid B^c] P(B^c).$$

Now consider an arbitrary random variable X defined on a probability space $(\Omega, \mathcal{F}, P)$, along with a σ-field $\mathcal{G} \subseteq \mathcal{F}$. Motivated by the above discussion, we define a *conditional expectation* $E[X \mid \mathcal{G}]$ to be a random variable such that

1. $E[X \mid \mathcal{G}]$ is measurable with respect to $\mathcal{G}$ and integrable.

2. For all $A \in \mathcal{G}$, $E[X \mid \mathcal{G}]$ satisfies the functional equation

$$E[1_A E[X \mid \mathcal{G}]] = E[1_A X]. \tag{1.27}$$

Setting $A = \Omega$ in (1.27), we obtain the *law of total expectation*: $E[X] = E[E[X \mid \mathcal{G}]]$. To establish the condition in (2) for all $A \in \mathcal{G}$, it suffices to show that (1.27) holds for all A belonging to a "π-system" that generates $\mathcal{G}$. A collection $\mathcal{P}$ of subsets of Ω is a π-system if it is closed under finite intersections: $A \cap B \in \mathcal{P}$ whenever $A, B \in \mathcal{P}$. In general, many random variables satisfy the conditions in (1) and (2) above, but any two such random variables differ only on a set of probability 0. Unless otherwise noted, by "the" conditional probability we mean an arbitrary member of the foregoing collection of random variables. For most purposes, the particular version of conditional expectation that is chosen is immaterial.

The following proposition gives some elementary properties of conditional expectation, many of which coincide with properties of ordinary expectation.

Proposition 1.28. *Let X and Y be random variables, and let a, b, and c be real-valued constants.*

(i) *If $X = c$ a.s., then $E[X \mid \mathcal{G}] = c$ a.s..*

(ii) *If X and Y are integrable with $X \leq Y$ a.s., then $E[X \mid \mathcal{G}] \leq E[Y \mid \mathcal{G}]$ a.s..*

(iii) *$|E[X \mid \mathcal{G}]| \leq E[|X| \mid \mathcal{G}]$ a.s..*

(iv) *If X and Y are integrable, then $E[aX + bY \mid \mathcal{G}] = aE[X \mid \mathcal{G}] + bE[Y \mid \mathcal{G}]$ a.s..*

The next two results give key properties of conditional expectation that we use repeatedly throughout the text.

Proposition 1.29. *Suppose that X is measurable with respect to $\mathcal{G}$ and that both Y and XY are integrable. Then*

$$E[XY \mid \mathcal{G}] = XE[Y \mid \mathcal{G}] \text{ a.s..}$$

Thus, if X is determined by the information in $\mathcal{G}$, it can be "pulled out" of a conditional expectation with respect to $\mathcal{G}$.

Proposition 1.30. *Let X be an integrable random variable, and let $\mathcal{G}_1$ and $\mathcal{G}_2$ be σ-fields such that $\mathcal{G}_1 \subseteq \mathcal{G}_2$. Then*

$$E[E[X \mid \mathcal{G}_1] \mid \mathcal{G}_2] = E[E[X \mid \mathcal{G}_2] \mid \mathcal{G}_1] = E[X \mid \mathcal{G}_1] \text{ a.s..}$$

The *conditional probability* $P(A \mid \mathcal{G})$ is defined by

$$P(A \mid \mathcal{G}) = E[1_A \mid \mathcal{G}]$$

for $A \in \mathcal{F}$, and the basic properties of conditional probability follow from the corresponding properties of conditional expectation. The reader may wonder whether, for a given σ-field $\mathcal{G} \subseteq \mathcal{F}$, we can choose a version of $P(A \mid \mathcal{G})$ for each $A \in \mathcal{F}$ such that, for fixed ω, the set function $P(\cdot \mid \mathcal{G})_\omega$ is a probability measure. In general, such a choice is impossible. For a given σ-field $\mathcal{G}$ and random variable X, however, we can define a function μ on $\mathcal{B} \times \Omega$ such that (1) $\mu(\cdot, \omega)$ is a probability measure for each $\omega \in \Omega$ and (2) $\mu(H, \cdot)$ is a version of $P\{X \in H \mid \mathcal{G}\}$ for each $H \in \mathcal{B}$. The probability measure $\mu(\cdot, \omega)$ is a *conditional distribution* of X, given $\mathcal{G}$.

If X is a random variable and $\mathcal{H}$ is a collection of random variables, then we use the notation $E[X \mid \mathcal{H}]$ to denote the conditional expectation $E[X \mid \mathcal{G}]$, where $\mathcal{G}$ is the σ-field generated by the random variables in $\mathcal{H}$. For example, $E[X \mid Y]$ is interpreted as $E[X \mid \sigma\langle Y \rangle]$.

All of the classical conditional probability formulas follow from the foregoing general framework. For example, suppose that the real-valued random vector (X, Y) has a joint density function f. Then $E[X \mid Y] = g(Y)$ a.s., where

$$g(y) = \frac{\int x f(x, y) \, dx}{\int f(x, y) \, dx}.$$

Since $g(Y)$ is clearly measurable with respect to $\sigma\langle Y \rangle$, showing that $g(Y)$ is a version of $E[X \mid Y]$ amounts to showing that $g(Y)$ satisfies the relation in (1.27). Thus it suffices to show that $E[1_A g(Y)] = E[1_A X]$ for all $A \in \sigma\langle Y \rangle$. Fix a set A of the form $A = \{Y \in E\}$, where E is a Borel set. Using Fubini's theorem, we have

$$E[1_A g(Y)] = \iint 1_E(y) g(y) f(x, y) \, dx \, dy$$

$$= \int 1_E(y) g(y) \left(\int f(x, y) \, dx \right) dy$$

$$= \iint 1_E(y) x f(x, y) \, dx \, dy$$

$$= E[1_A X],$$

and the desired result follows.

A.1.8 Stochastic Convergence

Limit theorems for random variables involve several different modes of convergence.

Definition 1.31. Let X and $\{X_n : n \geq 1\}$ be random variables defined on a common probability space $(\Omega, \mathcal{F}, P)$. Then X_n *converges with probability 1* to X if

$$P\{\lim_{n \to \infty} X_n = X\} = 1.$$

We also say that X_n converges to X *almost surely* (a.s.), and we often write "$X_n \to X$ a.s. as $n \to \infty$" or "$\lim_{n \to \infty} X_n = X$ a.s.." It can be shown that

$\lim_{n\to\infty} X_n = X$ a.s. if and only if, for all $\epsilon > 0$,

$$\lim_{m\to\infty} P\{|X_n - X| \le \epsilon \text{ for all } n \ge m\} = 1.$$

The following result, due to Kolmogorov, gives necessary and sufficient conditions for a.s. convergence of $\{S_n : n \ge 0\}$ when each S_n is a partial sum: $S_n = \sum_{i=1}^{n} X_i$. Denote by $X^{(c)}$ the random variable X truncated at c: $X^{(c)} = X1_{\{|X| \le c\}}$.

Proposition 1.32 (Three-series theorem). *Let $\{X_n : n \ge 1\}$ be a sequence of mutually independent random variables, and consider the three series*

$$\sum_n P\{|X_n| > c\}, \qquad \sum_n E[X_n^{(c)}], \qquad \sum_n \mathrm{Var}[X_n^{(c)}].$$

In order that $\sum_n X_n$ converge with probability 1, it is necessary that the three series converge for all positive c and sufficient that they converge for some positive c.

The next result relates a.s. convergence to convergence of moments.

Proposition 1.33. *Let $X, X_1, X_2, \ldots$ be random variables defined on a probability space $(\Omega, \mathcal{F}, P)$.*

(i) *(Monotone convergence) If each X_n is nonnegative and $X_n \uparrow X$ a.s., then $E[X_n] \to E[X]$.*

(ii) *(Dominated convergence) If $\sup_n |X_n| \le Y$ a.s. for some integrable random variable Y and $X_n \to X$ a.s., then X and the X_n are integrable and $E[X_n] \to E[X]$.*

(iii) *(Bounded convergence) If $\sup_n |X_n| \le c$ a.s. for some finite constant c, then X and the X_n are integrable and $E[X_n] \to E[X]$.*

Definition 1.34. Let X and $\{X_n : n \ge 1\}$ be random variables defined on a common probability space $(\Omega, \mathcal{F}, P)$. Then X_n *converges in probability* to X if

$$\lim_{n\to\infty} P\{|X_n - X| \le \epsilon\} = 1$$

for $\epsilon > 0$, and we write $X_n \xrightarrow{\text{pr}} X$.

We write $X \in L^p$ $(p \ge 0)$ if $E[|X|^p] < \infty$.

Definition 1.35. Let X and $\{X_n : n \ge 1\}$ be random variables defined on a common probability space $(\Omega, \mathcal{F}, P)$ such that $X, X_1, X_2, \ldots \in L^p$. Then X_n *converges in L^p* to X if

$$\lim_{n\to\infty} E[|X_n - X|^p] = 0.$$

For random variables X and $\{X_n : n \geq 1\}$, set $F(x) = P\{X \leq x\}$ and $F_n(x) = P\{X_n \leq x\}$.

Definition 1.36. A sequence $\{X_n : n \geq 1\}$ *converges in distribution* to X (or *converges weakly to X*) if

$$\lim_{n \to \infty} F_n(x) = F(x)$$

for all x at which the function F is continuous, and we write $X_n \Rightarrow X$.

Observe that the random variables involved in the foregoing definition need not be defined on the same probability space.

Proposition 1.37. *$X_n \Rightarrow X$ if and only if $E[h(X_n)] \to E[h(X)]$ for every bounded continuous function h.*

The above characterization can serve as a definition of weak convergence in settings more complicated than that of real-valued random variables. For example, we can easily extend the notion of weak convergence to the setting of $\Re^l$-valued random vectors—see also Section A.2.5.

Let μ and ν be probability measures defined on a common measurable space $(\Omega, \mathcal{F})$. The *total variation distance* between μ and ν, denoted by $\|\mu - \nu\|$, is defined as

$$\|\mu - \nu\| = \sup_{A \in \mathcal{F}} |\mu(A) - \nu(A)|.$$

The following definition applies to random variables $X, X_1, X_2, \ldots$, having respective distributions $\mu, \mu_1, \mu_2, \ldots$.

Definition 1.38. A sequence $\{X_n : n \geq 1\}$ *converges in total variation* to X if

$$\lim_{n \to \infty} \|\mu_n - \mu\| = 0,$$

and we write $X_n \overset{\text{tv}}{\to} X$.

Convergence in total variation can be viewed as a uniform version of convergence in distribution. If the random variables $X, X_1, X_2, \ldots$ take values in a finite or countably infinite state space, then the notions of convergence in distribution and total variation convergence coincide.

The following result gives some key relationships between the various modes of convergence. Recall that X is dominated by Y if $X \leq Y$ a.s.. We say that the sequence $\{X_n : n \geq 0\}$ is dominated by Y if each X_n is dominated by Y.

Proposition 1.39. *Let X and $\{X_n : n \geq 0\}$ be random variables, and let c be a real-valued constant.*

(i) If $X_n \to X$ a.s., then $X_n \overset{pr}{\to} X$.

(ii) If $X_n \overset{pr}{\to} X$, then $X_n \Rightarrow X$.

(iii) If $X_n \Rightarrow c$, then $X \overset{pr}{\to} c$.

(iv) If $X_n \overset{tv}{\to} X$, then $X_n \Rightarrow X$.

(v) If $X_n \to X$ in L^p, then $X_n \overset{pr}{\to} X$, and hence $X_n \Rightarrow X$.

(vi) If $X_n \overset{pr}{\to} X$ and $\{|X_n - X| : n \geq 1\}$ is dominated by some random variable $Y \in L^p$, then $X_n \to X$ in L^p.

Remark 1.40. The goal in a simulation experiment usually is to estimate some unknown constant c that quantifies the performance of the system under study. Suppose that $\{X_n : n \geq 1\}$ is a sequence of estimators of c, indexed by the length n of the simulation. If $X_n \to c$ a.s., then we say that "X_n is *strongly consistent* for c." If $X_n \Rightarrow c$ or, equivalently, $X_n \to c$ in probability, then we say that "X_n is *(weakly) consistent* for c." Finally, if $E[X_n] = c$ for each n, then we say that "X_n is *unbiased* for c."

The following propositions, which pertain specifically to convergence in distribution, are used frequently in the text.

Recall that if U is a random variable uniformly distributed on $[0,1]$ and F is a distribution function with inverse F^{-1} defined by

$$F^{-1}(u) = \inf\{x : F(x) > u\}$$

for $u \in [0,1]$, then the random variable $Z = F^{-1}(U)$ has distribution function F. Use of this trick leads to the following result.

Proposition 1.41 (Skorohod's theorem). *If $X_n \Rightarrow X$, then there exist random variables $X', X'_1, X'_2, \ldots$ defined on a common probability space $(\Omega, \mathcal{F}, P)$ such that X' is distributed as X, X'_n is distributed as X_n for $n \geq 1$, and $X'_n(\omega) \to X(\omega)$ for each $\omega \in \Omega$.*

The idea is to set $X'_n = F_n^{-1}(U)$ for $n \geq 1$, where F_n is the distribution function of X_n and U is uniformly distributed on $[0,1]$. Skorohod's theorem can be extended to random vectors $X, X_1, X_2, \ldots \in \Re^l$ ($l > 1$), but a more complicated argument is required.

The following result on continuous mappings follows almost immediately from Skorohod's theorem. Denote by $D(h)$ the set of discontinuity points for the real-valued function h, so that $x \in D(h)$ if and only if there exists a sequence $x_1, x_2, \ldots$ such that $x_n \to x$ but $h(x_n) \not\to h(x)$.

Proposition 1.42 (Continuous mapping theorem). *If $X_n \Rightarrow X$ and $P\{X \in D(h)\} = 0$, then $h(X_n) \Rightarrow h(X)$.*

Observe that $P\{X \in D(h)\} = 0$ trivially whenever h is a continuous function. It can be shown that the set $D(h)$ is at most countably infinite, and it follows that $P\{X \in D(h)\} = 0$ whenever the distribution function of X is absolutely continuous. Finally, if the state space S of the random variables $X, X_1, X_2, \ldots$ is finite or countably infinite, then $P\{X \in D(h)\} = 0$ automatically[4]. As with Skorohod's theorem, Proposition 1.42 can be extended to random variables taking values in $\Re^l$ for $l > 1$.

It can be shown that if $\{(X_n, Y_n) : n \geq 1\}$ is a sequence of random pairs such that $X_n \Rightarrow X$ and $Y_n \Rightarrow c$ for some random variable X and real-valued constant c, then $(X_n, Y_n) \Rightarrow (X, c)$. This fact, combined with the continuous mapping theorem, leads to the following result.

Proposition 1.43 (Slutsky's theorem). *If $X_n \Rightarrow X$ and if $Y \Rightarrow c$ for some real-valued constant c, then*

(i) $X_n + Y_n \Rightarrow X + c$;

(ii) $Y_n X_n \Rightarrow cX$; *and*

(iii) $X_n/Y_n \Rightarrow X/c$ *provided $c \neq 0$.*

The following important corollary to Slutsky's theorem is obtained by taking $Y_n = X'_n - X_n$, where $X_n - X'_n \Rightarrow 0$ and applying the result in (i).

Proposition 1.44 (Converging-together lemma). *If $X_n \Rightarrow X$ and $X_n - X'_n \Rightarrow 0$, then $X'_n \Rightarrow X$.*

Next, using Skorohod's theorem together with standard results for Taylor series, we obtain Proposition 1.45, which can be viewed as complementary to the continuous mapping theorem. Given a real-valued function f defined on $\Re^l$ ($l \geq 1$), we denote by $\nabla f(\alpha)$ the gradient of f at the point $\alpha \in \Re^l$. We assume that elements of $\Re^l$ are column vectors and denote by x^t the transpose of $x \in \Re^l$.

Proposition 1.45 (The delta method). *Let $X, X_1, X_2, \ldots$ be random vectors taking values in $\Re^l$ for some $l \geq 1$, and suppose that*

$$\gamma_n(X_n - \alpha) \Rightarrow X$$

for some constant $\alpha \in \Re^l$ and sequence $\{\gamma_n : n \geq 1\}$ of nonnegative constants such that $\gamma_n \to \infty$. Let f be a real-valued function that is differentiable at α. Then

$$\gamma_n\big(f(X_n) - f(\alpha)\big) \Rightarrow \nabla f(\alpha)^t X.$$

The "Cramér–Wold theorem" can be used to extend weak-convergence results for real-valued random variables to corresponding results for $\Re^l$-valued random vectors. In the following, $\{X_n : n \geq 0\}$ and Y are $\Re^l$-valued

---

[4]Here S is viewed as being endowed with the "discrete topology" in which all subsets of S are open and all functions defined on S are continuous. S can be metrized by defining a distance function ρ such that $\rho(x, y)$ equals 0 if $x = y$ and equals 1 otherwise.

random vectors for some $l > 1$, and we write $X_n = (X_{n,1}, X_{n,2}, \ldots, X_{n,l})$
for $n \geq 0$ and $Y = (Y_1, Y_2, \ldots, Y_l)$.

Proposition 1.46 (Cramér–Wold theorem). $X_n \Rightarrow Y$ *in* $\Re^l$ *if and
only if*

$$\sum_{i=1}^{l} u_i X_{n,i} \Rightarrow \sum_{i=1}^{l} u_i Y_i$$

in $\Re$ *for each* $(u_1, u_2, \ldots, u_l) \in \Re^l$.

Limit theorems in discrete time often can be converted to limit theorems
in continuous time by invoking the following result.

Proposition 1.47. *The stochastic process* $\{X(t) : t \geq 0\}$ *converges in distribution to* X *at* $t \to \infty$ *if and only if* $X(t_n) \Rightarrow X$ *for every subsequence*
$t_n \to \infty$.

The next two results relate convergence in distribution to convergence
of moments. The first result combines Skorohod's theorem with Fatou's
lemma (Proposition 1.10) to yield a version of Fatou's lemma for weak
convergence.

Proposition 1.48. *If* $X_n \Rightarrow X$, *then* $E[\|X\|] \leq \liminf_n E[\|X_n\|]$.

To state the second result, we need to introduce the notion of "uniform
integrability."

Definition 1.49. A sequence of random variables $\{X_n : n \geq 1\}$ is *uniformly integrable* if

$$\lim_{x \to \infty} \sup_{n \geq 0} E\left[|X_n| 1_{\{|X_n| > x\}}\right] = 0.$$

Sufficient conditions for uniform integrability are that

$$\sup_{n \geq 0} E[|X_n|^{1+\epsilon}] < \infty$$

for some $\epsilon > 0$ or that $\{X_n : n \geq 1\}$ is stochastically dominated by an
integrable random variable X.

Proposition 1.50. *If* $X_n \Rightarrow X$ *and* $\{X_n : n \geq 1\}$ *is uniformly integrable,
then* X *is integrable and* $E[X_n] \to E[X]$.

A.2   Limit Theorems for Stochastic Processes

A.2.1   Definitions and Existence Theorem

A stochastic process $\{X(t) : t \in T\}$ is an indexed collection of random variables taking values in a set S, where either $T = [0, \infty)$ or $T = \{0, 1, 2, \ldots\}$.

In the latter case, we usually use the notation $\{X_n: n \geq 0\}$ for a discrete-time process. A realization $\{x(t): t \in T\}$ of a stochastic process is called a *sample path* of the process. We focus throughout on processes with a state space $(S, \mathcal{S})$ that is a well-behaved subset of $(\Re^l, \mathcal{B}^l)$ for some $l \geq 1$; recall from Section A.1.2 that $\mathcal{B}^l$ are the l-dimensional Borel sets. More specifically, we require that S be a complete, separable metric space.[5]

A stochastic process $\{X(t): t \in T\}$ typically is defined in one of two ways. The first approach is to construct $\{X(t): t \in T\}$ in terms of a previously defined stochastic process. For example, in Chapter 3 the marking process of an SPN is defined in terms of an underlying general state-space Markov chain. Also in that chapter, a CTMC is constructed from the process $\{(Y_n, T_n): n \geq 0\}$, where $\{Y_n: n \geq 0\}$ is a DTMC and $\{T_n: n \geq 0\}$ is a sequence of exponential random variables with intensities that depend on $\{Y_n: n \geq 0\}$.

The other approach is to specify a set of finite-dimensional distributions and then show that there exists a process having these distributions. We outline this approach in the setting of discrete-time processes. Recall from Section A.1.6 the definition of the product space $(S^k, \mathcal{S}^k)$, where $k \geq 2$. For a discrete-time process $\{X_n: n \geq 0\}$, the finite-dimensional distributions are given by

$$P_{n_0 \cdots n_k}(A) = P\{(X_{n_0}, \ldots, X_{n_k}) \in A\},$$

where $k \geq 0$, $A \in \mathcal{S}^k$, and $n_0, \ldots, n_k$ are distinct indices. Proposition 2.1 below, known as the *Kolmogorov existence theorem*, gives some sufficient "consistency conditions" on the finite-dimensional distributions under which the existence of $\{X_n: n \geq 0\}$ is guaranteed.

The version of $\{X_n: n \geq 0\}$ guaranteed by the proposition is defined on the product space $(S^\infty, \mathcal{S}^\infty)$. Here the notion of product space extends the definition given in Section A.1.6. An element of S^∞ is a sequence $(\omega_0, \omega_1, \ldots)$ with each ω_i an element of S. In other words, an element of S^∞ is a possible sample path of the process. For $\omega \in S^\infty$ and $n \geq 0$, define the *coordinate projection function* $Z_n: S^\infty \mapsto S$ by $Z_n(\omega) = \omega_n$. Then $\mathcal{S}^\infty$ is defined as $\sigma\langle Z_n: n \geq 0\rangle$, that is, as the σ-field generated by the coordinate projection functions.

Proposition 2.1. *Suppose that for any integer $k \geq 0$, indices $0 \leq n_0 < n_1 < \cdots < n_k$, events $A_0, A_1, \ldots, A_k \in \mathcal{S}$, and permutation π of $\{0, 1, \ldots, k\}$,*

$$P_{n_0 \cdots n_k}(A_0 \times \cdots \times A_k) = P_{n_{\pi 0} \cdots n_{\pi k}}(A_{\pi 0} \times \cdots \times A_{\pi k})$$

---

[5]For a metric space S with metric ρ, recall that a *Cauchy sequence* $\{x_n: n \geq 1\}$ has the property that, for $\epsilon > 0$, $\rho(x_n, x_m) < \epsilon$ for all sufficiently large n and m. The space S is *complete* if every Cauchy sequence converges to a limit in S. If there exists a countable subset $S_0 \subseteq S$ such that every point in S is the limit of a sequence of points in S_0, then S is said to be *separable*. For the space $\Re$, the rationals play the role of the points in S_0.

and

$$P_{n_0 \cdots n_{k-1}}(A_1 \times \cdots \times A_{k-1}) = P_{n_0 \cdots n_{k-1} n_k}(A_1 \times \cdots \times A_{k-1} \times S).$$

Then there exists a probability measure P on the product space $(S^\infty, \mathcal{S}^\infty)$ such that the coordinate-projection process $\{ Z_n : n \geq 0 \}$ defined on $(S^\infty, \mathcal{S}^\infty, P)$ has the $P_{n_0 \cdots n_k}$ as its finite-dimensional distributions.

EXAMPLE 2.2 (Underlying chain of an SPN). To illustrate the application of Proposition 2.1, we consider in detail the construction of the underlying chain $\{ (S_n, C_n) : n \geq 0 \}$ for an SPN with marking set G and with M transitions; see the discussion in Section 3.1.1. The chain takes values in the measurable space $(\Sigma, \mathcal{S})$, where $\Sigma = \bigcup_{s \in G}(\{ s \} \times C(s))$ and $C(s)$ is the set of possible clock-reading vectors when the marking is s. In this setting, $\mathcal{S}$ is taken as the σ-field generated by sets of the form $\{ s \} \times (C(s) \cap B)$ with $s \in S$ and $B \in \mathcal{B}^M$. Recall that the underlying chain is specified by giving the initial distribution μ and the transition kernel[6] P. We can then define a set of finite-dimensional distributions $P_{n_0 \cdots n_k}$ as follows. Set

$$P_n^*(A) = \int_{A_0} \mu\big(d(s_0, c_0)\big) \int_{A_1} P\big((s_0, c_0), d(s_1, c_1)\big)$$

$$\cdots \int_{A_{n-1}} P\big((s_{n-2}, c_{n-2}), d(s_{n-1}, c_{n-1})\big) P\big((s_{n-1}, c_{n-1}), A_n\big)$$

for $n \geq 0$ and $A = A_0 \times A_1 \times \cdots \times A_n$. For $k \geq 0$, distinct indices $n_0, n_1, \ldots, n_k$ and event $A = A_0 \times A_1 \times \cdots \times A_k$, set $P_{n_0 \cdots n_k}(A) = P_m^*(B)$, where $m = \max(n_0, n_1, \ldots, n_k)$ and $B = B_0 \times B_1 \times \cdots \times B_m$ with $B_l = A_j$ if $l = n_j$ for some j and $B_n = \Sigma$ otherwise. It follows from standard measure-theoretic arguments that $P_{n_0 \cdots n_k}$ can be uniquely extended to a probability measure on $(\Sigma^k, \mathcal{S}^k)$—this probability measure constitutes the desired finite-dimensional distribution for the specified values of k and $n_0, n_1, \ldots, n_k$. The conditions of Proposition 2.1 hold almost by definition, so that there exists a measure P as in the conclusion of the proposition—write P_μ for P to emphasize the dependence on μ. Thus the chain $\{ (S_n, C_n) : n \geq 0 \}$ can be defined on the probability space $(\Sigma^\infty, \mathcal{S}^\infty, P_\mu)$ as the coordinate projection function.[7]

---

[6]In general, a *transition kernel* on a measurable space $(S, \mathcal{S})$ is a mapping P from $S \times \mathcal{S}$ to $[0, 1]$ such that, for fixed $x \in S$, the mapping $P(x, \cdot)$ is a probability measure and, for fixed $A \in \mathcal{S}$, the function $P(\cdot, A)$ is measurable.

[7]Other definitions of the underlying chain are possible. One common construction defines the chain on the probability space $([0, 1]^\infty, \mathcal{B}_0^\infty, \mu_\infty^{\text{Leb}})$, where $\mathcal{B}_0$ is the σ-field generated by the open subsets of $[0, 1]$ and $\mu_\infty^{\text{Leb}} = \mu_0^{\text{Leb}} \times \mu_0^{\text{Leb}} \times \cdots$ with μ_0^{Leb} equal to Lebesgue measure on $[0, 1]$. This probability space corresponds to the probabilistic experiment in which we generate a sequence $U_0, U_1, \ldots$ of mutually independent random

As discussed previously, the σ-field $\mathcal{F}_k = \sigma\langle X_0, X_1, \ldots, X_k \rangle$ represents "complete information" about the process $\{X_n : n \geq 0\}$ until time k. For a random variable K that is a stopping time with respect to $\{X_n : n \geq 0\}$, we define $\mathcal{F}_K$ as the σ-field generated by sets of the form $\{K \leq k\} \cap A_k$, where $k \geq 0$ and $A_k \in \mathcal{F}_k$. Informally, $\mathcal{F}_K$ represents "complete information" about $\{X_n : n \geq 0\}$ until the random time K.

A.2.2   I.I.D., O.I.D., and Stationary Sequences

Perhaps the simplest stochastic process is a sequence $\{X_n : n \geq 0\}$ of i.i.d. random variables. We now state the key limit theorems for such a process. In the following, set $\bar{X}_n = (1/n) \sum_{i=0}^{n-1} X_i$ and, as usual, denote by $N(0,1)$ a standard normal random variable.

Proposition 2.3 (Strong law of large numbers). *Let* $\{X_n : n \geq 0\}$ *be a sequence of i.i.d. random variables, and suppose that* $\mu = E[X_0] < \infty$. *Then*

$$\lim_{n \to \infty} \bar{X}_n = \mu \ a.s..$$

Proposition 2.4 (Central limit theorem). *Let* $\{X_n : n \geq 0\}$ *be a sequence of i.i.d. random variables with common mean* μ, *and suppose that* $\sigma^2 = \mathrm{Var}[X_0] < \infty$. *Then*

$$\sqrt{n}(\bar{X}_n - \mu) \Rightarrow \sigma N(0,1)$$

as $n \to \infty$.

An important variant of the foregoing result replaces the deterministic index n by the random index $N(t)$, where $\{N(t) : t \geq 0\}$ is an integer-valued stochastic process.

Proposition 2.5 (Random-index central limit theorem). *Let* $\{X_n : n \geq 0\}$ *be a sequence of i.i.d. random variables. Suppose that* $\sigma^2 = \mathrm{Var}[X_0] < \infty$ *and that* $N(t)/t \xrightarrow{pr} c$ *as* $t \to \infty$ *for some constant* $c \in (0, \infty)$. *Then*

$$\sqrt{N(t)}(\bar{X}_{N(t)} - \mu) \Rightarrow \sigma N(0,1)$$

as $t \to \infty$.

An easy application of the Cramér–Wold theorem (Proposition 1.46) extends Proposition 2.4 to $\Re^l$ for $l > 1$. Specifically, let $\{X_n : n \geq 0\}$ be a sequence of i.i.d. $\Re^l$-valued random vectors for some $l > 1$, and write

---

variables, each of which is uniformly distributed on $[0,1]$. This setup is characteristic of discrete-event simulation, where each U_n represents the output of a uniform random-number generator. We focus on the construction that uses Proposition 2.1 because it is better suited to our discussion of modelling power in Chapter 4.

$X_n = (X_{n,1}, X_{n,2}, \ldots, X_{n,l})$ for $n \geq 0$. Denote the common mean vector by $\mu = (\mu_1, \mu_2, \ldots, \mu_l)$ and the common covariance matrix by $\Sigma = \|\sigma_{ij}\|$; that is, $\mu_i = E[X_{0,i}]$ and $\sigma_{i,j} = \text{Cov}[X_{0,i}, X_{0,j}]$. As before, set $\bar{X}_n = (1/n) \sum_{i=0}^{n-1} X_i$.

Proposition 2.6 (Multivariate central limit theorem). *Suppose that* $\text{Var}[X_{0,i}] < \infty$ *for* $1 \leq i \leq l$. *Then*

$$\sqrt{n}(\bar{X}_n - \mu) \Rightarrow N(0, \Sigma)$$

as $n \to \infty$, *where* $\Rightarrow$ *denotes weak convergence in* $\Re^l$ *and* $N(0, \Sigma)$ *is an* *l-dimensional normal random vector with mean vector* $(0, 0, \ldots, 0)$ *and co-variance matrix* Σ.

A sequence of random variables $\{X_n : n \geq 0\}$ is *one-dependent* if X_{n+j} is independent of $\{X_0, X_1, \ldots, X_n\}$ for each $n \geq 0$ and $j > 1$. We now consider a stochastic process that comprises a sequence $\{X_n : n \geq 0\}$ of one-dependent and identically distributed (o.i.d.) random variables. By applying Proposition 2.3 separately to the odd and even terms of an o.i.d. sequence, we obtain the following result.

Proposition 2.7 (SLLN for o.i.d. sequences). *Let* $\{X_n : n \geq 0\}$ *be a* *sequence of o.i.d. random variables, and suppose that* $\mu = E[X_0] < \infty$. *Then*

$$\lim_{n \to \infty} \bar{X}_n = \mu \ a.s..$$

Definition 2.8. A sequence $\{X_n : n \geq 0\}$ is *(strictly) stationary* if $(X_0, X_1, \ldots, X_k)$ and $(X_n, X_{n+1}, \ldots, X_{n+k})$ are identically distributed for all $k, n \geq 0$.

For a stationary process $\{X_n : n \geq 0\}$ and $k, l \geq 0$, set $\mathcal{F}_0^k = \sigma\langle X_0, \ldots, X_k \rangle$ and $\mathcal{F}_l^\infty = \sigma\langle X_l, X_{l+1}, \ldots \rangle$. Then $\{X_n : n \geq 0\}$ is said to be ϕ-*mixing* if

$$\sup_{A \in \mathcal{F}_0^k, B \in \mathcal{F}_{k+n}^\infty} |P(B \mid A) - P(B)| \leq \phi_n$$

for $k, n \geq 0$, where $\lim_{n \to \infty} \phi_n = 0$. Heuristically, $\{X_n : n \geq 0\}$ is ϕ-mixing if the behavior of the process at widely separated points in time is approximately independent. The following result extends both the standard and random-index central limit theorems for i.i.d. random variables.

Proposition 2.9 (CLT for stationary sequences). *Let* $\{X_n : n \geq 0\}$ *be a stationary sequence of random variables. Suppose that* $\text{Var}[X_0] < \infty$ *and that the sequence is* ϕ-*mixing with* $\sum_n \phi_n^{1/2} < \infty$. *Then*

$$\sum_{k=1}^{\infty} |\text{Cov}[X_0, X_k]| < \infty$$

and

$$\sqrt{n}(\bar{X}_n - \mu) \Rightarrow \sigma N(0, 1)$$

as $n \to \infty$, *where* $\sigma^2 = \mathrm{Var}\,[X_0] + 2\sum_{k=1}^{\infty} \mathrm{Cov}\,[X_0, X_k]$. *If, moreover,* $N(t)/t \overset{\mathrm{pr}}{\to} c$ *as* $t \to \infty$ *for some constant* $c \in (0, \infty)$, *then*

$$\sqrt{N(t)}(\bar{X}_{N(t)} - \mu) \Rightarrow \sigma N(0, 1)$$

as $t \to \infty$.

An important special case of the above result is the following limit theorem for one-dependent stationary (o.d.s.) sequences, which are trivially ϕ-mixing.

Corollary 2.10 (CLT for o.d.s. sequences). *Let* $\{X_n : n \geq 0\}$ *be a stationary sequence of one-dependent random variables, and suppose that* $\sigma^2 = \mathrm{Var}\,[X_0] < \infty$. *Then*

$$\sqrt{n}(\bar{X}_n - \mu) \Rightarrow \sigma N(0, 1)$$

as $n \to \infty$, *where* $\sigma^2 = \mathrm{Var}\,[X_0] + 2\,\mathrm{Cov}\,[X_0, X_1]$. *If, moreover,* $N(t)/t \overset{\mathrm{pr}}{\to} c$ *as* $t \to \infty$ *for some constant* $c \in (0, \infty)$, *then*

$$\sqrt{N(t)}(\bar{X}_{N(t)} - \mu) \Rightarrow \sigma N(0, 1)$$

as $t \to \infty$.

Remark 2.11. Observe that o.d.s. sequences are a subclass of o.i.d. sequences: for the former class of sequences, the random vectors $\{(X_n, X_{n+1}, \dots, X_{n+k}) : n \geq 0\}$ must be identically distributed for each $k \geq 0$, whereas for the latter class of sequences, this requirement need only hold for $k = 0$.

A.2.3   Renewal Processes

Consider a machine that fails after a random time and is immediately replaced (renewed), and suppose that the successive lifetimes of the machines are i.i.d.. The associated "renewal counting process" counts the number of renewals in the interval $(0, t]$ and the "renewal process" is the sequence of random times at which the renewals occur. Formally, let $\{X_n : n \geq 0\}$ be a sequence of i.i.d. random variables (the lifetimes) with common mean μ and distribution function F. Set $S_0 = 0$ and form the partial sums $S_n = X_1 + \cdots + X_n$ for $n \geq 1$.

Definition 2.12. The process $\{S_n : n \geq 0\}$ is a *renewal process*, and the process $\{N(t) : t \geq 0\}$ defined by

$$N(t) = \sup\{n \geq 0 : S_n \leq t\}$$

is a *renewal counting process*.

In a *delayed* renewal process, the distribution of X_1, the time until the first renewal, may differ from the common distribution of $\{X_n : n \geq 2\}$.

The partial sum S_n is distributed as the n-fold convolution[8] F^{*n} of F with itself, and $P\{N(t) = n\} = F^{*n}(t) - F^{*(n+1)}(t)$ for $n, t \geq 0$. Set

$$m(t) = E[N(t)] = \sum_{n=1}^{\infty} F^{*n}(t)$$

for $t \geq 0$. The *renewal function* $m(t)$ is finite for $t \geq 0$.

Proposition 2.13 (Elementary renewal theorem).

$$\lim_{t \to \infty} \frac{m(t)}{t} = \frac{1}{\mu}.$$

Definition 2.14. Let h be a function defined on $\Re_+$, and let

$$\underline{m}_n(\delta) = \inf\{h(t) : (n-1)\delta \leq t \leq n\delta\}$$

and

$$\overline{m}_n(\delta) = \sup\{h(t) : (n-1)\delta \leq t \leq n\delta\}$$

for $\delta > 0$ and $n \in \{1, 2, \dots\}$. The function h is *directly Riemann integrable (d.R.i.)* if $\sum_{n=1}^{\infty} |\overline{m}_n(\delta)| < \infty$ and $\sum_{n=1}^{\infty} |\underline{m}_n(\delta)| < \infty$ for all $\delta > 0$ and

$$\lim_{\delta \to 0} \delta \sum_{n=1}^{\infty} \overline{m}_n(\delta) = \lim_{\delta \to 0} \delta \sum_{n=1}^{\infty} \underline{m}_n(\delta).$$

A d.R.i. function is bounded and continuous almost everywhere with respect to Lebesgue measure. A function h defined on $\Re_+$ is d.R.i. if (1) $h \geq 0$, (2) h is nonincreasing, and (3) $\int_0^{\infty} h(t)\, dt < \infty$. Other sufficient conditions for a function h that is bounded and continuous almost everywhere to be d.R.i. are (1) $h \leq g$ with g d.R.i., or (2) $h(t) = 0$ for all t outside of a bounded set.

Definition 2.15. A distribution function F is *spread out* if $F^{*n} \geq G$ for some $n \geq 1$, where G is nonnegative, not identically zero, and absolutely continuous.

In the terminology of Section 5.1.2, F is spread out if F^{*n} has a density component for some $n \geq 1$. If F is spread out, then F is aperiodic as in Definition 1.19 in Chapter 6.

---

[8]For distribution functions F and G of nonnegative random variables X and Y, the *convolution* of F and G, denoted $F * G$, is defined by $(F * G)(t) = \int_0^t F(t - x)\, dG(x)$ and is the distribution function of the random variable $X + Y$.

Proposition 2.16 (Key renewal theorem). *Let m be a renewal function for a renewal process with interrenewal-time distribution function F, and let h be a function defined on $\Re_+$. Suppose that either*

(i) *F is aperiodic and h is directly Riemann integrable; or*

(ii) *F is spread out and h is bounded and integrable with $\lim_{x \to \infty} h(x) = 0$.*

Then

$$\int_0^t h(t - x)\, dm(x) \to \frac{1}{\mu} \int_0^\infty h(x)\, dx$$

as $t \to \infty$, where $\mu = \int_0^\infty x\, dF(x)$.

A.2.4 Discrete-Time Markov Chains

To motivate the results for general state-space Markov chains that are given in the text, we briefly review some key results for chains evolving in discrete time and having a finite or countably infinite state space.

Definition 2.17. The stochastic process $\{X_n : n \geq 0\}$ taking values in a finite or countably infinite state space S is a *discrete-time Markov chain* (DTMC) if

$$P\{X_{n+1} = j \mid X_n, X_{n-1}, \ldots, X_0\} = P\{X_{n+1} = j \mid X_n\} \text{ a.s.}$$

for $n \geq 0$ and $j \in S$.

The above *Markov property* asserts that the future evolution of the process depends on its past and its present only through the current state. Define a vector μ whose ith entry is $\mu_i = P\{X_0 = i\}$ for $i \in S$. If, as we assume throughout, there exists a matrix $P = \|p_{ij}\|$ such that

$$P\{X_{n+1} = j \mid X_n = i\} \equiv p_{ij}$$

for $n \geq 0$ and $i, j \in S$, then the DTMC is said to be *time-homogeneous* with *transition matrix* P and *initial probability vector* μ. We often write P_μ for the probability law of the chain to emphasize the dependence on the initial probability vector μ. When $\mu_i = 1$ for some $i \in S$, then we write P_i instead of P_μ. Similarly, we write E_μ and E_i to denote expectations.

A DTMC is *irreducible* if any state can be reached from any other state in a finite number of state transitions: for each $i, j \in S$, there exists an integer $n = n(i, j) \in (0, \infty)$ such that $p_{ij}^n > 0$, where p_{ij}^n is the (i, j)th entry of the nth power P^n of the transition matrix P. State i is said to be *periodic* with period d if $p_{ii}^n = 0$ whenever n is not divisible by d, and d is the greatest integer having this property. A state with period 1 is *aperiodic*.

State i is *recurrent* if $P_i\{X_n = i \text{ i.o.}\} = 1$; otherwise, state i is *transient*. It can be shown that state i is recurrent if and only if $P_i\{\tau_i < \infty\} = 1$, where τ_i is the first hitting time of state i: $\tau_i = \inf\{n > 0 : X_n = i\}$. A recurrent state i is *positive recurrent* if $E_i[\tau_i] < \infty$; otherwise, state i is *null recurrent*. If a DTMC is irreducible, then all states are transient, or all states are null recurrent, or all states are positive recurrent; either all states are aperiodic or, if one state is periodic with period d, then all states are periodic with period d. If all states are transient, then the entire chain is said to be transient, and similarly for other properties.

Proposition 2.18 (Foster's criterion for recurrence). *Let* $\{X_n : n \geq 0\}$ *be an irreducible* DTMC *with state space* S *and let* S_0 *be a finite subset of* S. *Then the chain is positive recurrent if there exist a nonnegative real-valued function* g *defined on* S *and a constant* $\epsilon > 0$ *such that*

$$\sum_{j \in S} p_{ij}\, g(j) < \infty$$

for $i \in S_0$ *and*

$$\sum_{j \in S} p_{ij}\, g(j) < g(i) - \epsilon$$

for $i \in S - S_0$.

Observe that the two conditions in the theorem can be rewritten as

$$\sup_{i \in S_0} E_i\left[g(X_1) - g(X_0)\right] < \infty$$

and

$$E_i\left[g(X_1) - g(X_0)\right] \leq -\epsilon$$

for all $i \in S - S_0$.

A vector $\pi = \{\pi_i : i \in S\}$ is a *stationary distribution* of a DTMC with transition matrix P if π is a solution of $\pi P = \pi$ and $\sum_{i \in S} \pi_i = 1$.

Proposition 2.19. *Let* $\{X_n : n \geq 0\}$ *be an irreducible positive recurrent* DTMC *with state space* S. *Then there exists a unique stationary distribution* π *given by*

$$\pi_i = \frac{1}{E_i[\tau_i]}.$$

An irreducible, aperiodic, and positive recurrent DTMC is called *ergodic*.

Proposition 2.20. *Let* $\{X_n : n \geq 0\}$ *be an ergodic* DTMC *with state space* S *and stationary distribution* π. *Then*

$$\lim_{n \to \infty} p_{ij}^n = \pi_j$$

for $i, j \in S$.

Proposition 2.21. *Let* $\{X_n : n \geq 0\}$ *be an irreducible positive recurrent* DTMC *with state space* S *and stationary distribution* π. *Also, let* f *be a real-valued function defined on* S *such that*

$$\sum_{i \in S} \pi_i |f(i)| < \infty.$$

Then

$$\lim_{n \to \infty} \frac{1}{n} \sum_{i=0}^{n-1} f(X_i) = \pi(f) \ a.s.$$

for any initial distribution μ, *where* $\pi(f) = \sum_{i \in S} \pi_i f(i)$. *If, moreover,* $\sigma^2(f) < \infty$, *where*

$$\sigma^2 = \pi_i E_i \left[\left(\sum_{j=0}^{\tau_i - 1} \left(f(X_j) - \pi(f) \right) \right)^2 \right]$$

and i *is a fixed element of* S, *then*

$$\frac{1}{\sqrt{n}} \sum_{j=0}^{n-1} \left(f(X_j) - \pi(f) \right) \Rightarrow \sigma N(0,1)$$

as $n \to \infty$ *for any initial distribution* μ.

A.2.5   Brownian Motion and FCLTs

The central limit theorems given so far assert the convergence of a sequence of random variables to a limiting normal random variable. Typically, each random variable in the sequence is a suitably normalized partial sum. A *functional central limit theorem* (FCLT) is a generalization of a CLT and asserts the convergence of a sequence of random functions to a limiting random process. In our setting the limiting random process is always a Brownian motion. In this subsection we define what "convergence in distribution" means in this more general setting, give some basic properties of Brownian motion, and discuss Donsker's theorem. This latter result is the simplest FCLT and generalizes the central limit theorem for i.i.d. random variables. We also give extensions of Donsker's theorem to dependent random variables and random numbers of random variables.

Recall that for real-valued random variables $X, X_1, X_2, \ldots$ with corresponding distributions $\mu, \mu_1, \mu_2, \ldots$ we have $X_n \Rightarrow X$ if and only if

$$E[f(X_n)] \to E[f(X)]$$

for every bounded continuous real-valued function f. We now take this latter condition as the definition of convergence in distribution for random variables $X, X_1, X_2, \ldots$ that take values in an arbitrary metric space S. In this setting we say that the probability measures $\{\mu_n \colon n \geq 0\}$ *converge weakly* to μ and that the "random elements" $\{X_n \colon n \geq 0\}$ converge weakly to X.

For our purposes, the space S of interest is the space of continuous real-valued functions defined on $[0, 1]$—we denote this space by $C[0, 1]$. The space $C[0, 1]$ can be metrized by the uniform metric:

$$\rho(x, y) = \sup_{0 \leq t \leq 1} |x(t) - y(t)|$$

for $x, y \in C[0, 1]$. Thus a sequence of elements $x_1, x_2, \ldots \in C[0, 1]$ converges to $x \in C[0, 1]$ if and only if $\lim_{n \to \infty} \sup_{0 \leq t \leq 1} |x_n(t) - x(t)| = 0$. We denote by $\mathcal{C}[0, 1]$ the *Borel sets* of $C[0, 1]$, that is, the σ-field generated by the open[9] subsets of $C[0, 1]$ with respect to ρ.

Wiener measure is defined to be the unique probability measure W on $(C[0, 1], \mathcal{C}[0, 1])$ having the following properties.

1. $W\{x \colon x(0) = 0\} = 1$.

2. For all $t \in (0, 1]$ and $a \in \Re$,

$$W\{x \colon x(t) \leq a\} = \frac{1}{\sqrt{2\pi t}} \int_{-\infty}^{a} e^{-u^2/2t} \, du.$$

3. For $0 \leq t_0 \leq t_1 \leq \cdots \leq t_k \leq 1$ and $a_1, \ldots, a_k \in \Re$,

$$W\left(\bigcap_{i=1}^{k} \{x \colon x(t_i) - x(t_{i-1}) \leq a_i\}\right)$$
$$= \prod_{i=1}^{k} W\{x \colon x(t_i) - x(t_{i-1}) \leq a_i\}.$$

A random element taking values in $(C[0, 1], \mathcal{C}[0, 1])$ and having distribution W is called a one-dimensional *Brownian motion* or *Wiener process*. With a slight abuse of notation, we denote this process by $W = \{W(t) \colon 0 \leq t \leq 1\}$. It follows from the properties of Wiener measure that a Brownian motion W satisfies the following conditions.

1. $W(0) = 0$ with probability 1.

2. W has continuous sample paths.

3. $W(t)$ is distributed as $N(0, t)$ for each $t \in (0, 1]$.

---

[9]Recall that a set $A \subseteq S$ is *open* if for each $x \in A$ there exists $r > 0$ such that $B_r(x) \subseteq A$, where $B_r(x) = \{y \in S \colon \rho(x, y) \leq r\}$.

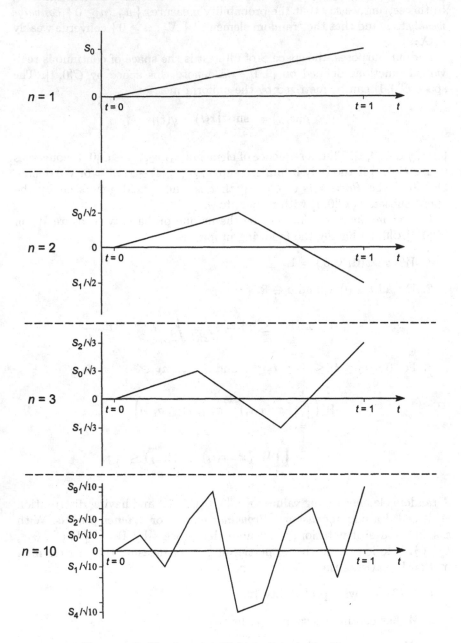

Figure A.1. The function $U_n(t)$ in Donsker's theorem.

4. For $0 \leq t_0 \leq t_1 \leq \cdots \leq t_k \leq 1$, the increments $W(t_1) - W(t_0), \ldots,$ $W(t_k) - W(t_{k-1})$ are mutually independent.

We are now ready to state Donsker's theorem. Let $\{X_n : n \geq 0\}$ be a sequence of i.i.d. random variables with common mean 0. Define a sequence of random functions, each with sample paths in $C[0, 1]$, by setting

$$U_n(t) = \frac{1}{\sqrt{n}} \int_0^{nt} X_{\lfloor u \rfloor} \, du \qquad (2.22)$$

for $0 \leq t \leq 1$ and $n \geq 1$, where $\lfloor x \rfloor$ is the greatest integer less than or equal to x. Setting $S_{-1} = 0$ and $S_i = X_0 + X_1 + \cdots + X_i$ for $i \geq 0$, we observe that $U_n(t) = S_{i-1}/\sqrt{n}$ for $t = i/n$ $(i = 0, 1, \ldots, n)$. If $i/n < t < (i+1)/n$ for some i, then the value of $U_n(t)$ is obtained by linearly interpolating between $S_{i-1}/\sqrt{n}$ and $S_i/\sqrt{n}$. The function U_n is sometimes expressed using a slightly more cumbersome notation:

$$U_n(t) = \frac{1}{\sqrt{n}} \sum_{i=0}^{\lfloor nt \rfloor - 1} X_i + (nt - \lfloor nt \rfloor) \frac{1}{\sqrt{n}} X_{\lfloor nt \rfloor}.$$

Some sample paths of U_n are shown in Figure A.1 for $n = 1, 2, 3,$ and 10, based on a hypothetical realization of $\{X_n : n \geq 0\}$.

Proposition 2.23 (Donsker's theorem). *Let $\{U_n : n \geq 1\}$ and W be defined as above. If $\sigma^2 = \mathrm{Var}\,[X_0] < \infty$, then $U_n \Rightarrow \sigma W$ as $n \to \infty$, where $\Rightarrow$ denotes weak convergence in $C[0, 1]$.*

The power of Donsker's theorem derives from that fact that many of the key results for convergence in distribution, such as the continuous mapping theorem, carry over to the setting of $C[0, 1]$. It can be shown, for example, that the coordinate projection mapping $x \mapsto x(1)$ is continuous on $C[0, 1]$, so that $U_n(1) \Rightarrow \sigma W(1)$, where $\Rightarrow$ denotes convergence in distribution for ordinary random variables. Since, as discussed above, $W(1)$ is distributed as $N(0, 1)$, we recover the central limit theorem for i.i.d. random variables. As another example, the limiting distribution of a quantity such as $\max_{0 \leq k \leq n} S_n$ can be obtained by analyzing the relatively tractable random variable

$$M = \sup_{0 \leq t \leq 1} W(t),$$

because the mapping $x \mapsto \sup_{0 \leq t \leq 1} x(t)$ is continuous on $C[0, 1]$.

Proposition 2.24 extends Donsker's theorem to ϕ-mixing sequences of stationary random variables. In the proposition, $\Rightarrow$ denotes weak convergence in $C[0, 1]$.

Proposition 2.24. *Let $\{X_n : n \geq 0\}$ be a stationary sequence of random variables with common mean 0, and define functions $\{U_n : n \geq 1\}$ as in (2.22). If $\mathrm{Var}\,[X_0] < \infty$ and $\{X_n : n \geq 0\}$ is ϕ-mixing with $\sum_n \phi_n^{1/2} < \infty$,* *then $\sum_{k=1}^{\infty} |\text{Cov}\,[X_0, X_k]| < \infty$ and $U_n \Rightarrow \sigma W$ as $n \to \infty$, where $\sigma^2 = \text{Var}\,[X_0] + 2\sum_{k=1}^{\infty} \text{Cov}\,[X_0, X_k]$.*

Donsker's theorem can also be extended to a sequence of $\Re^l$-valued i.i.d. random vectors $X_1, X_2, \ldots$ having common mean vector μ and common covariance matrix Σ. In this setting, convergence occurs in $C^l[0,1]$, the space of $\Re^l$-valued functions on $[0,1]$ having continuous sample paths. When Σ is equal to the $l \times l$ identity matrix $I^{(l)}$, the limiting random function is easily seen to be an l-dimensional standard Brownian motion

$$W^{(l)} = \left\{ \left(W_1^{(l)}(t), \ldots, W_l^{(l)}(t)\right) : 0 \le t \le 1 \right\}$$

in which the component processes $W_1^{(l)}, W_2^{(l)}, \ldots, W_l^{(l)}$ are mutually independent one-dimensional Brownian motions. For a general covariance matrix Σ, we can write $\Sigma = QQ^t$ for some matrix Q, and the limiting random function is given by $QW^{(l)}$. Proposition 2.24 can similarly be extended.

The following result can be used to extend both Donsker's theorem and Proposition 2.24 to permit a random index in the partial sums.

Proposition 2.25. *Suppose that*

$$U_n \Rightarrow QW^{(l)}$$

as $n \to \infty$, where $\{U_n : n \ge 0\}$ is a sequence of random elements of $C^l[0,1]$ for some $l \ge 1$ and $W^{(l)}$ is a standard l-dimensional Brownian motion. Also suppose that $N(t)/t \overset{pr}{\to} c$ as $t \to \infty$ for some constant $c \in (0, \infty)$. Then

$$U_{N(t)} \Rightarrow QW^{(l)}$$

as $t \to \infty$.

A.3 Terminology Used in the Text

In the main text we suppress measure-theoretic terminology wherever possible. To do this, the following conventions are used:

- Whenever, for instance, a result on a probability space $(\Omega, \mathcal{F}, P)$ is said to hold for "all subsets A" or "all functions" f, we mean that the result holds for all measurable subsets $A \in \mathcal{F}$ or all measurable functions f.

- As mentioned previously, a conditional expectation such as $E\,[X \mid Y]$ is interpreted as $E\,[X \mid \sigma\langle Y\rangle]$, and this convention carries over to conditional probabilities. For a random variable Y, a process $\{X_n : n \ge 0\}$, and a nonnegative random index K that is a stopping time with respect to $\{X_n : n \ge 0\}$, an expression such as $E[Y \mid X_0, X_1, \ldots, X_K]$

or $E[Y \mid K, X_0, X_1, \ldots, X_K]$ is interpreted as $E[Y \mid \mathcal{F}_K]$, where the σ-field $\mathcal{F}_K$ is defined as in Section A.2.1. As a special case, the notation $E[Y \mid X_0, X_1, \ldots, X_n]$ is interpreted to mean $E[Y \mid \mathcal{F}]$, where $\mathcal{F} = \sigma\langle X_0, X_1, \ldots, X_n \rangle$.

Notes

Billingsley (1986), Breiman (1968), Chung (1974), Durrett (1991), and Loéve (1977) provide excellent treatments of probability and measure at the level given in this Appendix. Texts on stochastic processes include Asmussen (1987a), Çinlar (1975), Doob (1953), Karlin and Taylor (1975), and Ross (1983). Gut (1988) studies sums of random numbers of i.i.d. random variables—in particular, the proof of Proposition 1.20 is contained in the proof of Gut's Theorem I.5.2. Discussions of weak convergence, Brownian motion, and FCLTs can be found in Billingsley (1968), Ethier and Kurtz (1986), and Whitt (2002). Glasserman and Yao (1994) discuss the use of the space $([0, 1]^\infty, \mathcal{B}_0^\infty, \mu_\infty^{\text{Leb}})$, defined in Example 2.2, for construction of stochastic processes associated with discrete-event systems. Serfling (1980) surveys basic limit theorems that arise in mathematical statistics, including many of the results discussed in this appendix.

References

Aggarwal, S., Garay, J., and Herzberg, A. (1995). Adaptive video-on-demand. In *Algorithms - ESA '95 (Proc. Third Annual European Symposium on Algorithms)*, Vol. 979 of *Lecture Notes in Computer Science*, 538–553. Springer-Verlag.

Ajmone Marsan, M., Balbo, G., Chiola, G., and Conte, G. (1987). Generalized stochastic Petri nets revisited: Random switches and priorities. In *Intl. Workshop Petri Nets Performance Models*, 44–53. IEEE Computer Society Press.

Ajmone Marsan, M., Balbo, G., Donatelli, S., and Franceschinis, G. (1995). *Modelling with Generalized Stochastic Petri Nets*. Wiley, New York.

Ajmone Marsan, M., Conte, G., and Balbo, G. (1984). A class of generalized stochastic Petri nets for the performance evaluation of multiprocessor systems. *ACM Trans. Comput. Sys.* **2**, 93–122.

Altman, E., Konstantopoulos, P., and Liu, Z. (1992). Stability, monotonicity and invariant quantities in general polling systems. *Queueing Sys. Theory Appl.* **11**, 35–57.

Anderson, T. W. (1971). *The Statistical Analysis of Time Series*. Wiley, New York.

Andradottir, S. (1998). Simulation optimization. In J. Banks (ed.), *Handbook of Simulation*. Wiley, New York.

Andradottir, S., Calvin, J. M., and Glynn, P. W. (1994). Increasing the frequency of regeneration for Markov processes. In *Proc. 1994 Winter Simulation Conference*, 320–323.

Archibald, G., Karabakal, N., and Karlsson, P. (1999). Supply chain vs. supply chain: Using simulation to compete beyond the four walls. In *Proc. 1999 Winter Simulation Conference*, 1207–1214.

Asmussen, S. (1987a). *Applied Probability and Queues*. Wiley, New York.

Asmussen, S. (1987b). Validating the heavy traffic performance of regenerative simulation. *Comm. Statist. Stochastic Models* **5**, 617–628.

Asmussen, S. and Foss, S. G. (1993). Renovation, regeneration, and coupling in multiple-server queues in continuous time. *Frontiers Pure Appl. Probab.* **1**, 1–6.

Athreya, K. B. and Ney, P. (1978). A new approach to the limit theory of recurrent Markov chains. *Trans. Amer. Math. Soc.* **245**, 493–501.

Baccelli, F. (1992). Ergodic theory of stochastic Petri networks. *Ann. Probab.* **20**, 375–396.

Baccelli, F. and Canales, M. (1993). Parallel simulation of stochastic Petri nets using recurrence equations. *ACM Trans. Model. Comput. Simul.* **3**, 20–41.

Baccelli, F., Cohen, G., Olsder, G. J., and Quadrat, J.-P. (1993). *Synchronization and Linearity: An Algebra for Discrete Event Systems*. Wiley, Chichester, England.

Baccelli, F., Foss, S., and Gaujal, B. (1996). Free-choice Petri nets— An algebraic approach. *IEEE Trans. Automatic Control* **41**, 1751–1778.

Baccelli, F. and Schmidt, V. (1996). Taylor series expansions for Poisson-driven (max, +)-linear systems. *Ann. Applied Probab.* **6**, 138–185.

Banks, J. (ed.) (1998). *Handbook of Simulation*. Wiley, New York.

Barlow, R. E. and Proschan, F. (1975). *Statistical Theory of Reliability and Life Testing: Probability Models*. Holt, Reinhart and Winston, New York.

Bergman, C. D. and Shedler, G. S. (1993). Estimation procedures for discrete event simulations using SPSIM. IBM Research Report RJ 9317, IBM Almaden Research Center, San Jose, CA.

Billingsley, P. (1968). *Convergence of Probability Measures*. Wiley, New York.

Billingsley, P. (1986). *Probability and Measure.* Wiley, New York.

Bobbio, A., Kulkarni, V. G., Puliafito, A., Telek, M., and Trivedi, K. S. (1995). Preemptive repeat identical transitions in Markov regenerative stochastic Petri nets. In *Proc. Sixth Intl. Workshop Petri Nets Performance Models*, 113–123. IEEE Computer Society Press.

Borovkov, A. A. (1984). *Asymptotic Methods in Queueing Theory.* Wiley, New York.

Borovkov, A. A. (1986). Limit theorems for queuing networks I. *Theory Probab. Appl.* **31**, 413–427.

Borovkov, A. A. and Foss, S. G. (1992). Stochastically recursive sequences and their generalizations. *Siberian Adv. Math.* **2**, 16–81.

Boucherie, R. J. (1994). A characterization of independence for competing Markov chains with applications to stochastic Petri nets. *IEEE Trans. Software Engrg.* **20**, 536–544.

Bratley, P., Fox, B. L., and Schrage, L. E. (1987). *A Guide to Simulation*, 2nd ed. Springer-Verlag, New York.

Breiman, L. (1968). *Probability.* Addison-Wesley, Reading, MA.

Brillinger, D. R. (1973). Estimation of the mean of a stationary time series by sampling. *J. Appl. Probab.* **10**, 419–431.

Brockwell, P. J. and Davis, R. A. (1987). *Time Series: Theory and Methods.* Springer-Verlag, New York.

Brown, M. and Ross, S. M. (1972). Asymptotic properties of cumulative processes. *SIAM J. Appl. Math.* **22**, 93–105.

Buzacott, J. A. and Shanthikumar, G. S. (1993). *Stochastic Models of Manufacturing Systems.* Prentice Hall, Englewood Cliffs, NJ.

Calvin, J. (1994). Return state independent quantities in regenerative simulation. *Oper. Res.* **42**, 531–542.

Calvin, J. and Nakayama, M. (2000). Simulation of processes with multiple regeneration points. *Probab. Engrg. Inform. Sci.* **14**, 179–201.

Campos, J., Colom, J. M., Chiola, G., and Silva, M. (1989). Tight polynomial bounds for steady-state performance of marked graphs. In *Proc. Third Intl. Workshop Petri Nets Performance Models*, 200–209. IEEE Computer Society Press.

Campos, J., Colom, J. M., Jungnitz, H., and Silva, M. (1994). Approximate throughput computation for stochastic marked graphs. *IEEE Trans. Software Engrg.* **20**, 526–535.

Carlstein, E. (1986). The use of subseries for estimating the variance of a general statistic from a stationary sequence. *Ann. Statist.* **14**, 1171–1179.

Cassandras, C. G. and LaFortune, S. (1999). *Introduction to Discrete Event Systems*. Kluwer Academic, Boston.

Chan, T. F., Golub, G. H., and LeVeque, R. J. (1983). Algorithms for computing the sample variance: Analysis and recommendation. *Amer. Statist.* **37**, 242–247.

Chien, C. (1989). Small sample theory for steady state confidence intervals. Tech. Rep. 37, Depatment of Operations Research, Stanford University, Stanford, CA.

Chien, C., Goldsman, D., and Melamed, B. (1997). Large-sample results for batch means. *Management Sci.* **43**, 1288–1295.

Chiola, G. (1991). Simulation framework for timed and stochastic Petri nets. *Intl. J. Comput. Simulation* **1**, 153–168.

Chiola, G. (1995). Characterization of timed well-formed Petri nets behavior by means of occurrence equations. In *Proc. Sixth Intl. Workshop Petri Nets Performance Models*, 127–136. IEEE Computer Society Press.

Chiola, G., Bruno, G., and Demaria, T. (1988). Introducing a color formalism into generalized stochastic Petri nets. In *Proc. 9th European Workshop Appl. Theory Petri Nets*.

Chiola, G., Dutheillet, C., Franceschinis, G., and Haddad, S. (1993). Stochastic well-formed colored nets and symmetric modeling applications. *IEEE Trans. Comput.* **42**, 1343–1360.

Choi, H., Kulkarni, V. G., and Trivedi, K. S. (1994). Markov regenerative stochastic Petri nets. *Performance Evaluation* **20**, 337–357. Erratum: *Performance Evaluation*, vol. 21, p. 271, 1995.

Chung, K. L. (1967). *Markov Chains with Stationary Transition Probabilities*, 2nd ed. Springer-Verlag, Berlin.

Chung, K. L. (1974). *A Course in Probability Theory*, 2nd ed. Academic Press, New York.

Çinlar, E. (1975). *Introduction to Stochastic Processes*. Prentice Hall, Englewood Cliffs, NJ.

Coleman, J. L. (1993). Algorithms for product-form stochastic Petri nets—A new approach. In *Proc. Fifth Intl. Workshop Petri Nets Performance Models*. IEEE Computer Society Press.

Conway, R. W. (1963). Some tactical problems in digital simulation. *Management Sci.* **10**, 47–61.

Cox, D. R. (1952). Estimation by double sampling. *Biometrika* **39**, 217–227.

Coyle, A. J. and Taylor, P. G. (1995). Tight bounds on the insensitivity of generalized semi-Markov processes with a single generally distributed lifetime. *J. Appl. Probab.* **32**, 63–73.

Crane, M. A. and Iglehart, D. L. (1975). Simulating stable stochastic systems: III, regenerative processes and discrete event simulation. *Oper. Res.* **23**, 33–45.

Crane, M. A. and Lemoine, A. J. (1977). *An Introduction to the Regenerative Method for Simulation Analysis*. Lecture Notes in Control and Information Sciences. Springer-Verlag, New York.

Dai, J. G. (1995). On positive Harris recurrence of multiclass queueing networks: A unified approach via fluid limit models. *Ann. Applied Probab.* **5**, 49–77.

Damerdji, H. (1991). Strong consistency and other properties of the spectral variance estimator. *Management Sci.* **37**, 1424–1440.

Damerdji, H. (1994). Strong consistency of the variance estimator in steady-state simulation output analysis. *Math. Oper. Res.* **19**, 494–512.

Damerdji, H. (1995). Mean-square consistency of the variance estimator in steady-state simulation output analysis. *Oper. Res.* **43**, 282–291.

Damerdji, H. and Goldsman, D. (1995). Consistency of several variants of the standardized time series area variance estimator. *Naval Res. Logist. Quart.* **42**, 1161–1176.

Doob, J. L. (1953). *Stochastic Processes*. Wiley, New York.

Durrett, R. (1991). *Probability: Theory and Examples*. Wadsworth and Brooks/Cole, Pacific Grove, CA.

Efron, B. and Tibshirani, R. J. (1993). *An Introduction to the Bootstrap*. Chapman and Hall, London.

Eswaran, K. P., Hamacher, V. C., and Shedler, G. S. (1978). Collision-free access control for computer communication bus networks. *IEEE Trans. Software Engrg.* **SE-7**, 574–582.

Ethier, S. N. and Kurtz, T. G. (1986). *Markov Processes: Characterization and Convergence*. Wiley, New York.

Ferscha, A. and Richter, M. (1997). Time Warp simulation of timed Petri nets: Sensitivity of adaptive methods. In *Proc. Seventh Intl. Workshop Petri Nets Performance Models*, 205–216. IEEE Computer Society Press.

Foss, S. G. and Kalashnikov, V. (1991). Regeneration and renovation in queues. *Queueing Sys. Theory Appl.* **8**, 211–224.

Fox, B. L. and Glynn, P. W. (1985). Discrete-time conversion for simulating semi-Markov processes. *Oper. Res. Lett.* **5**, 191–196.

Fox, B. L. and Glynn, P. W. (1987). Estimating time averages via randomly-spaced observations. *SIAM J. Appl. Math.* **47**, 186–200.

Fox, B. L. and Glynn, P. W. (1989). Simulating discounted costs. *Management Sci.* **35**, 1297–1315.

Fox, B. L., Goldsman, D., and Swain, J. J. (1991). Spaced batch means. *Oper. Res. Lett.* **10**, 255–263.

Gaeta, R. and Chiola, G. (1995). Efficient simulation of SWN models. In *Proc. Sixth Intl. Workshop Petri Nets Performance Models*, 137–146. IEEE Computer Society Press.

Genrich, H. J. and Lautenbach, K. (1981). System modelling with high-level Petri nets. *Theoret. Comput. Sci.* **13**, 109–136.

Glasserman, P. (1991). *Gradient Estimation via Perturbation Analysis*. Kluwer Academic, Boston.

Glasserman, P. and Glynn, P. W. (1992). Gradient estimation for regenerative processes. In *Proc. 1992 Winter Simulation Conference*, 280–288.

Glasserman, P. and Yao, D. D. (1994). *Monotone Structure in Discrete-Event Systems*. Wiley, New York.

Glynn, P. W. (1982a). Asymptotic theory for nonparametric confidence intervals. Tech. Rep. 63, Department of Operations Research, Stanford University, Stanford, CA.

Glynn, P. W. (1982b). Simulation output analysis for general state space Markov chains. Ph.D. Dissertation, Department of Operations Research, Stanford University, Stanford, CA.

Glynn, P. W. (1987). A low bias steady-state estimator for equilibrium processes. Tech. Rep. 47, Department of Operations Research, Stanford University, Stanford, CA.

Glynn, P. W. (1989a). The covariance function of a regenerative process. Tech. Rep. 49, Department of Operations Research, Stanford University, Stanford, CA.

Glynn, P. W. (1989b). A GSMP formalism for discrete event systems. *Proc. IEEE* **77**, 14–23.

Glynn, P. W. (1989c). Likelihood ratio derivative estimators for stochastic systems. In *Proc. 1989 Winter Simulation Conference*, 374–380.

Glynn, P. W. (1994). Some topics in regenerative steady-state simulation. *Acta Appl. Math.* **34**, 225–236.

Glynn, P. W. and Haas, P. J. (2002a). Consistent estimation of the variance in steady-state simulation output analysis. IBM Research Report, IBM Almaden Research Center, San Jose, CA.

Glynn, P. W. and Haas, P. J. (2002b). Laws of large numbers and central limit theorems for discrete-event systems. IBM Research Report, IBM Almaden Research Center, San Jose, CA.

Glynn, P. W. and Heidelberger, P. (1990). Bias properties of budget constrained simulations. *Oper. Res.* **38**, 801–814.

Glynn, P. W. and Heidelberger, P. (1992). Jackknifing under a budget constraint. *ORSA J. Comput.* **4**, 226–234.

Glynn, P. W. and Iglehart, D. L. (1986a). Consequences of uniform integrability for simulation. Tech. Rep. 15, Department of Operations Research, Stanford University, Stanford, CA.

Glynn, P. W. and Iglehart, D. L. (1986b). Estimation of steady state central moments by the regenerative method of simulation. *Oper. Res. Lett.* **5**, 271–276.

Glynn, P. W. and Iglehart, D. L. (1987). A joint central limit theorem for the sample mean and regenerative variance estimator. *Ann. Oper. Res.* **8**, 41–55.

Glynn, P. W. and Iglehart, D. L. (1988). Simulation methods for queues: An overview. *Queueing Sys. Theory Appl.* **3**, 221–256.

Glynn, P. W. and Iglehart, D. L. (1989). Smoothed limit theorems for equilibrium processes. In T. W. Anderson, K. B. Athreya, and D. L. Iglehart (eds.), *Probability, Statistics, and Mathematics: Papers in Honor of Samuel Karlin*. Academic Press, New York.

Glynn, P. W. and Iglehart, D. L. (1990). Simulation output analysis using standardized time series. *Math. Oper. Res.* **15**, 1–16.

Glynn, P. W. and Iglehart, D. L. (1993). Conditions for the applicability of the regenerative method. *Management Sci.* **39**, 1108–1111.

Glynn, P. W. and L'Ecuyer, P. (1995). Likelihood ratio gradient estimation for stochastic recursions. *Adv. Appl. Probab.* **27**, 1019–1053.

Glynn, P. W. and Meyn, S. P. (1996). A Lyapunov bound for solutions of Poisson's equation. *Ann. Probab.* **24**, 916–931.

Glynn, P. W. and Sigman, K. (1992). Uniform cesaro limit theorems for synchronous processes with applications to queues. *Stochastic Process. Appl.* **40**, 29–44.

Glynn, P. W. and Whitt, W. (1986). A central-limit-theorem version of $L = \lambda W$. *Queueing Sys. Theory Appl.* **2**, 191–215.

Glynn, P. W. and Whitt, W. (1987). Sufficient conditions for the functional-central-limit-theorem versions of $L = \lambda W$. *Queueing Sys. Theory Appl.* **1**, 279–287.

Glynn, P. W. and Whitt, W. (1988). Ordinary CLT and WLLN versions of $L = \lambda W$. *Math. Oper. Res.* **13**, 693–710.

Glynn, P. W. and Whitt, W. (1989). Indirect estimation via $L = \lambda W$. *Oper. Res.* **37**, 82–103.

Glynn, P. W. and Whitt, W. (1992a). The asymptotic efficiency of simulation estimators. *Oper. Res.* **40**, 505–520.

Glynn, P. W. and Whitt, W. (1992b). The asymptotic validity of sequential stopping rules for stochastic simulation. *Ann. Applied Probab.* **2**, 180–198.

Gnedenko, B. V. and Kovalenko, I. N. (1974). *Introduction to Queueing Theory*, German ed. Akademie-Verlag, Berlin.

Goldsman, D., Meketon, M. S., and Schruben, L. W. (1990). Properties of standardized time series weighted area variance estimators. *Management Sci.* **36**, 602–612.

Goldsman, D. and Nelson, B. L. (1998). Comparing systems via simulation. In J. Banks (ed.), *Handbook of Simulation*. Wiley, New York.

Goldsman, D. and Schruben, L. (1990). New confidence interval estimators using standardized time series. *Management Sci.* **36**, 393–397.

Goyal, A., Shahabuddin, P., Heidelberger, P., Nicola, V., and Glynn, P. (1992). A unified framework for simulating Markovian models of highly dependable systems. *IEEE Trans. Comput.* **41**, 36–51.

Grenander, U. and Rosenblatt, M. (1984). *Statistical Analysis of Stationary Time Series*, 2nd ed. Chelsea, New York.

Gut, A. (1988). *Stopped Random Walks: Limit Theorems and Applications*. Springer-Verlag, New York.

Haas, P. J. (1999a). Estimation methods for non-regenerative stochastic Petri nets. *IEEE Trans. Software Engrg.* **25**, 218–236.

Haas, P. J. (1999b). Estimation of delays in non-regenerative stochastic Petri nets. IBM Research Report RJ 10138, IBM Almaden Research Center, San Jose, CA.

Haas, P. J. (1999c). On simulation output analysis for generalized semi-Markov processes. *Comm. Statist. Stochastic Models* **15**, 53–80.

Haas, P. J. and Shedler, G. S. (1985a). Regenerative simulation methods for local area computer networks. *IBM J. Res. Develop.* **29**, 194–205.

Haas, P. J. and Shedler, G. S. (1985b). Regenerative simulation of stochastic Petri nets. In *Intl. Workshop Timed Petri Nets*, 14–23. IEEE Computer Society Press.

Haas, P. J. and Shedler, G. S. (1986). Regenerative stochastic Petri nets. *Performance Evaluation* **6**, 189–204.

Haas, P. J. and Shedler, G. S. (1987a). Recurrence and regeneration in non-Markovian networks of queues. *Comm. Statist. Stochastic Models* **3**, 29–52.

Haas, P. J. and Shedler, G. S. (1987b). Regenerative generalized semi-Markov processes. *Comm. Statist. Stochastic Models* **3**, 409–438.

Haas, P. J. and Shedler, G. S. (1987c). Stochastic Petri nets with simultaneous transition firings. In *Intl. Workshop Petri Nets Performance Models*, 24–32. IEEE Computer Society Press.

Haas, P. J. and Shedler, G. S. (1988). Modelling power of stochastic Petri nets for simulation. *Probab. Engrg. Inform. Sci.* **2**, 435–459.

Haas, P. J. and Shedler, G. S. (1989a). Stochastic Petri net representation of discrete event simulations. *IEEE Trans. Software Engrg* **15**, 381–393.

Haas, P. J. and Shedler, G. S. (1989b). Stochastic Petri nets with timed and immediate transitions. *Comm. Statist. Stochastic Models* **5**, 563–600.

Haas, P. J. and Shedler, G. S. (1991). Stochastic Petri nets: Modeling power and limit theorems. *Probab. Engrg. Inform. Sci.* **5**, 477–498.

Haas, P. J. and Shedler, G. S. (1992). Simulation methods for manufacturing systems using stochastic Petri nets. IBM Research Report RJ 8672, IBM Almaden Research Center, San Jose, CA.

Haas, P. J. and Shedler, G. S. (1993a). Estimation of passage times via equilibrium processes with one-dependent cycles. IBM Research Report RJ 9520, IBM Almaden Research Center, San Jose, CA.

Haas, P. J. and Shedler, G. S. (1993b). Passage times in colored stochastic Petri nets. *Comm. Statist. Stochastic Models* **9**, 31–79.

Haas, P. J. and Shedler, G. S. (1995). One-dependent cycles and passage times in stochastic Petri nets. In *Proc. Sixth Intl. Workshop Petri Nets Performance Models*, 191–202. IEEE Computer Society Press.

Haas, P. J. and Shedler, G. S. (1996). Estimation methods for passage times using one-dependent cycles. *Discrete Event Dynam. Sys. Theory Appl.* **6**, 43–72.

Hack, M. (1975). Decidability questions for Petri nets. Ph.D. Dissertation, Department of Electrical Engineering, Massachusetts Institute of Technology, Cambridge, MA.

Hall, P. and Heyde, C. C. (1980). *Martingale Limit Theory and Its Application*. Academic Press, New York.

Heidelberger, P. (1979). A variance reduction technique that increases regeneration frequency. In N. Adam and A. Dogramaci (eds.), *Current Issues in Computer Simulation*. Academic Press, New York.

Heidelberger, P. and Iglehart, D. L. (1979). Comparing stochastic systems using regenerative simulation with common random numbers. *Adv. Appl. Probab.* **11**, 804–819.

Heidelberger, P. and Lewis, P. A. W. (1981). Regression-adjusted estimates for regenerative simulations, with graphics. *Comm. ACM* **24**, 260–273.

Heidelberger, P., Shahabuddin, P., and Nicola, V. (1994). Bounded relative error in estimating transient measures of highly dependable non-Markovian systems. *ACM Trans. Model. Comput. Simul.* **4**, 137–164.

Heidelberger, P. and Welch, P. D. (1981). A spectral method for confidence interval generation and run length control in simulation. *Comm. ACM* **24**, 233–245.

Henderson, S. G. and Glynn, P. (1999a). Can the regenerative method be applied to discrete-event simulation? In *Proc. 1999 Winter Simulation Conference*, 367–373.

Henderson, S. G. and Glynn, P. (1999b). Derandomizing variance estimators. *Oper. Res.* **47**, 907–916.

Henderson, S. G. and Glynn, P. (2001). Regenerative steady-state simulation of discrete-event stochastic systems. *ACM Trans. Model. Comput. Simul.* **11**.

Hordijk, A., Iglehart, D. L., and Schassberger, R. (1976). Discrete time methods for simulating continuous time Markov chains. *Adv. Appl. Probab.* **8**, 772–788.

Iglehart, D. L. (1975). Simulating stable stochastic systems, V: Comparison of ratio estimators. *Naval Res. Logist. Quart.* **22**, 554–565.

Iglehart, D. L. (1976). Simulating stable stochastic systems, VI: Quantile estimation. *J. ACM* **23**, 347–360.

Iglehart, D. L. (1977). Simulating stable stochastic systems, VII: Selecting the best system. In M. F. Neuts (ed.), *Algorithmic Methods in Probability*, 37–50. North Holland, Amsterdam.

Iglehart, D. L. and Lewis, P. A. W. (1979). Regenerative simulation with internal controls. *J. ACM* **26**, 271–282.

Iglehart, D. L. and Shedler, G. S. (1980). *Regenerative Simulation of Response Times in Networks of Queues*, Vol. 26 of *Lecture Notes in Control and Information Sciences*. Springer-Verlag, Berlin.

Iglehart, D. L. and Shedler, G. S. (1983). Simulation of non-Markovian systems. *IBM J. Res. Develop.* **27**, 472–480.

Iglehart, D. L. and Shedler, G. S. (1984). Simulation output analysis for local area computer networks. *Acta Inform.* **21**, 321–338.

Iglehart, D. L. and Stone, M. (1983). Regenerative simulation for estimating extreme values. *Oper. Res.* **31**, 1145–1166.

Ingalls, R. G. and Kasales, C. (1999). CSCAT: The Compaq supply chain analysis tool. In *Proc. 1999 Winter Simulation Conference*, 1201–1206.

Jančar, P. (2000). Bouziane's algorithm for the Petri net reachability problem is incorrect. Unpublished manuscript, Technical University of Ostrava, Czech Reublic.

Jensen, K. (1981). Coloured Petri nets and the invariant-method. *Theoret. Comput. Sci.* **14**, 317–336.

Jochens, H. W. and Shedler, G. S. (1989). Modelling and simulation of stochastic systems with SPSIM. IBM Research Report RJ 6825, IBM Almaden Research Center, San Jose, CA.

Kalashnikov, V. (1994). *Topics on Regenerative Processes*. CRC Press, Boca Raton, FL.

Karlin, S. and Taylor, H. M. (1975). *A First Course in Stochastic Processes*, 2nd ed. Academic Press, New York.

Kaspi, H. and Mandelbaum, A. (1992). Regenerative closed queueing networks. *Stochastics Stochastics Rep.* **39**, 239–258.

Kohlas, J. (1982). *Stochastic Methods of Operations Research*. Cambridge University Press, Cambridge, England.

König, D., Matthes, K., and Nawrotzki, K. (1967). *Verallgemeinerungen der Erlangschen und Engsetschen Formeln*. Akademie-Verlag, Berlin.

König, D., Matthes, K., and Nawrotzki, K. (1974). Unempfindlichkeitseigenshaften von Bedienungsprozessen. Appendix to Gnedenko and Kovalenko (1974).

Kosaraju, S. (1973). Limitations of Dijkstra's semaphore primitives and Petri nets. Tech. rep., Computer Science Program, Johns Hopkins University, Baltimore, MD.

Kosten, A. E. and Tchoudaikina, S. V. (1998). Yet another reachability algorithm for Petri nets. *SIGACT News* **29**, 98–110.

Lavenberg, S. and Sauer, C. H. (1977). Sequential stopping rules for the regenerative method of simulation. *IBM J. Res. Develop.* **21**, 545–558.

Law, A. M. and Kelton, D. W. (2000). *Simulation Modeling and Analysis*, 3rd ed. McGraw-Hill, New York.

Lin, C. and Marinescu, D. C. (1988). Stochastic high-level Petri nets and applications. *IEEE Trans. Comput.* **37**, 815–825.

Lindemann, C. and Shedler, G. S. (1996). Numerical analysis of deterministic and stochastic Petri nets with concurrent deterministic transitions. *Performance Evaluation* **27/28**, 565–582.

Loéve, M. (1977). *Probability Theory*, 4th ed. Springer-Verlag, New York.

Loucks, W. M., Hamacher, V. C., and Preiss (1982). Performance of short packet local area rings. Tech. Rep. 25, Departments of Electrical Engineering and Computer Science, University of Toronto, Toronto.

Matthes, K. (1962). Zur Theorie der Bedienungsprozessen. In *Trans. 3rd Prague Conf. Inform. Theory Statist. Decision Functions*, 39–42.

Meketon, M. S. and Heidelberger, P. (1982). A renewal theoretic approach to bias reduction in regenerative simulations. *Management Sci.* **28**, 173–181.

Meketon, M. S. and Schmeiser, B. (1984). Overlapping batch means: Something for nothing? In *Proc. 1984 Winter Simulation Conference*, 227–230.

Meyn, S. P. and Tweedie, R. L. (1993a). *Markov Chains and Stochastic Stability*. Springer-Verlag, London.

Meyn, S. P. and Tweedie, R. L. (1993b). Stability of Markov processes II: Continuous time processes and sampled chains. *Adv. Appl. Probab.* **25**, 487–517.

Meyn, S. P. and Tweedie, R. L. (1993c). Stability of Markov processes III: Foster–Lyapunov criteria for continuous time processes. *Adv. Appl. Probab.* **25**, 518–548.

Miller, D. R. (1972). Existence of limits in regenerative processes. *Ann. Math. Statist.* **43**, 1275–1282.

Miller, D. R. (1974a). Limit theorems for path-functionals of regenerative processes. *Stochastic Process. Appl.* **2**, 141–161.

Miller, R. G. (1974b). The jackknife—A review. *Biometrika* **61**, 1–15.

Minh, D. L. (1987). Simulating GI/G/k queues in heavy traffic. *Management Sci.* **33**, 1192–1199.

Miyazawa, M. (1993). Insensitivity and product-form decomposability of reallocatable GSMP. *Adv. Appl. Probab.* **25**, 415–437.

Molloy, M. K. (1981). On the integration of delay and throughput measures in distributed processing models. Ph.D. Dissertation, Department of Computer Science, University of California, Los Angeles, CA.

Molloy, M. K. (1982). Performance analysis using stochastic Petri nets. *IEEE Trans. Comput.* **C-31**, 913–917.

Morozov, E. V. (1994a). Regeneration of closed queueing networks. *J. Math. Sci.* **69**, 1186–1192.

Morozov, E. V. (1994b). Wide sense regenerative processes with applications to multi-channel queues and networks. *Acta Appl. Math.* **34**, 189–212.

Morozov, E. V. (1998). Weak regenerative structure of open Jackson queueing networks. *J. Math. Sci.* **91**, 2956–2961.

Morozov, E. V. and Sigovtsev, S. G. (2000). Queueing process simulation based on weak regeneration. Unpublished manuscript.

Motwani, R. and Raghavan, P. (1995). *Randomized Algorithms*. Cambridge University Press, Cambridge, England.

Muñoz, D. F. and Glynn, P. W. (1997). A batch means methodology for estimation of a nonlinear function of a steady-state mean. *Management Sci.* **43**, 1121–1135.

Muñoz, D. F. and Glynn, P. W. (2001). Multivariate standardized time series for steady-state simulation output analysis. *Oper. Res.* **49**, 413–422.

Muppala, J. K., Trivedi, K. S., Mainkar, V., and Kulkarni, V. G. (1994). Numerical computation of response time distributions using stochastic reward nets. *Ann. Oper. Res.* **48**, 155–184.

Murata, T. (1989). Petri nets: Properties, analysis, and applications. *Proc. IEEE* **77**, 541–580.

Nakayama, N. K. and Shahabuddin, P. (1998). Likelihood ratio gradient estimation for finite-time performance measures in generalized semi-Markov processes. *Management Sci.* **44**, 1426–1441.

Natkin, S. (1980). Les réseaux de Petri stochastiques et leur application a l'evaluation des systemes informatiques. Thèse de Docteur Ingénieur, Conservatoire National des Arts et Metier, Paris.

Natkin, S. (1985). Les réseaux de Petri stochastiques. *Technique et Science Informatiques* **4**, 143–160.

Nummelin, E. (1978). A splitting technique for Harris recurrent chains. *Z. Wahrscheinlichtkeitstheorie und Ver. Geb.* **43**, 309–318.

Nummelin, E. (1984). *General Irreducible Markov Chains and Non-Negative Operators*. Cambridge University Press, Cambridge, England.

Ólafsson, S. and Shi, L. (1999). Optimization via adaptive sampling and regenerative simulation. In *Proc. 1999 Winter Simulation Conference*, 666–672.

Peterson, J. L. (1981). *Petri Net Theory and the Modeling of Systems*. Prentice Hall, Englewood Cliffs, NJ.

Prisgrove, L. A. and Shedler, G. S. (1986). Symmetric stochastic Petri nets. *IBM J. Res. Develop.* **30**, 278–293.

Puliafito, A., Scarpa, M., and Trivedi, K. S. (1998). Petri nets with K simultaneously enabled generally distributed timed transitions. *Performance Evaluation* **32**, 1–34.

Pyssysalo, T. and Ojala, L. (1995). A high-level net model of a video on demand system. Research Report A36, Digital Systems Laboratory, Helsinki University of Technology, Otaniemi, Finland.

Reisig, W. (1985). *Petri Nets: An Introduction*. EATCS Monographs on Theoretical Computer Science. Springer-Verlag, Berlin.

Roberts, G. O. and Tweedie, R. L. (1999). Bounds on regeneration times and convergence rates for Markov chains. *Stochastic Process. Appl.* **80**, 211–229.

Ross, S. M. (1983). *Stochastic Processes*. Wiley, New York.

Rubinstein, R. and Melamed, B. (1998). *Modern Simulation and Modeling*. Wiley, New York.

Rubinstein, R. and Shapiro, A. (1993). *Discrete Event Systems: Sensitivity Analysis and Stochastic Optimization by the Score Function Method*. Wiley, New York.

Schmeiser, B. (1982). Batch size effects in the analysis of simulation output. *Oper. Res.* **30**, 556–568.

Schruben, L. W. (1983). Confidence interval estimation using standardized time series. *Oper. Res.* **31**, 1090–1108.

Seila, A. F. (1982). A batching approach to quantile estimation in regenerative processes. *Management Sci.* **28**, 573–581.

Serfling, R. J. (1980). *Approximation Theorems of Mathematical Statistics*. Wiley, New York.

Shedler, G. S. (1987). *Regeneration and Networks of Queues*. Springer-Verlag, New York.

Shedler, G. S. (1993). *Regenerative Stochastic Simulation*. Academic Press, New York.

Shedler, G. S. (1994). Estimation of delay distributions with SPSIM. IBM Research Report RJ 9874, IBM Almaden Research Center, San Jose, CA.

Sigman, K. (1990a). Correction: Notes on the stability of closed queueing systems. *J. Appl. Probab.* **27**, 735.

Sigman, K. (1990b). One-dependent regenerative processes and queues in continuous time. *Math. Oper. Res.* **15**, 175–189.

Sigman, K. and Wolff, R. (1993). A review of regenerative processes. *SIAM Rev.* **2**, 269–288.

Smith, W. L. (1955). Regenerative stochastic processes. *Proc. Roy. Soc. London Ser. A* **232**, 6–31.

Smith, W. L. (1958). Renewal theory and its ramifications. *J. Roy. Statist. Soc. Ser. B* **20**, 243–302.

Song, W. T. and Schmeiser, B. W. (1993). Variance of the sample mean: Properties and graphs of quadratic-form estimators. *Oper. Res.* **41**, 501–517.

Song, W. T. and Schmeiser, B. W. (1995). Optimal mean-squared-error batch sizes. *Management Sci.* **41**, 110–123.

Symons, F. J. W. (1978). Modelling and analysis of communication protocols using numerical Petri nets. Ph.D. Thesis, Department of Electrical Engineering Science, University of Essex, Essex, England.

Symons, F. J. W. (1980). The description and definition of queueing systems by numerical Petri nets. *Austral. J. Telecom. Res.* **13**, 20–31.

Thorisson, H. (2000). *Coupling, Stationarity, and Regeneration*. Springer-Verlag, New York.

van der Aalst, W. M. P. (1998). The application of Petri nets to workflow management. *J. Circuits Sys. Comput.* **8**, 21–66.

Viswanadham, N. and Narahari, Y. (1988). Stochastic Petri net models for performance evaluation of automated manufacturing systems. *Inform. Decision Tech.* **14**, 125–142.

Viswanadham, N. and Raghavan, N. S. (2000). Performance analysis and design of supply chains: A Petri net approach. *J. Operational Res. Soc.* **51**, 1158–1169.

Welch, P. D. (1987). On the relationship between batch means, overlapping batch means, and spectral estimation. In *Proc. 1987 Winter Simulation Conference*, 320–323.

Whitt, W. (1980). Continuity of generalized semi-Markov processes. *Math. Oper. Res.* **5**, 494–501.

Whitt, W. (2002). *Stochastic-Process Limits*. Springer-Verlag, New York.

Xie, A., Kim, S., and Beerel, P. A. (1999). Bounding average time separations of events in stochastic timed Petri nets with choice. In *Proc. 5th Intl. Sympos. Adv. Res. Asynchronous Circuits Systems (ASYNC '99)*, 94–107. IEEE Computer Society Press.

Zenie, A. (1985). Colored stochastic Petri nets. In *Proc. Intl. Workshop Timed Petri Nets*, 262–271. IEEE Computer Society Press.

Index